Chemistry
of
Pesticides

Chemistry of Pesticides

N.K. Roy

Emeritus Scientist,
Former Head, Professor & Project Coordinator,
All India Co-ordinated Research Project on Pesticide Residues
Division of Agricultural Chemicals
Indian Agricultural Research Institute
New Delhi - 110012 (India)

CBS

CBS Publishers & Distributors Pvt. Ltd.

New Delhi • Bengaluru • Chennai • Kochi • Kolkata • Mumbai
Hyderabad • Nagpur • Patna • Pune • Vijayawada

Chemistry of Pesticides

ISBN: 978-81-239-0854-0

First Edition: 2002
Reprint: 2006, 2010, 2013, 2017, 2021

Published by **Satish Kumar Jain** and produced by **Varun Jain** for

CBS Publishers & Distributors Pvt Ltd

4819/XI Prahlad Street, 24 Ansari Road, Daryaganj, New Delhi 110 002, India.
Ph: 011-23289259, 23266861, 23266867   Website: www.cbspd.com
Fax: 011-23243014   e-mail: delhi@cbspd.com;
      cbspubs@airtelmail.in.

Corporate Office: 204 FIE, Industrial Area, Patparganj, Delhi 110 092
Ph: 011-4934 4934   Fax: 011-4934 4935
      e-mail: publishing@cbspd.com; publicity@cbspd.com

Branches

- **Bengaluru:** Seema House 2975, 17th Cross, K.R. Road, Banasankari 2nd Stage, Bengaluru 560 070, Karnataka, India
  Ph: +91-80-26771678/79   Fax: +91-80-26771680   e-mail: bangalore@cbspd.com
- **Chennai:** 7, Subbaraya Street, Shenoy Nagar, Chennai 600 030, Tamil Nadu, India
  Ph: +91-44-26680620, 26681266   Fax: +91-44-42032115   e-mail: chennai@cbspd.com
- **Kochi:** 42/1325, 1326, Power House Road, Opp KSEB, Ernakulum, Kochi 682 018, Kerala, India
  Ph: +91-484-4059061-65,67   Fax: +91-484-4059065   e-mail: kochi@cbspd.com
- **Kolkata:** 6/B, Ground Floor, Rameswar Shaw Road, Kolkata-700014 (West Bengal), India
  Ph: +91-33-2289-1126, 2289-1127, 2289-1128   e-mail: kolkata@cbspd.com
- **Mumbai:** PWD Shed, Gala no 25/26, Ramchandra Bhatt Marg, Next to JJ Hospital Gate no. 2, Opp. Union Bank of India, Noorbaug Mumbai-400009, Maharashtra, India
  Ph: +91-22-66661880/89   e-mail: mumbai@cbspd.com

Representatives

• Hyderabad	0-9885175004	• Jharkhand	0-9811541605	• Nagpur	0-9421945513
• Patna	0-9334159340	• Pune	0-9623451994	• Uttarakhand	0-9716462459

Printed at Rashtriya Printers, Dilshad Garden, Delhi, India

Foreword

During the last sixty years, pesticide science has evolved into an integrated discipline of crop protection and public health, involving chemistry, biochemistry, toxicology, chemical technology and biotechnology. The essential criteria of a pesticide has also expanded encompassing aspects of safety of the environment. Now there is a growing interest in the greening not only of pesticidal molecules but even the various stages of technology of manufacture to application. Persistence and toxic pesticides like DDT, BHC, aldrin, endrin etc. have been phased out through a well orchestrated mechanism of feedback and redressal.

The challenges before farm scientists in augmenting the production of food, fodder and fibre at various levels commensurate with the requirements of burgeoning population in the coming decades are stupendous. Crop protection would have to play a major role in ensuring higher productivity. The crop protection methods and agents must also conform to the standards of safety to environment and ecology. Integrated Pest Management modules must be developed suiting to the needs of cropping system under various agro-climatic zones.

Biorational approaches have been weaved into the design of new molecules in addition to computer assisted molecular modeling (CAMM) and quantitative structure activity relationship (QSAR). The subject of pesticide chemistry covering all aspects of production, formulation, application, metabolism, analysis and registration requirements has become more complex and sophisticated. Rapid advances have taken place in the related frontier areas such as genetic engineering and biotechnology.

The subject is being taught at different levels in the universities including IARI, a deemed university. Prof. N.K. Roy has been involved in the research and teaching of pesticide chemistry for about thirty years. Prof. Roy, through his book on 'Chemistry of Pesticides' has now brought out a comprehensive treatise including recent developments in the major chemical classes of synthetic and botanical pesticides, intellectual property rights, WTO agreement, phytosanitary measures etc. I congratulate Prof. Roy for his excellent work. I am sure that this book will serve as an excellent source of information to all those interested in the area of pesticide science, especially students, teachers, researchers and plant protection scientists concerned with public health programme and policy makers.

(R.S. Paroda)
Secretary, Department of Agricultural Research & Education, and
Director General, Indian Council of Agricultural Research,
Ministry of Agriculture, New Delhi-110001

Preface

In spite of constraints of natural resources such as land and water, the future demand of food, feed and fibre in the next two decades can be met only by ensuring higher productivity for which crop protection more than crop improvement holds the key.

During the last decade or so, many universities have been dealing with crop protection, including application of recombinant DNA technology, plant breeding for durable resistance to pests and diseases. Deployment of biocontrol agents have also been harnessed to ensure effective and ecofriendly crop protection. In spite of all these efforts, chemical pesticides continue to be the kingpin of crop protection. This situation is unlikely to change at least for another two decades.

The demand for green or soft pesticides may never be quenched. Sophistication in design and discovery, formulation and application will be on the rise. The subject of pesticide chemistry is being taught in several agricultural universities and a few traditional universities both at undergraduate and postgraduate levels. But unfortunately, in India no books suitable to the syllabi are available on the pesticide chemistry. The need for a book on the topic has been felt by the author.

In this book, an attempt has been made to give a holistic scenario of pesticide chemistry. The present book on 'Chemistry of Pesticides' comprises eleven chapters dealing with major chemical classes having bearing on practical application.

The chapter on 'Botanical Pesticides' provides an overview of recent advances on natural pesticides with a special reference to azadirachtins, active constituents of neem an emphasis on the discovery of lead molecules as templates of potential synthetics.

The chapter on 'Synthetic Pesticides' deals with the chemistry of insecticides, nematicides, rodenticides, fungicides and herbicides, incorporating the newer developments in this area. There is a brief discussion on the mode of action, metabolism and photodegradation of these chemical groups as supplementary tools in understanding aspects of safety to non-target systems.

The chapter on 'Improved Formulations' gives an account of new generation of formulations vis-a-vis conventional formulations emphasizing the essentialities of quality in delivery. The accumulation of pesticide residues in the environment is of grave concern to all. This aspect has been discussed thoroughly including phytosanitation measures in the light of WTO agreement.

Biotechnology, an emerging applied field in the pest management programme, has been highlighted. Quantitative structure activity relationship (QSAR) and computer assisted molecular modelling (CAMM) help in optimising the design of potent pesticides. These are elaborated with copious examples.

A chapter is devoted to alternative methods of pest control using pheromones, hormones, antifeedants, repellants. The future pesticide scenario is depicted in the background of Intellectual Property Rights (IPR) and Indian patent system following WTO agreement. The regulatory aspects of pesticides—Insecticides Act, 1968 and safety measures in handling and use of pesticides are discussed. The book is intended to serve as a source book of knowledge to students specialising in pesticide chemistry, agricultural chemistry and plant protection science as well as a reference book for those dealing with pesticide science, technology and managements.

The author wishes to thank Human Resource Unit, Department of Science & Technology, Government of India for providing financial grant to support the project (HR/UR/01/1997, titled "Recent Advances in Agrochemicals"). Finally, I would like to thank all my colleagues in the Division of Agricultural Chemicals particularly Drs. C. Devakumar, Suresh Walia, Prem Dureja, S.K. Goel, R.L. Kalra and P.K. Ramdas for their valuable suggestions and help in the preparation of this manuscript. I am grateful to Dr. R.S. Paroda, Secretary, DARE and D.G., Indian Council of Agricultural Research, New Delhi for Foreword note in this book. My gratitude is also due to the Director, Indian Agricultural Research Institute, New Delhi and Head of the Division of Agricultural Chemicals, IARI for their inspiration, cooperation and providing all facilities to complete this work.

Special thanks are due to my loving children Ms. Sucharita for editing the manuscript with great patience and Narendra for inspiration and my wife Shyamali for constant encouragement and kind cooperation. I thank Shri Dipankar and Shri Trithankar, loving Ph.D. students, and Shri Avinash for word processing the manuscript.

N.K. Roy

Contents

Foreword v
Preface vii
Abbreviations used in the text xi

1. Introduction 1
2. Pesticides of Botanical Origin 9
3. Synthetic Pesticides 44
   (i) Chemistry of Chlorinated Insecticides 44
   (ii) Chemistry of Pyrethroid Insecticides 58
   (iii) Chemistry of Miscellaneous Insecticides 71
   (iv) Chemistry of Organophosphorus Insecticides 76
   (v) Chemistry of Carbamate Insecticides 118
   (vi) Chemistry of Nematicides 130
   (vii) Chemistry of Rodenticides 136
   (viii) Chemistry of Fungicides 148
   (ix) Chemistry of Herbicides 198
4. New Improved Pesticide Formulations 242
5. Biotechnology in Pest Management 257
6. Pesticide Residues and Their Environmental Implications 265
7. Alternative Methods of Insect Pests Control 281
8. Quantitative Structure Activity Relationship (QSAR) and Computer Assisted
   Molecular Modelling (CAMM) in Pesticide Design 293
9. Suggestions for Future Development, Intellectual Property Rights and
   Indian Patent System 305
10. Regulatory Aspects of Pesticides—Insecticides Act, 1968 314
11. Safety Measures in Handling and Uses of Pesticides 322

Appendix : Important pesticides registered for use in the country with their common name
including warning symbol, chemical name, trade name and manufacturers 331
Subject Index 341

Abbreviations Used in the Text

a.i.	Active ingredient
ACS	American Chemical Society
abs	Abstract
Aq	aqueous
ASE	Accelerated Solvent Extracts
bp	Boiling point
CCPR	Codex Committee on Pesticide Residues
C	Celcius
CA	Chemical abstract
cf.	compare
Conf.	Conference
Congr.	Congress
cm	centimeter
CS	Encapsulated suspension
C(S)	C=S
CE	Capillary electrophoresis
d	Day(s)
ED_{50}	Lethal dose required to kill 50% population of test animals
ed.	Editor(s)
EO	Emulsion water in soil
Ed.	Edition
EW	Emulsion, oil in water
ELISA	Enzyme linked immunosorbent assay
Fam.	Family
ft.	feet
GAP	Good Agricultural Practices
GLP	Good Laboratory Practices
GC/GLC	Gas liquid chromatography
g	gramme(s)
GR	Granules
HPLC	High performance liquid chromatography
HPTLC	High performance thin layer chromatography
h	hour(s)
ha	hectare

het.	Hetrocyclic
Inf	Information
Internatn.	International
i.e.	that is
I.D.	Internal diameter
ICI	Imperial Chemical Industries
IUPAC	International Union of Pure and Applied Chemistry
IPM	Integrated Pest Management
J	Journal
Kg	kilogramme(s)
LD_{50}	Lethal dose required to kill 50% test animal population
LC_{50}	Lethal concentration required to kill 50% test insect population
mg	milligramme(s)
mt	metric tonnes
mp	Melting point
min	minute
MS	Mass spectrometer
mm	millimeter
MRL	Maximum residue levels
oxd.	Oxidation
OP	Organophosphorus
OC	Organochlorine
Proc.	Proceedings
PFA	Prevention of Food Adulteration Act
pat.	Patent
Prot.	Protection
ppm	Parts per million
ppb	Parts per billion
P(O)/P(S)	P=O; P=S
Rep.	Report
Red.	Reduction
rt.	Room temperature
SC	Suspension concentrates (Flowable concentrate)
sec.	seconds
SG	Suspoemulsion
Symp.	Symposium
SFE	Supercritical fluid extraction
Sr.	Series
t.	Tertiary
UV	Ultraviolet
ULV	Ultra Low Volume
Vp	Vapour pressure
Wt	Weight
WDP	Water dispersible powder
WG	Water dispersible granule
WP	Wettable powder

CHAPTER 1

Introduction

- Classification of pesticides
- Historical development of pesticides
- Advances in the area of pesticides
- Current Indian situation in the area of food production cropwise *vis-a-vis* the world
- Pesticides registered, banned and under review in India
- Characteristics of ecofriendly pesticides
- *References*

Agrochemicals have an important role in ensuring food supply and better health for a growing world population. The rapid advancement in science and technology in the last four decades has projected several new pesticidal molecules, formulations, application technology and also brought revolutionary changes in farm practices.

The term 'agrochemicals' includes plant nutrients like fertilizers as well as other alternate chemicals like plant growth regulators, pheromones, hormones, attractants, repellents, chemosterilants and bio-organisms.

Plant protection chemicals (pesticides) are chemicals or mixture of chemicals used for killing, repelling, mitigating, phasing or even regulating pests (depending on the stage of the pest infestation) with a view to minimise the damage caused to crops. Pesticides in general can be categorised as insecticides, fungicides and herbicides (weedicides). There are minor pesticides as well, such as nematicides, acaricides, rodenticides, molluscicides and larvicides. Insecticides may again be classified based on their use as a) crop, b) veterinary, c) public health, d) stored grain, e) household insecticides. Sometimes insecticides may also be grouped based on their mode of entry as a) contact, b) stomach, c) systemic, d) fumigant. The pesticides of first and second generations are mostly contact, non-systemic types, were easier to discover and are free from problem of development of resistance and rapid detoxification compared to the case with systemics where the chemical comes in close contact with the tissues of the host plant and develops resistance within a short-time, resulting in restriction in their use. However, a large number of systemic pesticides like antibiotics, benomyl, hexaconazole (fungicides), schradan (organophosphorus insecticide) and phenoxy-acetic acids (herbicides) are still being used as agents of crop protection.

Development of pesticides : Historical background

In early days, very little plant protection chemicals were used and these were mainly inorganic chemicals like sulphur, calcium arsenate, copper, mercury, barium salts preparations, and botanical like tobacco, derris and pyrethrum extracts. Some of these find application even today. Among the modern synthetic pesticides, alkyl thiocyanate, salicylanilide and dithiocarbamate fungicides were first marketed in the year 1930. After World War-II, the advent of synthetic organic pesticides began with the discovery of BHC and DDT, followed by a large number of chlorinated cyclodiene insecticides. Many of these chemicals were potent contact insecticides and possessed low mammalian toxicity. However, all of them proved to be stable in the environment and thus have high persistence They accumulate in animal tissues through agroecological cycle posing threat to human health and environment. With increasing environmental awareness, the focus has subsequently shifted towards development of bio-degradable types such as organophosphorus and carbamate group of compounds that were introduced in mid-fifties. Many of these chemicals were more potent than organochlorines and would not accumulate in the environment, hence safe. However, a few of them are extremely toxic to mammals, hazardous and non-selective in nature.

The pyrethrum extracts containing pyrethrins were used as insecticide since 1850. However due to their thermal and photolability, they could not be used for outdoor agricultural application. Dr. Elliott and his colleagues at Rothamsted Experimental Station, England [1] first time synthesized pyrethroids which are more potent and photostable insecticides than natural pyrethrins. Since the last two decades, several potent synthetic pyrethroids have been developed, commercialized and are recommended as safe insecticides in the control of insect pests of cereals, vegetables, fruit crops, pulses, oilseeds and only a few g/ha dose is adequate. These may be called as eco-friendly insecticides.

In the year 1943, scientists of I.C.I. England [2] discovered the herbicidal activity of 2,4-dichloro (2,4-D) and 2-methyl 4-chloro (MCPA) phenoxyacetic acids. These compounds are translocated in plants and are extremely valuable for the selective control of broad-leaved weeds in cereal crops. Later on, bipyridylium compounds like diquat and paraquat herbicides were introduced, which were also absorbed and translocated in plants causing desiccation of the foliage. Then came herbicides like urea, triazine, carbamate, pyridine, pyrimidine and organophosphorus compounds, the latest being sulfonylurea derivatives which are the most effective and require only a few g/ha dose without apparently causing environmental hazard.

In the area of fungicides too, significant advances have been made. N-Trichloromethyl thiotetrahydrophthalimide (Captan) was commercialised in the year 1951, followed by antibiotics as plant chemotheraputic agents to control fungi as well as bacteria. The systemic fungicides like oxathin, benzimidazole, thiophanate, pyridine, pyrimidine, morpholine and organophosphorus compounds were introduced subsequently later, while the latest group being added is triazole compounds. Rapid advances in biotechnology and genetic engineering have led to novel microbial products as substitutes for synthetic pesticides as well as generation of new durable resistant crop varieties.

Advances in formulation and application technology have been trying with time to meet the ever-changing demands of farmers and crop requirements under different agroclimatic conditions.

Some valuable pesticides, being very toxic to mammals, should be formulated safely to minimise the hazard associated with their application. Safety considerations from view points of consumer and public health are very pertinent since pesticides after application can contaminate soil, water and commodity. In order to evolve the safety index, the crops, soil and water samples were regularly monitored by following the protocols of residue analysis using sophisticated instruments like GC., GC-MS, HPLC, HPLC-MS, HPLC-MS-MS. ELISA techniques are being applied now-a-days in pesticide residue analysis of the crop produce, soil, water and fodders to generate accurate and reliable data. The nature of metabolites with mechanism of degradation of the active ingredients in animals, including birds, fish and plants and the behaviour of the pesticides in the environment including ground water warrant a close scrutiny before these are recommended. In order to minimise the hazards due to synthetic pesticides, alternate methods of pest control employing semiochemicals like pheromones, juvenile hormones (insect growth regulators), chitin synthesis inhibitors, repellents, antifeedants, chemosterilants have been tried but these were not so popular to the farmers due to inherent constraints like application technology.

The pesticide research and development process is a long journey and involves multidisciplinary approach. The first step is to design and then synthesize. A series of compounds after preparation are subjected to a preliminary screening to establish whether they show activity against a targetted organism like insects, fungi and herbs. On the basis of SAR and QSAR studies, the structure of the most active molecule is further optimised, and then subjected to detailed bioassay study both *in vitro* and *in vivo*. Analytical methods are then developed for qualitative and quantitative analysis of the commercially promising active ingredient both in formulations, and on various food commodities including soil and water samples. This is followed by toxicological studies like chronic and acute toxicity, mutagenicity, carcinogenicity, metabolism in plant and soil and related studies like photodegradation. Thus the total period for commercializing an active pesticidal molecule might require atleast 8 to 10 years with an expenditure of about 50 million US dollars.

Loss of food grains due to insect pests, diseases and weeds in the field as well as in storage condition due to insect pests, rodents, mites and fungi are considerably high. The loss of food crops in countries having tropical climate is generally more than that in countries having temperate climate. The increasing use of pesticides in developing and underdeveloped countries is thus absolutely vital for greater supply of food to feed their ever increasing population. This would mean that there has to be more food production and feed supply to fulfil domestic and export projections. The current Indian situation in the area of food production, cropwise as compared to world and its future projections are listed in Tables 1.1 to 1.4. A glimpse of the consumption of pesticides *vs.* food production and population growth in India has been projected in Fig. 1.1. The list of pesticides registered for use, banned and under review are given in Tables 1.5 to 1.7. The use of pesticides vary from country to country due to the nature of crops grown and climatic conditions. It has been observed that the consumption of herbicides is maximum in developed countries than in developing countries where insecticide consumption is more.

Today, about one thousand pesticides are available in the market. The use pattern of pesticides varies from country to country. The consumption of pesticides is the highest for Japan (12 kg/ha) followed by USA, and the European countries while India uses only 400 g. The modern trend is to increase the biological efficacy with low application rates thereby minimising the environmental load i.e. pollution effects. There are already a few efficient chemicals available today in the market, like triazole group of fungicides whose dose is about 80-100 g/ha, the synthetic pyrethroid requires

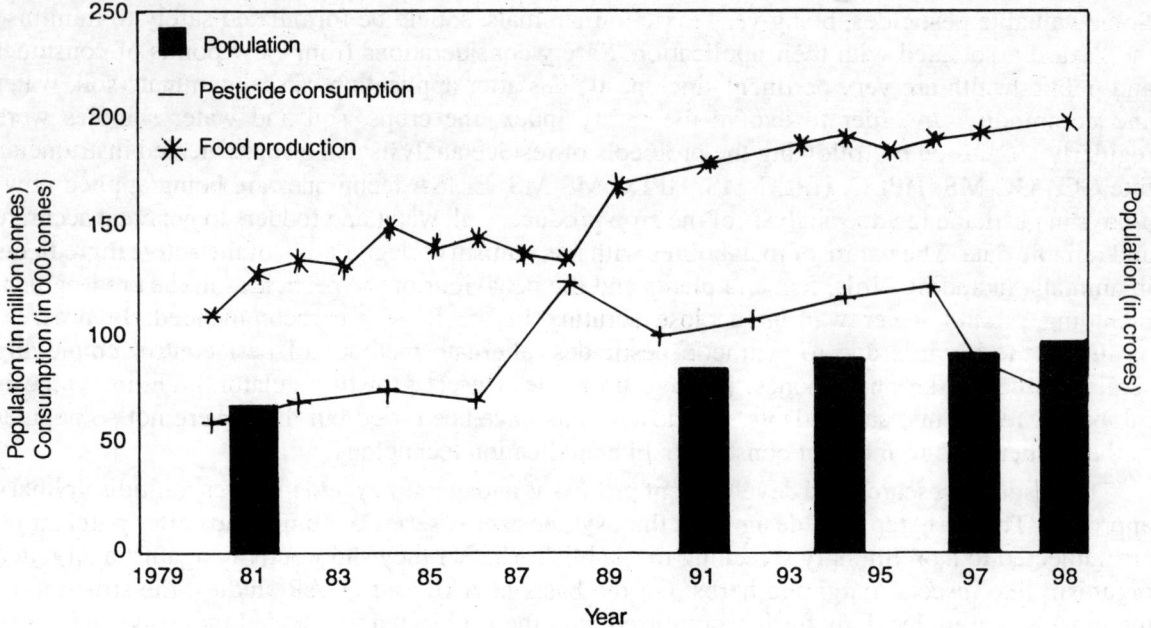

Fig. 1.1. Pesticide consumption, food grain production and population growth in India.

Table 1.1. Crop yield in India during 1996-97 as a percentage of world's average

Food crop	Rice (%)	Wheat (%)	Maize (%)	Bajra (%)	Jowar (%)	Pulses (%)	Oil seeds (%)
Average yield per ha	49	98	38	72	57	57	73
The highest yield per ha	30	31	19	20	20	18	31

Non-food crop	Potatoes (%)	Groundnuts (%)	Sugarcane (%)	Sunflower (%)	Rapeseed (%)	Seed cotton (%)
Average yield per ha	98	76	109	51	65	59
The highest yield per ha	42	36	70	27	27	34

Source : Economic Survey, 1997-98

Table 1.2. Cotton : A global scenario

S. No.		Average yield per hectare in kgs	
	Country	1990	1996
1.	Israel	1656	1452
2.	China	814	879
3.	U.S.A.	711	602
4.	Pakistan	615	879
5.	India	267	316

Table 1.3. Domestic demand (million tonnes) for food grains in India

Items	1991-92 Base year	1996-97	2001-02	2006-07
Plan end	VII	VIII	IX	X
1. Rice	72.6	81.2	89.8	98.8
2. Wheat	57.8	65.0	72.5	80.7
3. Coarse cereals	28.9	30.7	32.6	34.4
4. Total cereals	158.0	175.7	194.9	213.1
5. Pulses	13.4	15.9	18.4	21.5
6. Food grains (total)	171.3	191.6	213.3	234.5

Table 1.4. Targets for food grains production in 2020

Crop	1996-97			2019-20		
	Area million ha	Yield Kg/ha	Production millon tonnes	Area million ha	Yield Kg/ha	Production million tonnes
Rice	43.3	1878	81.3	43.5	4900	213.2
Wheat	25.9	2675	69.3	28.5	6400	182.4
Coarse grains	32.1	1069	34.3	21.5	3200	68.8
Pulses	23.2	621	14.4	31.5	1142	36.0
Total	124.5	1601	199.3	125.0	4144	500.4

Source : IARI Economists, New Delhi, India

about 20 g/ha while the sulfonylurea group of herbicide requires only 5-10 g/ha as compared to a few kg/ha needed by conventional pesticides. Now-a-days integrated pest management (IPM) technology comprising the use of biopesticides, biocontrol agents and eco-friendly synthetic pesticides in low volume coupled with cultural, physical and mechanical methods are utilized in an integrated manner to reduce the use of hazardous pesticides. The IPM technologies are now available for crops such as rice, cotton, sugarcane and vegetables, and have been successfully adapted particularly in South-East Asia especially Indonesia and China. It appears that there exists a strong demand for eco-friendly pesticides with respect to specificity of action with no side effects on non-target species like human beings and animals. The characteristics of eco-friendly pesticides are given in Table 1.8.

This shows that our productivity is low in most of the crops as compared to the world average. The country, therefore, needs to put more efforts to improve productivity substantially by effectively controlling the crop losses due to insect pests.

Table 1.3 indicates that the demand for food grains would be approximately 213.3 million tonnes in 2001-02 and 234.5 million tonnes in the year 2006-07. The area under the largest grown crops *viz*. rice and wheat (Table 1.4) would increase marginally but the production targets required to be achieved from the same acreage are almost three times higher. A similar situation is prevalent in non-food grains like edible oilseeds, vegetables, fruits, sugar and cotton. The intensification of production and quality of foodgrains, non-food grains and non-crop commodities should be future

Table 1.5. Pesticides registered for use in India under Section 9(3) of the Insecticides Act, 1968.

Name of the Pesticide	Name of the Pesticide	Name of the Pesticide
1. 2,4-Dichlorophenoxy Acetic Acid	40. Cypermethrin	76. Fosetyl-Al
2. Acephate	41. Cyphenothrin	77. Gibberellic Acid
3. Alachlor	42. D-trans allethrin	78. Glyphosate
4. Aldicarb	43. Dalpon	79. Glufosinate-ammonium
5. Allethrin	44. Decamethrin (Deltamethrin)	80. Hexaconazole
6. Alphacypermethrin	45. Diazinon	81. Iprodione
7. Alphanaphthyl Acetic Acid	46. Dichloro Diphenyl Trichloroethane (DDT)	82. Isoprothiolane
8. Aluminium Phosphide		83. Isoproturon
9. Anilofos	47. Dichloropropene and Dichloro-propanes mixture (DD Mixture)	84. Kasugamycin
10. Atrazine		85. Kitazin
11. Aureofungin		86. Lamdacyhalothrin
12. Bacillus thuringiensis (B.t.)	48. Dichlorovos (DDVPS)	87. Lime Sulphur
13. Barium Carbonate	49. Diclofop-methyl	88. Lindane
14. Benomyl	50. Dicofol	89. Linuron
15. Benthiocarb (Thiobencarb)	51. Dieldrin	90. Malathion
16. Bitertanol	52. Diflubenzuron	91. Maleic Hydrazide (MH)
17. Bromadiolone	53. Dimethoate	92. Mancozeb
18. Butachlor	54. Dinocap	93. Metalaxyl
19. Captafol	55. Dithianon	94. Metaldehyde
20. Captan	56. Diuron	95. Methabenzthiazuron
21. Carbaryl	57. Dodine	96. Methomyl
22. Carbendazim	58. Edifenphos	97. Methoxy Ethyl Mercury Chloride (MEMC)
23. Carbofuran	59. Endosulfan	
24. Carbosulfan	60. Ethephon	98. Methyl Bromide
25. Carboxin	61. Ethion	99. Methyl Chlorophenoxy Acetic Acid (MCPA)
26. Cartap Hydrochloride	62. Ethofenprox (Etofenprox)	
27. Chlorfenvinphos	63. Ethylene Dibromide (EDB)	100. Methyl Parathion
28. Chlorimuron ethyl	64. Ethylene Dibromide and Carbon Tetrachloride mixture (EDCT Mixture)	101. Metasulfuron methyl
29. Chlormequat Chloride (CCC)		102. Metolachlor
30. Chlorobenzilate		103. Metoxuron
31. Chlorothalonil	65. Fenarimol	104. Metribuzin
32. Chlorpyriphos	66. Fenitrothion	105. Monocrotophos
33. Copper Oxychloride	67. Fenobucarb (BPMC)	106. Myclobutanil
34. Copper Hydroxide	68. Fenpropathrin	107. Neem Products
35. Copper Sulphate	69. Fenthion	108. Nickel chloride
36. Coumachlor	70. Fenvalerate	109. Oxadiazon
37. Coumatetralyl	71. Ferbam	110. Oxidiargyl
38. Cuprous Oxide	72. Fipronil	111. Oxycarboxin
39. Cyfluthrin	73. Fluchloralin	112. Oxydemeton-methyl
	74. Fluvalinate	113. Oxyfluorfen
	75. Formothion	

Name of the Pesticide	Name of the Pesticide	Name of the Pesticide
114. Paradichlorobenzene	128. Propetamphos	142. Thiram
115. Paraquat dichloride	129. Propiconazole	143. Tridimefon
116. Penconazole	130. Propoxur	144. Triallate
117. Pendimethalin	131. Pyrethrins (Pyrethrum)	145. Trizophos
118. Permethrin	132. Quinalphos	146. Trichlorfon
119. Phenthoate	133. Simazine	147. Trichloro Acetic Acid (TCA)
120. Phorate	134. Siramate	148. Tricyclazole
121. Phosalone	135. Sodium Cyanide	149. Tridemorph
122. Phosphamidon	136. Streptomycin + Tetracycline	150. Trifluralin
123. Pirimiphos-methyl	137. Sulphur	151. Validamycin
124. Prallethrin	138. Temephos	152. Warfarin
125. Pretilachlor	139. Thiodicarb	153. Zine Phosphide
126. Profenphos	140. Thiometon	154. Zineb
127. Propanil	141. Thiophanate-Methyl	155. Ziram

Table 1.6. Pesticides banned

Name of the Pesticide	Name of the Pesticide
1. Aldrin	12. Menazon
2. Benzene Hexachloride (BHC)	13. Nicotine Sulphate (for export only)
3. Calcium cyanide	14. Nitrofen
4. Chlordane	15. Paraquat dimethyl sulphate
5. Copper Acetoarsenite	16. Pentachloro nitrobenzene (PCNB)
6. Dibromochloropropane (DBCP)	17. Pentachlorophenol (PCP)
7. DDT	18. Phenyl Mercury Acetate (PMA)
8. Endrin	19. Sodium Methane Arsonate (MSMA)
9. Ethyl Mercury Chloride	20. Tetradifon
10. Ethyl Parathion	21. Toxaphene
11. Heptachlor	22. Methomyl 24% formulation

Table 1.7. List of pesticides under review by Government of India

Name of the Pesticide	Name of the Pesticide	Name of the Pesticide
1. Alachlor	6. Oxyflourfen	11. Monocrotophos
2. Benomyl	7. Phosphamidon	12. Ziram
3. Diuron	8. Thiometon	13. Zineb
4. Fenarimol	9. Triazophos	
5. Methomyl	10. Tridemorph	

Table 1.8. Characteristics of eco-friendly pesticides

1. Non-phytotoxic and non-injurious to plants.
2. Selective action in killing insect pests (selectivity)
3. Quick toxic action (with knock down effect)
4. High toxicity to pests and less toxicity towards mammals.
5. Stability (moderate) on treated surface (from safety angle)
6. Toxicity to as many stage of insects (egg, larvae, adults) as possible
7. Harmless to higher animals/human being.
8. Compatibility with other group of pesticides.
9. Harmless to beneficial soil fauna and flora and its nonaccumulation in soil
10. Non-tainting of edible plant products
11. Effective in small doses per unit of surface area or material covered.
12. Reasonable stability in storage and transport (i.e. shelf life)
13. Freedom from obnoxious odours and from irritating action on human skin and system.
14. Availability for use in as many form as desired such as D,WP, GR., EC at economic prices.
15. Non-inflammability and non-corrosive action on metals.

priorities of the nation which can be achieved only by efficient crop protection methodologies under integrated pest management (IPM) paradigm.

REFERENCES

1. Elliott, M. and Janes, N.F. *Recent structure-activity correlations in synthetic pyrethroids.* In: *Advances in Pesticide Chemistry* (H. Geissbuhler, ed.), Part 2, Pergamon Press, Oxford, 166 (1978).
2. Fryer, J.D. and Makepeace, R.J. (eds.). *Weed Control Handbook*, 8th edn., Blackwell, Oxford, (1978).

CHAPTER 2

Pesticides of Botanical Origin

- Introduction
- Natural insecticides
- Neem, Chemistry and effects
- Minor insecticides of plant origin
- Ryania, sabadilla,
- Unsaturated amides
- Light induced insecticides
- Insect antifeedant and growth regulators
- Miscellaneous insecticides

- Insecticides of marine origin
- Nematicides of plant origin
- Fungicides of plant origin
- Natural products of microbial origin
- Herbicides of plant origin
- Plant growth regulators
- Allelochemicals
- *References*

Plants afford a rich source of chemicals with diverse biological activities. A host of new pest control chemicals like insecticides, fungicides, herbicides, PGR and antifeedants have been isolated for use in agricultural crops and public health. So far over 2000 plants species belonging to 60 families are known to possess insecticidal properties (Table 2.1).

1. Natural insecticides

Some of the well known plant products possessing insecticidal activity are 'pyrethrum' from dried flowers of *Chrysanthemum cinerariaefolium* (Fam. Compositae), 'rotenoids' from *Derris* (Fam. Leguminosae) and 'nicotinoids' from *Nicotiana* (Fam. Solanaceae). The active principles of pyrethrum are pyrethrin-I (1), -II (2), cinerin-I (3), -II (4) and jasmolin-I (5), -II (6) which are derivatives of (+) trans-chrysanthemic acid (7) and (+) *trans*-pyrethric acid (8) (Fig. 2.1a).

Pyrethrum is used to control storage and house hold pests but it is not as effective against agricultural pests under field conditions because of its photolability. However, based on the structure of natural pyrethrins, more effective synthetic pyrethroids have been designed and prepared. These are used extensively against agricultural pests as a substitute for natural pyrethrum.

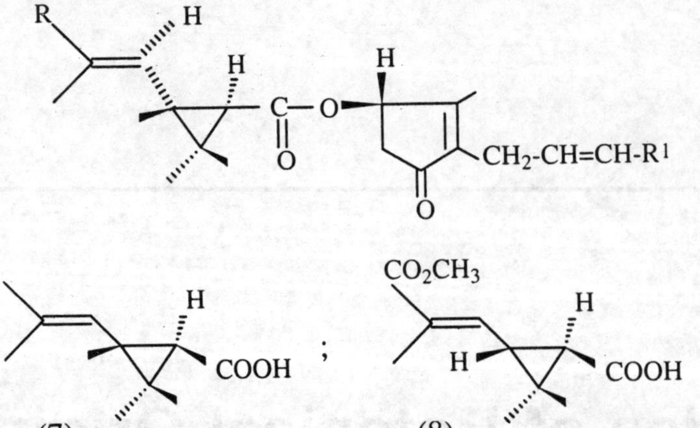

Fig. 2.1a. Natural insecticides.

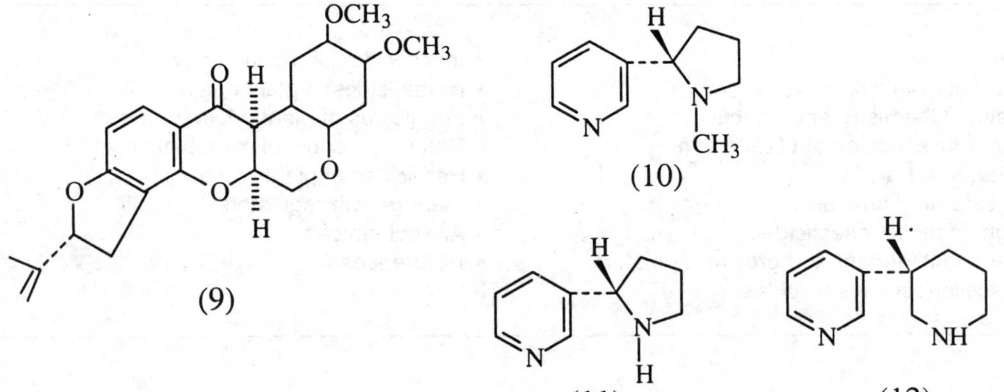

R =
Pyrethrin-I (1) CH_3,
Pyrethrin-II (2) CO_2CH_3,

R^1 =
-CH = CH
-CH = CH

Cinerin-I (3) - CH_3,
Cinerin-II (4) - CO_2CH_3,
-CH_3
-CH_3

Jasmolin-I (5) - CH_3 ,
Jasmolin-II (6) - CO_2CH_3,
-C_2H_5
-C_2H_5

Fig. 2.1b. Structure of natural insecticides.

Rotenoids (9), another group of natural insecticides are also not used in agriculture because of its high fish toxicity and photo-instability. Similar is the case with nicotine (10), which possesses unpleasant odour and high mammalian toxicity. Structurally related two alkaloids, nornicotin (11) and anabasine (12) also exhibit similar activity (Fig. 2.1b). Detailed information relating to structure activity relationships of this group of compounds is very well reviewed by Jacobson & Crosby [1], Jacobson [2a] and Elliger *et al.* [2b].

2. Neem

Azadirachtin

Extracts of seeds of the neem tree *Azadirachta indica* A. Juss, (Fam. Meliaceae) yield a large number of tetranortriterpenoids, one of which is azadirachtin A having insect antifeedant and growth retarding properties [3]. The chief chemical constituents, as a class, identified in different

Table 2.1. Plant families with insecticidal importance

No.	Family	No.	Family	No.	Family
1.	Acanthaceae	21.	Ebenaceae	41.	Myristicaceae
2.	Agavaceae	22.	Ericaceae	42.	Myrtaceae
3.	Annoanaceae	23.	Euphorbiaceae	43.	Papaveraceae
4.	Apocynaceae	24.	Flacourtiaceae	44.	Piperaceae
5.	Araceae	25.	Guttiferae	45.	Poaceae
6.	Aristolochiaceae	26.	Helleboraceae	46.	Polygonaceae
7.	Asclepiadaceae	27.	Hippocastanaceae	47.	Polypodiaceae
8.	Balanitaceae	28.	Hypericaceae	48.	Ranunculaceae
9.	Berberidaceae	29.	Illiciaceae	49.	Rosaceae
10.	Boraginaceae	30.	Juglandaceae	50.	Rubiaceae
11.	Brassicaceae	31.	Labiatae (Lamiaceae)	51.	Rutaceae
12.	Burseraceae	32.	Lauraceae	52.	Sapindaceae
13.	Capparaceae	33.	Leguminosae (Fabaceae)	53.	Sapotoceae
14.	Capparidaceae	34.	Liliaceae	54.	Simaroubaceae
15.	Celastraceae	35.	Loganiaceae	55.	Solanaceae
16.	Chenopodiaceae	36.	Lycopodiaceae	56.	Stemonaceae
17.	Compositae (Asteraceae)	37.	Magnoliaceae	57.	Taxaceae
18.	Convolvulaceae	38.	Malvaceae	58.	Theaceae
19.	Cucurbitaceae	39.	Meliaceae	59.	Umbelliferae (Apiaceae)
20.	Dioscoreaceae	40.	Menispermaceae	60.	Verbenaceae

Source : Sukh Dev and Opender Koul, Insecticides of Natural Origin, Harwood Academic Publishers, Germany, India (1997).

Table 2.2. Neem parts yielding insecticidal compounds

Plant part	Composition
Seeds	Lipids, limonoids, organosulphur compounds
Stem bark	Phenolic diterpenoids
Root bark	Polysaccharides
Heartwood	Nimbolins A & B
Fruit coats	Protolimonoids
Leaves	Protolimonoidolides
Green twigs	C-*seco*-meliacinolides
Flowers	Flavonoids and their glycosides
Gum exudate	Complex condensate of protein and hetero-polysaccharides

parts of the tree are mentioned in Table 2.2 [4]. The important insect pests affected by the neem products are summarized in Table 2.3 [4].

Various behavioural and physiological effects of neem on insects

Neem products disrupt the development of eggs, larvae and pupae, metamorphosis, mating and sexual communication in insects. Besides blocking the moulting of larvae or nymphs, these prod-

Table 2.3. Organisms affected by neem products

Insects (major class)	Examples	Others
Coleoptera	Desert locust	Nematodes
Diptera	Brown plant hopper	Snails
Heteroptera	Cockroach	Crustaceans
Hymenoptera	Stored product insect	Fungi
Homoptera	Army worm	Plant viruses
Lepidoptera	Leafminers	Soil bacteria
Thysanoptera	European corn borer	
	Mosquitoes	
	Aphids	
	Fruitflies	
	Gypsymoth	
	Hornflies	
	Blowflies	

ucts also have repellent action and act as oviposition and feeding deterrent. They inhibit chitin synthesis and can be used as insect sterilant.

It is established that the seed kernel is the rich storehouse of tetranortriterpenoids or the meliacins. Over 125 compounds of this class have been characterised from kernel alone. Neem kernel constituents can be broadly classified under a) protolimonoids, b) limonoids with all the four rings intact and a modified side chain in the form of furan ring, such as azadirone (13), c) limonoids in which 'D' ring is expanded such as gedunin (14), d) pentacyclic limonoids like vepinin (15), limbocinin (16) and e) C-secolimonoids such as nimbin (17), salanin (18) (Fig. 2.2) and azadirachtins (19). Till today, neem products comprise of 23 natural congeners of this class but only ten have been called azadirachtin such as A, B, C, D, E, F. G, H., I, & K. The absolute stereochemical configurations of azadirachtin A, B, D, E, F and G (Fig. 2.3a) and H, I, K, L, vepaol (20) and isovepaol (21) are shown in Fig. 2.3b.

Rembold [5a, 5b] reviewed the ability of azadirachtin to interfere with neuroendocrinal control of metamorphosis in susceptible insects. Although the intake of azadirachtin by insects was very low yet it initiated many endocrinal and morphological changes, leading to growth retardation of the larvae which refused to eat the untreated leaves and ultimately died.

Lee *et al.* [6] gave considerable evidence of similarity of activity for azadirachtin with phytoecdysones and synthetic ecdysone antagonists. Structural similarities between deesterified azadirachtin and ecdysterone (Fig. 2.3c) suggest that despite the differences at the A/B ring junction for these compounds, the hydroxyl group in the 3-position is axial for azadirachtin and equatorial for ecdysteriods, and might occupy same relative space with respect to the rest of the molecule. Thus, either of the carbon frame work may hold good for the C-3 hydroxyl group in space so that interaction with an appropriate receptor can occur.

Important reaction products of azadirachtin [7]

Catalytic hydrogenation of azadirachtin (i) for 3.5 and 24 h in presence of Pd/C furnished dihydro (ii) and tetrahydro (iii) derivatives respectively in 75% and 53% yields. Dihydroderivative (ii) on treatment with $KMnO_4$, $NaIO_4$ in presence of Na_2CO_3 gave detigloylated product (iv) (36%) which also can be obtained by ozonolysis of (ii) followed by treatment with Na_2CO_3. The reaction of (i) with HOAc for 36 h gave addition product (va & vb) (95%). Bromination of (i) in presence of different alcohols yielded (vi. Fig. 2.4). The antifeedant activity of (ii) & (iii) is shown in Table 2.4.

Table 2.4. Antifeedant Index [(C-T)/(C+T)]%

Compounds	Mean (SE)	
	Spodoptera littoralis	Helicoverpa armigera
Azadirachtin	99 (1.1)	85 (2.7)
Dihydroazadirachtin	100 (0.0)	74 (8.4)
Tetrahydroazadirachtin	79 (10.6)	64 (4.6)

Synthesis of model compounds of azadirachtin

SAR studies on azadirachtin showed that the hydrofuran acetal portion of the molecule played an important role in determining antifeedant activity. This is borne out by the dramatic change in the activity due to structural variations effected at the 22, 23 double bond. In order to assess the level of antifeedant activity of different portions of the azadirachtin, a few model compounds were synthesized (Fig. 2.5). Three out of four fragments relating to hydrofuran acetal portion of azadirachtin yielded interesting biological activity (Table 2.5), i.e. compounds based on a & b models gave excellent antifeedant activity as compared to dihydroazadirachtin model (c).

Table 2.5. Antifeedant Index [(C-T)/(C+T)]%

Model compounds	Spodoptera littoralis	Helicoverpa virescens
Azadirachtin model, (a)	99 (1.1)	72 (4.8)
	66 (2.3)	54 (6.8)
Epoxide model, (b)	85 (6.9)	—
Dihydroazadirachtin model, (c)	100 (0.0)	67 (12.2)
	55 (7.3)	6 (11.6)

Total synthesis of azadirachtin

The synthesis of these model fragments as well as decalin fragment has been reported by Ley *et al.* [8a, 8b]. The decalin fragment did not show much antifeedant activity. During the early steps in the synthesis of decalin fragment, intramolecular Diels-Alder reaction was used to constitute two rings of the system with a high degree of stereochemical control. Studies on model compounds yielded valuable information concerning functional group manipulation in this complex molecule to establish the desired 1, 3-diaxial diol arrangement in ring A. Although these compounds were inactive, they contained eight of the stereogenic centres common to the natural product. These studies atleast paved the way to find an alternative convergent approach towards synthesis of azadirachtin, where these two fragments form the C.8-C.14 bond at a fairly late stage in the synthesis. A new decalin synthesis (Fig. 2.6) employing a silicon substituted dienophile, a key control element in the intramolecular Diels-Alder reaction was established. This group facilitated introduction of the C-3 hydroxyl group via a silyl Baeyer-Villiger process. The new route also

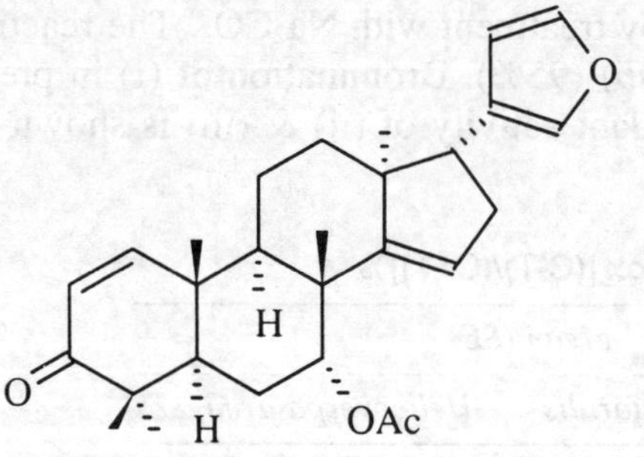

Fig. 2.2. Limonoids.

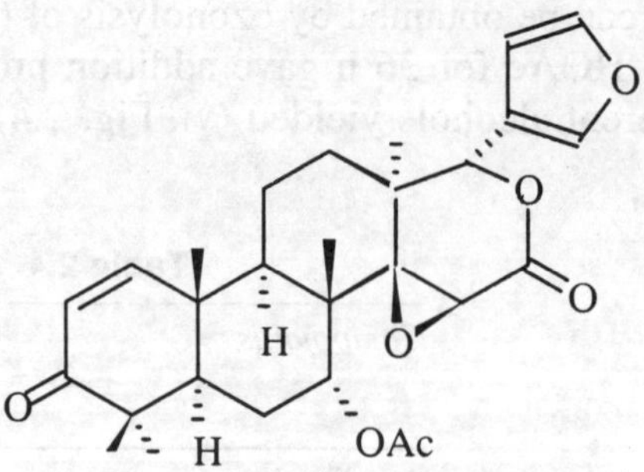

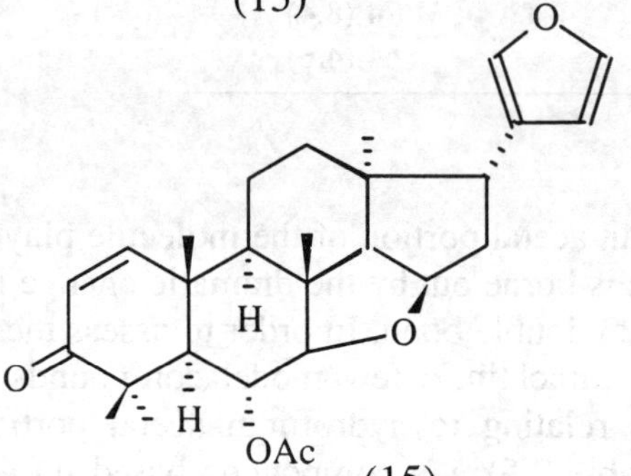

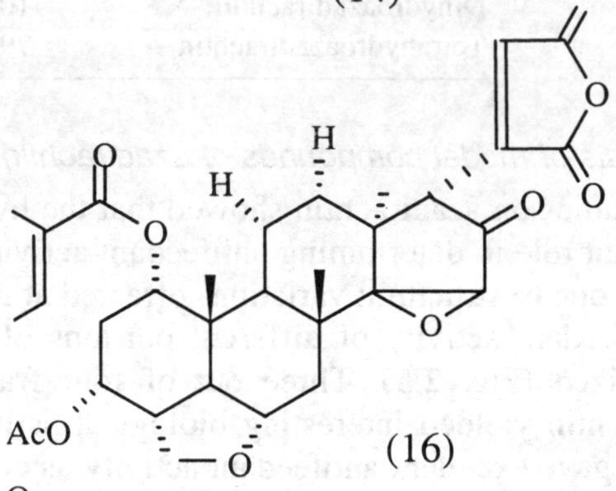

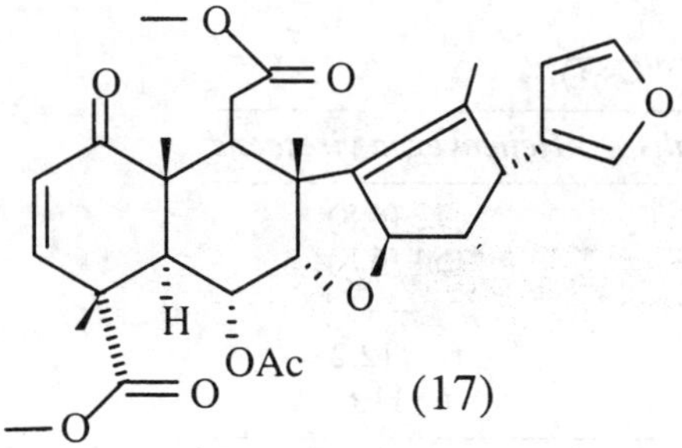

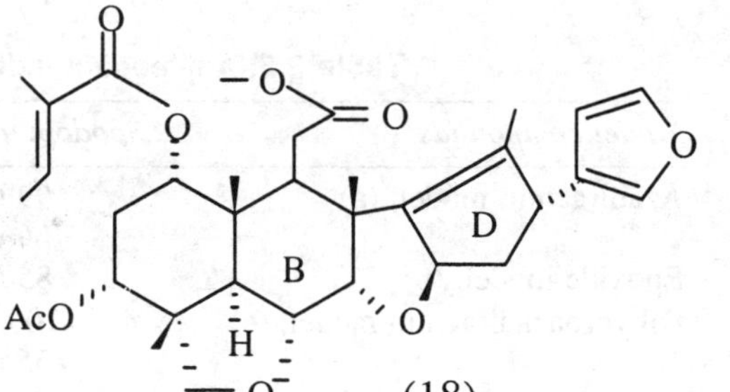

gave an access to a C-19 oxidised product which was otherwise difficult to achieve by the earlier route.

Finally, a concise route to the synthesis of remaining right hand acetal fragment in its correct optically active form was also achieved. The total synthesis may provide many more analogues and products which will shed more light in understanding the mechanism of action within various insect species and may lead to new opportunities in insect pest control.

The starting material for the preparation of acetal fragment was (−)-3-endo-bromotricyclo [3.2.0.0^{2,7}] heptane-6-one (i) which on reduction with sodium borohydride followed by oxidation with tetra-n-propylammoniumperruthenate (TRAP) gave bromide (ii). Reaction of (ii) with silver trifluoroacetate (CF$_3$COOAg, acetone/H$_2$O) gave ketoalcohol (iii) which on oxidation with m-chloroperbenzoic acid in presence of p-toluene sulphonic acid afforded (iv). Oxidation of (iv) with

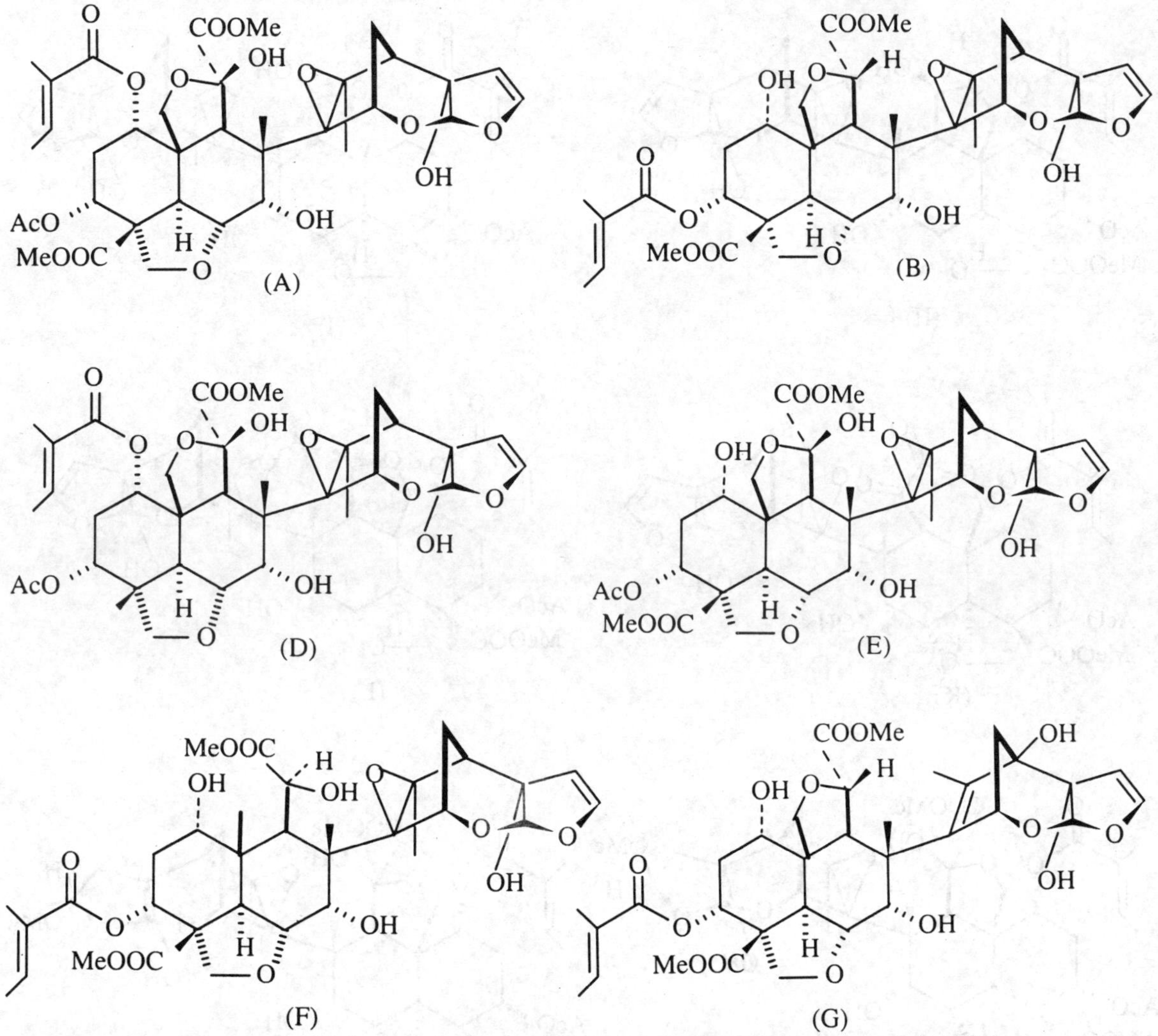

Fig. 2.3a. Azadirachtins (19).

LDA/molybdenum peroxide reagent (MoOPH) gave (v). Reaction of (v) with *t*-butyldimethylsilyl chloride in presence of imidazole followed by treatment with KDA in HMPA and allyl bromide yielded (vi). Reduction of (vi) with diisobutylaluminium hydride (DIBAL) gave (viia) which on ozonolysis yielded tricyclic lactol (viib). Methylation and selective deprotection with acetic acid yielded the desired product (viii) with high yield (Fig. 2.7).

MINOR INSECTICIDES OF PLANT ORIGIN

A number of plants have been investigated for the presence of various kinds of pest control chemicals. However, these products were not commercialized because of drawbacks of photolability and economical constraints. A few of these are discussed in this chapter.

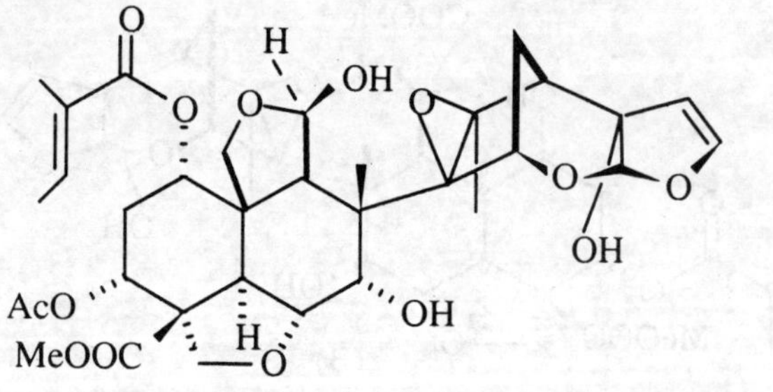

(H)

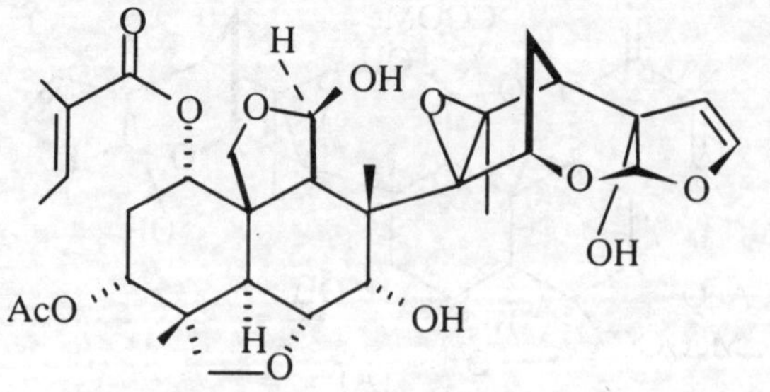

(I)

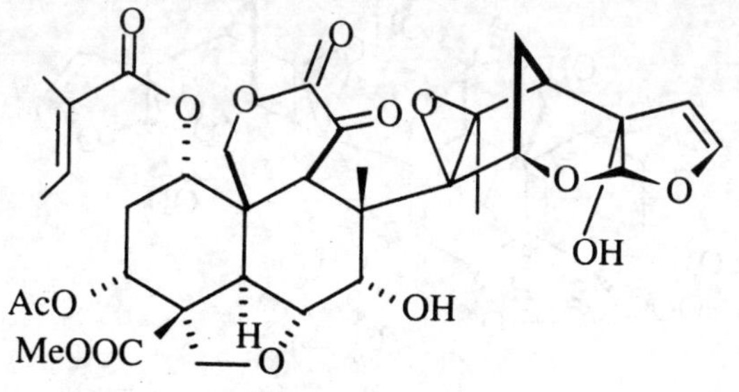

(K)

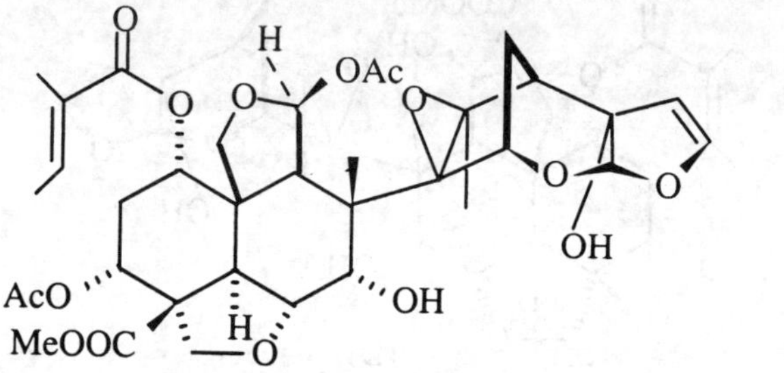

(L)

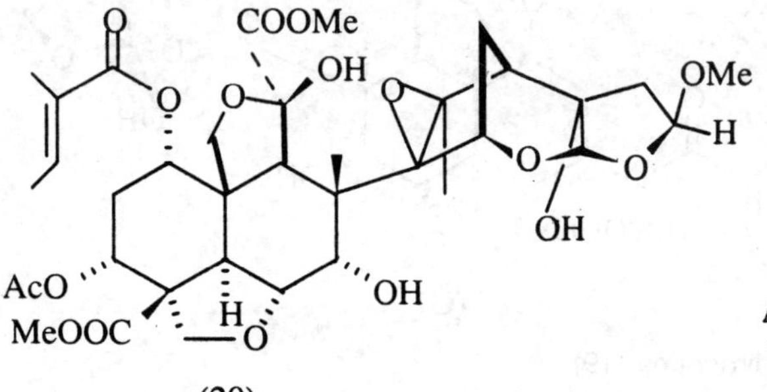

(20)

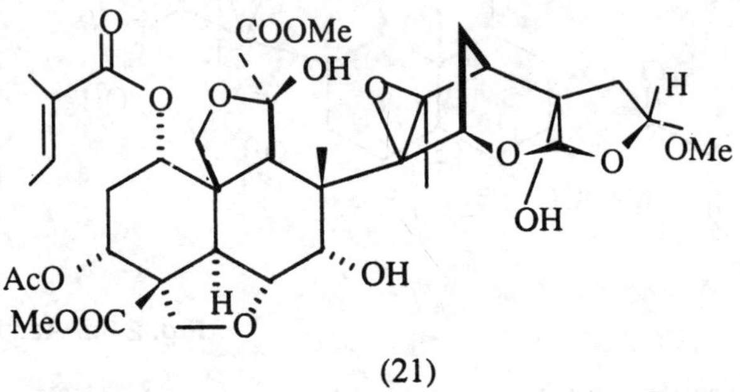

(21)

Fig. 2.3b. Azadirachtins.

3. Ryania

A review on Ryania is given by Crosby [9]. The Casida group utilizing radial TLC and preparative HPLC were able to isolate 10 out of 11 compounds, identified earlier and provided a procedure for monitoring different lots of plant samples [10].

The most active components of *Ryania speciosi* plant are Ryanodine (22) and its di-dehydro derivative (23). A pyridine-3-carboxylate analogue (24) of ryanodine showed little activity but its degradation products ryanodol (25) and di-dehydro ryanodol (26) were found to have good knock down property. A total synthesis of ryanodol has been reported [11].

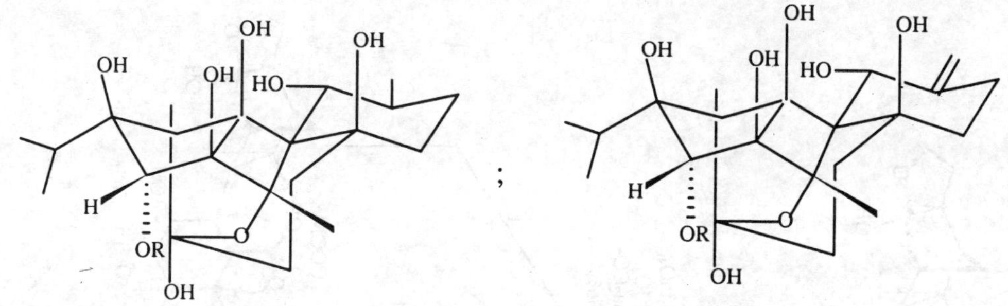

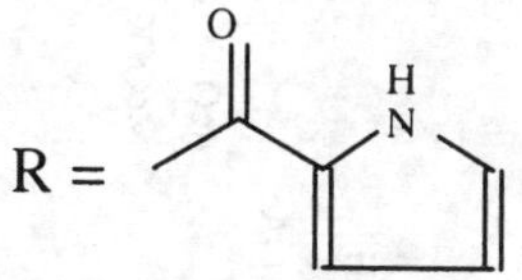

R = (22) ; R = (23)

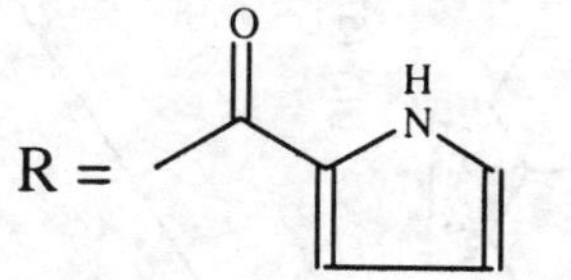

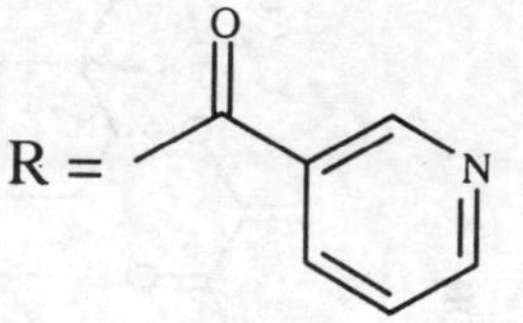

R = (24) ;

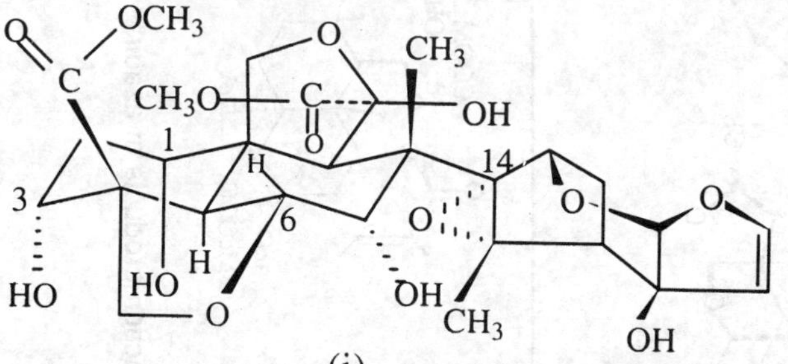

R = H (25)
R = H (26)

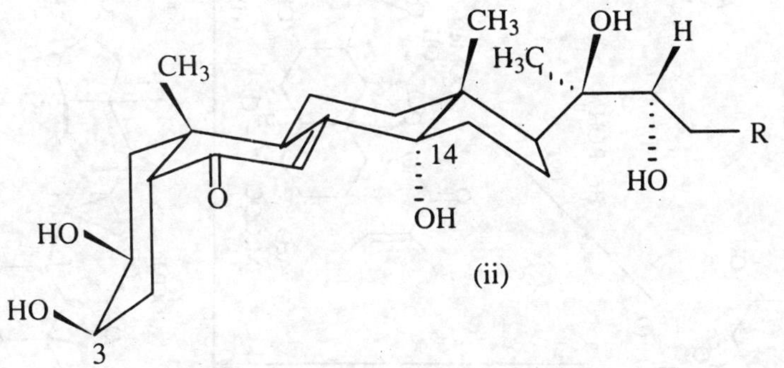

(i)

(ii)

R = (iii) R =

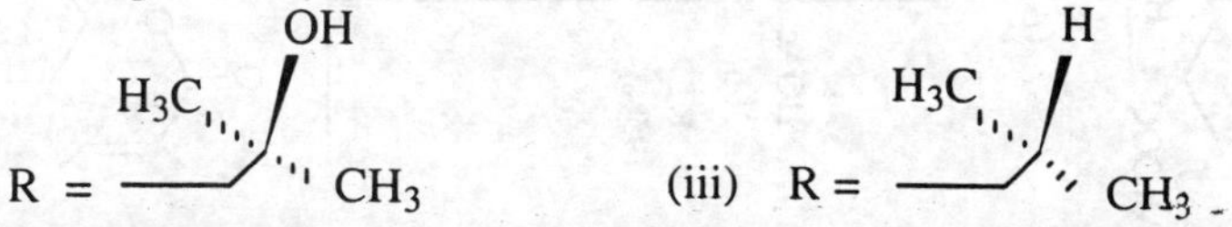

Fig. 2.3c. Stereostructural relationships between (i) azadirachtin skeleton, deesterified (1,3-dideacylate), (ii) ecdysterone (the insect moulting hormone, (20-hydroxyecdysone) and (iii) the phytoecdysone, ponasterone A. [Adapted from Agrochemicals from Natural Products by CRA Godfrey, 1994, 7]

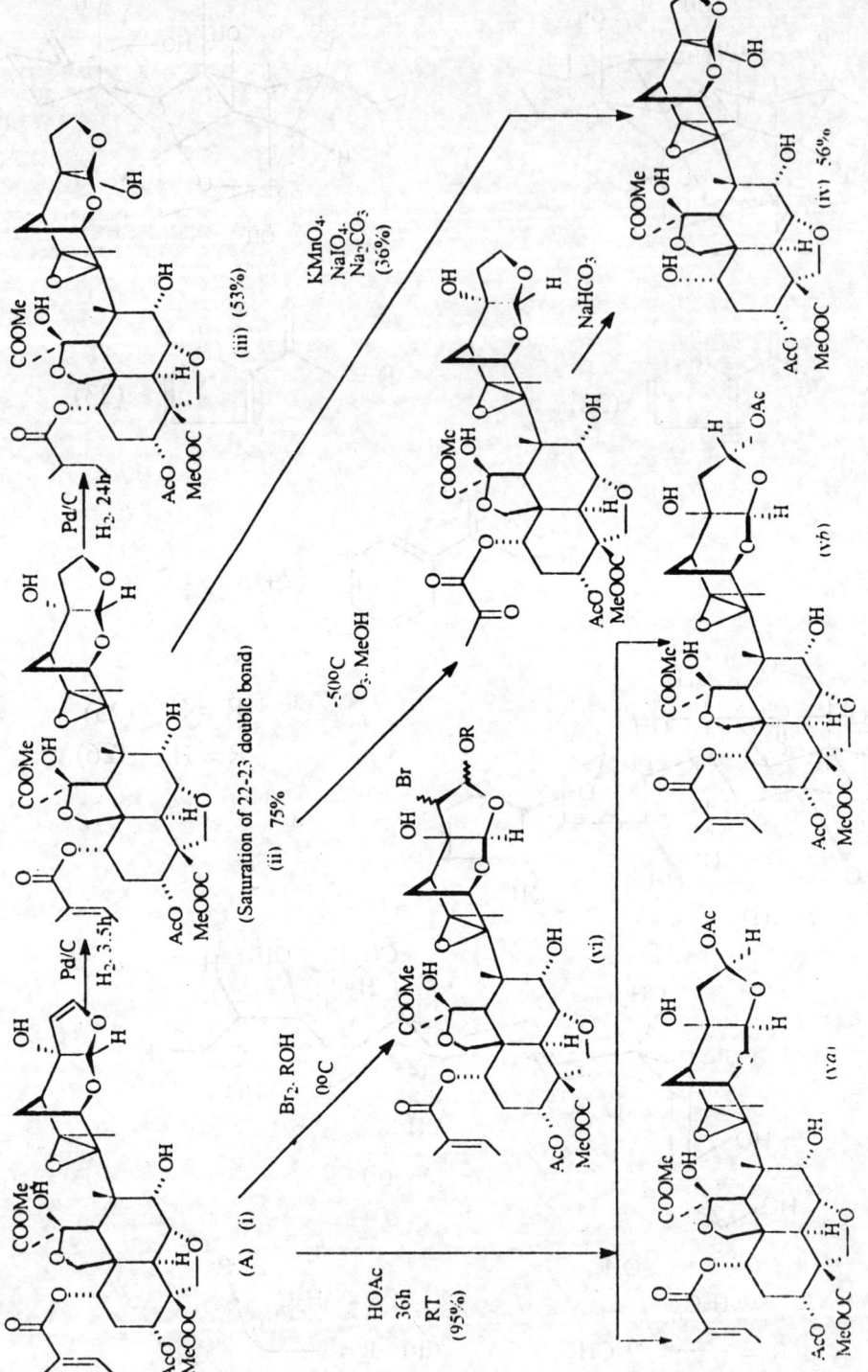

Fig. 2.4. Reaction products of azadirachtins [Adapted from Ley, 1990, 7].

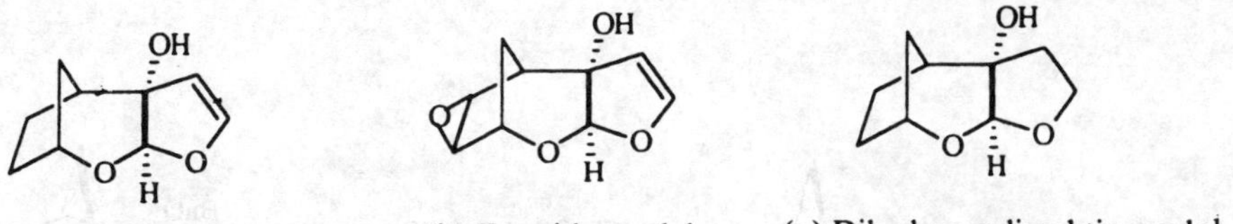

(a) Azadirachtin model   (b) Epoxide model   (c) Dihydroazadirachtin model

Fig. 2.5. Model compounds synthesised.

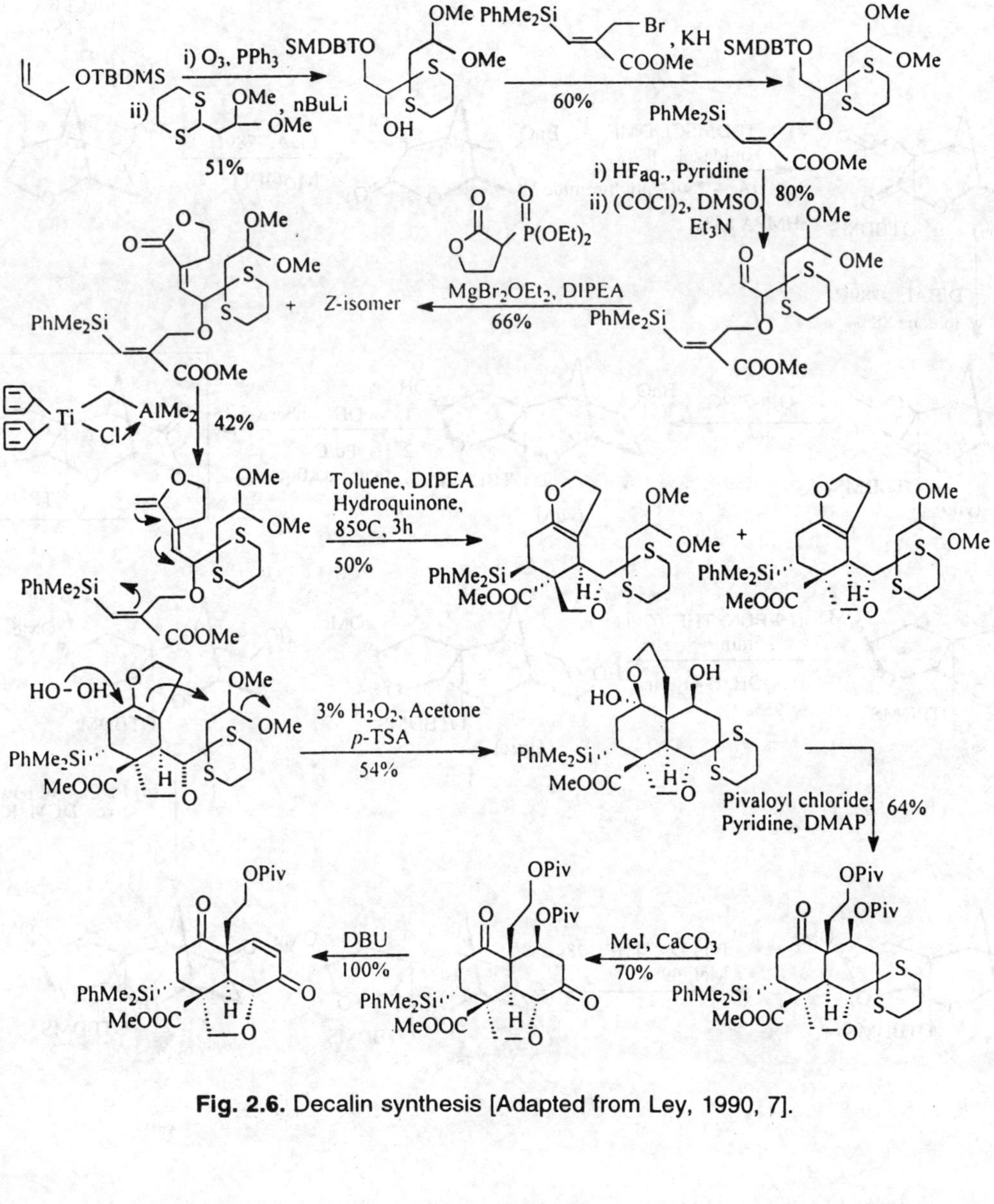

Fig. 2.6. Decalin synthesis [Adapted from Ley, 1990, 7].

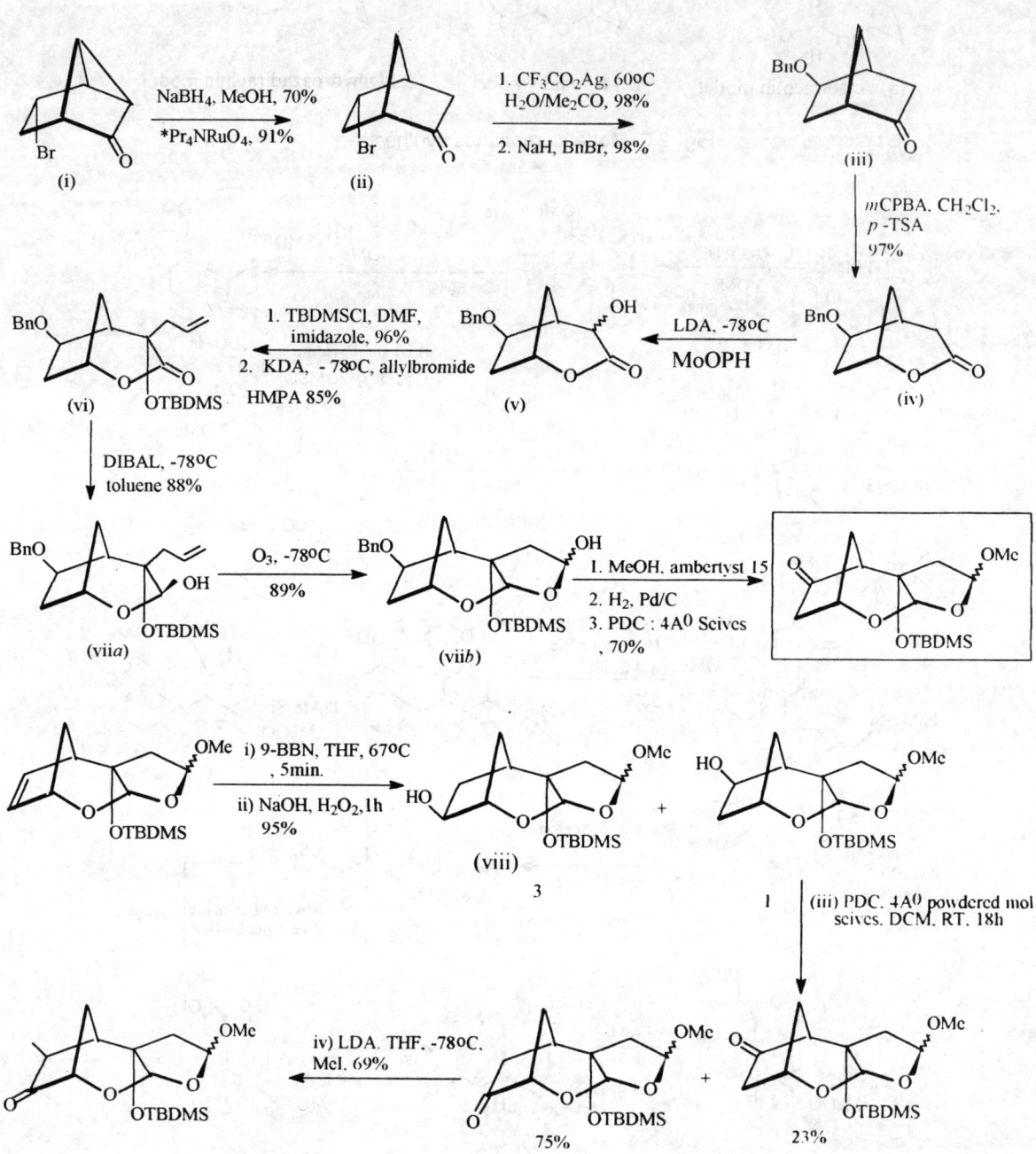

Fig. 2.7. Acetal fragment synthesis [Adapted from Ley, 1990, 7].

4. Sabadilla

The insecticidal property of Sabadilla is due to the presence of veratridine (27), the 3-veratroyl and cevadine (28), the 3-angenoyl esters of veracevine (29). These alkaloids act as neurotoxins which affect sodium ion channels in excitable membranes [12a].

   Studies on the effect of changing the nature of acyl group attached to 3-position of veracevine on insecticidal property showed that among benzoyl esters, 3, 5-dimethoxybenzoyl derivative (30) was most active [12b].

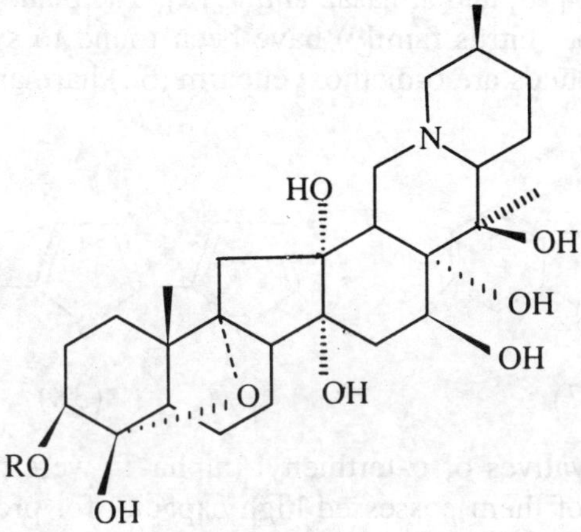

R = 3,4-$(CH_3O)_2$PhCO   (27)

R = 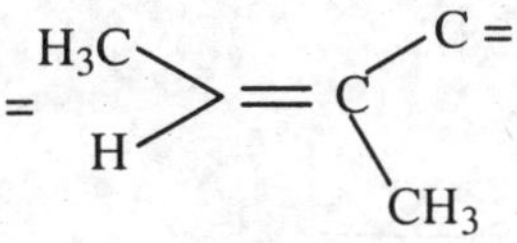 (28)

R = H   (29)

R = 3,5 - $(CH_3O)_2$ PhCO   (30)

5. Unsaturated amides

The insecticidal activity of unsaturated isobutylamides isolated for the first time from genera belonging to families Compositae, Piperaceae and Rutaceae has been reviewed by Jacobson [13]. The plant derived compounds pellitorine (31) and affinin (32) possessed good insecticidal properties. A potent compound dihydro-pipericide (33) was also isolated from fruits of black pepper, but due to their unstability, these compounds did not find practical application as pest control agents.

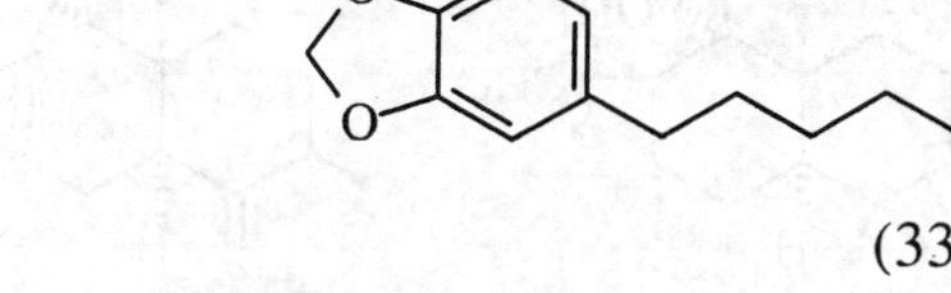

(31)

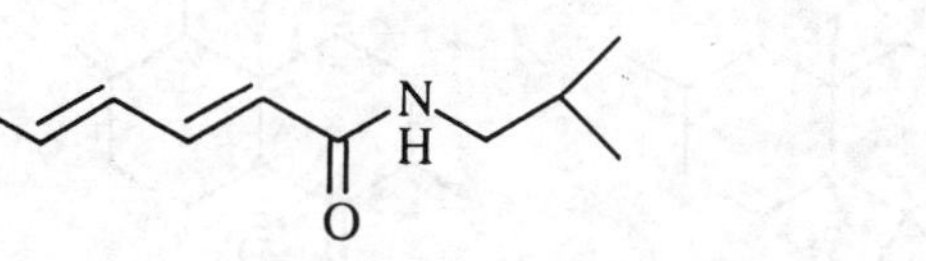

(32)

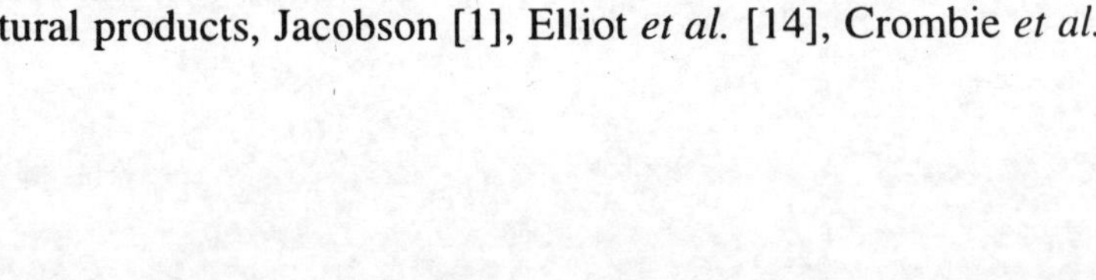

(33)

   A large number of synthetic analogues (34, 35) were prepared and tested but they were not as effective as the natural products, Jacobson [1], Elliot *et al.* [14], Crombie *et al.* [15].

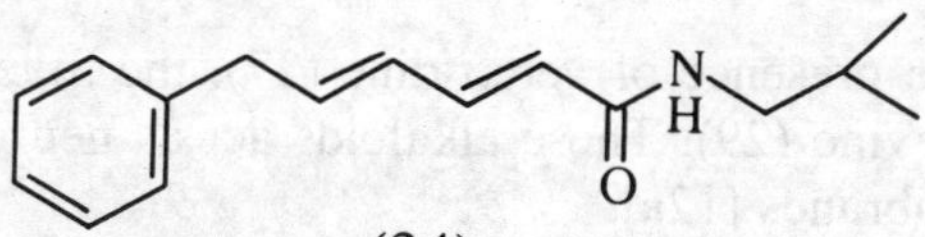

(34) 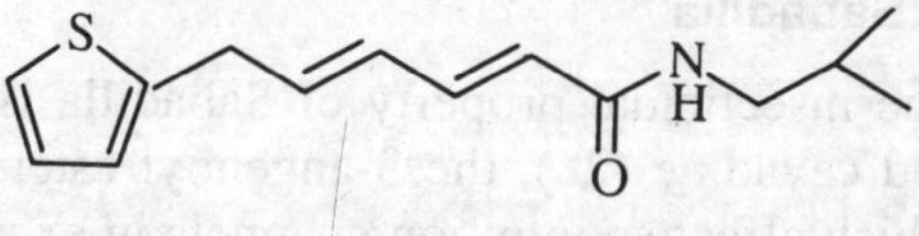   (35)

6. Light activated insecticides (Photoinduced toxins)

Plants yield a number of chemicals which can be excited by sunlight to induce toxicity against bacteria, fungi, nematodes and insects, Towers, [16] and Amasan *et al.* [17]. The plant Asteraceae (family Compositae) and members of Rutaceae (citrus family) have been found to synthesize a number of photoinducible toxins. Some compounds are 6-methoxyeuparm (36) harmane (37) and α-terthienyl (38).

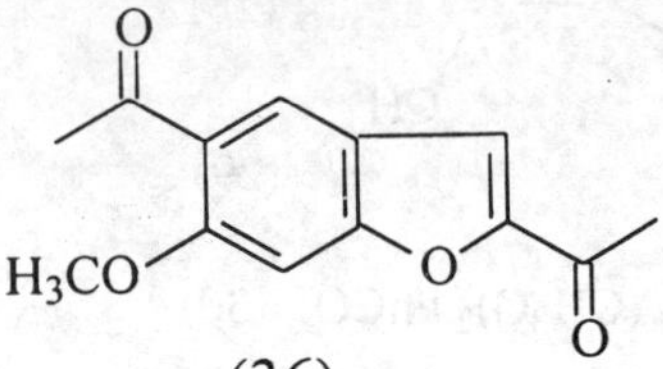

(36) 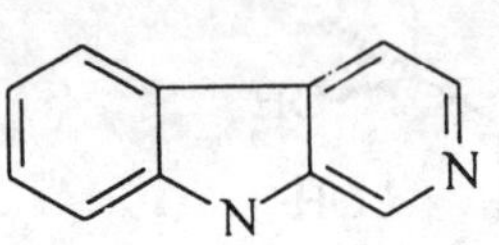   (37) 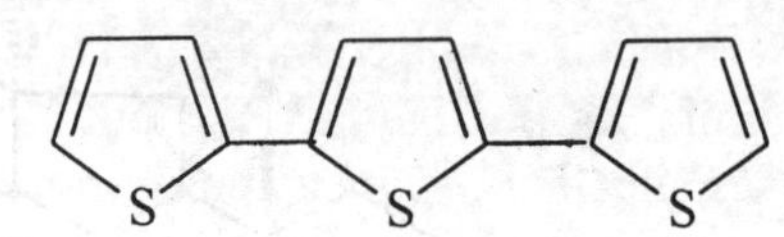   (38)

A number of synthetic analogues and derivatives of α-terthienyl (alpha-T) were prepared to evaluate structure-activity relationship. Many of them possessed high capacity for production of toxic singlet O_2 responsible for toxicity [Knox and Dodge, 18]. Further work showed that besides singlet O_2, partition coefficient or Hammett constants alone or alongwith the complex biological interactions between phototoxins and target, play an important role in the expression of phototoxicity.

7. Insect antifeedants and growth regulators

i) Drimane sesquiterpenoids [Jansen and de Groot, 19]

Extracts of the bark of *Drimys* spp. show pesticidal activity. Chemically they comprise sesquiterpenoids (39, 40) belonging to the group drimanes. Drimanes occur in nine plants belonging to four genera of the family Cannellaceae. Of these, *Warburgia* is endemic to East Africa, *Cinnamosma* to Madagascar and *Winterana* and *Cinnamodendron* to South America. The rich source of drimanes is the marsh pepper, *Polygonum hydropiper* L. (Polygonaceae). Several fungi such as *Penicillium purpurogenum* Stoll and *Mycocalia reticulata* Petch also produce drimanes. Drimanes are the versatile compounds showing antibacterial, antifungal, antifeedant, PGR, cytotoxic, phytotoxic, piscicidal and molluscicidal activity.

(39)   (40)   (41)   (42)

The most potent compound is polygodial (41) which can kill the fungi like *Saccharomyces cerevisiae* within 10 min at a concentration of 50 g/ml. It inhibits germination in rice and promotes root elongation. Other analogues exhibit similar type of PGR activity. Muzigadial (42) at 5 ppm acts as an effective helicocide against *Binmphalaria pfeiffer* and *B. glabratus*, the casual organisms of schistosomes and bilharzia. It also exhibits piscicidal and molluscicidal activity [Kubo and Nakanishi, 20]. It was observed that the activity was species-specific and analogue dependent.

ii) Clerodanes [Hanson, 21; Vadar, 22].

Clerodane belongs to a group of diterpenes having general structure (43). Till date more than 500 clerodanes have been isolated from numerous species of many plant families, micro-organisms and marine fauna. These compounds exhibit broad spectrum of biocidal activity such as antimicrobial, piscicidal; antipeptic-ulcer and as insect antifeedant. SAR studies on these compounds indicate that their activity is reduced if a) α-hydroxyl at C-2 is epimerised or acetylated or b) the ester groups are hydrolysed and c) the stereochemistry of epoxide is altered.

Further studies on ajugarin (44) indicated that the ajugarin analogues (44a, 44b and 44c) showed similar antifeedant activity but 44d & 44e were inactive. Pickett *et al.* (1987) [23] showed that 44a was inactive against aphids but active as antifeedant against the diamond-back moth, *Plutella xylostella* at 250 ppm.

	R_1	R_2	R_3	R_4
a,	CH_2	O	OAc	OAc
b,	CH_2	O	OAc	OH
c,	CH_2OH	OH	OAc	OAc
d,	H	COOCH_3	H	OAc
e,	CH_2	O	H	OAc

(43)   ;   (44)

iii) Chromenes

Chromenes (benzopyrans) and benzofurans of diverse structures were isolated from many species of higher plants such as Asteraceae. The biological activities along with chemical configurations of 167 isolated compounds have been reviewed by Proksch *et al.* [24]. Chromene derivatives, the precocene I and II (45a, 45b) isolated by Bowers and co-workers [25] from the plant *Ageratum honstonianum* prevent juvenile hormone synthesis. The desert sunflowers of the genus *Encelia* (Asteraceae) contain three major acetyl chromenes like encecalin (46a), demethylencecalin (46b), and demethoxyencecalin (46c) possessing antifeedant activity. In another experiment, the isolates of four tarweed *Hemizonia fitchii* A. Gray (Asteraceae) comprising some volatile compounds (46a, 46b) and 6-vinyl-7-methoxy 2,2-dimethylchromene (47) were found to reduce the population of *Culex pipiens*.

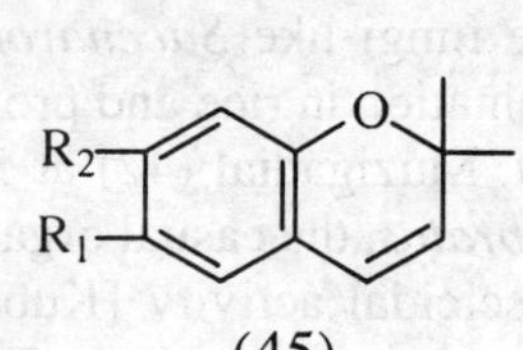

(45)

a) $R_1 = H$, $R_2 = OCH_3$
b) $R_1 = R_2 = OCH_3$

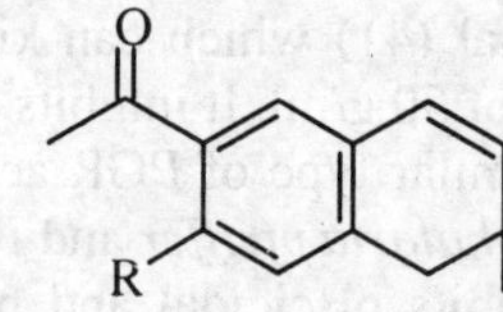

(46)

a) $R = OCH_3$
b) $R = OH$
c) $R = H$

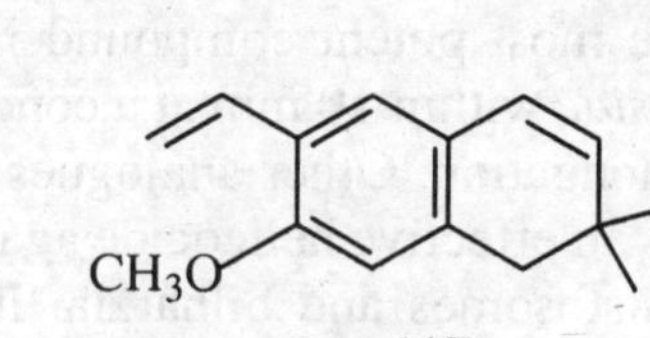

(47)

iv) Acetogenins

Asimisin, a trihydroxy-bis-tetrahydrofuran fatty acid γ-lactone (48a) was isolated from an extract of the bark of paw tree *Asimina triloba* Dunal (Annonaceae) [26]. Bioassay showed that asimisin and its analogue annonin I (48b) possessed insecticidal activity at 10-15 ppm against Mexican bean beetle.

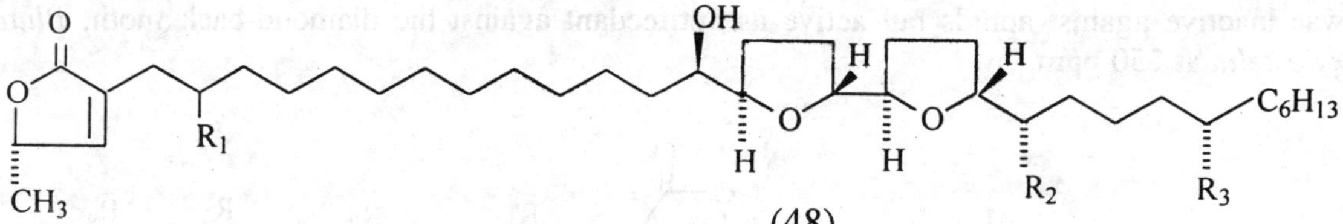

(48)

	R_1	R_2	R_3
a)	OH	⟍ OH	H
b)	H	---- OH	OH

Similarly, the seeds of *Annona squamosa* L. were reported to contain several acetogenins such as annonin (48b) asimisin (48a) and related compounds [27] which exhibit ovicidal and larvicidal activity. Evidences suggest that the insecticidal activity of the acetogenins may be due to their ability to inhibit electron transport at complex 1 of the mitochondrial respiratory chain [28].

v) Miscellaneous insecticides

The cultures of *Streptomyces antibiotics* strain DSM 1951 produced two cyclic phosphates (49a, 49b) [29] whose activity was comparable to the synthetic insecticides monocrotophos and carbofuran. Similarly the sponge *Ulosa ruetzleri* was reported to have hydantoin phosphonate (50) active against tobacco hornworm, *Manduca sexta* [30]. The petroleum ether extract of South African plant *Aloe pluridens* (Liliacae) exhibited insecticidal activity against variety of mosquito species and the southern armyworm (*Spodoptera eridania*). The most active compound was identified as pluridane (51). Several phenols and aromatic ethers such as *trans*-anethole (52) eugenol (53) and estragole (54) were isolated from anise plants (*Pimpinella anisum* L.) having insecticidal activity against houseflies [31]. Synthetic analogues of phenolic compounds (55) isolated from the Panamanian hardwood *Dalbergia retussa* noted for resistance towards fungi and other microorganisms, yielded compounds with fly-sterilant and mosquito growth inhibitory properties [32]. The mushroom *Clitocybe inversa* contains nitro containing nucleoside clitocine (56) showed strong activity against the pink bollworm (*Pectinophora gossypiella*) [33].

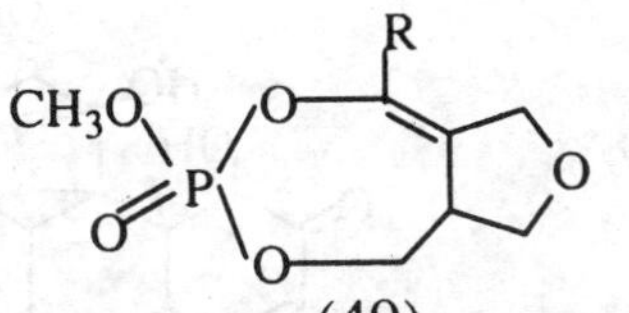

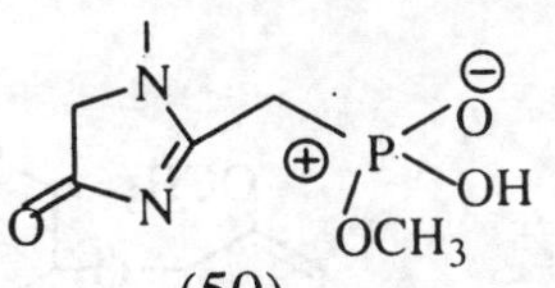

(49)

a, R = CH3; b, R = n - C3H7

(50)

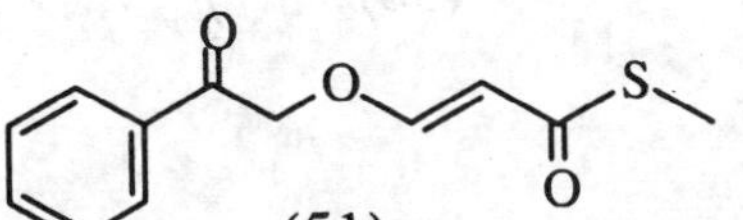

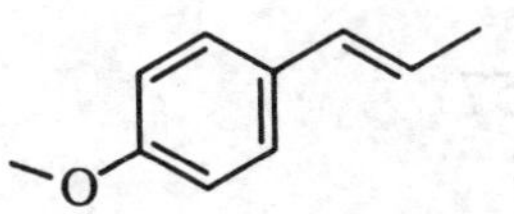

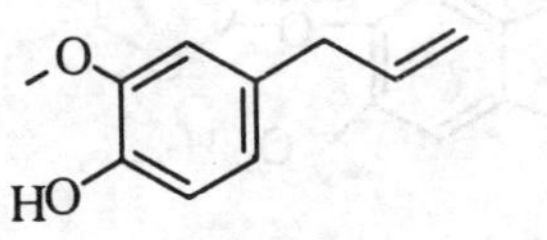

(51) ; (52) ; (53)

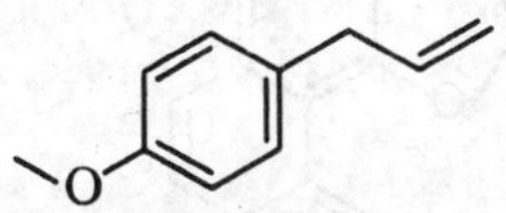

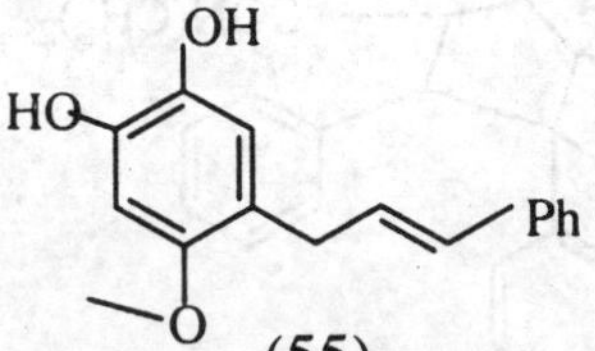

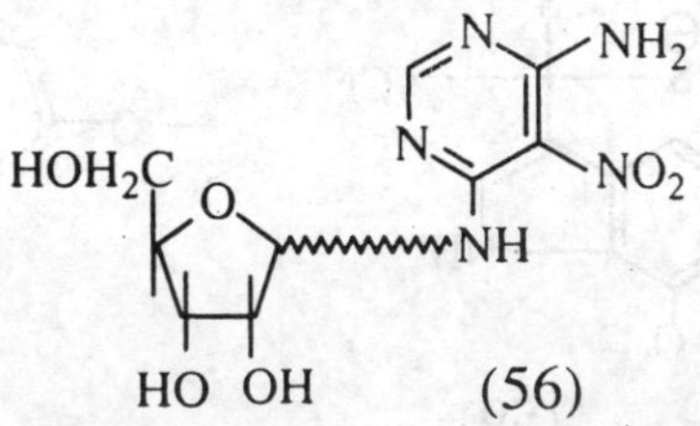

(54)   (55)   (56)

The extract of roots of *Phryma leptostachya* L. contains a number of lignans. One of them is sesquilignan haedoxan A (57) which imparts resistance against several lepidopterous insects and houseflies. Compared to this the synthetic analogues and other natural lignans exhibit reduced activity [34]. The extracts of *Simaba multiflora* A. Juss and *Soulamea soulameoides* Nooteboom were mixed with the feed for examining their growth inhibitory and insecticidal activity against tobacco budworm (*Helicoverpa virescens*) and antifeedant activity against *H. virescens* and the fall armyworm (*Spodoptera frugiperda*). The most effective compound identified was 6-senecioyloxychaparrinone (58) [35]. The seed extracts of *Dithyrea waslizenii* (Cruciferae) possessed antifeedant property and the active ingredient was characterised as dithreanitrile (59). The dried twigs of tropical Asian tree *Aglaia odoratalour* extracts contained mocaglamide (60) and three analogues having insecticidal property [36]. The root extracts of *Aristolochia albida* yielded aristolochic acid showing antifeedant activity [37]. The leaf extract of cocoa plant containing the alkaloid cocaine exhibited both antifeedant and insecticidal activities which were attributed to the blockage of uptake of the neurotransmitter octopamine [38].

INSECTICIDES OF MARINE ORIGIN

8. Nereistoxin

The insecticide 4-dimethylamino-1,2-dithiolane (62) known as nereistoxin was isolated from the marine annelid *Lumbriconeren heteropoda* by Mareuz in 1934. The structural determination and synthesis of this compound was done out by Konishi [39]. A large number of analogues were synthesised but none of them were found active.

The structure activity relationship studies showed that dimethylamino group in the 4-position was essential for imparting high activity. The activity was probably due to blocking action in the central nervous system [40]. Recent studies indicated it to be a partial agonist of nicotinic acetyl-

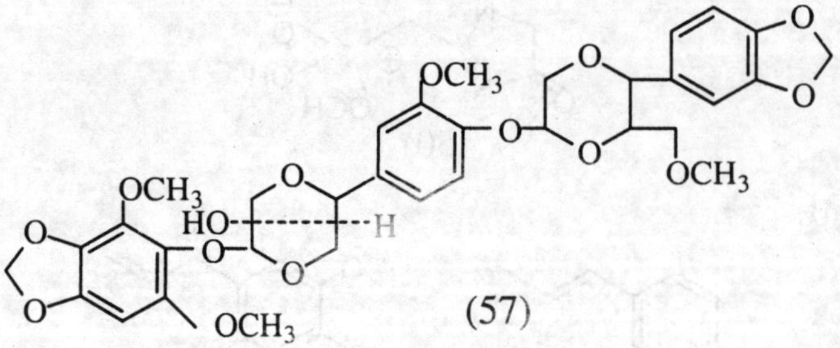

(57) 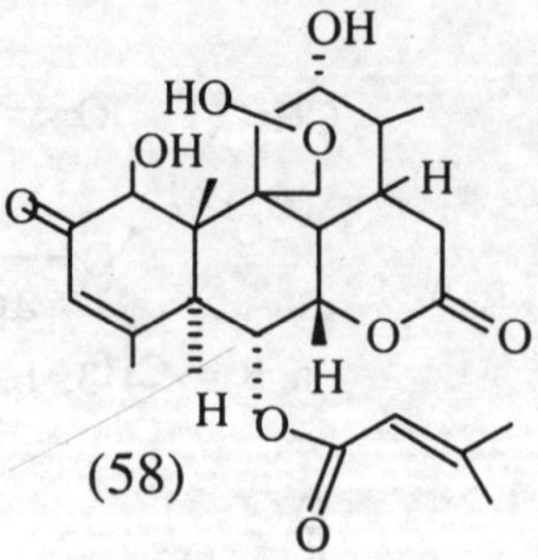 (58)

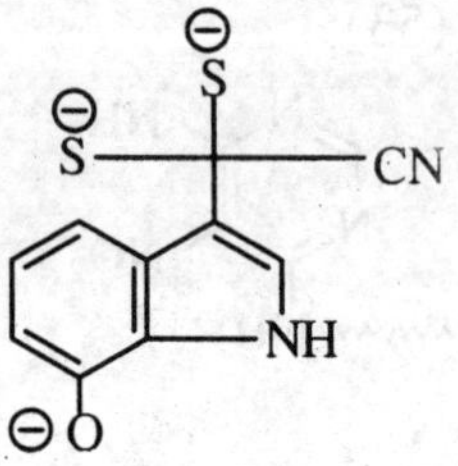

(59) 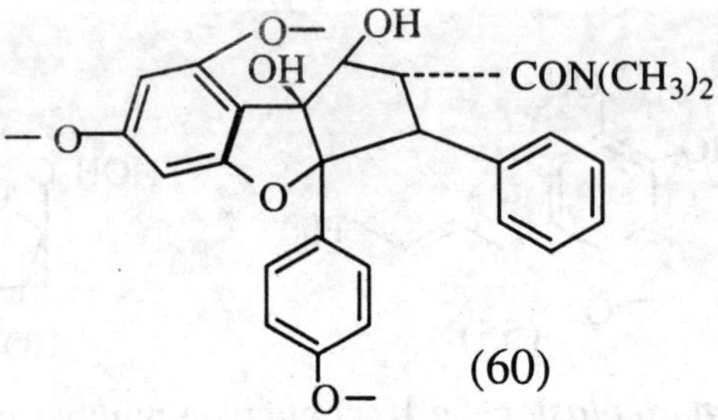 (60) 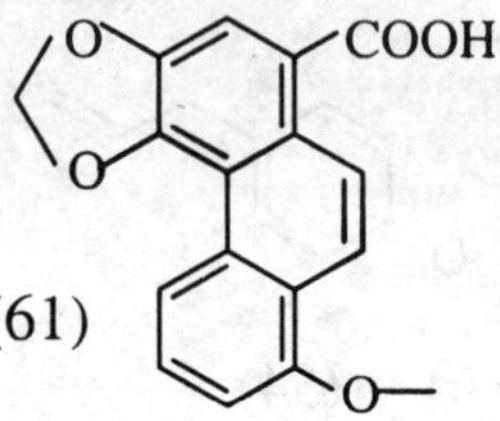 (61)

choline receptors at low concentration [41] and inhibitor of the receptor channel at higher concentrations.

Synthesis of a large number of analogues and SAR study identified three synthetic analogues the binamide, cartap, (63a) most effective against rice stem borer (*Chilo suppressalis*); the bisbenzenesulfonate, bensultap (63b) effective against the colorado potato beetle (*Leptinotarsa decemlineata*) and the thiocyclam (*Chara globularis*) (64). Similar effective insecticidal preparations are commercially available.

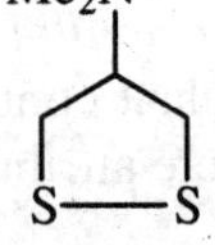

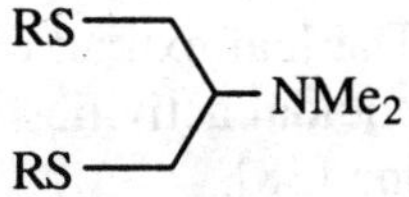

a, R = H$_2$N - CO

b, R = C$_6$H$_5$SO$_2$

(62) (63) (64)

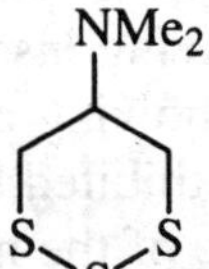

Domoic acid

The insecticidal activity of sea weeds *Chondria armata* and *Dignea simplex* was due to domoic acid (65) and α-kainic acid (66) and was comparable to that of pyrethroid phenothrin, DDT and BHC etc. A number of synthetic analogues of domoic acid and α-kainic acid have been prepared and evaluated for their bioactivity [42].

Studies on American cockroaches showed that very low concentrations of iomoic acid increases the sensitivity of insect neuromuscular junction to glutamic acid causing contraction of the excised hindgut. The natural neuropeptide proctolin also causes the same effect suggesting that domoic acid acts by binding to proctolin receptors.

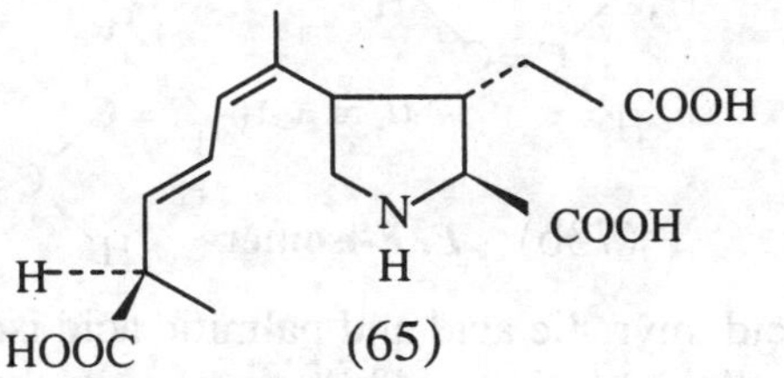

'Insecto', a diatom frustule based commercial preparation effectively controls weevils, moths and beetles. The silica present in the cell walls of diatoms penetrates the exoskeleton of insects leading to swift death due to dehydration. The diatoms are unicellular algae which form marine bloom and on death form large deposits on the sea bed.

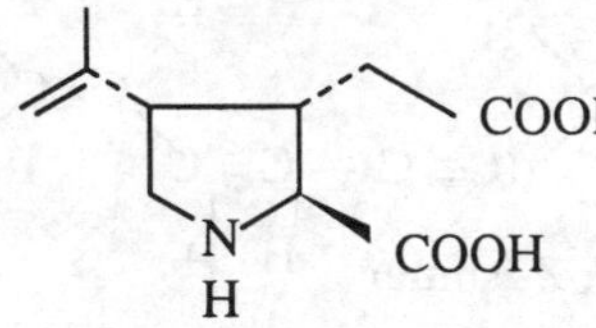

NEMATICIDES OF PLANT ORIGIN

A wide variety of plants and root exudates possess characteristic property to control specific nematodes. Marigolds (*Tagetes el. ta* L.) effectively control *Meloidogyne, Pratylenchus* and other nematode genera. The bioactive chemicals present in the roots of merigolds have been identified as α-terthienyl (38) and 5-(3-buten-1-ynyl) 2,2'-bisthienyl (67). Two compounds, 1-tridecanene-3,5,7,9,11-pentyne (68) and 2,3-dihydro-3-methylene-6-methylbenzofuran (69) having nematicidal activity were identified from the roots of *Helenium* hybrid. Asparagusic acid (70) isolated from the roots of *Asparagus* inhibited the growth of *Heterodera rostochiensis* and *Meloidogyne horpla*. Similarly the roots of *Bocconia cordata* contained alkaloid bocconine (71) which exhibited nematicidal activity. Angular pyranocoumarin (72) extracted from *Angelica archanangelica* also exhibited good nematicidal activity [43a]. The flowers, stems and roots of the safflower plant *Carthamus tinctorius* L. were found to contain polyacetylene compounds like 3-*cis*, 11-*trans* trideca-1,3,11-triene-5,7,9-triyne (73a) and 3-*trans*, 11-*trans* trideca-1, 3,11-triene-5, 7,9-triyne (73b), having nematicidal property [43b]. Many plants of the family *Cruciferae* when damaged by disease or herbivorous attack, release volatile *iso*thiocyanates by enzymatic action on glucosinolates stored in the plant, which have nematicidal action comparable to synthetic dazomet [43c].

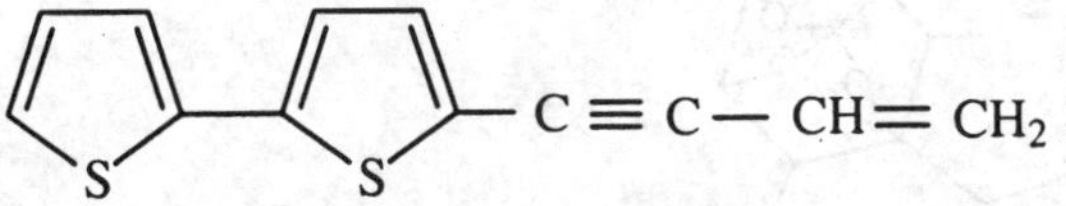

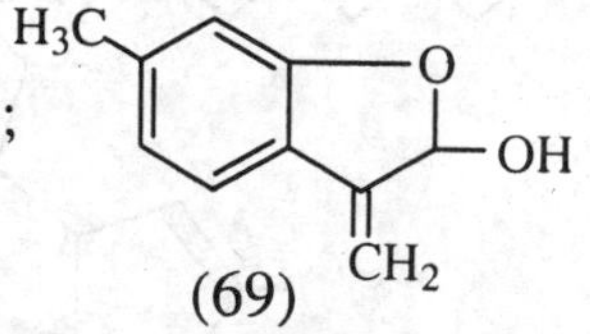

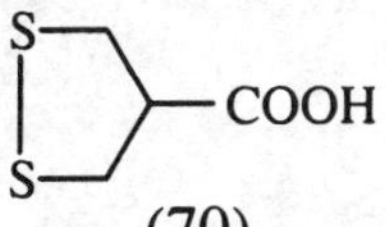

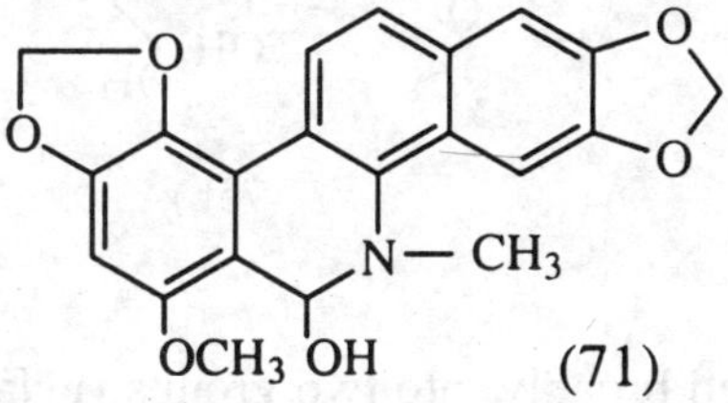

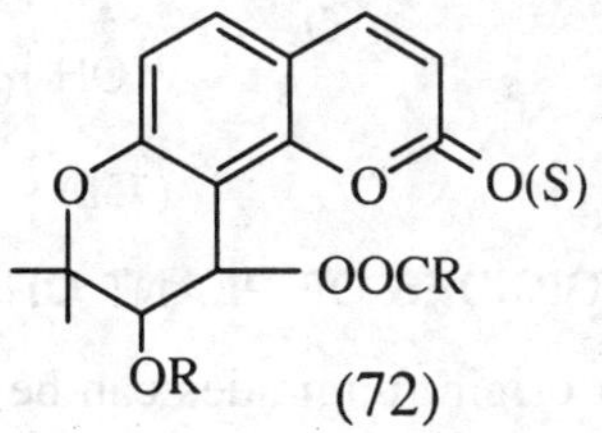

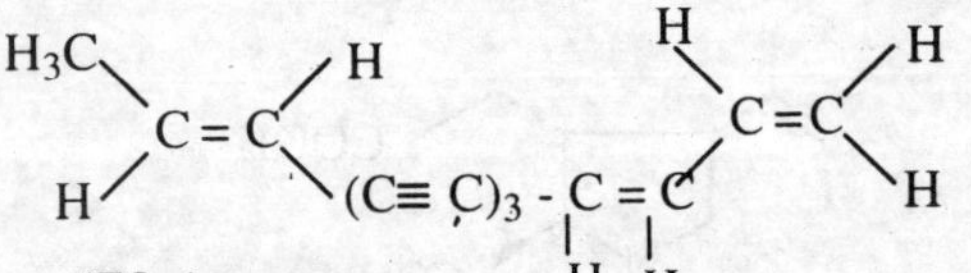

(73a) *E, Z*-isomer 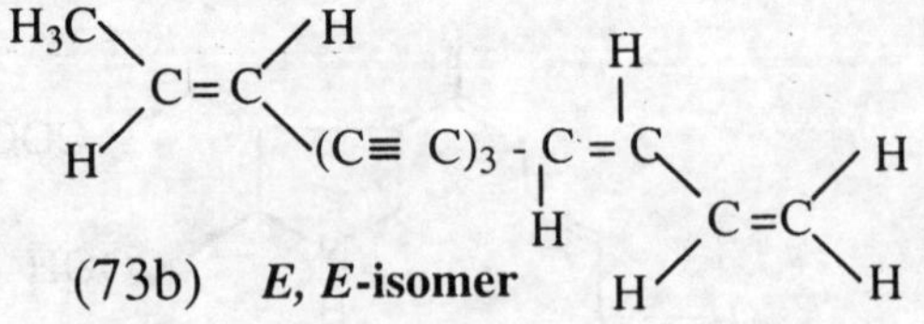   ;   (73b) *E, E*-isomer

A number of fatty acids such as undecanoic acid, myristic acid and palmitic acid isolated from roots of *Iris japonica* Thumb showed strong nematicidal activity [43d].

The alkaloids matrine (74a) and sophocarpine (74b) [44] isolated from the epigeal part of *Sophora flavescens* were found to be active against pine wood nematode (*Bursaphelenchus xylosphilus*).

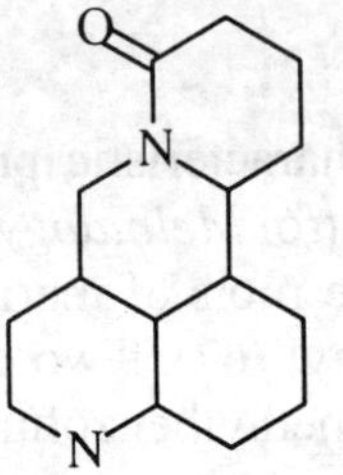

(74a) 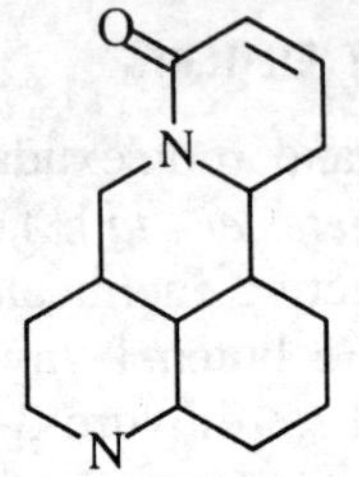   (74b)

Eleven neem constituents such as nimbin, solanin, azadirachtin and related compounds isolated from neem seed kernels exhibited activity against root knot nematode *M. incognita* [45a]. Two more diterpenes such as odoracin (75a) and odoratrin (75b) were isolated from the roots of *Daphne odora* [45b, 45c] having nematicidal properties.

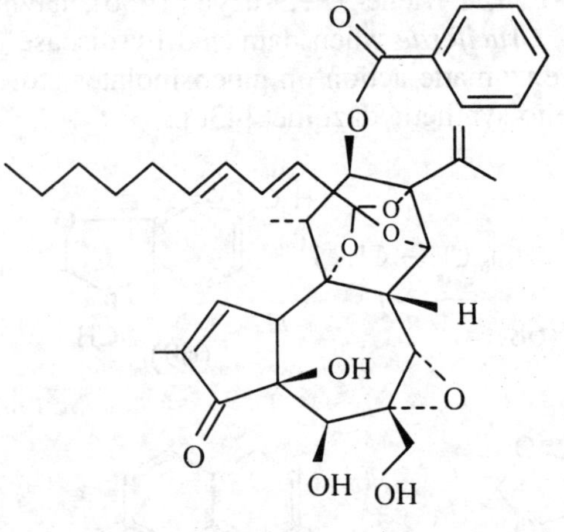

(75a) 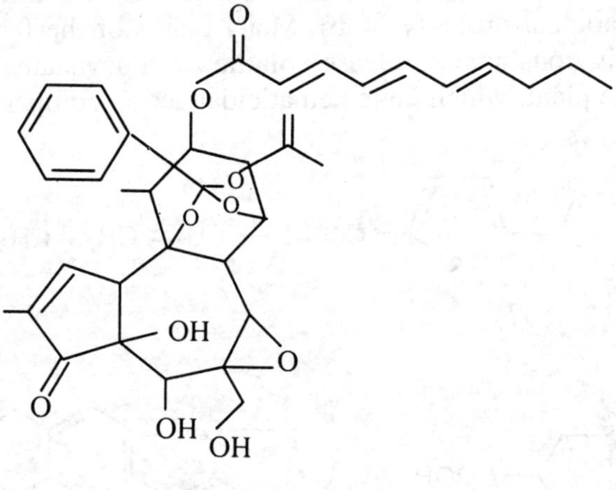   (75b)

FUNGICIDES OF PLANT ORIGIN

Plant origin fungicides can be classified broadly into two groups i) Natural, ii) Phytoalexins.

Natural fungicides are those plant products which are responsible for the natural resistance of plants towards fungal pathogens. These are metabolic products of plants whose concentration varies with plant growth and environmental factors.

Phytoalexins are produced by plant cells in response to an infecting organism causing metabolic interaction between host and the infectant. However, in recent years, phytoalexins have been detected in traces in healthy tissues also indicating that their presence is not always dependent on infection. A large number of earlier known natural products [46a, 46b] like capillin (76), lupulone (77), pinosylvin (78), juglone (79), α, β and γ-thujaplicin benzoxazolones possess varying degree of fungicidal activity.

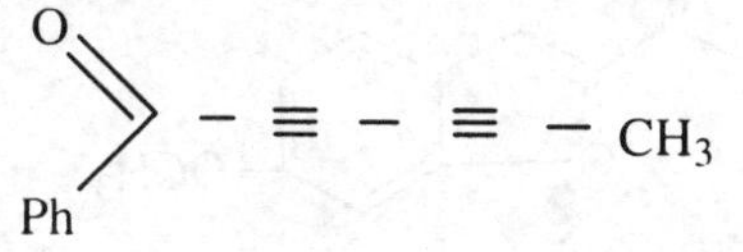

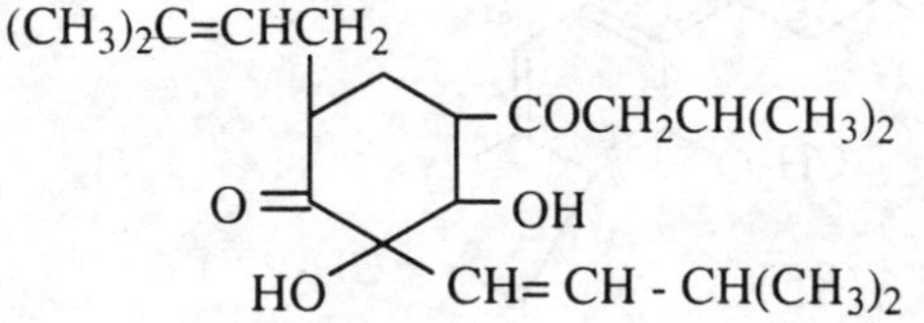

(76)   (77)

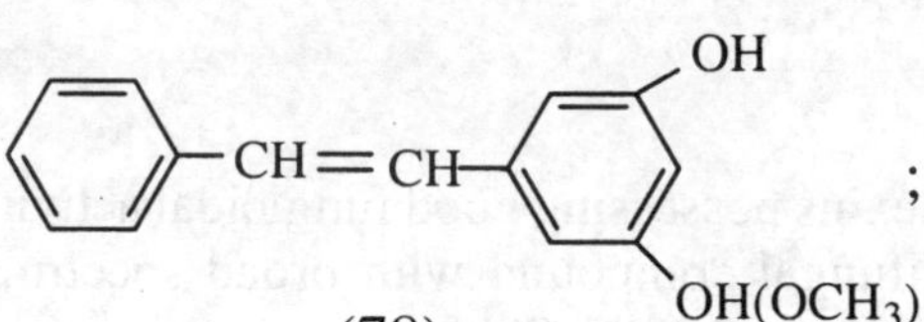

(78) 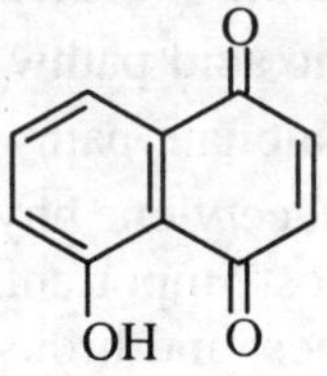   (79) 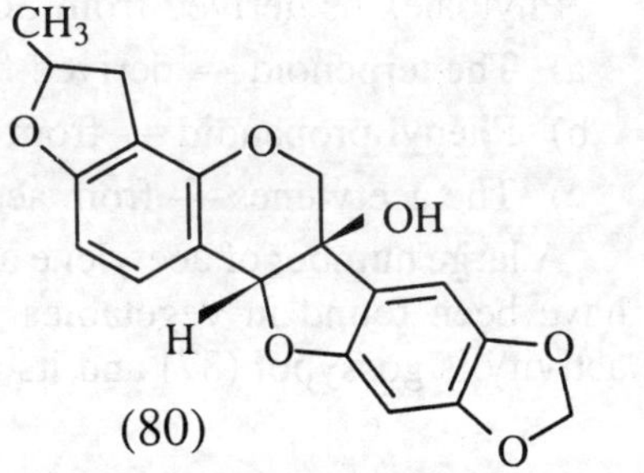   (80)

The families Solanaceae and Leguminosae have been investigated thoroughly to form a source of large number of phytoalexins of economic importance. The first phytoalexin identified was (+) pisatin (80).

Excellent reviews on phytoalexins are available [47, 48] in literature. Various classes of phytoalexins viz. isoflavans, isoflavanones, isoflavones, pterocarpans, coumestans and pterocarpenes have been identified and their toxicity to fungi has been ascertained [49]. A few most active antifungal compounds like genistein (81), luteone (82), pterocarpan (83), coumestrol (84), sativan (85) whose ED_{50} values (concentration of chemicals required to cause 50% inhibition) are below 50 µg/ml, have been identified. Some of the legumes also produce non-isoflavavanoid phytoalexins such as vignafuran (86).

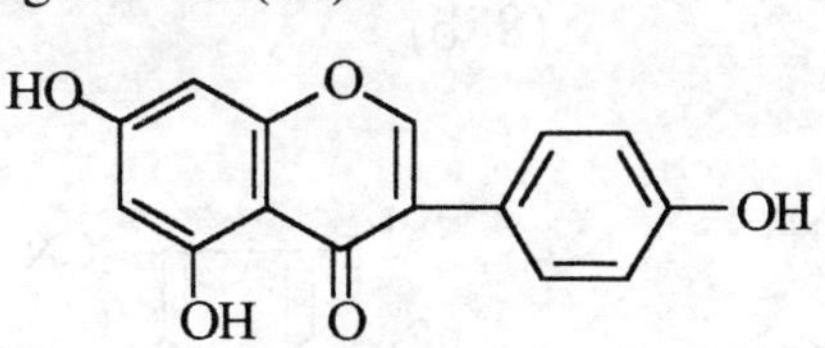

(81) 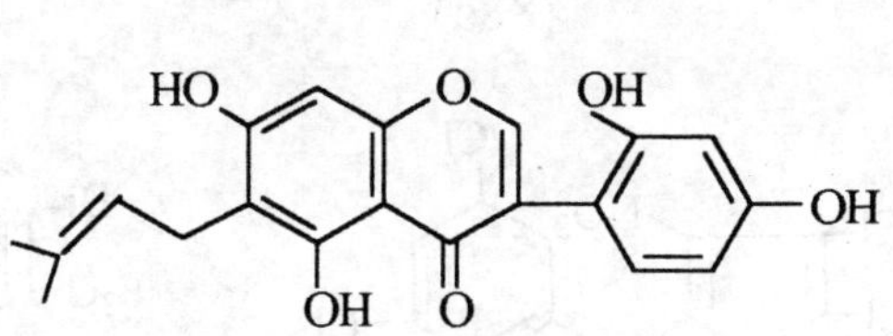   (82)

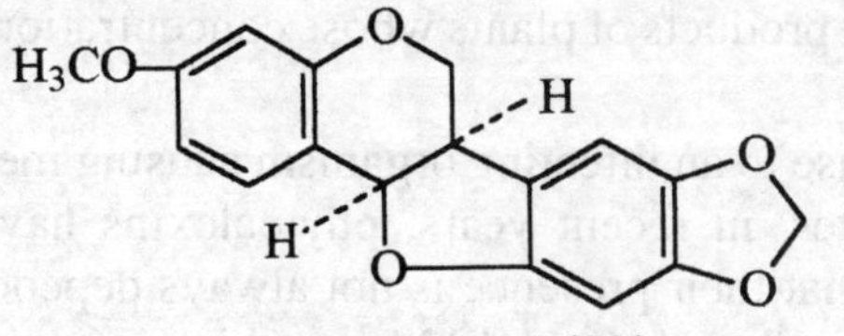

(83)

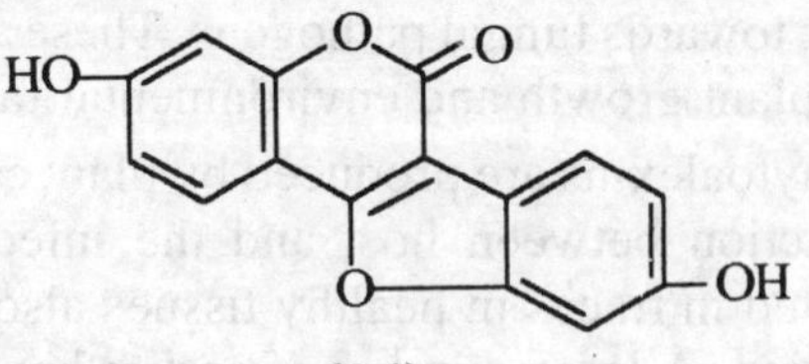

(84)

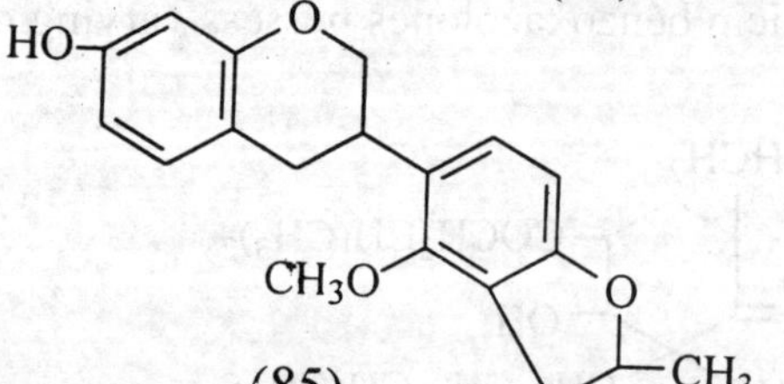

(85)

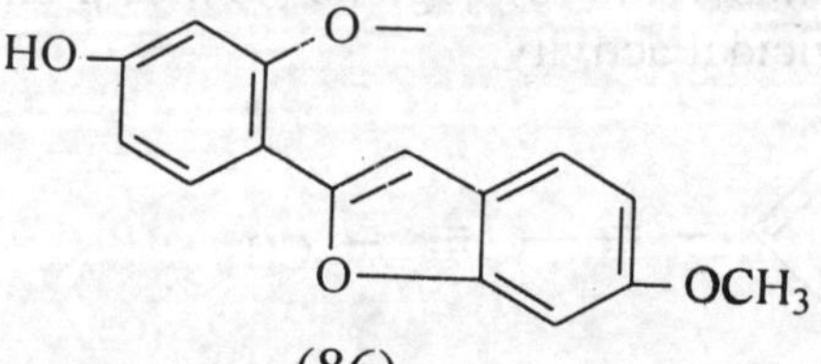

(86)

Phytoalexins derived from Solanaceae are classified into three categories :

a) The terpenoid — derived from acetate-mevalonate pathway.

b) Phenyl propanoid — from shikimic acid pathway.

c) The acetylenes — from acetate-malonate pathway.

A large number of acetylene and polyacetylene phytoalexins possessing good fungicidal activity have been found in vegetables. The most important antifungal compound with broad spectrum activity is gossypol (87) and its analogues, and broussonin A & B (88a, 88b).

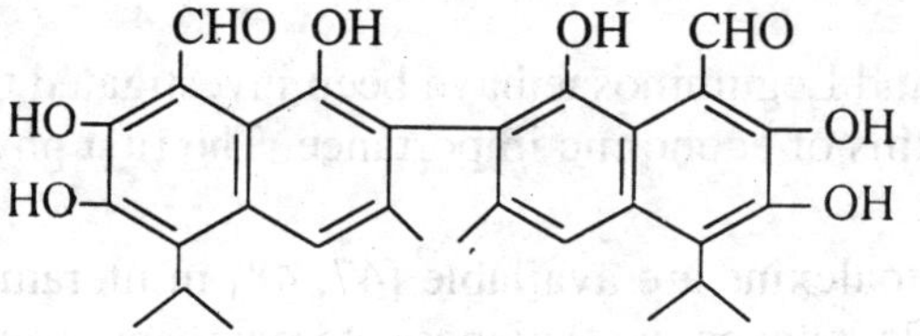

(87)

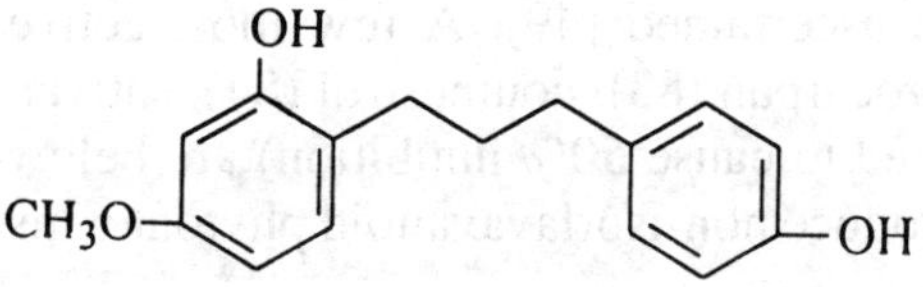

(88a)

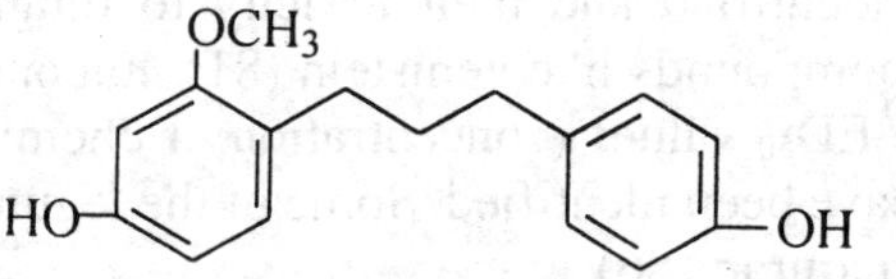

(88b)

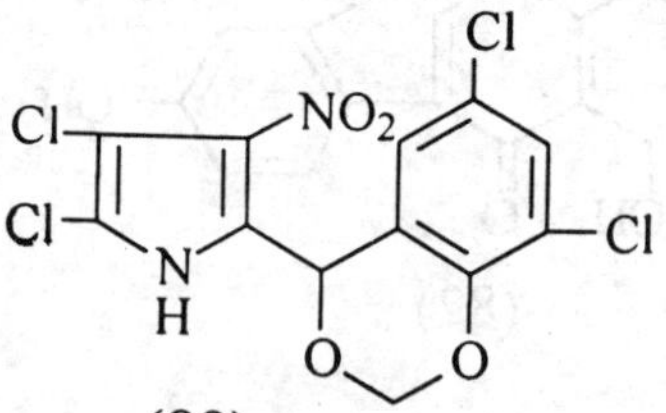

(89)

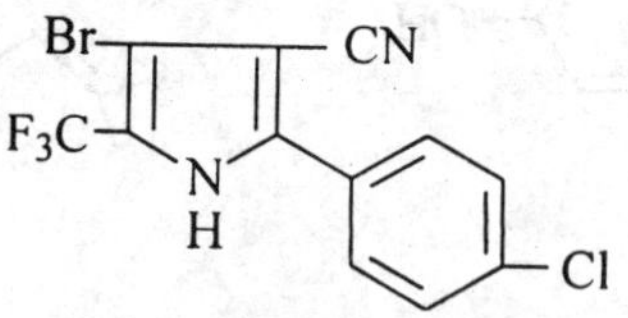

(90)

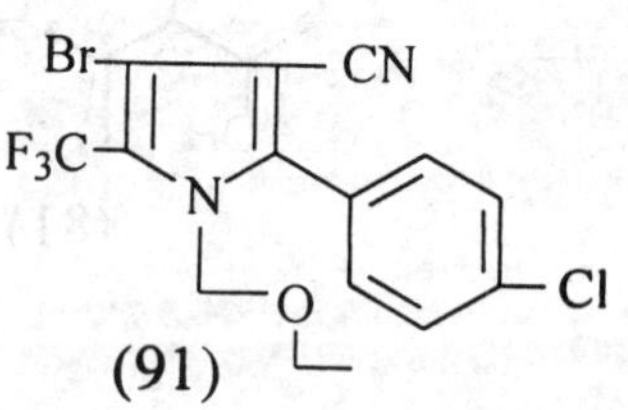

(91)

A wide variety of phytoalexins can control a number of diseases and do not show plant-host specificity. Many possess low fungicidal activity comparable to synthetic ones. It is necessary, therefore, to investigate their systematic structure-activity relationship using most active phyto-alexin molecules as the lead compound, in order to obtain most effective synthetic analogues. Alternatively modern biotechnology methods can be applied to evolve disease resistant races of crops.

Natural products of microbial origin

i) Insecticidal activity

A large number of insecticidal compounds have been isolated and identified from a variety of micro-organisms. Quite a few of them have not been commercialized because they are toxic to human beings. The compound dioxapyrolomycin, (89) a product of fermentation activity of *Strep-tomyces* strains, possesses insecticidal activity. SAR studies on a large number of synthetic ana-logues proved that the presence of electronegative group CN in the 3-position in place of nitro group and CF_3 in the 5-position make them effective broad-spectrum insecticides and acaricides (90). Ethoxy methylated derivative designated as AC 303,630 (91) possessed broad spectrum activity and has been registered for use on a variety of crops and ornamentals [50].

A few promising compounds like avermectins (92a) and milbemycins (92b), the complex macrocyclic lactones having high activity against mites, insects and nematodes, were isolated from *Streptomyces avermitilis* and *S. hygroscopicus* respectively. Amongst the isomers of avermectins, avermectin B_1 is the most active and widely commercialized for application against agricultural pests under the trade name Abamectin Milbemycins. It structurally differs from avermectins in lacking carbohydrate side-chain at C-13 and possess activity against aphids, mites and tent cater-pillars. Tetranacitin (93) with low mammalian toxicity and LD_{50} (mouse oral) 15 mg/kg, is an effective insecticide. Destruxins (94a) metabolites of *Metarrhizium anisopliae* and piericidin-A (94b) isolated from *Streptomyces mobaraensis* are not suitable as practical insecticides due to their instability and high mammalian toxicity. Similar properties are also associated with rofidin-A (95) and rubratoxin-B (96).

ii) Antifungal and antibacterial activity

A comprehensive review of natural products with antifungal and antibacterial activity was pub-lished by Worthington [51]. Natural products derived from fermentation broths are mostly complex molecules, present in micro quantities making it difficult to purify on a large scale. These products are photoinducible and non-phytotoxic with low persistance and do not exhibit mammalian toxicity. The commercially exploited antibiotics can be classified into two major groups, a) the aminoglycosides like streptomycin (97), validamycin-A (98) and kasugamycin (99), and b) nucle-osides like polyoxins B (100) and D (101) and blasticidin S (102). All these compounds are being manufactured commercially in Japan using different species of *Streptomyces*. Other structurally unrelated antibiotics like cycloheximide, actidione (103) and cellocidin (104) are also being pro-duced and used as fungicides in Japan.

Streptomycin

The first natural product used in agriculture was streptomycin (97), which is still used for the treatment of tuberculosis and as an adjunct to many other antibiotics. It is produced from the cultures of *Streptomyces griseus*. It has very low acute oral toxicity, LD_{50} (mice) > 10,000 mg/kg

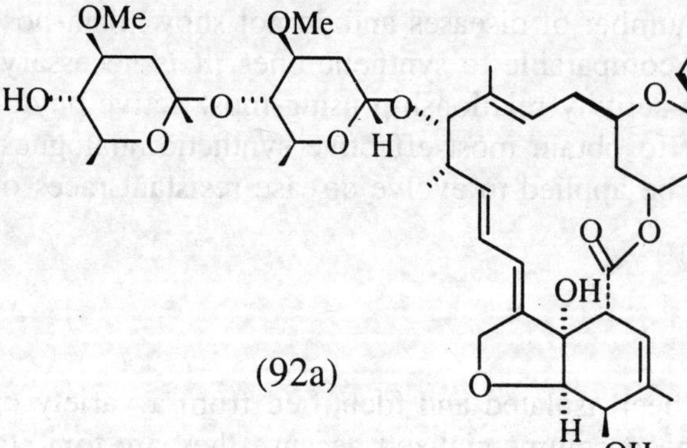

(92a)

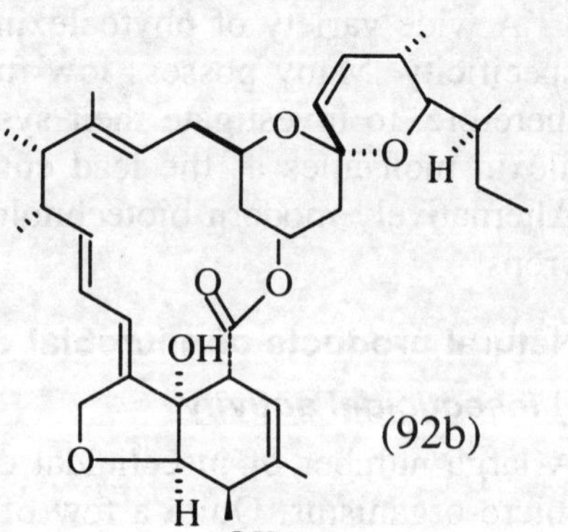

;

(92b)

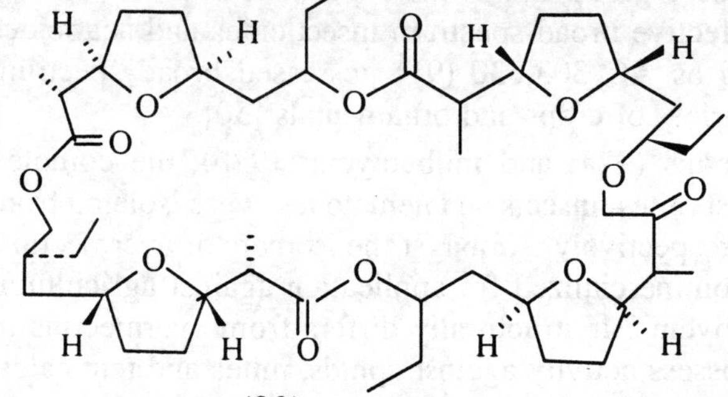

(93)

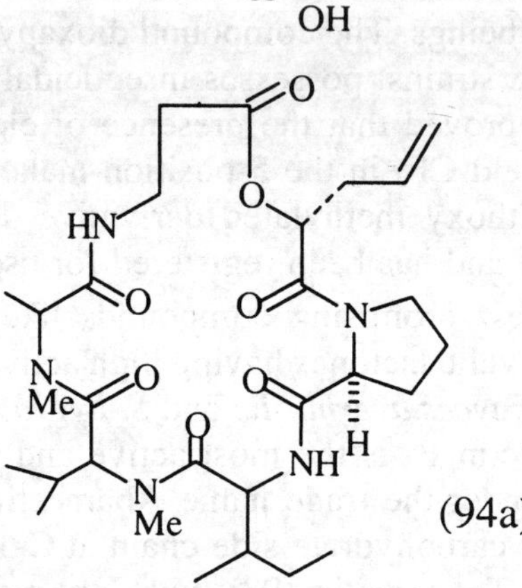

(94a)

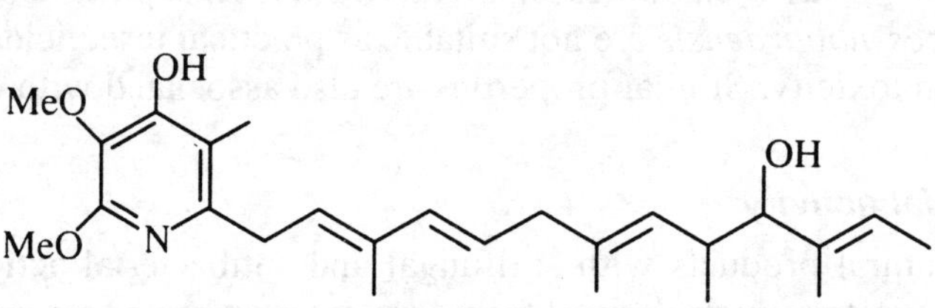

(94b)

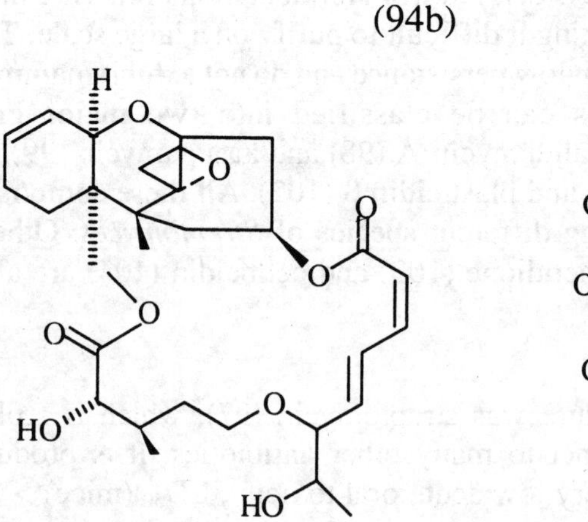

(95)

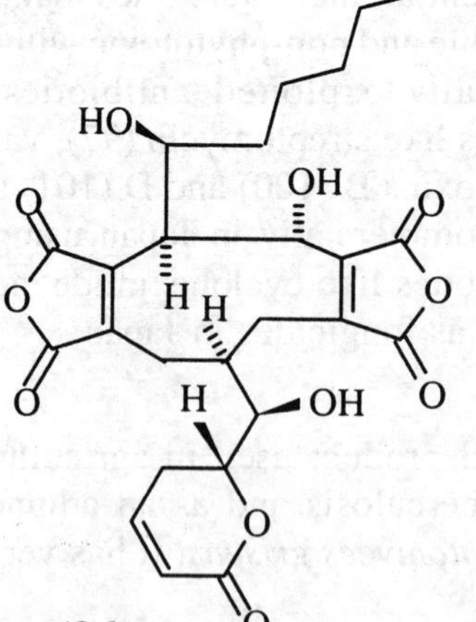

(96)

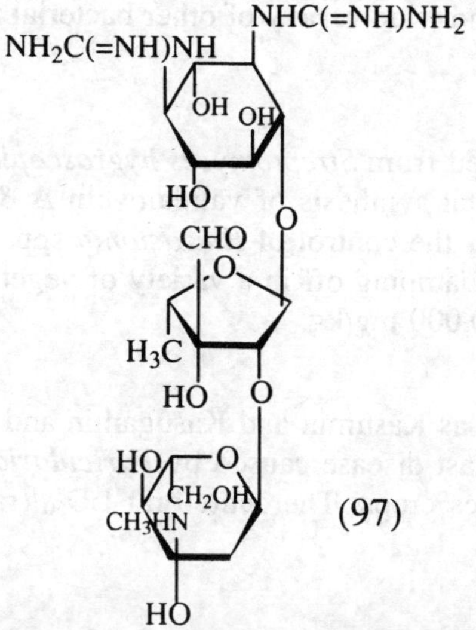

(97)

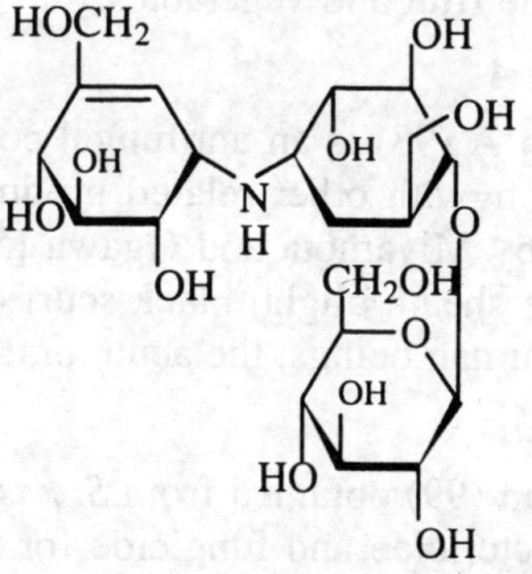

(98)

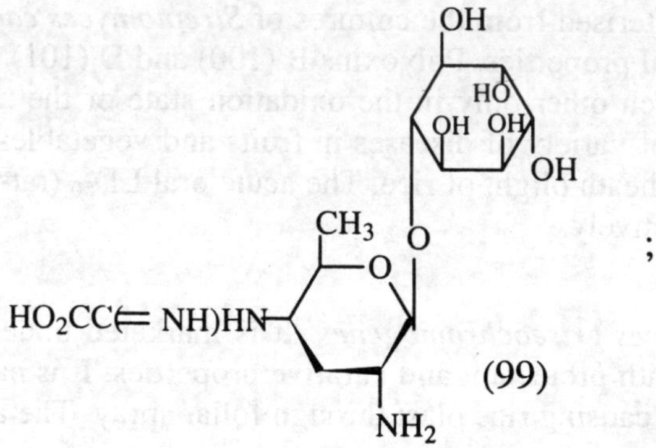

(99)

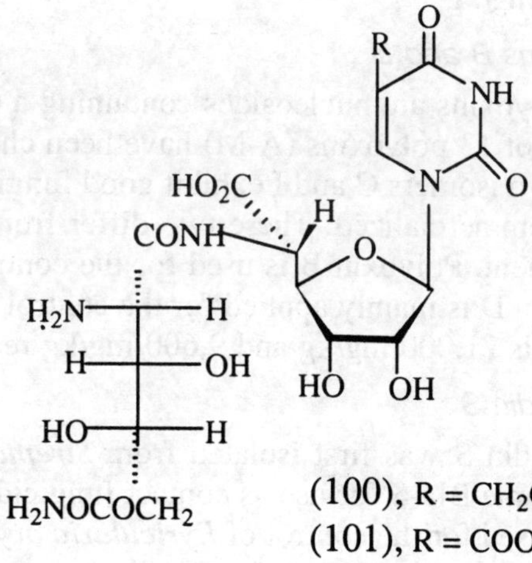

(100), $R = CH_2OH$

(101), $R = COOH$

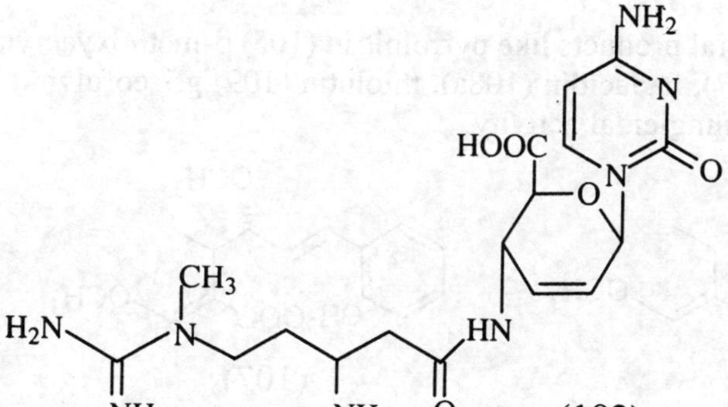

(102)

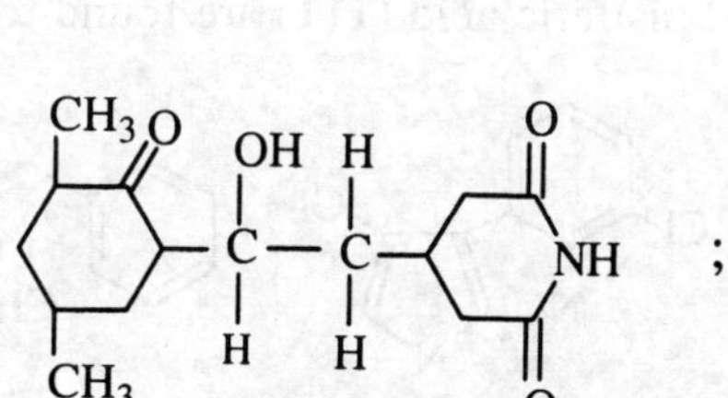

(103)

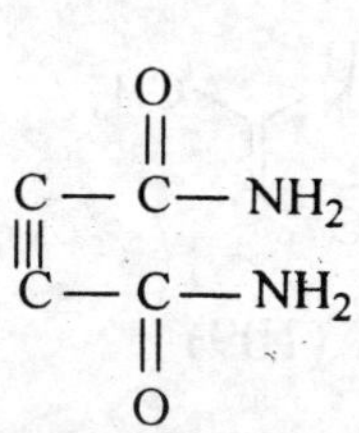

(104)

and is used for the control of apple and pear fireblight and wide variety of other bacterial rots and wilts on stone fruit and vegetables.

Validamycin A

Validamycin A (98) is an antifungal compound produced from *Streptomyces hygroscopicus* var. *limoneus* alongwith other related products B-G. The total synthesis of validamycin A & E was carried out by Miyamota and Ogawa [52]. It is used for the control of *Rhizoctonia* spp. causing diseases like sheath blight, black scurf on potatoes and damping off in a variety of vegetables. It is safe for human beings, the acute oral LD_{50} (rats) > 20,000 mg/kg.

Kasugamycin

Kasugamycin (99) obtained from *S. kasugaensis* is sold as Kasumin and Kasugamin and act as a systemic bactericide and fungicide for the control of blast disease caused by *Pyricularia oryzae* and variety of bacterial diseases of fruits and vegetables crops. The acute oral LD_{50} (rats) is > 22,000 mg/kg.

Polyoxins B and D

The polyoxins are nucleosides containing a peptide chain substituted at C-4 of the furanose ring. A total of 13 polyoxins (A-M) have been characterised from the cultures of *Streptomyces cacaoi*. Only two isomers C and I exhibit good fungicidal properties. Polyoxins B (100) and D (101) have been commercialized. These two differ from each other only in the oxidation state of the uracil component. Polyoxin B is used for the control of variety of diseases in fruits and vegetables and polyoxin D is mainly applied for the control of sheath blight of rice. The acute oral LD_{50} (rats) for B & D is 21,000 mg/kg and 9,600 mg/kg respectively.

Blasticidin S

Blasticidin S was first isolated from *Streptomyces griseochromogenes*. It is marketed under the trade name Bla-S for use as contact fungicide with protectant and curative properties. It is particularly used for the control of *Pyricularia oryzae* causing rice blast through foliar spray. The acute oral LD_{50} (rat) is 39.5 mg/kg.

Some synthetic analogues based on natural products like pyrrolnitrin (105) β-methoxyacrylates, strobilurin-A (106) and oudemansin A (107), hadacidin (108a), thiolutin (109) griseofulvin (110) and pisiferic acid (111) are found to have fungicidal activity.

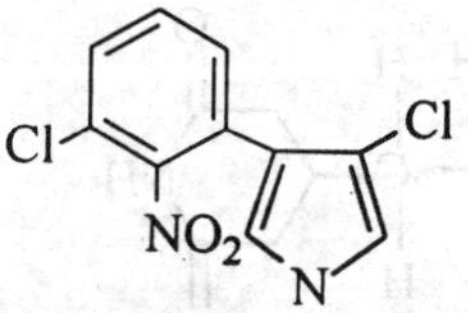

(105) H   (106) 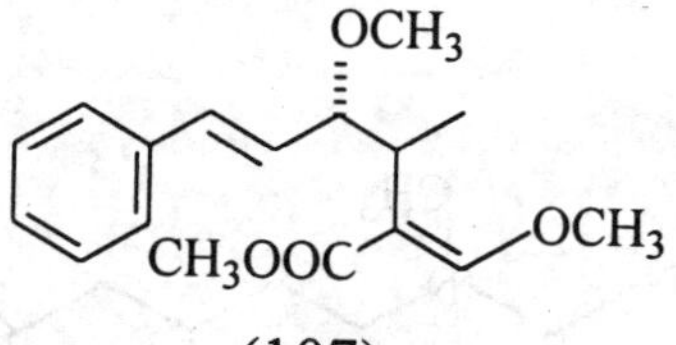   (107)

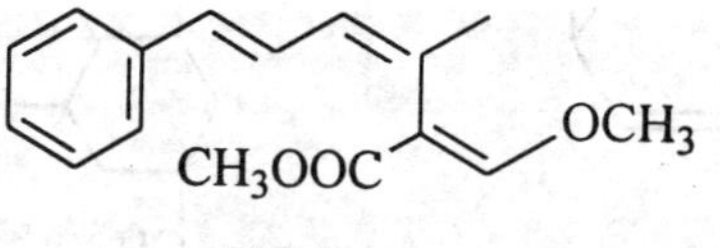

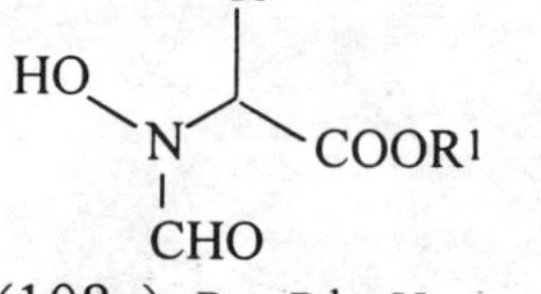

(108a), R = R¹= H
(108b), R = CH₃, R¹ = Na   (109) 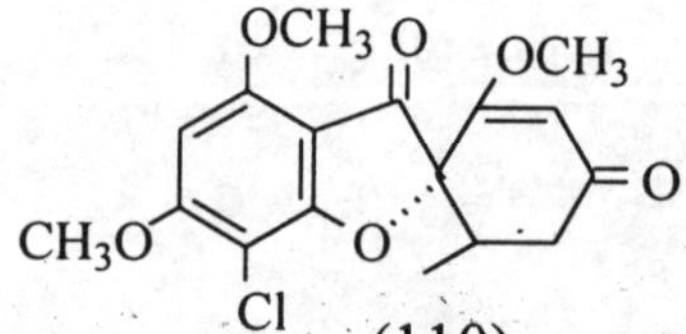   (110)

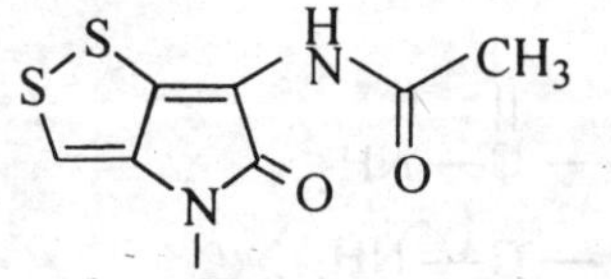

Pyrrolnitrin (105) isolated from *Pseudomonas pyrrocinia* is found very active as fungicide. But its less active synthetic analogue fenpiclonil (112) which controls snow mould and bunt and was reported in 1988 [53].

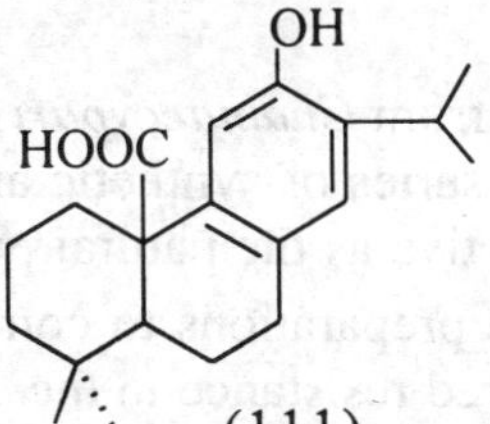

Later on the Ciba company developed a more active derivative CGA-173506 (113) to use as an antifungal seed dressing agent in USA under the trade name Saphire.

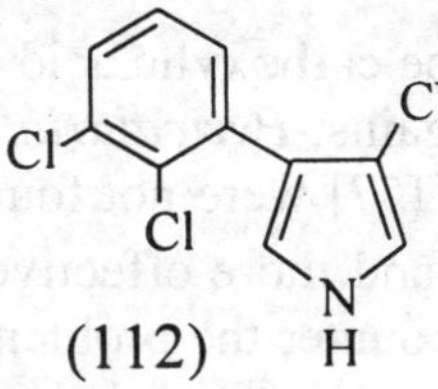

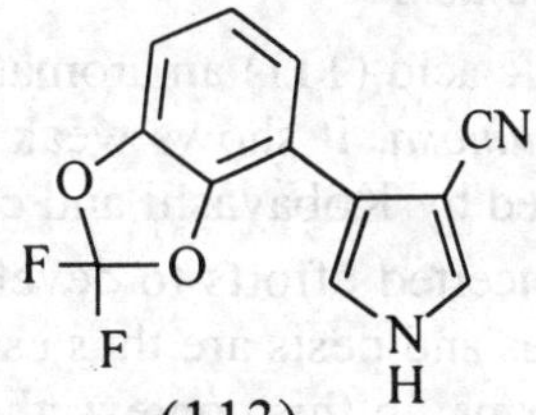

The β-methoxyacrylates

The synthetic analogues with fungicidal activity were prepared based on natural products strobilurin A (106) and oudemansin A (107) having low acute toxicity to mammals. A few synthetic strobilurin analogues (114a, 114b, 114c, 114f, 114h) were prepared which can control a large number of fungal parasites especially on rice, vegetables and fruit crops [54].

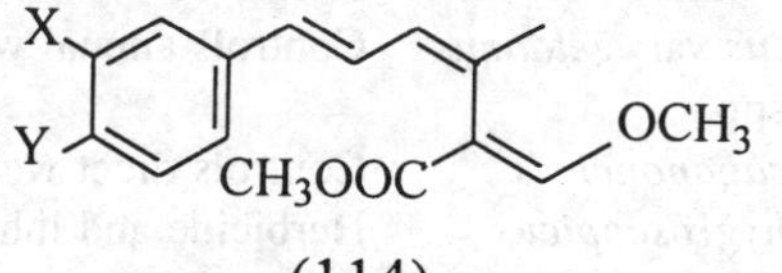

a, X = Y = H
b, X = OCH$_3$;   Y = Cl
c, X = H;   Y = OCH$_3$
f, X = OH;   Y = H
h, X = OCH$_3$;   Y = H

(114)

Hadacidin

A novel compound WL 87353 (108b) derived from natural product hadacidin (108a) was reported to be effective as post harvest treatment against downy mildew disease of vine [55].

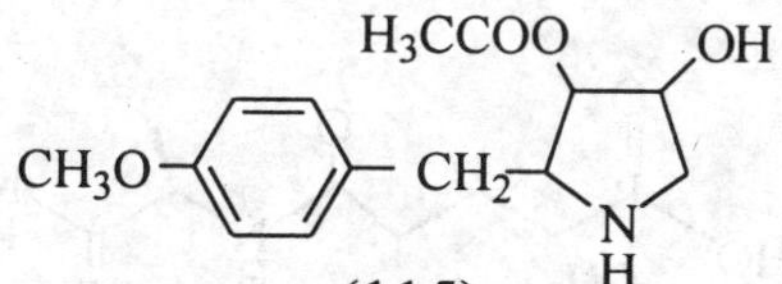

Thiolutin

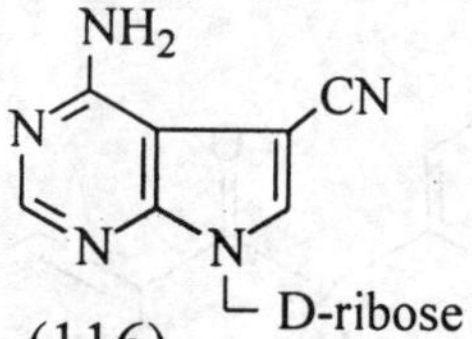

The natural product thiolutin (109) of the pyrrothine family, obtained from *Streptomyces* paved way for the synthesis of analogues which are effective as fungicides and bactericides. The compound thiolutin effectively controls black rot and fire blight of apples, tobacco and wilt on tomatoes. The optimised synthetic analogues showed good activity against many fungi [55].

Griseofulvin

Griseofulvin (110) was first isolated from *Penicillium griseofulvum*. It is used for the control of plant diseases like early blight of tomato, blossom blight of apples and canker of melons. A series of semisynthetic and synthetic analogues of griseofulvin have been prepared but do not possess good fungicidal property [56] as compared to the natural one.

Pisiferic acid

Pisiferic acid (111) an aromatic diterpene carboxylic acid is derived from *Chamaecyparis pisifera* var. *plumosa*. It shows weak activity against *Pyricularia oryzae*. A series of synthetic analogues prepared by Kobayashi and co-workers [57] were not found as effective as the natural product.

Concerted efforts to develop newer and more effective synthetic preparations to combat crop diseases and pests are thus essential to counter the problem of acquired resistance in the causative organisms. In this context, the natural products and their synthetic analogues hold out promise as they are more effective and eco-friendly and do not carry the risk of inducing resistance in the targetted organisms [58a, 58b].

iii) Herbicides of plant origin

Several microbial metabolites which exhibit herbicidal [59,60] activity are recorded in Table 2.6.

Table 2.6. Herbicides of microbial origin

Herbicide	Organism	Use
Cycloheximide (103)	*Streptomyces griseus*	Control perennial shrubs and weeds of rice. Inhibitor of seed germination.
Teutoxin (cyclic tetrapeptide)	*Alternaria alternata*	Herbicide.
Annisomycin (115)	*S. hygroscopicus* var. *geldanus*	Controls annual weeds.
Toyocamycin (116)	*Streptomyces* spp.	
Herbicidin A & B	*Streptomyces saganonensis*	Controls dicot weeds.
Herbimycins A & B (117a, 117b)	*Streptomyces hygroscopicus*	Herbicide and inhibitor of seed germination.
Trialaphos (118)	*S. hygroscopicus*	Broad spectrum herbicide.
Phosalacine (119)	*Kitasatosporia phosalacinea*	Herbicide; enzyme inhibitor.
Phosphinothricylalanylalanine (120) (bialaphos)	*S. hygroscopicus*	Herbicide.

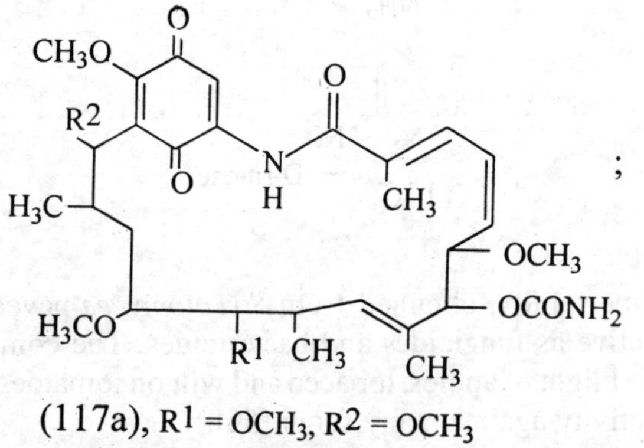

(117a), R1 = OCH₃, R2 = OCH₃
(117b), R1 = OH, R2 = H

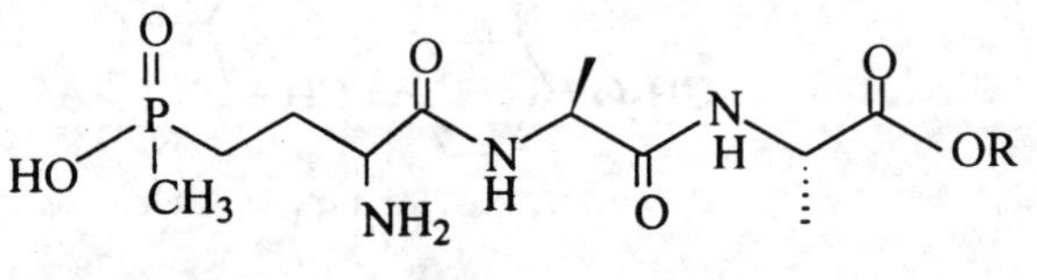

(118), R = L - alanine
(119), R = L - leucine
(120), R = H

Besides these compounds, many more microbially derived phytotoxic compounds have been reported in literature [61]. Based on the natural products, many synthetic analogues were prepared and commercialized. 'Basta' a synthetic product of Meiji Seika (Herbiace) and Hoechst A.G., whose active ingredient is phosphinothricin (121) is commercially available and used as a broad spectrum post emergent herbicides. Similarly glyphosate, N-(phosphonomethylglycine) (122), a synthetic analogue is widely used as broad specturm herbicide.

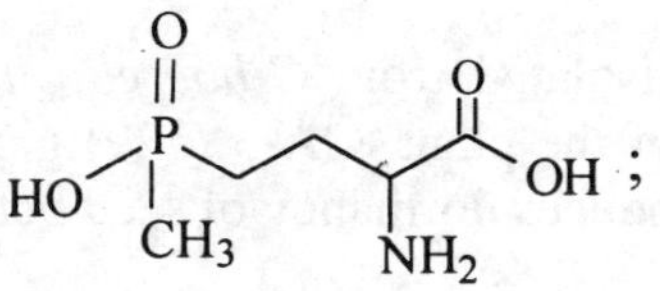

$$HOOCCH_2NHCH_2P(OH)_2$$

(121) 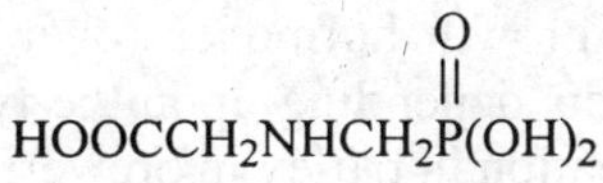   (122)

Plant growth regulators [62]

Some synthetic derivatives of naturally occuring phosphonates exhibit plant growth regulatory activity. Ethephon, 2-chloroethylphosphonic acid (123) is used for ripening fruits, induce flowering, promote abscission, enhance production and improve quality and colour. Glyphosine, [N, N-bis-(phosphonomethyl) glycine] (124) is used commercially for ripening sugarcane but at higher concentration it acts as a herbicide. The other PGR compounds are propylphosphonic acid and ethylhydrogenpropyl phosphonate (NIA-106 37) (125).

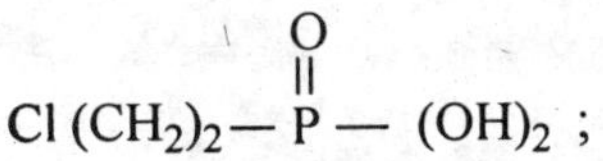

$$Cl(CH_2)_2 - P - (OH)_2$$

(123) 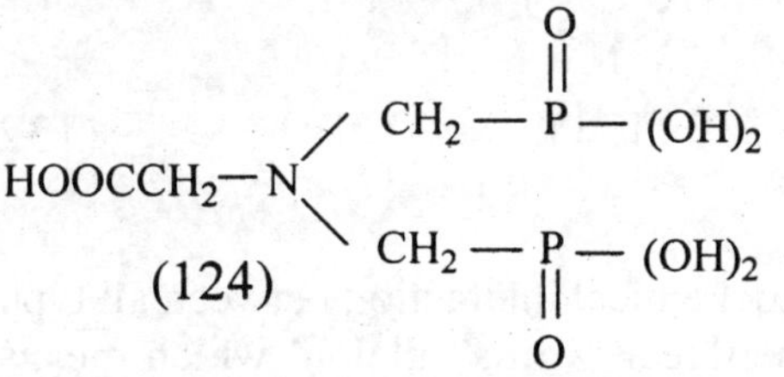   (124) 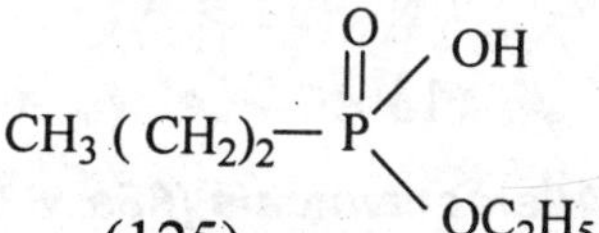   (125)

Paclobutrazol, (2R,3R+2S,3S)-1-(4-chlorophenyl)-4,4-dimethyl-2-(1H-1,2,4-triazol-1-yl)pentan-3-ol, (126) introduced by Zeneca ICI Agro Chemicals Ltd. is a plant growth regulator affecting various growth and development processes through inhibition of gibberellin biosynthesis. It is used to control vegetative growth and also induce flowering in commercial mango orchards.

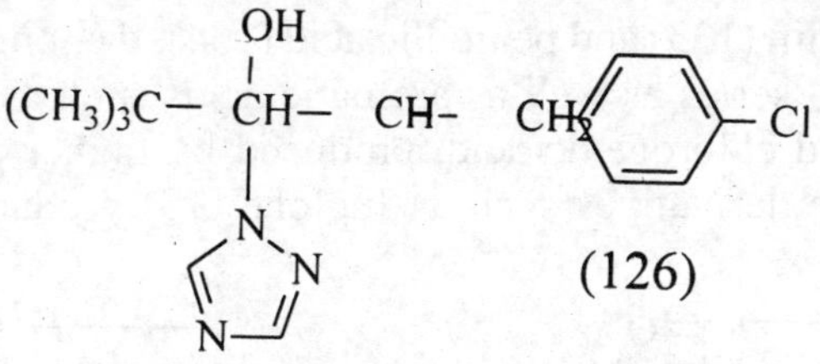

$$(CH_3)_3C - CH - CH - CH_2 \text{—} Cl$$

(126)

Auxins

The naturally occuring plant auxin is indol-3-yl acetic acid (IAA) (127a) and IBA, 4-indol-3-yl butyric acid (127b) which promotes cell elongation in shoots and accelerates rooting of cuttings. The synthetic analogues used in root growth in plant cuttings are α-naphthyl acetic acid (128).

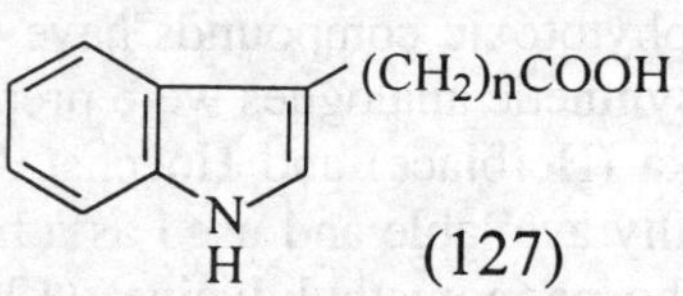

(127)

a, n=1; b, n = 3

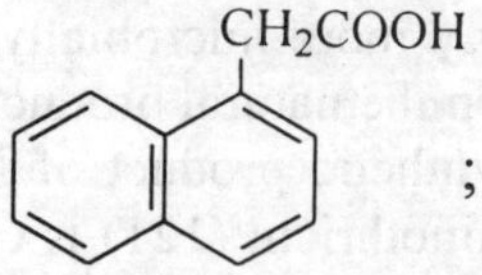

(128)

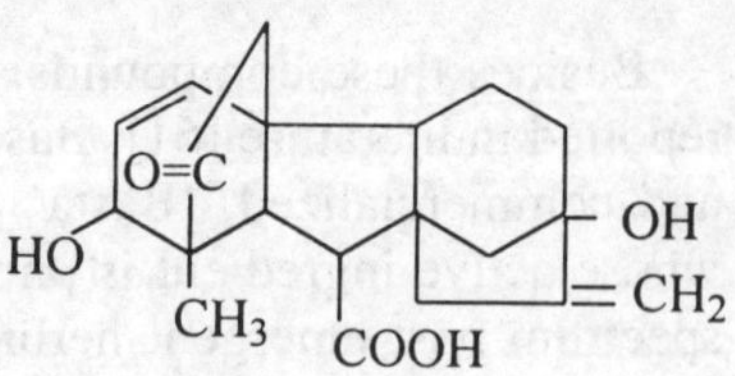

(129)

Gibberellins

Gibberellins, plant growth hormones (over 30) isolated from *Gibberella fujikuroa* have varied morphological effects depending upon the type of the plant. GA-3 (129) gibberellic acid is used for inducing germination in barley in breweries, induces dormancy of seed potatoes and sometimes for mobilizing sugars in certain plant organs.

Cytokinins

The naturally occuring cytokinins are zeatin (130), 6-(γ,γ-dimethylallyl amino) (131), 6-benzyl amino (132a) and 6-furfurylamino (132b) purine. Cytokinins help in foliation, photo response, ageing and induction of cell division.

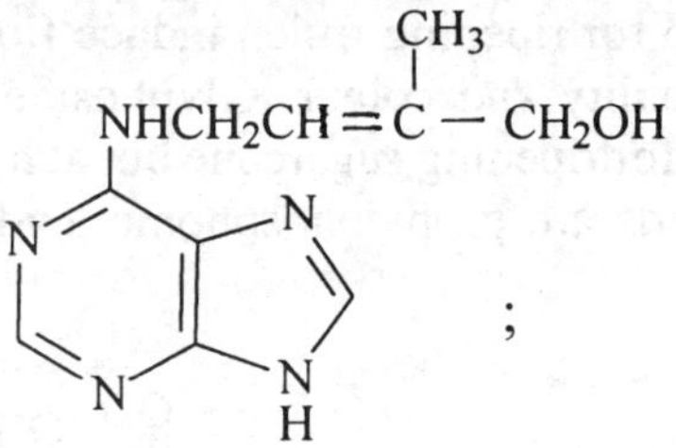

(130)

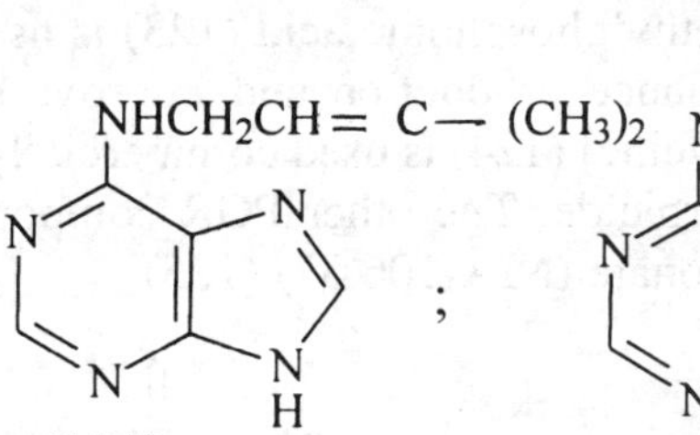

(131)

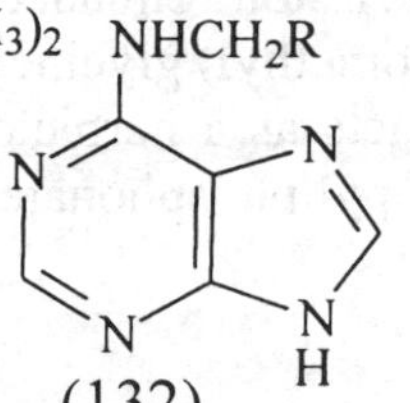

(132)

a) R = C_6H_5

b) R =

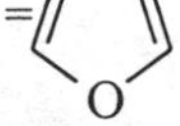

Allelochemicals [63a, 63b]

The term allelopathy refers to biochemical interaction between all types of plants including micro-organisms. It is derived from the Greek words 'allelon' which means to each other and 'pathos' referring to suffer. Allelo chemicals in general are synthesised almost in all plants and are released through volatilization, leaching, root exudation, lysis and decomposition of the plant residues.

A number of organic compounds such as organic acids of an aliphatic series are exuded as toxins by roots of cucumber and tomato plants. Butyric acid was among the toxins produced during decomposition of wheat straw, rye and corn residues. The lactones such as coumarins and their derivatives, flavonoids, patulin (133) and penicillic acid (134), the long chain fatty acids or mixture of fatty acids of C_{16} and C_{18} series as well as aromatic acids such as cinnamic acids, gallic acid, caffeic acid, ferulic acid and chlorogenic acid, produced by higher plants also act as inhibitors. Napthaquinone and quinone derivatives such as Juglone (79) produced by higher plants exhibit allelopathic effect.

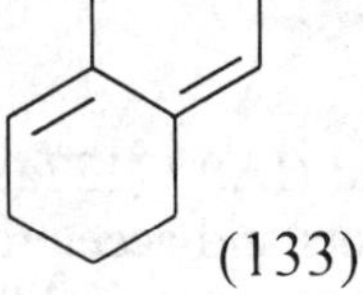

(133)

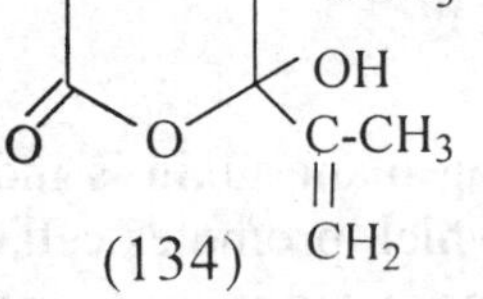

(134)

Terpenoids such as cineole, α-phellandrene, α-pinene and β-pinene the volatile inhibitors produced by *Eucalyptus camaldulensis* are adsorbed on to the soil particles in significant quantity. Abscisic acid, a sesquiterpene also exhibits allelopathic activity.

Similar allelopathic activity as seed germination inhibitor is observed with alkaloids such as cocaine (135), physostigmine (136) and quinine (137).

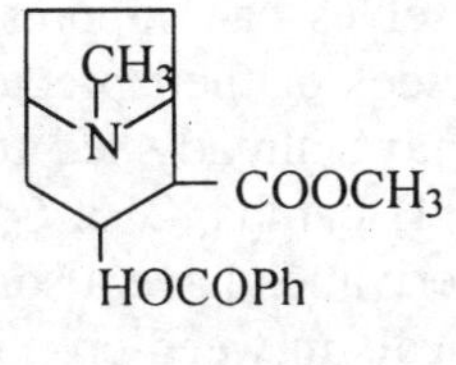

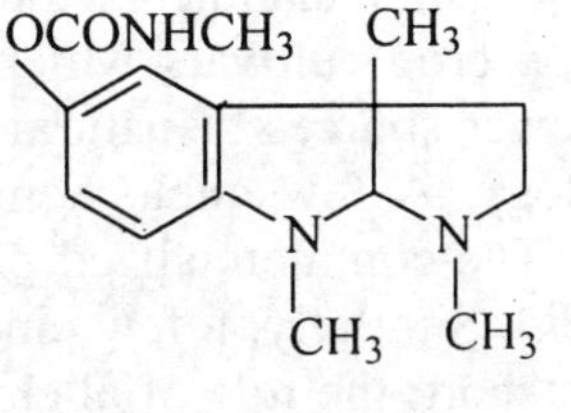

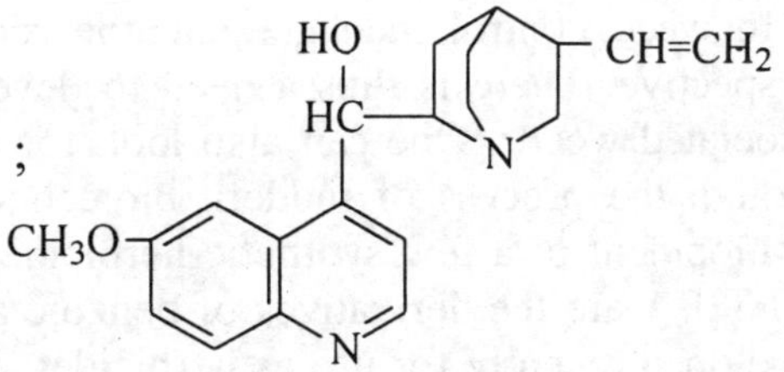

(135)   (136)   (137)

The sulphur and nitrogen compounds such as allylisothiocyanate, allyl thiocyanate and hydrolysed products of mustard oil and glycosides such as sinigrin, are potent inhibitors of seed germination and microbial growth.

The naturally occurring purines and nucleosides also show allelopathic activity. The only compounds in higher plants exhibited activity are caffeine (138) and theobromine found in coffee [Chou and Waller, 64].

Allelopathic effects in weed control management

The alleopathy can be used in weed control management in several ways. One way is to use allelopathic agents produced by various plants as herbicides. The another way is to apply residues of allelopathic weeds or crop plants as mulches or grow an allelopathic crop in a rotational sequence and allow the residues to remain in the field. It has been reported that mulches of sorghum and sudangrass when applied to apple orchards in early spring reduced weed biomass by 90% and 85% respectively. The allelopathic effect can also be utilized by growing a companion plant which has selective allelopathic activity against certain weeds and does not interfere with the crop growth.

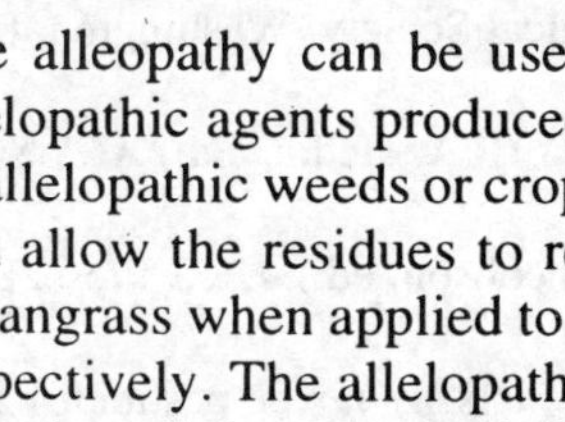

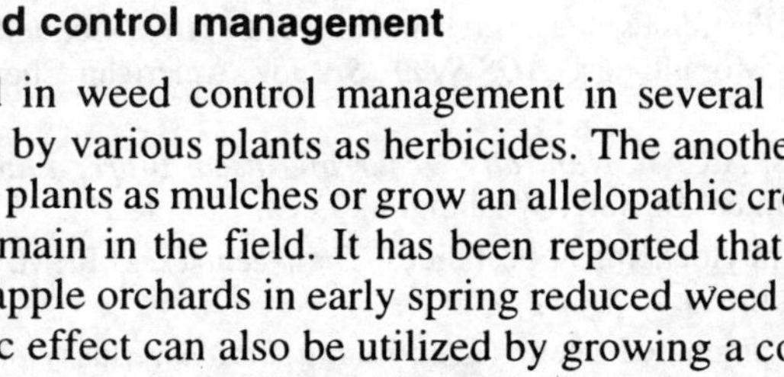

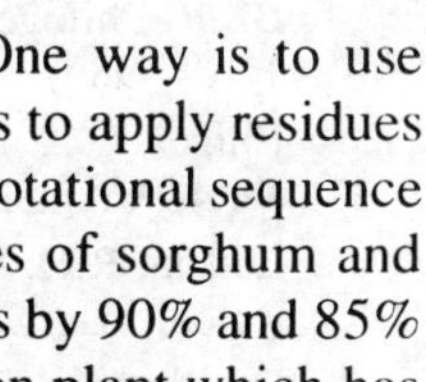

(138)   (139)   (140)

Strigol (139) isolated from root exudates of cotton is also a potent seed germination stimulant of root parasite witch weed and is active against the root parasites affecting warm season grasses and important crops such as maize, sorghum and sugarcane. SAR studies with synthetic analogues of strigol showed that the activity of even smaller fragments such as 3-hydroxy-2, 6,6-trimethylcyclohex-1-ene-1-carboxaldehyde (140) [Vail *et al.*, 65] was at par with strigol.

With voluminous information available on the exudation of allelochemicals by crop residues, none of the low tillage practices is found successful in certain situations such as wheat soybean crop rotation. In another way, the effectiveness and selectivity of allelochemicals may be enhanced by mixing with synthetic herbicides.

In weed control management, the role of allelochemicals must be recognised with proper perspective. There is thus a need to develop crop cultivars which themselves can suppress the associated weeds. One can also look for newer sources of natural herbicides or their precursors through the process of modern biotechnology. A few allelochemicals have already led to the development of a few synthetic herbicides. The commercialized product Tricamba, 2, 3, 6-TBA and TIBA are the derivatives of benzoic acid. Similarly, a few cineole derivaties have also been developed recently for use as herbicides. In short, the role of allelochemicals in weed-crop management programme should be duly recognised and developed further in order to get the desired results. Based on natural allelopathic compounds, suitable synthetic eco-friendly and more effective weedicides may be developed for practical use which can considerably minimize the crop losses due to weeds.

REFERENCES

1. Jacobson, M. and Crosby, D.G. (eds.). *Naturally Occurring Insecticides*, Marcel Dekker, New York (1971).
2a. Jacobson, M. Botanical Pesticides : Past, Present and Future. In : *Insecticides of Plant Origin* (J.T. Arnason, B.J.R. Philogene and P. Morand eds.) *ACS Symp. Sr. 387*, American Chemical Society, Washington, D.C., 1-10 (1989).
2b. Elliger, C.A. and Waiss, A.C. Jr. *Naturally Occurring Insect Bioregulators* (P.A. Hedin, Ed.) *ACS Symp. Sr. 449*, American Chemical Society, Washington, D.C., 210 (1991).
3. Larson, R.O. Focus on phytochemical pesticides, *The Neem Tree*, **I**, (M. Jacobson, ed.), CRC Press, Boca Raton, FL., USA, 155 (1989).
4. Devakumar, C. Development of neem (*Azadirachta indica* A.Juss) products as novel pesticides : In. *Agrochemicals and Sustainable Agriculture* (N.K. Roy, ed.) APC Publications Pvt. Ltd., New Delhi, 109-121 (1996).
5a. Rembold, H. Focus on phytochemical pesticides, *The Neem Tree, 1*, (M. Jacobson, ed.) CRC Press, Boca Raton, FL., USA, 47 (1989).
5b. Rembold, H. *Insecticides of Plant Origin* (J.T. Arnason, B.J. Philogene, and P. Morand, eds.), *ACS Symp. Sr. 387*, American Chemical Society, Washington, D.C., 150 (1980).
6. Lee, S.M., Klocke, J.A., Barnby, M.A., Yamasaki, R.B. and Balandrin, M.F. *Naturally occurring Pest Bioregulators* (P.H. Hedin, ed.), *ACS Symp. Sr. 449*, American Chemical Society, Washington, D.C., 239 (1991).
7. Ley, S.V. Synthesis of insect antifeedants in Pesticide Chemistry. *Proc. Seventh Internat. Cong. Pestic. Chem.*, Hamburg, (H. Frehse ed.), abs. 97-107 (1990).
8a. Ley, S.V., Santafianos, D., Blaney, W.M., Simmonds, M.S.J. *Tetrahedron Lett.* **28**, 221-224 (1987).
8b. Ley, S.V., Abad-Somavilla, A., Broughton, H.B., Craig, D., Slawin, A.M.Z., Toogood, P.L. and Silliams, D.J. *Tetradedron* **45**, 2143-2164 (1989).
9. Crosby, D.G. *Naturally Occuring Insecticides* (M. Jacobson and D.G. Crosby eds.). Marcel Dekkar, New York, 198 (1971).

10. Jefferies, P.R., Toia, R.F., Branunigan, B., Pessah, I. and Casida, J.E. Ryania insecticide : Analysis and biological activity of 10 natural Ryanoids. *J. Agric. Food Chem.* **40**, 142 (1992).

11. Deslongchamps, P., Belanger, A., Berney, D.J.F., Borschberg, H.J., Brouseau, R, Dontheau, A., Durand, R., Katayama, H., Lapalme, R., Leturc, D., Liao, C.C., MacLachlan, F.N., Maffrand, J.P., Marazza, F., Martino, R., Moreau, C., Saint-Laurent, L., Saintonge, R. and Soucy, P. The total synthesis of (+) ryanodol, p. iv, Preparation of (+) ryanodol from (+) anhydroryanodol. *Can. J. Chem.* **68**, 186 (1990).

12a. Ujvary, I., Eya, B.K., Grendell, R.L., Toia, R.F. and Casida, J.E. Insecticidal activity of various 3-acetyl and other derivatives of veracevine relative to the veratrum alkaloids veratridine and cevadine. *J. Agric. Food Chem.* **39**, 1875 (1991).

12b. Addov, R.W., Babcock, T.T., Black, B.C., Brown, D.G., Diehl, R.E., Fureh, J.A., Kameswaran, V., Kamhi, V.M., Kremer, K.A., Kuhn, D.G., Lovell, J.B., Trotto, S.H. and Wright, D.P. Jr. *Synthesis and Chemistry of Agrochemicals III* (D.R. Baker, J.J. Fenyes and J.L. Steffens eds.). *ACS Symp. Sr. 504*, American Chemical Society, Washington D.C., 283 (1992).

13. Jacobson, M. *Naturally Occurring Insecticides* (M. Jacobson and D.C. Crosby eds.) Marcel Dekker, New York, 137 (1971).

14. Elliott, M., Farnham, A.W., Jones, N.F., Johnson, D.M. and Pullman, D.A. *Pestic. Sci.* **18**, a, 191 (Part 1); b, 203 (Part 2); c, 211 (Part 3); d, 223 (Part 4); e, 229 (Part 5); f, 239 (Part 6) (1987).

15. Crombie, L., Horshamand, M.A., Blade, R.J. Synthetic approaches to *iso*butylamides of insecticidal interest. *Tetrahedron Lett.* **28**, 4879 (1987).

16. Towers, G.H.N. Interaction of light with phytochemicals in some natural and novel systems. *Can. J. Bot.* **62**, 2900-2911.

17. Amason, J.T., Philogene, B.J.R., Morand, P., Imrie, K., Iyenger, S., Duval, F., Soucy-Breau, C., Scaiano, J.G., Werstinu, N.H., Hasspieler, B. and Downe, A.E.R. Naturally occurring and synthetic thiophenes as photoactivated insecticides. In : *Insecticides of Plant Origin*, (J.T. Amason, B.J.R. Philogene and P. Morand eds.). *ACS Symp. Sr.* **387**, American Chemical Society, Washington D.C., 164-172 & 213 (1989).

18. Knox, J.P. and Dodge, A.D. Singlet oxygen and plants. *Phytochem* **24**, 889-896 (1985).

19. Jansen, B.J.M. and DeGroot, A. The Occurrence and biological activity of drimane sesquiterpenoids. *Nat. Prod. Rep.* **8**, 309-318 (1991).

20. Kubo, I. and Nakanishi, K. Some terpenoid insect antifeedant from tropical plants. In : Advances in Pesticide Science [H. Geissbuhler, ed.) *Proc. Fourth Internat. Congr. Pestic. Chem.* Pergamon Press, Oxford, part-2, 284-294 (1978).

21. Hanson, J.R. Diterpenoids; *Nat. Prod. Rep.* **1**, 533 (1984).

22. J. Vader. Studies towards the Total Synthesis of Insect antifeedant Clerodanes, *Ph.D. thesis*, University of Wageningen, Netherland (1989).

23. J.A. Pickett, G.W. Dawson, D.C. Griffiths, A. Hassanali, L.A. Merritt, A. Mudd, M.C. Smith, L.J. Wadhams, C.M. Woodcock, Z. Zhong-ning, Development of plant derived antifeedants for crop protection. In : *Pesticide Science and Biotechnology* (R. Greenhalgh and T.R. Roberts, eds.), IUPAC, Well Scientific Publication, Oxford, 125-126 (1987).

24. Proksch, P. and Rodriguez, E. Chromenes and benzofurans of the Asteraceae, their chemistry and biological significance. *Phytochemistry* **22**, 2335 (1983).

25. Bowers, W.S., Ohta, T., Cleere, J.S., and Marsella, P.A. Discovery of antijuvenile hormones in plants. *Science* **193**, 542 (1976).

26. Rupprechit, J.K., Hui, Y.H. and McLaughin, J.L. Annonaceous acetogenins : A review. *J. Nat. Prod.* **53**, 237 (1990).

27. Born, L., Lieb, F., Lorentzen, J.P., Moeschler, H., Nonfan, M., Sollner, R. and Wendisch, D. The relative configuration of acetogenins isolated from Annona squamosa : Annonin I (Squamocin) and Annonin VI *Planta Med.* **56**, 312-316 (1990).

28. Hollingworth, R.M., Ahammadsahib, K.I., Gadelhak, G.C. and McLaughlin, J.L. 203rd. *American Chemical Society National Meeting*, San Francisco, USA, abs 156 (1992).

29. Neumann, R. and Peter, H.H. Insecticidal organophosphates : natural made them fast. *Experientia*, **3**, 1235 (1987).

30. Cardellina, J.H. 11, Biologically Active Natural Products, Potential use in Agriculture (H.G. Cutler, ed.) *ACS Symp. Sr.* **380**, American Chemical Society, Washington D.C., 305 (1988).

31. Marcus, C. and Lichtenstein, E.P. Biologically active components of Anise : toxicity and interaction with insecticides in insects. *J. Agric. Food Chem.* **23**, 217 (1979).

32. Jurd, L. and Manners, G.D. Wood extractives as models for the development of new types of pest control agents. *J. Agric. Food Chem.* **28**, 183 (1980).

33. Kamikawa, T., Fujie, S., Yamagiwa, Y., Kim, M., and Kawaguchi, H. Synthesis of clitocine, a new insecticidal nucleoside from the mushroom clitocybe in versa. *Chem. Commun.*, 195 (1988).

34. Yamauchi, S. and Taniguchi, E. Synthesis and insecticidal activity of lignananalogues (11) *Biosci. Biotech. Biochem.* **56** (3), 412-417 (1992).

35. Klocke, J.A., Arisawa, M., Handa, S.S., Kinghorn, A.D., Cordell, G.A. and Farnsworth, N.R. Growth inhibitores, insecticidal and antifeedant effects of some antilekemic and cytotoxic quassinoids on two species of agricultural pests. *Experientia*, **41**, 379 (1985).

36. Satasook, C., Isman, M.B. and Wiriyachitra, P. Activity of Rocaglamide an insecticidal natural product against variegated cutworms, *Peridroma sucia* (Lepidoptera Noctuidae). *Pestic. Sci.* **36**(1), 53-58 (1992).

37. Lajide, L., Escoubas, P. and Mizutani, J. Antifeedant activity of metabolites of *Aristolochia albida* against the tobacco/cutworm, *Spodoptera liture. J. Agric. Food Chem.* **41**, 669 (1993).

38. Nathanson, J.A., Hunnicutt, E.J., Kantham, L. and Scavone, C. Cocaine as a naturally occuring insecticide. *Proc. Nat. Acad. Sci, U.S.A.* **90**, 9645-9648 (1993).

39. Konishi, K., Nereistoxin and its relatives. In. *Insecticides*, **I**, *Proc. Second Internatn. Congr. Pestic. Chem.* (A.S. Tahori ed.), Gordon and Breach Science Publishers, New York, 179-189 (1972).

40. Sakai, M. *Rev. Plant Prot. Res.,* **2**, 17 (1969).

41. Worthing, C.R. (ed.). The Pesticide Mannual, 9th Ed. British Crop Prot. Council, Surrey, 132 (1991).

42. Maeda, M., Kodama, T., Tanaka, T., Uhfune, Y., Nomota, K.N., Nishimura, K. and Fujita, T. Insecticidal and neuromuscular activities of domoic acid and its related compound. *J. Pestic. Sci.* **9**, 27 (1984).

43a. Bertram, H.J., Hartwig, J. and Sollner, R. Nematicidal activity of some angular coumarins. *Proc. Seventh Internatn. Congr. Pestic. Chem.,* Hamburg, abs. 254 (1990).

43b. Kogiso, S., Wada, K. and Munakata, K. Isolation of nematicidal polycetylenes from *Cathamus tinctorius* L. *Agr. Biol. Chem.* **40**, 2085 (1976).

43c. Hick, A.J., Dawson, G.W., Pickett, J.A. Compounds which mimic *iso*thiocyanate release from glucosinolates. *Proc. Seventh Internatn. Congr. Pestic. Chem.*, Hamburg, abs. 255 (1990).

43d. Munakata, K. *Natural Products for Inovative Pest Management* (D.L. Whitehead & W.S. Bowers eds.), Pergamon Press, Oxford, 299 (1983).

44. Matsuda, K., Yamada, K., Kimura, M. and Hamada, M. Nematicidal activity of matrine and its derivatives against pine wood nematodes. *J. Agric. Food Chem.* **39**, 189-191 (1991).

45a. Devakumar, C. and Goswami, B.K. Nematicidal principles from Neem (*Azadirachta indica* A. Juss) Part III Isolation and Bioassay of some Neem Meliacins. *Pestic. Res. J.* **4** (2), 81-86 (1992).

45b. Hick, A.J., Dawson, G.W., Pickett, J.A. Odoracin, a nematicidal constituent from *Dphne odora. Agr. Biol. Chem.*, **40**, 2119 (1976).

45c. Munakata, K. Nematicidal substances from plants. *Fourth Internatn. Congr. Pestic. Chem. Zurich*, abs. Book III, Part 2 (1978).

46a. Mitra, S.R., Choudhuri, A and Adityachaudhary, N. Production of antifungal compounds by higher plants— a review of recent researches. *Pl. Physiol. Biochem.* **11** (1), 533-77 (1984).

46b. Parmar, B.S. and Devakumar, C. *Botanical and Biopesticides*, Westvill Publishing House, New Delhi, 27-44 (1993).

47. Perrin, D.R. and Bottomley, W. Studies on phytoalexins, V. The structure of pisatin from *Pisum sativum* L. *J. Am. Chem. Soc.* **84**, 1919-1992 (1962).

48. Brooks, J.W. and Watson, D.G. Terpenoid phytoalexins. *Nat. Prod. Rep.* 8(4), 367-390 (1991).

49. Ingham, J.L. Phytoalexins from Leguminosae. In : *Phytoalexins* (J.A. Baily and J.W. Mansfield, eds.). Blackie, London, 21-80 (1982).

50. Miller, T.P., Treacy, M.F., Gard, I.E., Lovell, J.B., Wright, D.P. Jr., Addor R.W., and Kamhi, V.M. Brighton Crop Protection Conference—Pests and Diseases 1990, **1**, British Crop Prot. Council Surrey, 43 (1991).

51. Worthington, P.A. Antibiotics with antifeedant and antibacterial activity against plant diseases. *Nat. Prod. Rep.* **47** (1988).

52. Miyamota, Y. and Ogawa, S. Synthetic studies on antibiotic validamycins. Part-13, Total synthesis of (+) - validamycins A and E and related compounds. *J. Chem. Soc. Parkin* **1**, 1013 (1989).

53. Ammerman, E., Lorenz, G., Schelberger, K., Wenderoth, B., Sauter, H. and Rentzea, C. BAS 490F—a broad spectrum fungicides with a new mode of action. *Brighton Crop Prot. Conf. Pests and Diseases*-1992, **1**, British Crop Prot. Council, Farnham, England, 403-410 (1992).

54. Dunn, C.L. and Klein, S.P. Fungicides for crop protection-100 years of progress, British Crop Prot. Council, Farnham, England, 407, 455 (1985).

55. Dell, I., Godfrey, C.R.A. and Wadsworth, D.J. *Synthesis and chemistry of Agrochemicals* **111** (D.R. Baker, J.G. Fenyes and J.J. Steffens eds.) *ACS Symp. Sr.* **504**. American Chemical Society, Washington DC, 384 (1992).

56. Ko, B.S., Oritani, T. and Yamashita, K. Synthesis and biological activities of Griseofulvin analogues. *Agric. Biol. Chem.* **54**, 2199 (1990).

57. Kobayashi, K., Nishino, C., Tomita, H. and Fukushima, M. Antifungal activity of pisiferic acid derivatives against the rice blast fungus. *Phytochemistry* **26**, 3175 (1987).

58a. Evans, D.A. and Lawson, K.R. Crop protection chemicals-research and development perspectives and opportunities. *Pestic. Outlook*, **3**, 10 (1992).

58b. D.A. Evans and K.R. Lawson. *Milestones in 150 years of the Chemical Industry* (P.J.T. Morris, W.A. Campbell, and H.L. Roberts eds.). Royal Society of Chemistry, London, 68 (1992).

59. Stonard, R.J. and Miller-Wideman, M.A. *Herbicides and plant growth regulators.* Agrochemicals from natural products (C.R.A. Godfrey ed.). Marcel Dekker, Inc, New York, 285-310 (1994).

60. Cutler, H.G. (ed.) *Biologically active natural products, Potential use in agriculture. ACS Symp. Sr. 380,* American Chemical Society, Washington D.C., 198 (1987).

61. Campbell, W.C. (ed.) *Ivermectin and Abarmectin*, Springer-Verlag, New York, 1989.

62. Cremlyn, R. J. Naturally occuring insecticides. In : *Agrochemicals.* John, Wiley & Sons, N.Y., 166-183 (1990).

63a. Rice, E.L. *Allelopathy*, Academic Press, New York (1984).

63b. IAS Internatn. allelopathy society constitution, First World Cong. on allelopathy, a science for the future. Cadiz, Spain (1996).

64. Chou, C.H. and Walles, G.R. (eds.). *Phytochemical Ecology : Alleochemicals, Mycotoxins and Insect Pheromones and Allomones.* Institute of Botany Academia Sinica, Taipei (1989).

65. Vail, S.L., Dailey, O.D. Jr., Connick, W.J. Jr. and Pepperman, A.B. Jr. Strigol syntheses and related structure—bioactivity studies. In *Chemistry of Alleopathy : Biochemical interactions among plants* (A.C. Thompson Ed.), *ACS Symp. Sr. 268*, American Chemical Society, Washington D.C., 445-458 (1985).

CHAPTER 3

Synthetic Pesticides

Section	
I. Chemistry of Chlorinated Insecticides	VI. Chemistry of Nematicides
II. Chemistry of Pyrethroid Insecticides	VII. Chemistry of Rodenticides
III. Chemistry of Miscellaneous Insecticides	VIII. Chemistry of Fungicides
IV. Chemistry of Organophosphorus Insecticides	IX. Chemistry of Herbicides
V. Chemistry of Carbamate Insecticides	

Section I.   Chemistry of Chlorinated Insecticides

- Introduction
- DDT and its analogues
- Metabolism of DDT
- Mode of action
- Analytical methods
- Biodegradable DDT analogues
- DDT pyrethroid compound
- Hexachlorocyclohexane

- Mode of action
- Analytical methods
- Cyclodienes
- Stereochemistry
- Formulation and application
- Mode of action
- Photochemistry
- *References*

Zeidler [1] during his Ph.D. programme in the year 1873-74 synthesized a number of chlorinated hydrocarbons including dichloro-diphenyl trichloroethane, also known as DDT, p-p'-DDT, 4,4'-DDT or 1,1,1-trichloro-2,2-bis (*p*-chlorophenyl) ethane or 1,1-bis (*p*-chlorophenyl)-2,2,2-trichloroethane (1). The insecticidal property of compounds remained unknown till 1939-40 when Paul Muller of Switzerland [2] revealed the effectiveness of DDT for the control of insects and commercialized it as gesarol. This became a miracle insecticide during the second World War when the allied powers realised its high effectiveness against deadly diseases like malaria and typhoid. It has a wide spectrum of insecticidal action, cost effective preparation protocol, moderate mammalian toxicity with LD_{50} (oral, rats) 150 mg/kg and safe to handle. DDT is stable and has high persistence in plant, soil and water. It is true that DDT has boosted up agricultural production in many countries because of its high toxicity action to many insect pests but unfortunately its indiscriminate use has led to the development of resistance in many insect pests and caused environmental pollution and an ecological imbalance. As a result, it has now been banned from agricultural use in India and all developed countries but still its importance is being realised for the control of insect vectors.

Due to its high effective insecticidal properties, a large number of analogous organochlorine compounds were synthesized but only a few of them were found as effective as DDT (1). These compounds are DFDT (2), methoxychlor (3), DDD, 1,1-dichloro-2,2-di-(p-chlorophenyl) ethane (4), perthane (5), prolan (6) and bulan (7). The compounds methoxychlor (3), and DDD (4) have the advantage over DDT being less persistent and hence non-polluting and thus may be found useful for food crops.

Preparation of DDT and its analogues

The preparation procedure of DDT and its analogues is depicted in Fig. 3.1.

Chemical reaction of DDT

A few important chemical reactions products of DDT are DDE (8), dicofol (9) tetranitro DDT (10), unsymmetrical and symmetrical tetrachloroethanes (11, 12), dichlorobenzil (13), benzilic acid (14), ethyl chlorobenzilate (15) and benzophenone (16) as shown in Fig. 3.2.

Dehydrochlorination of DDT and their analogues

The dehydrochlorination is an *E2*-type elimination which is regulated by the availability of electrons at the benzylic carbon atom. The electron release is also influenced by the presence of aromatic substituents and heavy metals. Electron withdrawing substituents in the 4-and $4^{/}$-positions of DDT accelerate the process of dehydrochlorination of DDT and its analogues while electron donating groups decrease it.

a) With the symmetrical $4,4^{/}$ dihalogen substituted compounds, the dehydrochlorination rate decreases in the order I < Br < Cl < F.

b) The dehydrochlorination rate is reduced greatly in the DDT analogue lacking any aromatic substituents and $4,4^{/}$-dialkoxy and 4-$4^{/}$-dialkyl substituents.

c) A change of the trichloromethyl group also affects the dehydrochlorination rate. Thus in DDD (4), the rate of dehydrochlorination is half of that of DDT since the electron withdrawing power is reduced in the former.

d) If $4,4^{/}$-Cl atoms are further replaced by alkyl groups, in the DDD structure like perthane ($4,4^{/}$-diethyl analogue), the dehydrochlorination rate is reduced 180 times.

e) The dehydrochlorination rate also reduced when the $4,4^{/}$-Cl atoms of DDT are replaced by alkyl or alkoxy groups. This is of great significance in relation to insect resistance which is caused by the enzymatic dehydrochlorination of DDT. It is observed that where DDT is ineffective, compounds having $4,4^{/}$-alkoxy substituents retain considerable activity towards these resistant insects.

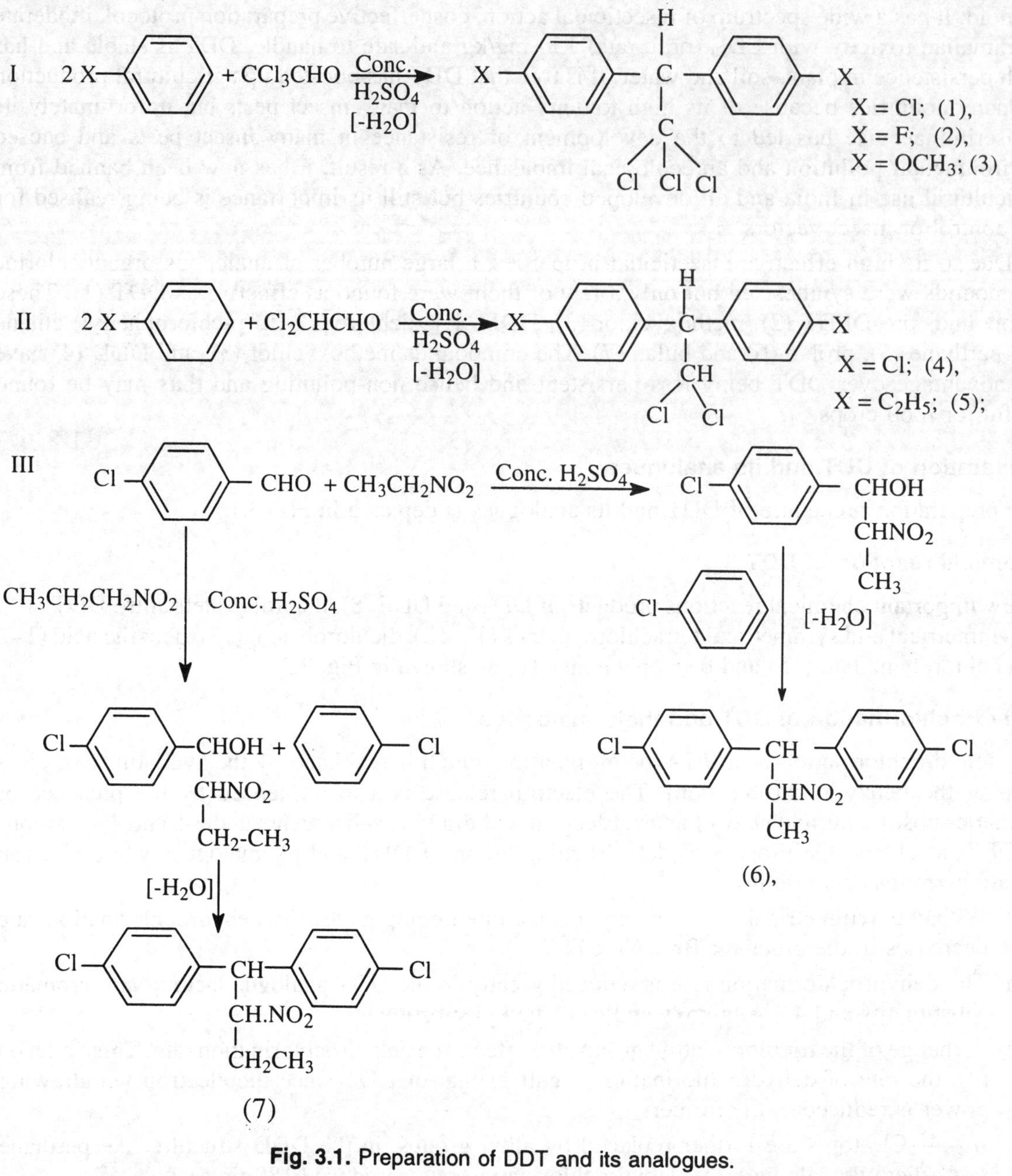

Fig. 3.1. Preparation of DDT and its analogues.

The dehydrochlorination reaction of DDT is affected in alkaline condition, while DDT can withstand high temperature with 20% aqueous NaOH solution for 24 h, its dehydrochlorination occurs quite rapidly in alcoholic-alkali solution. If an alcoholic solution of DDT is treated with 0.1 to 0.2 M alcoholic KOH solution at room temperature, the theoretical amount of HCl is liberated in 15 to 20 min. which can be titrated with $AgNO_3$ solution. When DDT is boiled with an alcoholic

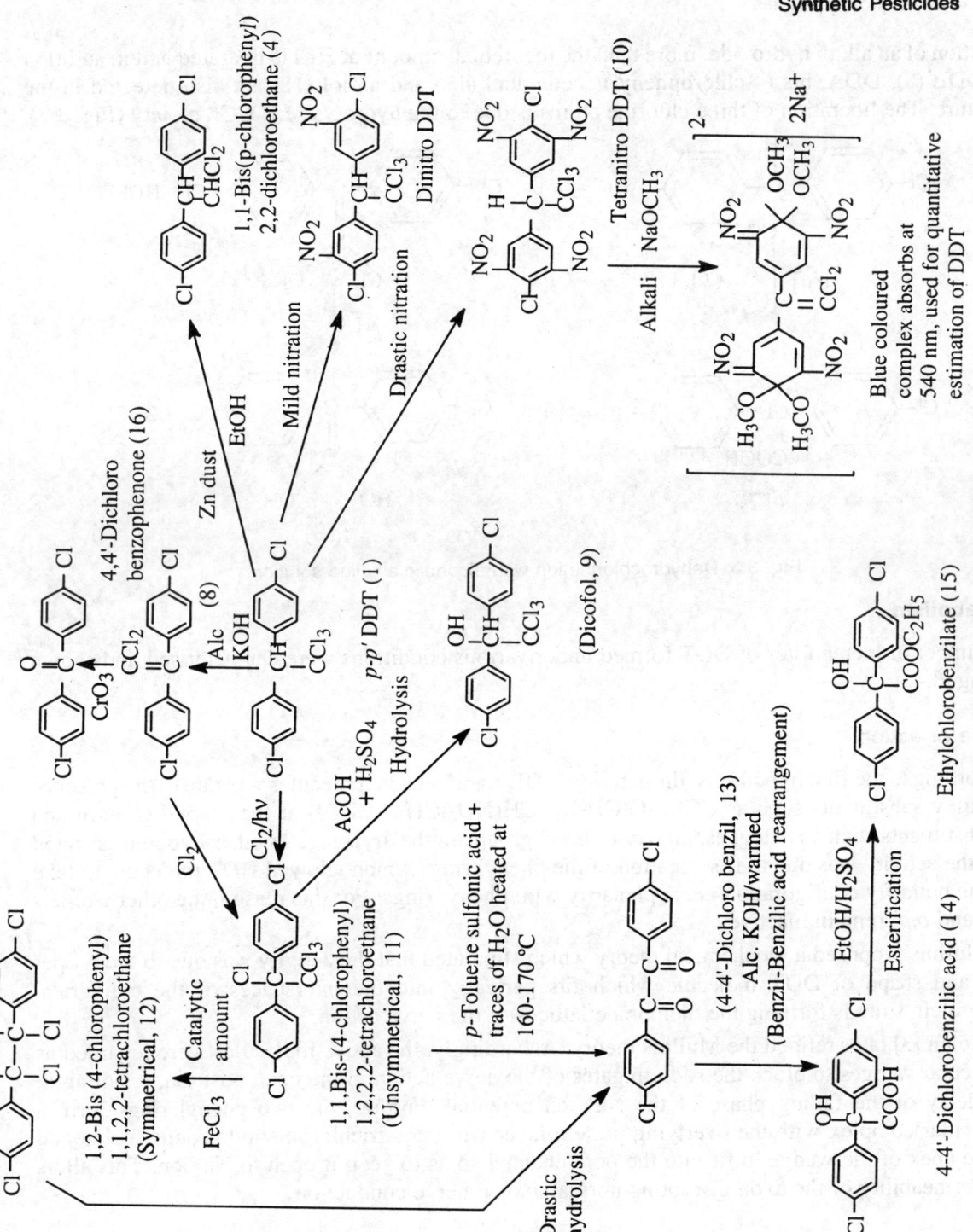

Fig. 3.2. Chemical reactions of DDT molecule.

solution of an alkali hydroxide, more than the theoretical amount of HCl is liberated and in addition to DDE (8), DDA, bis (4-chlorophenyl) acetic acid (17) and a diol (18) are also detected in the mixture. The liberation of three chlorine atoms is due to the hydrolysis of $-CCl_3$ moiety (Fig. 3.3).

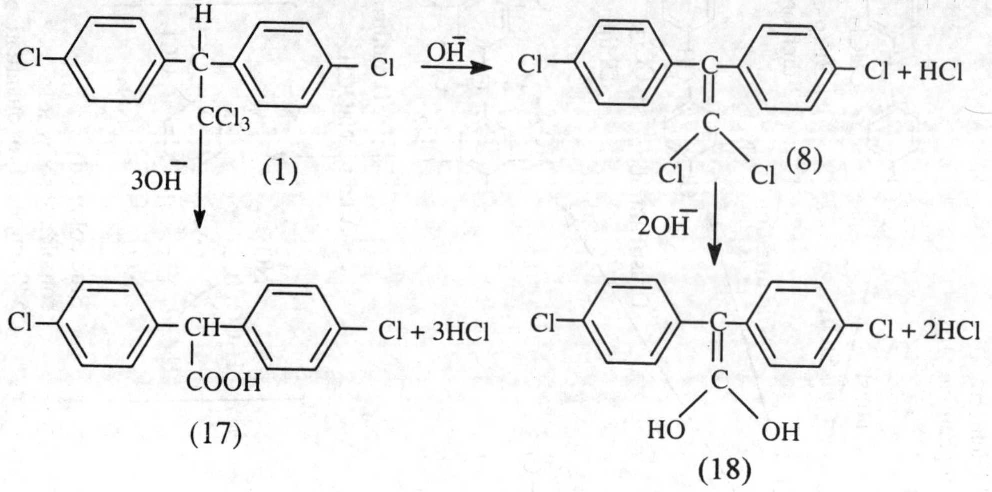

Fig. 3.3. Dehydrochlorination with alcoholic alkaline solution.

Metabolism

A number of metabolites of DDT formed under various conditions were characterised and shown in Fig. 3.4.

Mode of action

According to the first hypothesis, the activity of DDT and related molecules was due to the presence of bulky subsituents such as CCl_3, $-C(CH_3)_3$, $-CH(NO_2)CH_3$, $-CHCl_2$ at the second C-atom and p-substituents such as halogens, alkoxy or alkyl groups in the aryl ring. Another hypothesis stated that the activity was due to free rotation of the phenyl rings which allowed DDT molecule to take up the butterfly configuration i.e. coplanarity with phenyl rings. For this reason, the other isomers o,o' and o,p' remain inactive.

Mullins proposed a biochemical theory which stipulated that the activity was due to the proper size and shape of DDT molecule which fits perfectly into the interspaces of the cylindrical lipoprotein strands forming the membrane lattices of the nerve action.

Holan [3] later refined the Mullins theory. Accordingly, the active molecules were regarded as molecular wedges to block the sodium gates of the nerve action in the open position, resulting in the delay of the falling phase of the Na^+ ion potential. Further, the two phenyl rings form a molecular complex with the overlying protein layer while the trichloromethyl groups functioned as the apex of the wedge to fit into the pore channel so as to keep it open to Na^+ ion. This alters the permeability of the axon disrupting normal axonal nerve conduction.

Analytical methods

Gas liquid chromatography (GLC) using electron capture detector along with conventional volu-

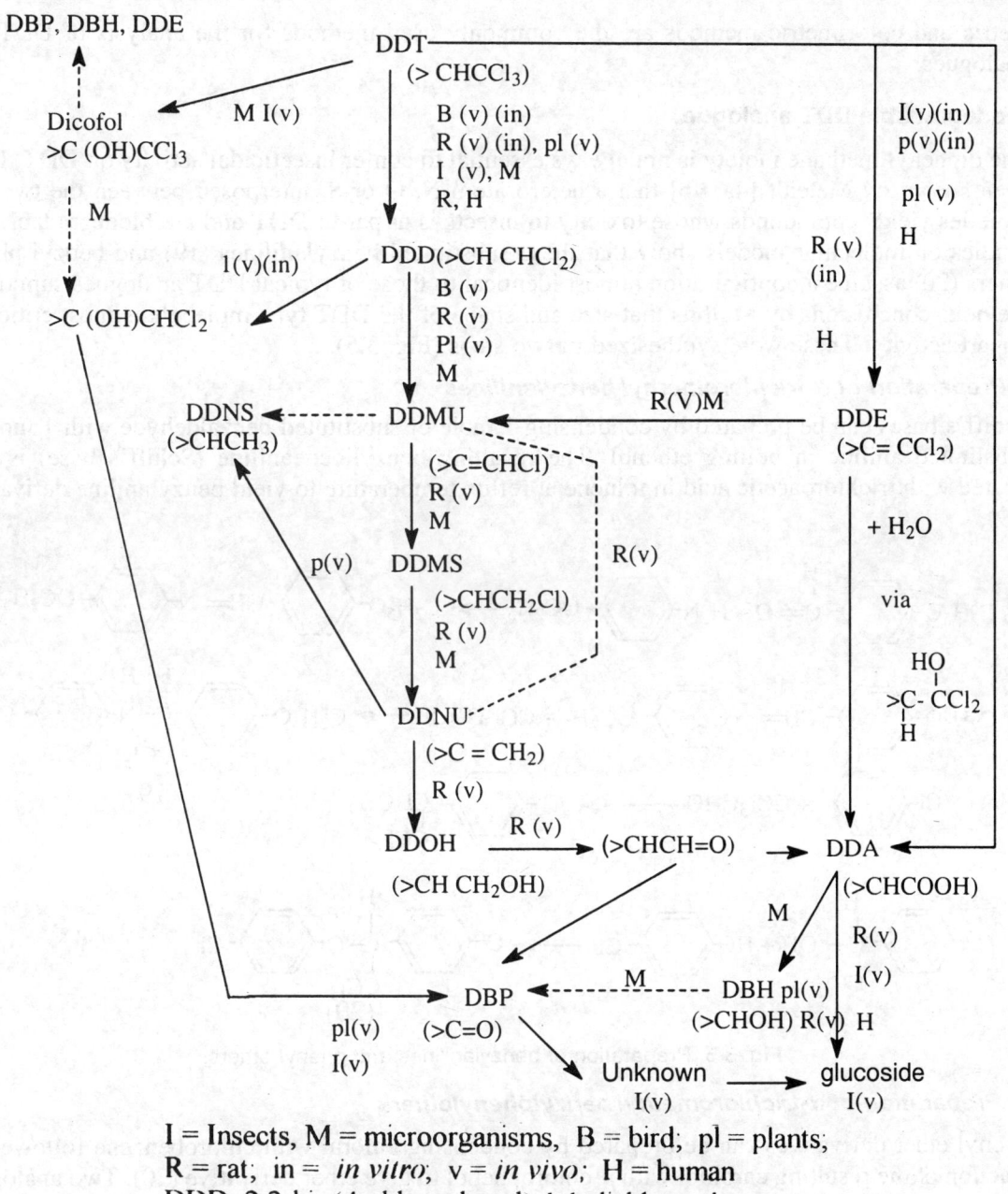

Fig. 3.4. DDT metabolism in some living organisms [Adapted from Chlorinated insecticide (GT Brooks ed. II, 1976)].

metric and colorimetric methods are the commonly used methods for the analysis of DDT and analogues.

Biodegradable DDT analogue

The diphenyl methane moiety is not always essential to confer insecticidal activity on DDT. It has been shown by Metcalf [4a, 4b] that a hetero atom N, O or S, interposed between the two aryl moieties yields compounds whose toxicity to insects is at par to DDT and are biodegradable too. Studies on molecular models show that these substituted benzylanilines (19) and benzyl phenyl ethers (20) assume a configuration almost identical to those of typical DDT analogues supporting previous conclusions by Mullins that size and shape of the DDT type molecule is most critical to impart activity. These were synthesized in two steps (Fig. 3.5).

i) Preparation of α-trichloromethyl benzylanilines

Schiff's bases can be prepared by condensing 1 mole of substituted benzaldehyde with 1 mole of substituted aniline in boiling ethanol. The resulting benzylideneaniline (Schiff's base) is then treated with trichloroacetic acid in toluene at reflux temperature to yield benzylaniline derivatives (19).

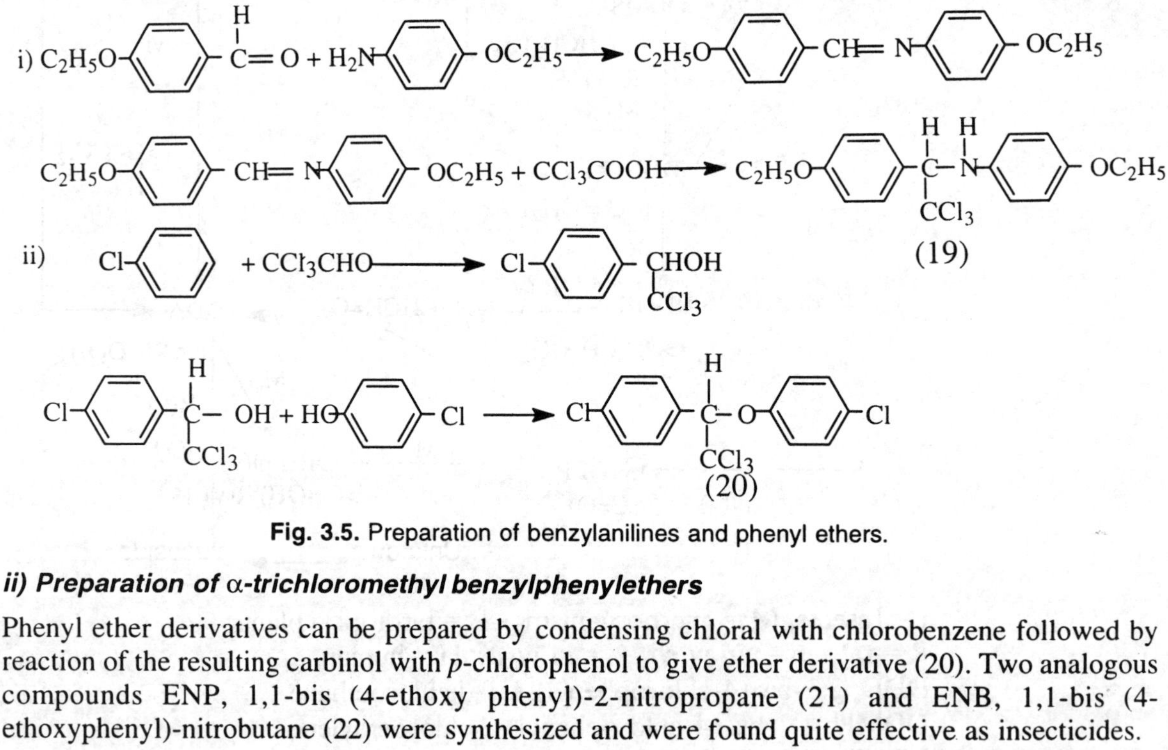

Fig. 3.5. Preparation of benzylanilines and phenyl ethers.

ii) Preparation of α-trichloromethyl benzylphenylethers

Phenyl ether derivatives can be prepared by condensing chloral with chlorobenzene followed by reaction of the resulting carbinol with p-chlorophenol to give ether derivative (20). Two analogous compounds ENP, 1,1-bis (4-ethoxy phenyl)-2-nitropropane (21) and ENB, 1,1-bis (4-ethoxyphenyl)-nitrobutane (22) were synthesized and were found quite effective as insecticides.

DDT-pyrethroid type compounds

Holan et al. [5] reported a DDT pyrethroid compound (23) whose activity was comparable to that of DDT and pyrethroids. SAR studies were carried out to determine i) the effect of substituents at

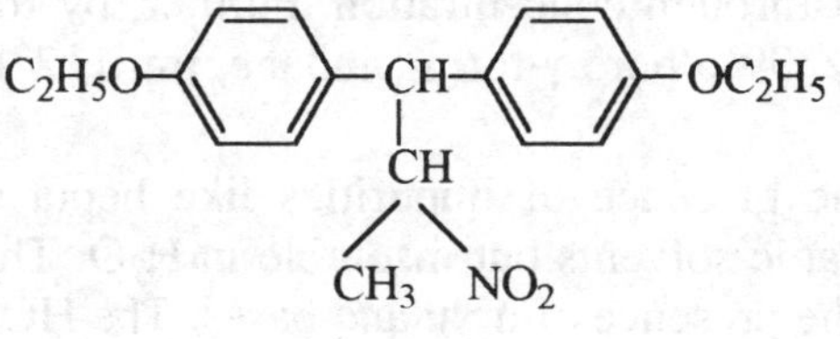

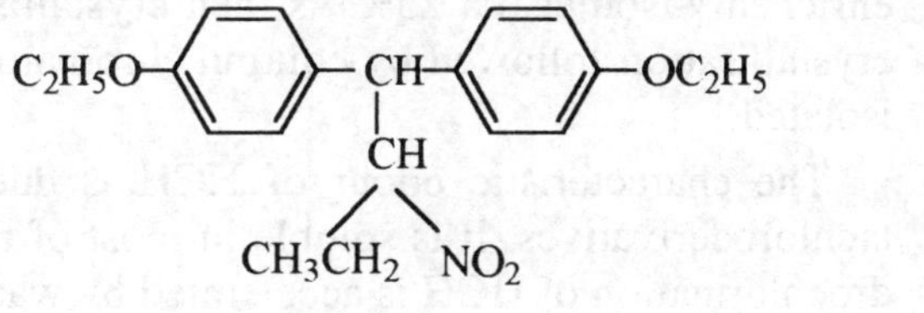

C-1 of phenyl ring and C-2 position, ii) the influence of the relative stereochemistry of substituents around the cyclopropane ring, and iii) the effect of different alcohols on the activity. Studies on the effect of substitution at C-2 showed that 2, 2-difluoro, 2-chloro-2-fluoro and 2,2-dichloro derivatives possessed maximum activity. Similarly, the effect of substitution on C-1 phenyl ring indicated that 4-ethoxy or 3,4-methylenedioxy group exhibited very good activity. Among the various alcohol moieties, 3-phenoxybenzyl, and 4-fluoro-3-phenoxybenzyl groups imparted better activity.

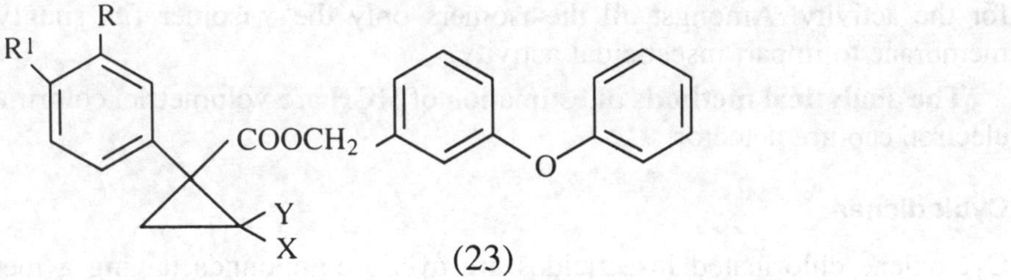

Hexachlorocyclohexane (HCH)

Hexachlorocyclohexane, popularly known as BHC, was discovered independently by a group of British and French chemists in 1942 [6a, 6b]. The compound HCH is a mixture of eight stereoisomers α, β, γ, δ, ε, ζ, η, and φ of which γ-isomer (lindane, 24) is the most active.

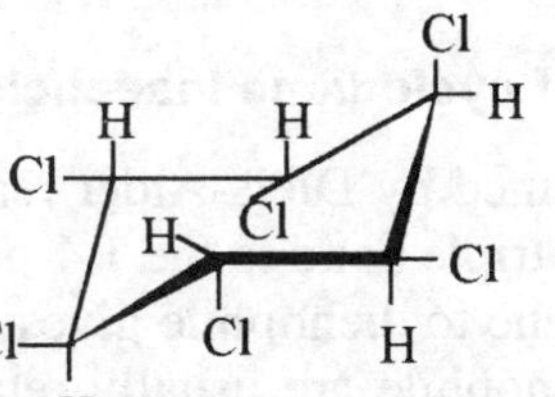

HCH is prepared by chlorination of benzene in presence of initiators, organic peroxide (benzoyl peroxide) and UV irradiation or hard γ-radiation from radioactive cobalt (Co-60).

Commercially available BHC is a mixture of HCH isomers and some related products although the reaction is carried out in excess benzene solvent or diluents like CH_2Cl_2, $CHCl_3$ and CCl_4. The reaction is temperature and time dependent. The γ-isomer is predominantly formed at higher concentration of chlorine and at a temperature range of 40-60°C. After chlorination, the solvent is distilled off and the residual liquid on cooling separates out less soluble α and β-isomers. The enriched γ-isomer (ca 25%) is then crystallised out through concentration. Further, by fractional crystallisation followed by column chromatography, 99% pure γ-form (lindane, mp 112°) can be isolated.

The characteristic odour of HCH is due to the presence of impurities like hepta and octachloroderivatives. It is soluble in most of the organic solvents but insoluble in H_2O. The dehydrochlorination of HCH is accelerated by water in the presence of light and bases. The HCHs, like DDT, are chemically stable under sunlight. They however, do not lose HCl easily. Boiling water/steam liberates only traces of HCl but at higher temperature (Ca 200°C) in a sealed tube, 1,2,4-trichloro benzene is formed. Amongst all isomers, β-HCH is most stable. On treatment with alkali, the HCH gives a mixture of trichlorobenzenes.

HCH is formulated into wettable powders, emulsifiable concentrates and dust. It is used both in agriculture and in public health programmes to control a large number of insects. Because of its high persistence and non-biodegradability, the use of HCH in agriculture has now been banned in India.

The mode of action of HCH is similar to that of DDT. The size and shape of HCH is responsible for the activity. Amongst all the isomers only the γ-isomer fits snugly in the pores of nerve membrane to impart insecticidal activity.

The analytical methods of estimation of HCH are volumetric, colorimetric and GLC using the electron capture detector.

Cyclodienes

Cyclodiene chlorinated insecticides are cyclic compounds having a methano bridged structure substituted with chlorine atoms. These dienes are prepared by Diels-Alder reaction. The starting material for the preparation of cyclodienes is hexachlorocyclopentadiene (HCCP, 25). A large number of compounds belonging to cyclodiene group of compounds are used as insecticides [6c] of which, a few important compounds have been discussed in this chapter. The preparation of chlordene (26) chlordane (27), heptachlor (28), its epoxide (29), aldrin (30), dieldrin (31), isodrin (32), endrin (33), endosulfan (34) and allodan (35) are depicted in Figs. 3.6*a* & 3.6*b*.

Stereochemistry nomenclature of cyclodiene insecticides

The cyclodiene compounds are formed by Diels-Alder reaction which involves the addition of a compound containing a double or a triple bond to the 1,4-position of a conjugated system. According to D.A. rules the addition of diene to dienophile gives a '*cis*' addition product and the relative positions of substituents in the dienophile are usually retained in the adduct. This means that a *cis*-dienophile will not give a trans-adduct and trans dienophile will not give '*cis*' adduct. However, under experimental conditions '*cis*' adduct can isomerize to trans-adduct. Diels-Alder reaction is governed by the Woodward-Hoffmann orbital symmetry rules and is identified by an endo-transition state. The endo positions are those which lie below the plane containing the methylene bridge (methanobridge) carbon of norbornene nucleus and the two carbon atoms carrying the exo-hydrogens at the ring junction. The reaction of norbornadiene nucleus (formed by the reaction of cyclopentadiene and acetylene) with cyclopentadiene gives products which have basic carbon skeleton of dimethano naphthalene series of cyclodienes. It is found that a planar representation is inadequate apart from the possibility of both *exo*- and *endo*-addition. Two orientations are possible for each of these modes of fusion. Thus the methanobridge skeleton associated with endofusion corresponds to the isodrin and aldrin series of cyclodiene compounds. The molecule heptachlor is called 1 *exo*-4, 5, 6, 7, 8, 8-heptachloro-3a, 4, 7, 7a, tetrahydro-4-7 methanoindene (28). The C-1 chlorine of heptachlor is in *exo*-position and should be expressed as 1-*exo*. Similarly the prefix *endo*- in *endo*-4,7 methanoindene can explain the stereochemistry of the ring junction.

(i)

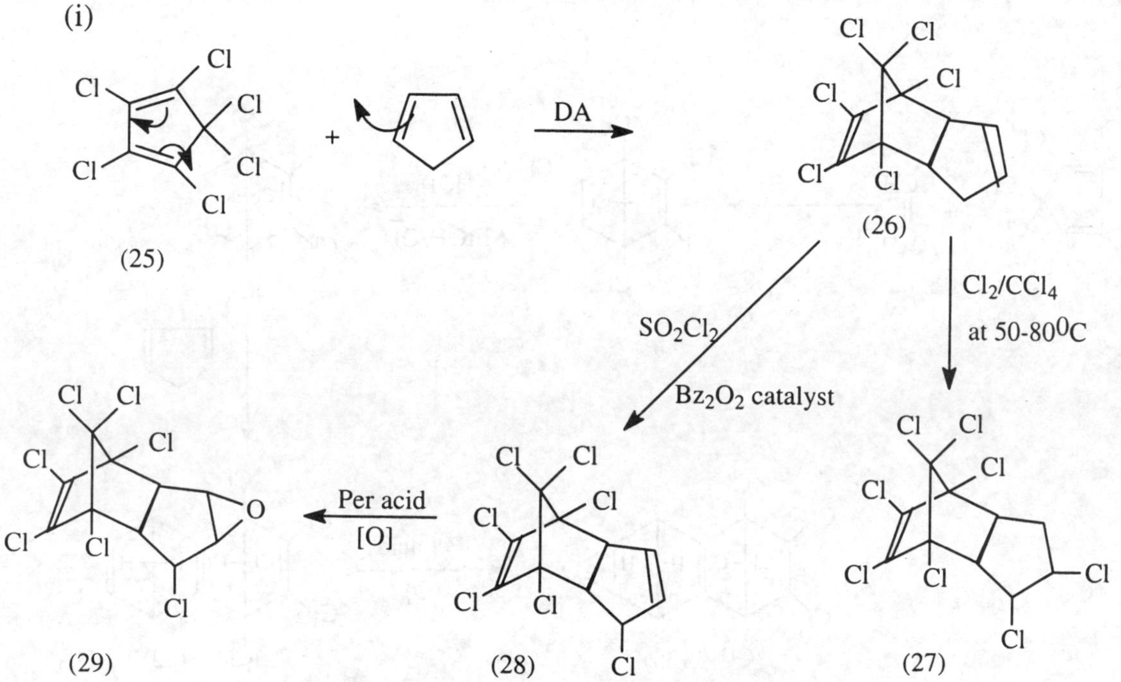

Fig. 3.6a. Preparation of cyclodiene insecticide.

In the British system, the prefix *exo*- means outside and *endo*- means inside. The first prefix refers to the methanobridge of the lowest numbered nucleus carrying the chlorine atoms while the second prefix refers to the methanobridge in the highest numbered nucleus. In aldrin (30), the 1,4-dichloromethanobridge exhibit *exo*- while the 5,8-methanobridge shows *endo*-configuration. Thus the aldrin molecule can be represented as 1, 2, 3, 4, 10, 10 hexachloro-1, 4, 4a, 5, 8, 8a-hexahydro-1, 4-*exo*, *endo*-5,8-dimethanonaphthalene (or HHDN). In isodrin (32, aldrin stereo-isomer) both methanobridges are outside the bent cage structure (*exo*-) and the molecule is called HHDN with 1, 4, *exo*, *exo*-5, 8, orientation.

In the American system, the prefixes *exo*- and *endo*- have different meaning. The *exo*- or *endo*-bonds of each of the norbornene nuclei are used to fuse the two ring systems together and each prefix therefore defines the stereochemistry of the norbornene nucleus to which it applies. Accordingly, HHDN is represented as 1, 2, 3, 4,10,10-hexachloro-1,4,4a,5,8,8a-hexahydro-1,4-*endo*, *exo*-5, 8-dimethanonaphthalene and isodrin (HHDN) with 1,4-*endo*, *endo*-5, 8 orientation.

The molecule dieldrin (31, HEOD) is designated as 1,2,3,4,10, 10-hexachloro-6,7-*exo*-epoxy 1,4,4a,5,6,7,8,8a-octahydro-1,4-*endo*, *exo*-5, 8-dimethanonaphthalene. Similarly endrin (β-HEOD) is the *endo*-*endo*-stereoisomer.

(ii)

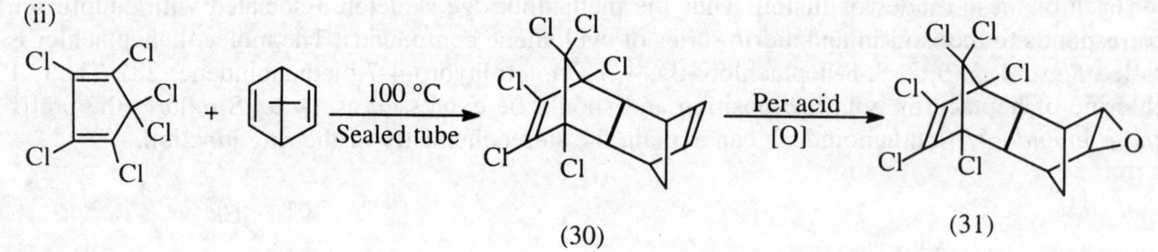

(30)

(31)

(iii)

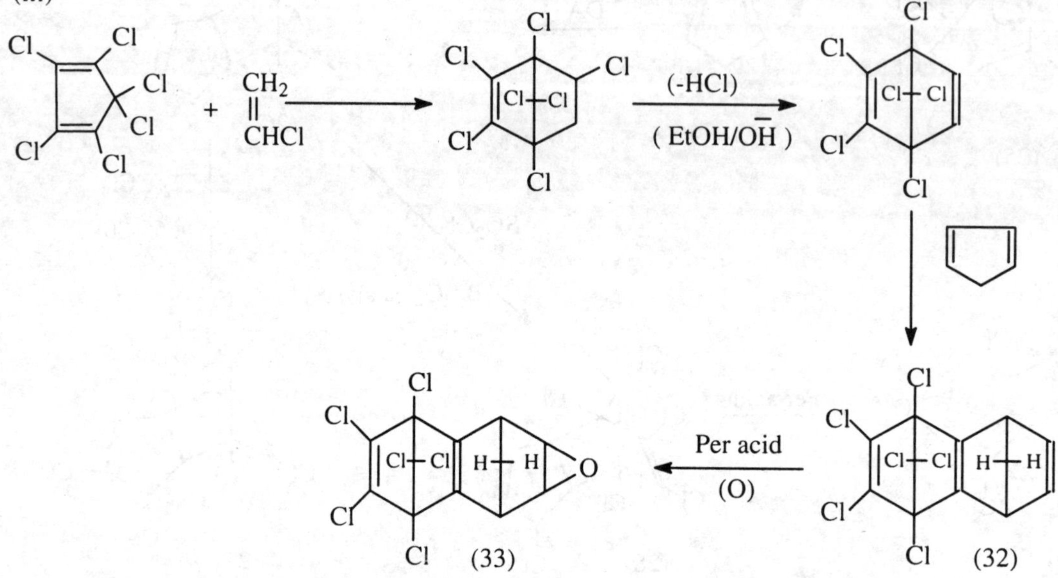

(32)

(33)

(iv)

(35)

(34)

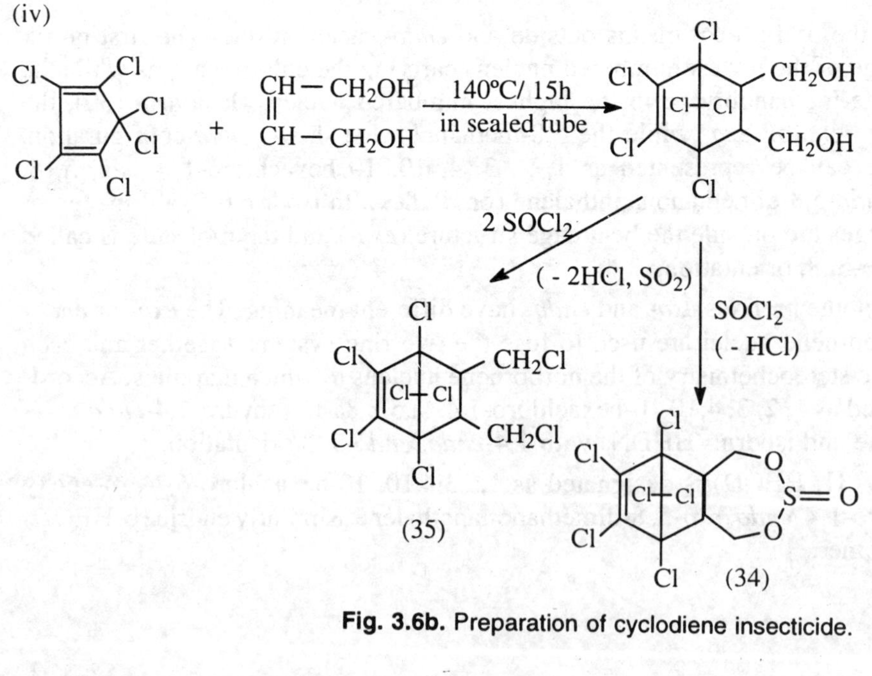

Fig. 3.6b. Preparation of cyclodiene insecticide.

The three-dimensional structure is rather complicated in case of planar heptachlor structure where stereochemistry of ring junction at position 4a and 8a is undefined. The methanobridge of the unchlorinated norbornene nucleus may be in one of two different configurations (*endo-* or *exo-*) relative to the dichloromethano bridge.

Formulation and application

All cyclodiene insecticides are available in conventional liquid and solid formulations. These are non-systemic and highly persistent having high contact and stomach activity towards many insects. Because of leaving toxic residues on consumer food, soil and water, these are totally banned by Government of India for use in agriculture except endosulfan which is less persistent and relatively safe as it volatilises rapidly from plant surfaces, and does not leave any toxic residue on the harvested crops.

Mode of action

The molecular shape and size appears to be responsible for the activity of cyclodiene group of insecticides. In general, these compounds act on central nervous system. Insect pests treated with cyclodiene insecticides, react violently followed by trembling, convulsions, and paralysis finally leading to death. The cyclodienes probably release excessive acetylcholine which is responsible for loss of nerve co-ordination.

Photochemistry

The fact that sunlight possesses sufficient energy to effect photochemical transformation in organic compounds, has created a great deal of interest in the study of environmental photochemistry of pesticides. Chlorinated pesticides are generally considered to be highly stable and persistent in the environment. But these undergo some phototransformation at wavelengths between 285-450 nm which are of general interest for pesticide photolysis in sunlight [7].

Since pesticides undergo leaching by water, volatilization, adsorption to surfaces and partitioning into organic films, their photochemical behaviour in each of these phenomena is important.

Photodegradation of pesticides involves two steps :

 (i) absorption of radiation (direct or sensitized) and

 (ii) transformation of the electronically excited species by chemical processes.

Use of photosensitizers

There are many pesticides which directly absorb light and get transformed slowly. Photosensitizers are compounds that facilitate the transfer of the light energy into receptor chemicals. Various photosensitizers facilitate photolysis such as benzophenone, amino compounds and anthraquinone. These are mixed with pesticides in minute qunatities before exposing them to sunlight. Other factors affecting photochemical reaction are the medium and the solvent in which the reacting substrate is dissolved or suspended.

Photolysis of DDT

Photochemical studies on DDT in organic solvents, water, air and also in vapour phase under U.V.

and sunlight were conducted by Crosby [8]. The products identified were DDE, epoxidation products of DDE, DDD and benzophenone. The mechanism of the conversion is still not clear.

Photolysis of cyclodienes

The photochemistry of cyclodiene compounds (Fig. 3.7) was widely studied under various conditions [9]. The intramolecular rearrangement process is common during transformation. Longer

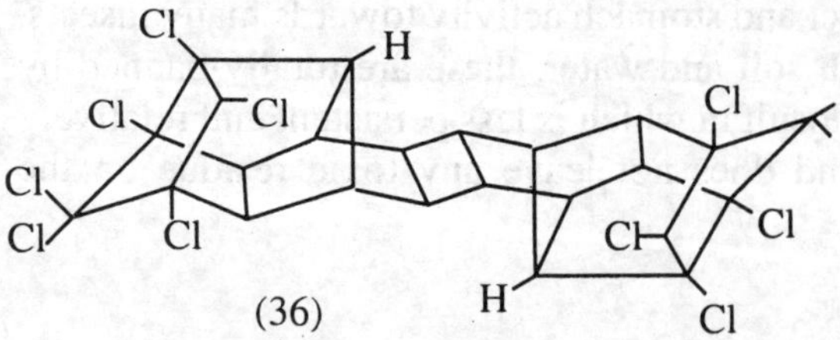

(36)

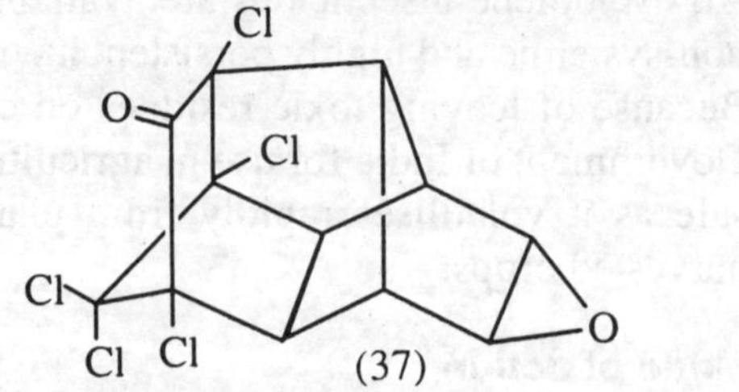

(37)

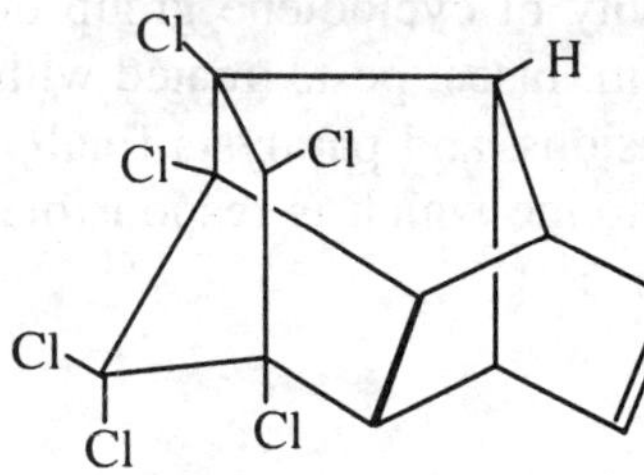

(38)

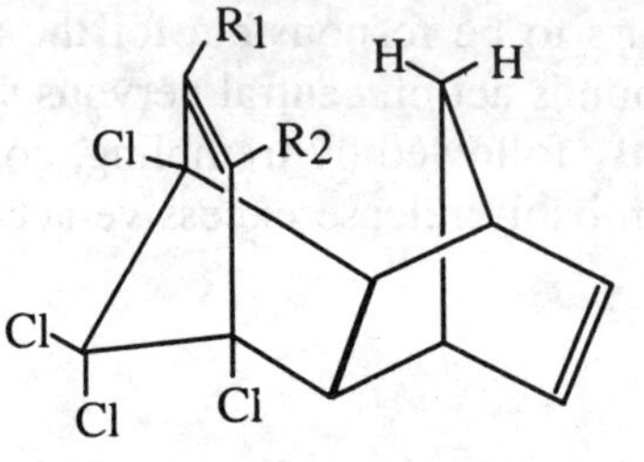

(39 = R1=R2=H)
(40 = R1=Cl, R2=H)

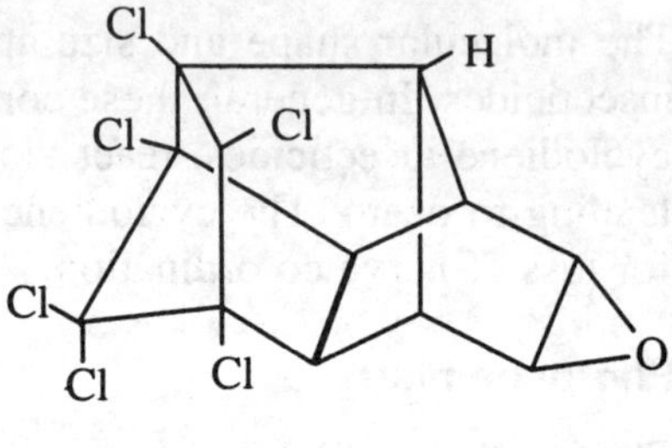

(41)

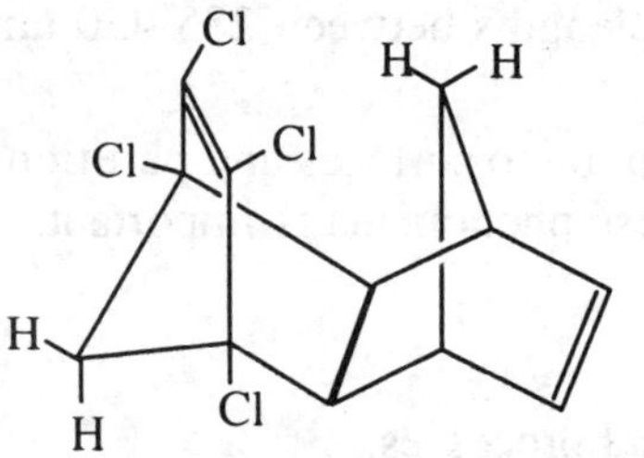

(42)

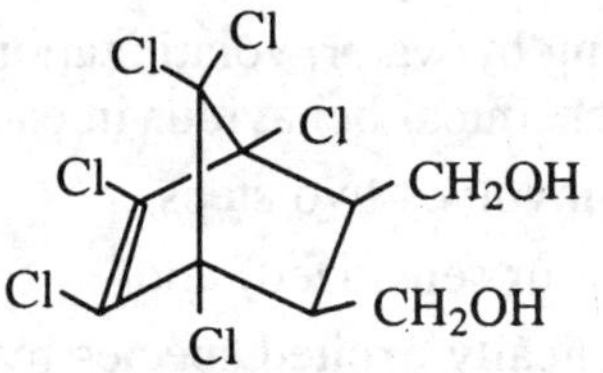

(43)

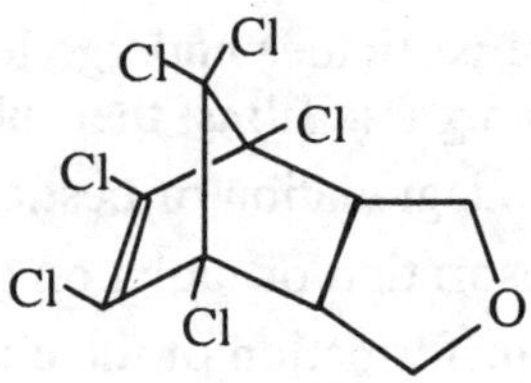

(44)

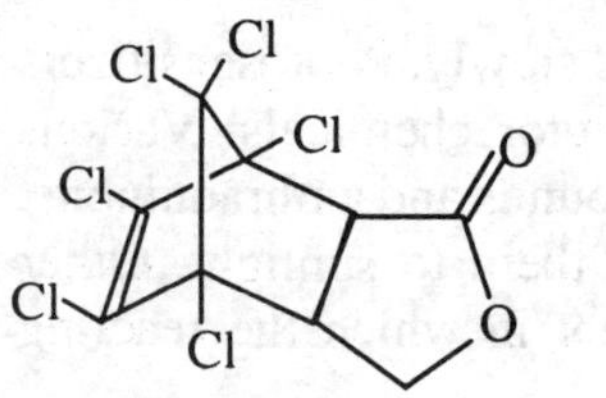

(45)

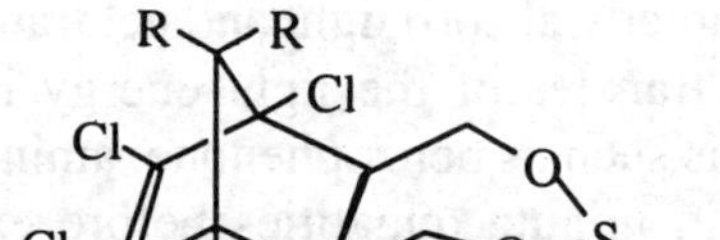

(46, R=Cl)
(48, R=H)

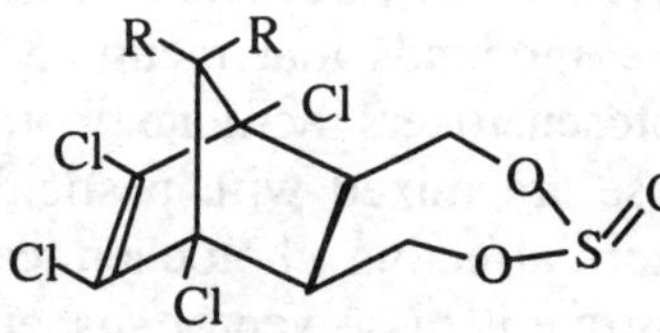

(47, R=Cl)
(49R=H)

Fig. 3.7. Photoproducts of cyclodiene insecticide.

irradiation leads to dimerised products (36). Other photochemical reactions included photooxidation and photomineralization. In case of dieldrin and aldrin, cage isomers are formed on irradiation [10] under UV light. Similarly, endrin on photolysis gives ketoendrin (37) the quality of which increases with storage period.

Photolysis of aldrin

Aldrin on irradiation with UV light in presence of acetone as sensitizer gives a dechlorinated product along with photoaldrin (38), tetrachloroaldrin (39), pentachloroaldrin (40) and an acetone adduct. On exposure to sunlight, a thin film of aldrin on a glass plate gets transformed to predominantly dieldrin and photodieldrin (41). The related cyclodiene insecticides dieldrin and endrin behave similarly like aldrin and yield photoisomers as principal products such as photodieldrin and photoheptachlor along with some mono and di-dechlorinated photoproducts (42). Treatment of cyclodiene insecticides with superoxide yield anti-dechlorinated products.

Although amine induced photodehalogenation of organochlorine insecticides has been observed earlier and mechanism of the formation of charge-transfer complexes proposed, the stereospecificity in the dehalogenation of cyclodienes observed is rather novel. Photodieldrin and photoaldrin gave novel products respectively on attempted anti-dechlorination with potassium superoxide [11].

Phoytolysis of endosulfan

Amongst the cyclodiene insecticides, only endosulfan has retained its importance in Indian agriculture as it is one of the environmentally safe insecticide. It has been earlier reported to form photoproducts by loss of one chlorine atom from the methanobridge position. In addition, it gives photoproducts like the diol (43), ether (44) and lactone (45) as a result of cleavage of the sulphite ester. Dureja and Mukherjee [12] however, isolated both (α, β) isomers of endosulfan (46, 47) as sole products by dechlorination from the methanobridge. The formation of the corresponding tetrachloro endosulfans (48, 49) was more pronounced in the presence of photosensitizers. On the other hand, in the presence of protic solvents like methanol, the sulphite ester cleavage was the major photoreaction. Since endosulfan diol (43) and endosulfan ether (44) did not undergo methanobridge dechlorination, the participation of S=O in the intramolecular energy transfer mechanism to form these photoproducts was envisaged. Like other cyclodienes, both the isomers of endosulfan in presence of triethylamine gave mainly, syn- dechlorinated photo-product along with traces of anti-dechlorinated products. The major photoreaction of endosulfan of some concern is the isomerisation of α-endosulfan (46) to its more stable β-isomer (47) under field conditions in sunlight and more rapidly in UV. Since β-endosulfan is less toxic to bees, the prevention of this photoisomerisation should enhance environmental safety of technical endosulfan rich in β-endosulfan.

REFERENCES

1. Zeidler, O. Dissertation of Strasbourg (1873). *Ber Deut. Chem. Ges.* **7**, 1180 (1974).
2. Muller, P. Swiss Pat. 226180 (1940). DDT das Insektizid. Dichlordiphenyltrichlorathan und seine Bedeutung **1**, Birkhauser, Basel-Stuttgart.
3. Holan, G. Halocyclopropane insecticides and the mode of action of DDT. *Nature* (London) **221**, 1025-29 (1969).

4a. Hirwe, A.S., Metcalf, R.L. and Kapoor, I.P. α-Trichloromethyl benzylanilines and α-trichloromethyl benzyl phenyl ethers with DDT like insecticidal action. *J. Agric. Food Chem.* **20**, 818 (1972).

4b. Melcalf, R.L. Century of DDT. *J. Agric. Food Chem.* **21**, 511 (1973).

5. Holan, G., O'Keefe, D.F., Ricks, K., Walser, R. and Virgona, C.T. New Insecticides, combined DDT-isosteres and pyrethroid structures. In : *Advances in Pesticide Science,* (H. Geissbuhler, ed.) Part 2, Pergamon Press, N.Y. 201 (1978).

6a. Brooks, G.T. *Chlorinated insecticides.* **I & II**, C.R.C. Press, USA (1976).

6b. Matolcsy, G. Anti-insect agents. In : *Pesticide chemistry*, (G. Matolcsy, M. Nadasy, V. Andriska, eds.). Elsevier, New York, 47-85 (1988).

6c. Synthetic insecticides I : Miscellaneous and organochlrine compounds. In : *Agrochemicals* (R.J. Cremlyn, ed.), John Wiley & Sons, New York, 79-103 (1990).

7. Matsumara, F. *Toxicology of Insecticides*, Plenum Press, New York, 598 (1985).

8. Crosby, D.G. The significance of light induced pesticide transformations. In : *Advances in Pesticide Science* (H. Geissbuhler, ed.), Part 3, Pergamon Press, New York, 568-576 (1979).

9. Parlar, H. and Korte, F. Photoreactions of cyclodiene insecticides under simulated environmental conditions—A review. *Chemosphere* **10**, 665-705 (1977).

10. Rosen, J.D. and Sutherland, D.J. The nature and toxicity of the photoconversion products of aldrin. *Bull Environ. Contam. Toxicol.* **2** (1-9) (1967).

11. Alley, E.G., Layton, B.R. and Minyard, Jr., J.P. Photoreduction of mirex in aliphatic amines. *J. Agric. Food Chem.*, **22**, 727 (1974).

12. Dureja, P. and Mukherjee, S.K. Photoinduced reaction P-N, Studies on photochemical fate of 6,7,8,9,10,10a-hexachloro-1,5,5a,6,9,9a-hexahydro-6, 9-methano-2,4,3-benzo(e)-dioxathiopin-3-oxide (endosulfan) an important insecticide. *Indian J. Chem.* **21B**, 411 (1982).

Section II. Chemistry of Pyrethroid Insecticides

- Introduction
- Pyrethroids
- Non ester pyrethroids
- Mode of action

- Metabolism
- Photochemistry
- *References*

Natural pyrethroids (cf. Chapter 2) are the most ideal and safer insecticides among pesticides of natural origin. The production of natural pyrethrum extract is limited because of the shortage of *Chrysanthemum cinerariaefolium* plant which requires specific agroclimatic conditions to grow. Due to the high demand of pyrethrins, the chemists have all along been interested in developing synthetic analogues with high potency, selectivity and photostability, and low mammalian toxicity.

Pyrethroids

Schechter *et al.* [1] first synthesized 3-allyl-2-methyl-4-oxo-cyclopent-2-en-1-yl ester of (±) - (*E,Z*)-chrysanthemic acid popularly known as **allethrin** (1, Fig. 3.8a). Elliott *et al.* [2] of Rothomstead Experimental Station, U.K. and scientists at Sumitomo Industries, Japan demonstrated that a large number of variations in the pyrethrin molecule are possible which led to the discovery of many active molecules which are now being commercialized.

Chemical and spectroscopic evidences showed that cyclopentenyl ring in the pyrethrin molecule

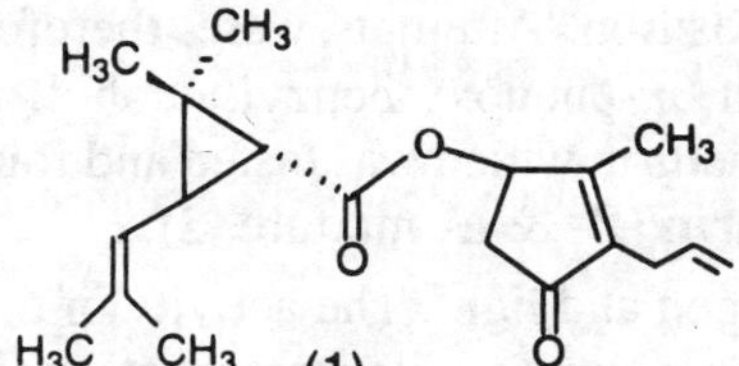

Fig. 3.8a. Synthetic pyrethroids.

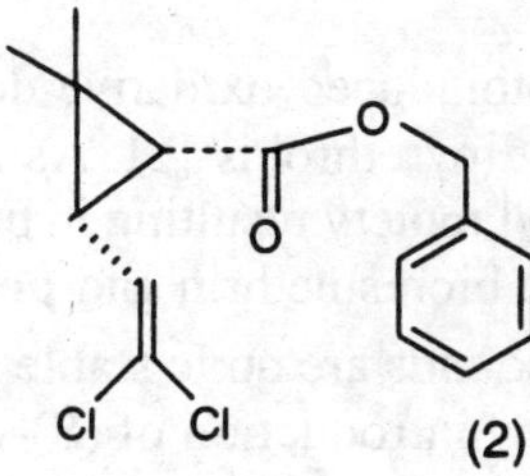

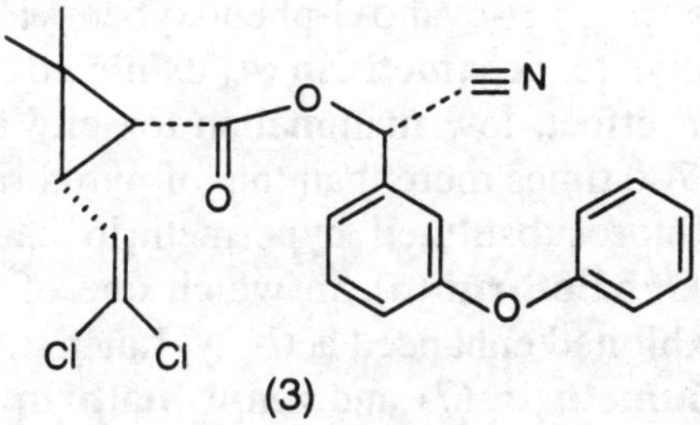

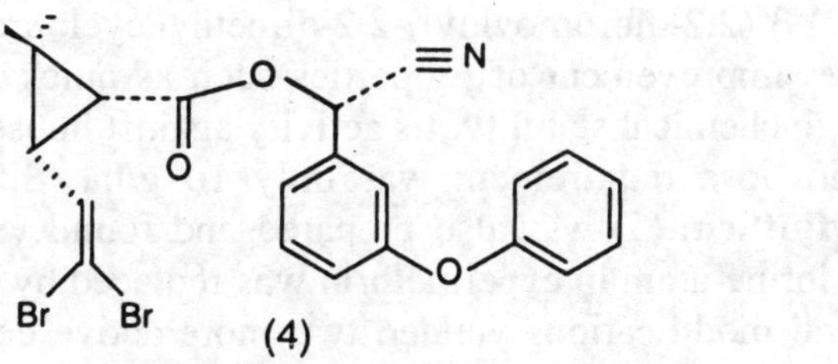

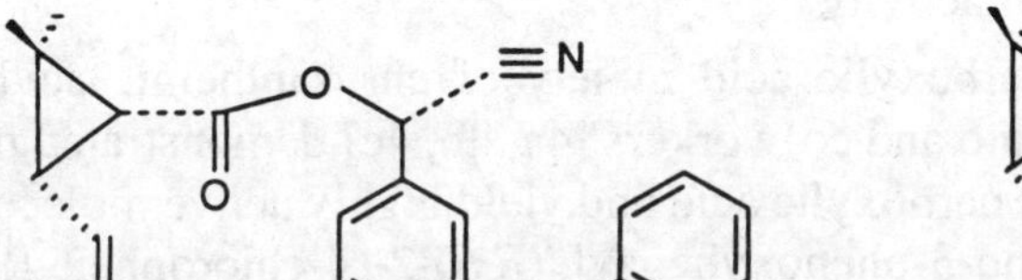

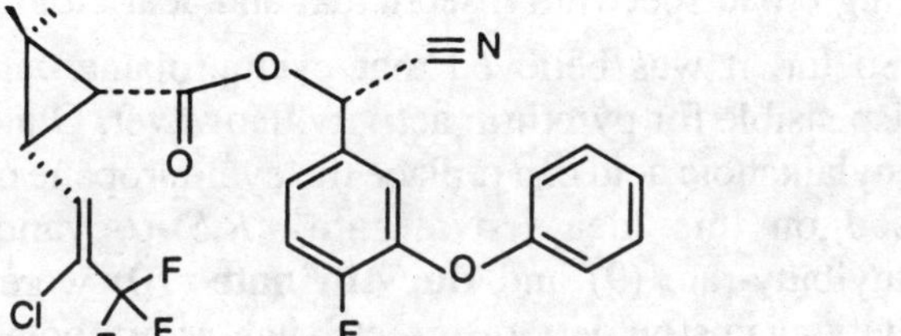

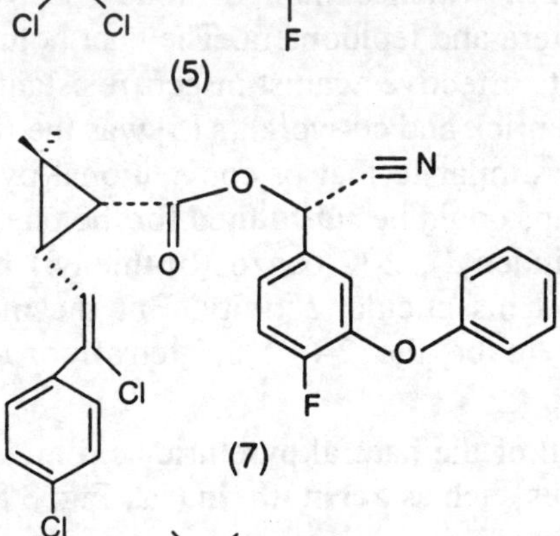

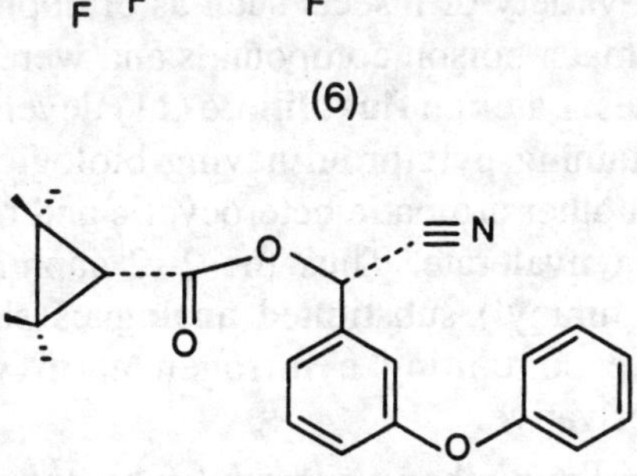

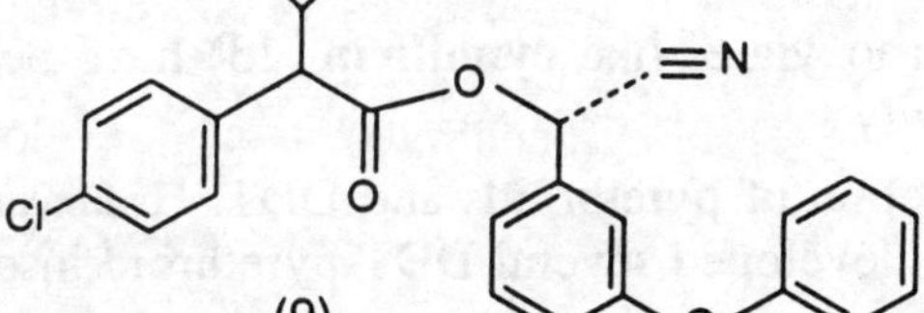

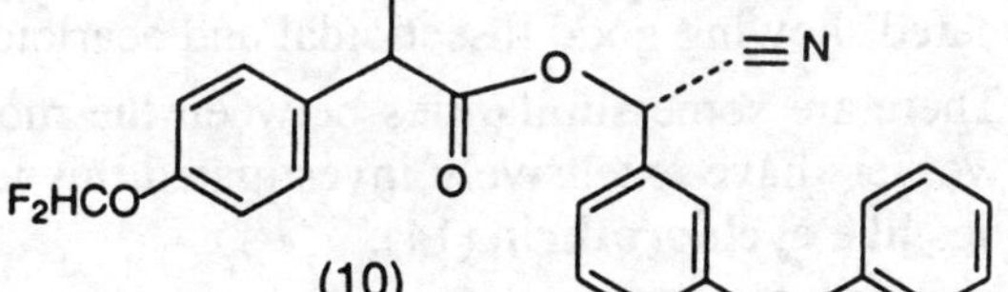

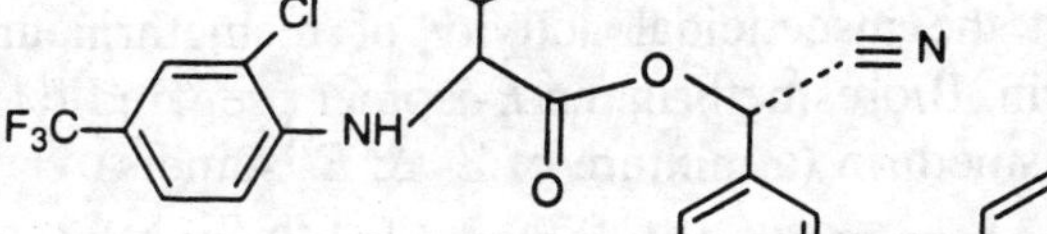

was the most probable site of photoinduced oxidative decomposition. Attempts were, therefore, made to discover more photostable pyrethroids [2]. As a result, m-phenoxy benzyl alcohol was substituted in place of cyclopentenyl moiety resulting in biopermethrin which was tested and found to be 2.5 times more effective than bioresmethrin and **permethrin** (Z- & E- mixture, 2).

It was observed that these compounds are quite stable to oxygen and light. The activity of these compounds was further enhanced by introduction of (S)-α-cyano group to give **cypermethrin** (3) [3]. The replacement of chloro group by bromo group as in (S)-α-cyano-3-phenoxybenzyl($1R$, cis)-3-(2,2-dibromovinyl)-2,2-dimethyl cyclopropane carboxylate, **decamethrin** (4) exhibited further improvement of properties such as quick knock down effect, low mammalian toxicity and high chemical stability. Its activity against houseflies was 1700 times more than that of pyrethrin-I and dose requirement was only 10 g/ha. Similarly, 4-fluorosubstituted cypermethrin called **cyfluthrin** (5) was also prepared and found very active. **Cyhalothrin** (6), in which one of the chlorine-atom in cypermethrin was replaced by CF_3, also exhibited enhanced activity. Later structural modifications yielded two more active compounds **flumethrin** (7) and **fenpropathrin** (8) having broad spectrum insecticidal and acaricidal activity.

So far, it was believed that cyclopropane carboxylic acid moiety of chrysanthemic acid is indispensible for pyrethrin activity. However, Ohno and co-workers [4a, 4b, 4c] demonstrated that phenylalkanoic acid can replace the cyclopropane carboxylic acid and yield highly active molecule. Based on this idea **fenvalerate**, (R,S)-α-cyano-3-phenoxybenzyl (R,S)-2-(4-chlorophenyl)-3-methylbuty-rate (9) and **flucythrinate** (10) were synthesized which exhibited broad spectrum activity against variety of insects such as orthoptera, hemiptera and lepidoptera. They act both as contact and stomach poison compounds and were found quite effective against insects resistant to other insecticides. Later on **fluvalinate** (11) developed by Henrick and co-workers [5] was the first aminoacid containing pyrethroid having biological activity similar to that of conventional pyrethroids. Several other aromatic heterocyclic and related groups could be substituted for the phenyl ring at C-2 in fenvalerate. Thus the 2-(2-naphthyl), 2-(2-indenyl), 2-(2-benzo [b]-thienyl) and 2-(2-benzo [b]-furanyl) substituted analogues showed good insecticidal activity. But the most active analogue containing a nitrogen atom was found to be the 2-(4,5,6,7-tetrafluoro-2-isoindolinyl) derivative.

Variations in the methoxy carbonyl substituted side chain of the natural pyrethric acid moiety in pyrethrin-II has yielded several active pyrethroid molecules such as **acrinathrin** (12, Fig. 3.8b) having good acaricidal and insecticidal activity.

Several aromatic spiro-substituted cyclopropane carboxylates like **cypothrin** (13) have been prepared showing good insecticidal and acaricidal activity.

There are some similarities between the mode of action of pyrethroids and DDT. Holan and co-workers have extensively investigated this idea and developed several DDT-pyrethroid insecticides like **cycloprothrin** (14).

Variation in alcohol moiety led to the preparation of **dimethrin** (15), **barthrin** (16), **resmethrin** (17) and **tetramethrin** (18). It was observed that the insecticidal activity of resmethrin and tetramethrin was two times more than that of allethrin. Bioresmethrin, an E-isomer prepared from (±) - (E) -chrysanthemic acid is more active than resmethrin (a mixture of Z- & E-isomers).

The alcohol moiety such as 3-benzyl phenyl group gave very active insecticide **bifenthrin** (19) which was as active as cyhalothrin. Earlier it was thought that a bridging atom between the two

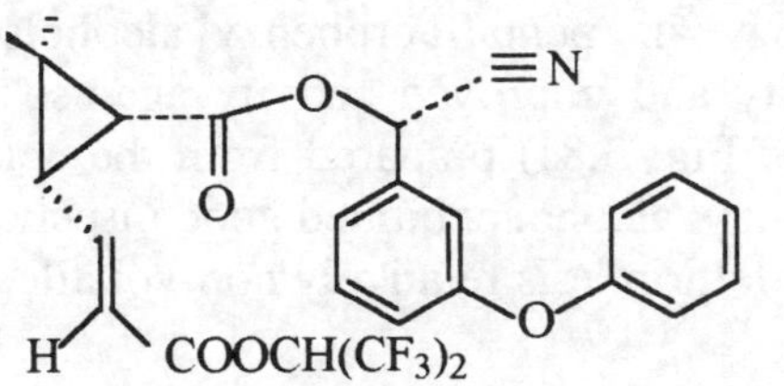

(12)

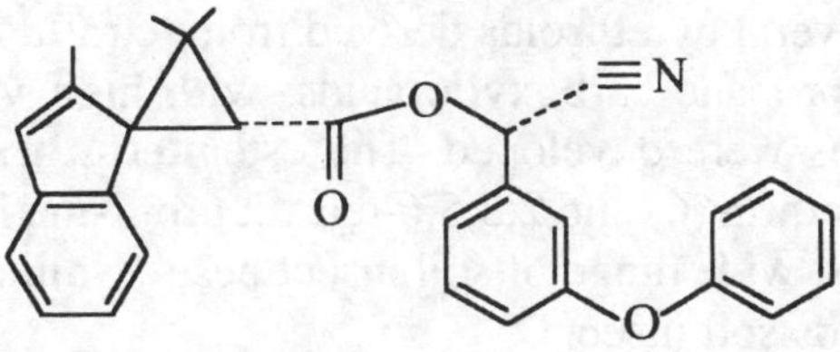

(13)

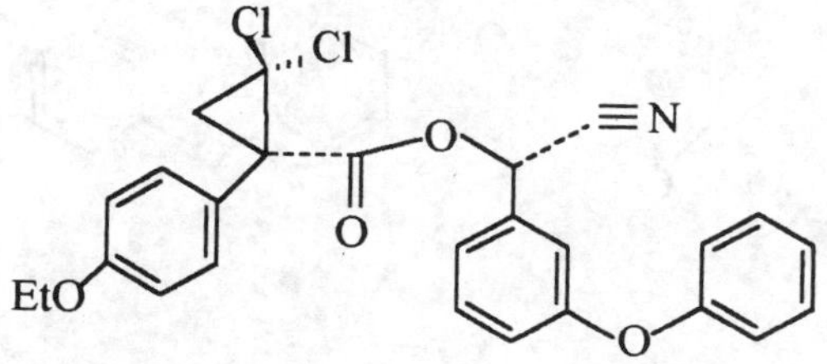

(14)

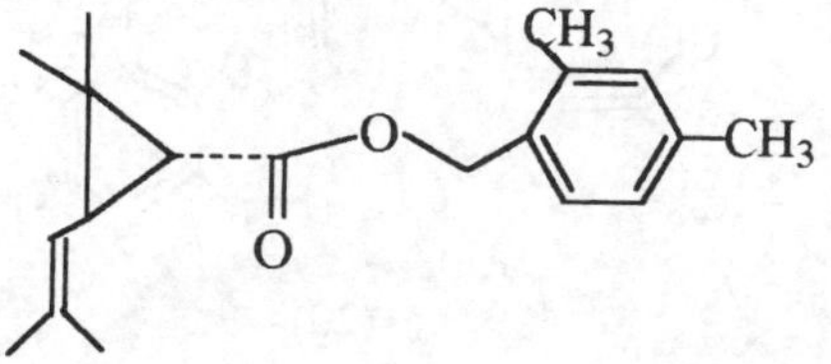

(15)

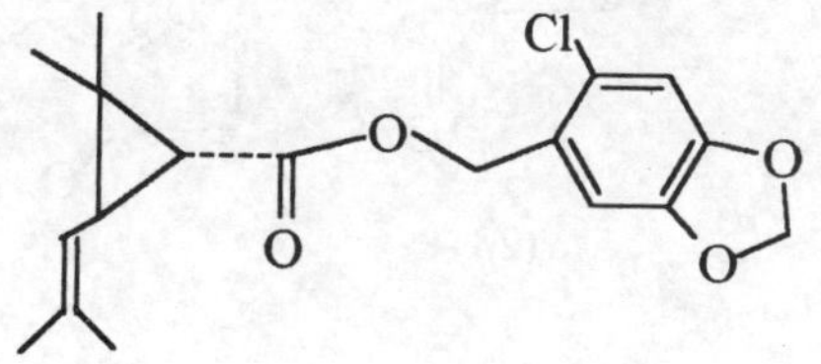

(16)

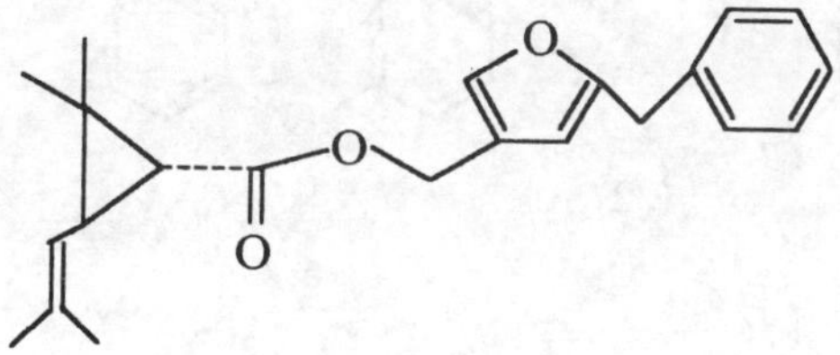

(17)

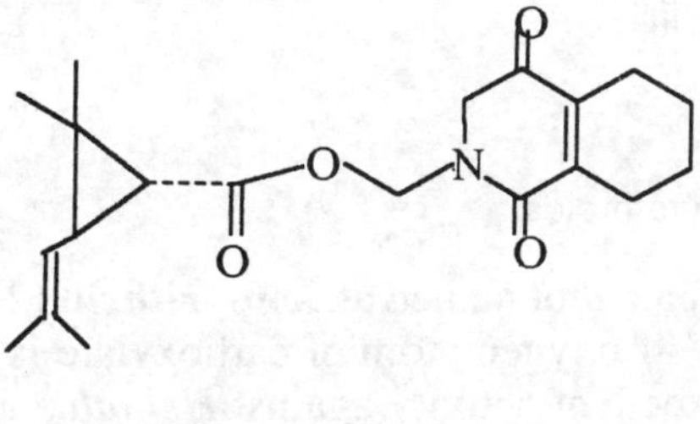

(18)

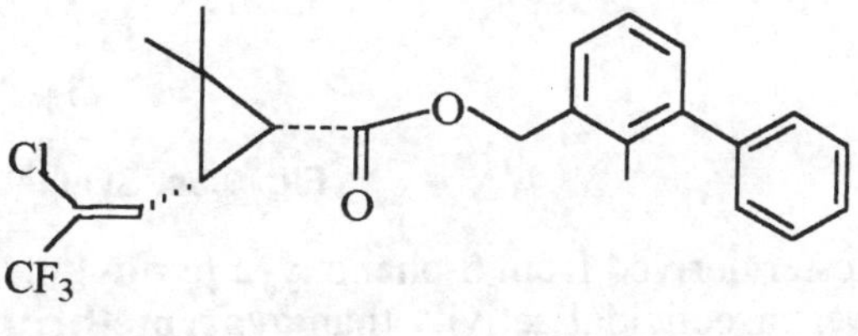

(19)

Fig. 3.8b. Synthetic pyrethroids.

centres of unsaturation in the alcoholic moiety was essential for the activity. However, Plummer [6] later demonstrated that a single methyl group at C-2 of the phenyl-benzyl ring induced a conformational effect similar to the bridging oxygen atom of the phenoxy benzyl moiety. Thus 2-methyl-3-phenyl benzyl derivative is more active than **permethrin** (2).

A few cyclic benzyl esters such as 4-allyl-1-indanyl chrysanthemate and 7-substituted-2, 3-dihydrobenzo [b] furan-3-yl chrysanthemates have been synthesised and shown to be more active than allethrin (1).

Several pyrethroids derived from tetrafluorobenzyl and pentafluorobenzyl alcohols and various cyclopropane carboxylic acids, with high volatility and improved activity against a few insect species were developed. The ester **tefluthrin** (20, Fig. 3.8c) prepared from the acid·moiety of cyhalothrin (6) and 2,3,5,6-tetrafluoro-4-methyl benzyl alcohol exhibited good insecticidal activity against wide range of soil insect pests. Unlike cyhalothrin, it is relatively non-volatile and inactive towards soil insects.

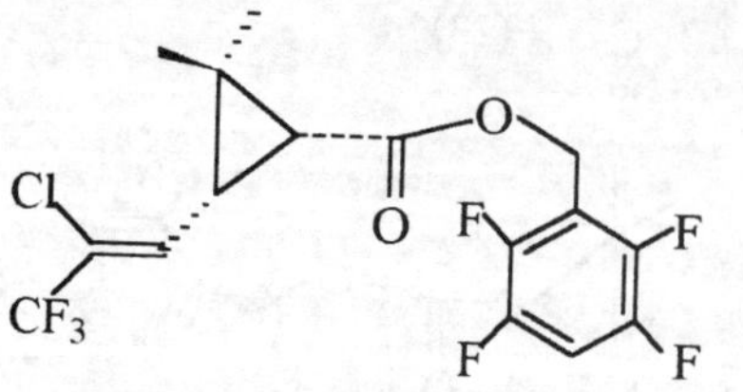

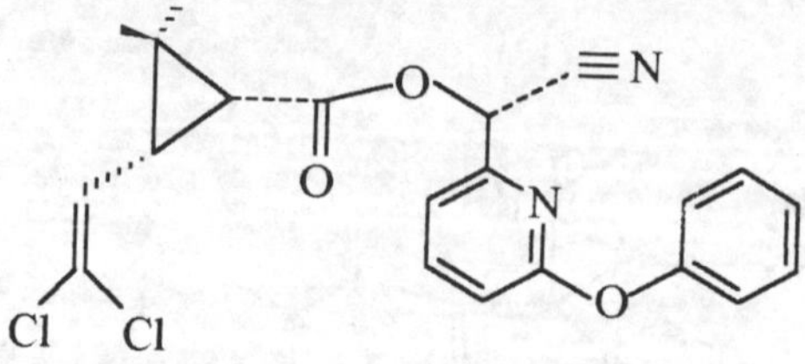

Fig. 3.8c. Synthetic pyrethroids.

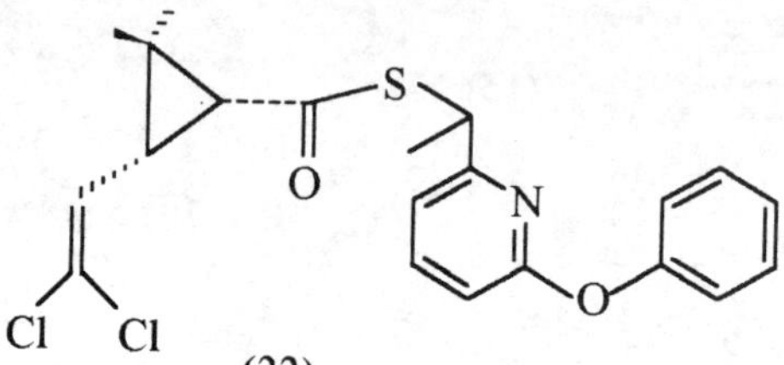

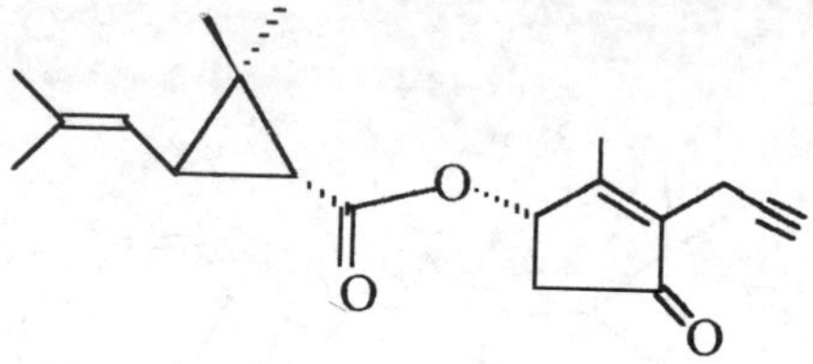

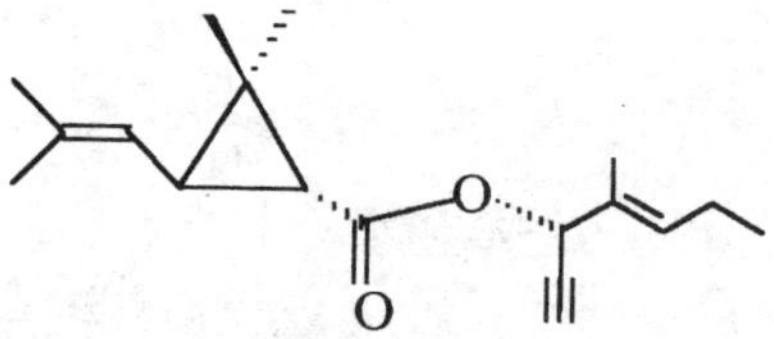

The ester derived from 6-phenoxy-2-pyridyl methyl alcohol named as **fenpyrithrin** (21) exhibited lower insecticidal activity than **cypermethrin** (3). If oxygen atom of carboxylate is replaced by sulphur, S-alkyl **thioester** (22) obtained showed excellent activity against *Heliothis virescens*. Analogue esters have also been prepared in the fluvalinate series but their activity was diminished.

A number of pyrethroids have been developed using simple alcohol with acetylenic moieties in order to produce high volatility and knock down activity against houseflies and mosquitoes. Thus **prallelthrin** (23) and **empenthrin** (24) synthesised have been found to be superior than bioallethrin as household and public health insecticide.

Non-ester pyrethroids

The first non-ester pyrethroids discovered with good insecticidal activity were the alkyl ketone **oxime-O-ethers** (25, Fig. 3.9) of which only the *E*-isomer showed good activity. These products are however, photolabile in nature.

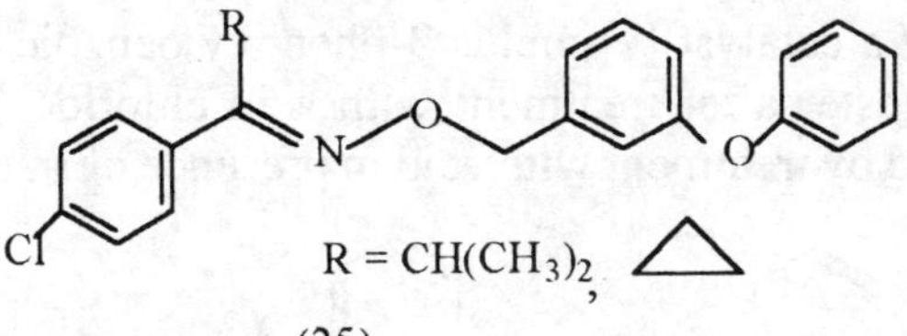

R = CH(CH$_3$)$_2$,

(25)

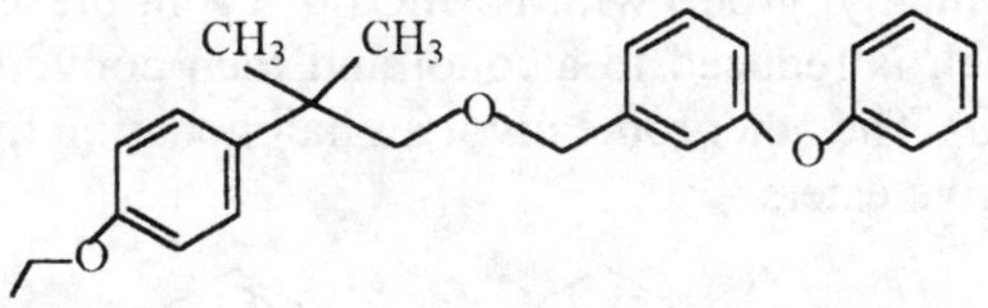

(26)

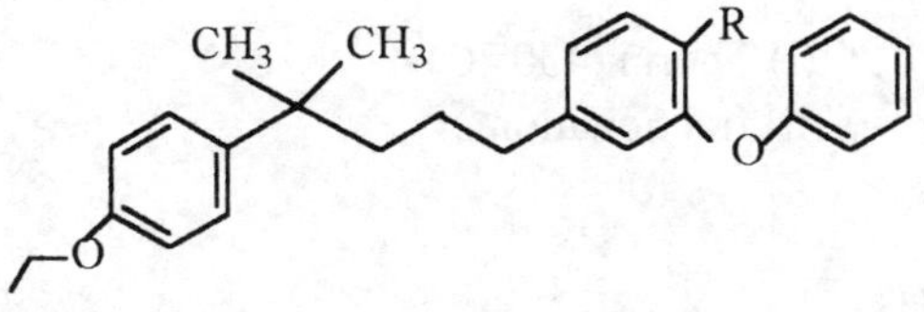

(27)

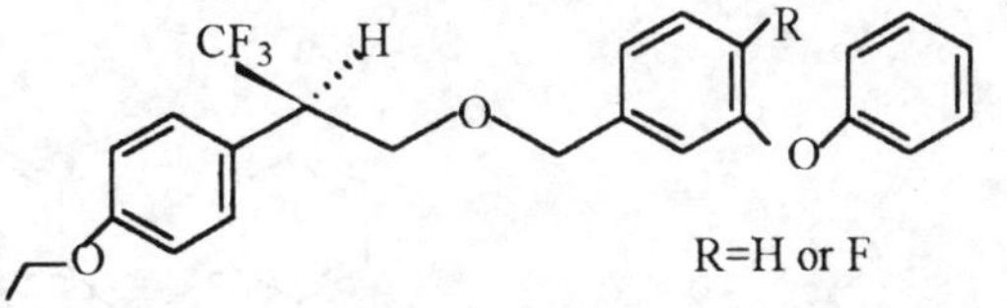

(28)

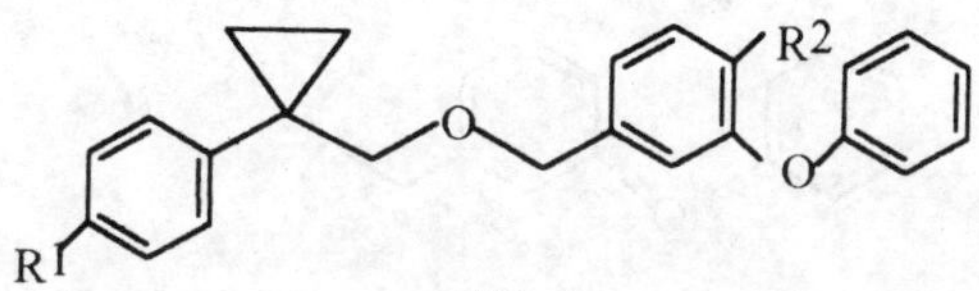

R=H or F

(29)

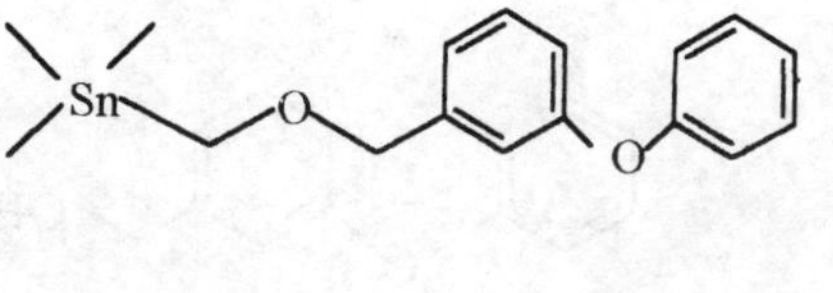

(30b)

R^1 = CH$_3$CH$_2$O ; R^2 = H
R^1 = CH$_3$CH$_2$O ; R^2 = F
R^1 = Cl　　　 ; R^2 = F
R^1 = Cl　　　 ; R^2 = F

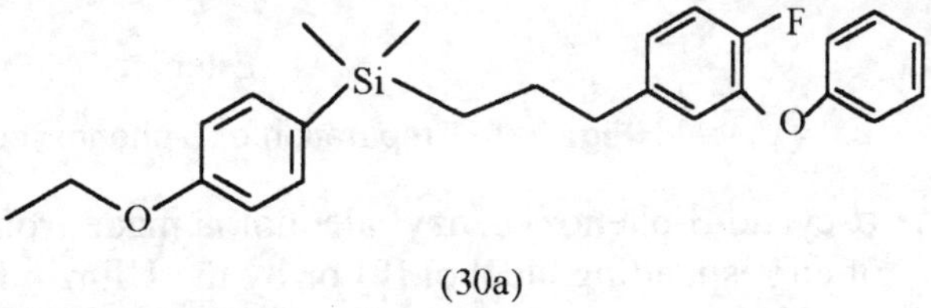

(30a)

Fig. 3.9. Synthesis of non-ester pyrethroids.

Other interesting compounds such as **etofenprox** (MT1-500, 26) and MT1-800 (27) have been developed with good insecticidal activity and low mammalian toxicity. A large number of non ester pyrethroids with variations in alkyl aryl ketones (ether) and the cyclopropyl group were obtained. If gem-dimethyl group in etofenprox is replaced by **fluoro** analogue (28) or cyclopropyl group (29), more active molecules effective against rice insects are produced. Further research has shown that silicon derivative **silaflunofen** (30a) possesses better activity than conventional pyrethroids, except trimethylstanniomethyl ether (30b) which shows very good activity against the Asiatic rice borer, *Chilo suppressalis* and housefly.

Synthesis of components of pyrethroid esters

A. 3-phenoxybenzyl alcohol and esters

These are the most important alcoholic components of photostable synthetic pyrethroids. The alcoholic component can be prepared by several routes, the most common one involves the reaction of 3-phenoxytoluene prepared by condensing potassium cresate with bromobenzene. Oxidation of the methyl group with $KMnO_4$ or O_2 in presence of a catalyst [7] yields 3-phenoxy benzoic acid which is reduced to alcohol and then converted to ester after treatment with acid chloride (Fig. 3.10). The other route involves halogenation followed by treatment with acid in presence of *t*-amine to give esters.

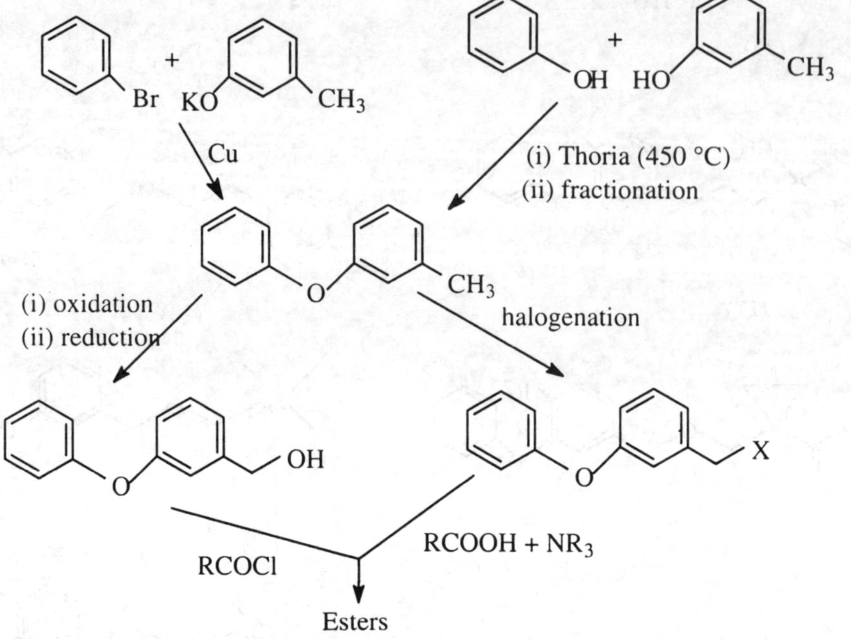

Fig. 3.10. Preparation of 3-phenoxybenzylalcohol and esters.

The α-cyano-3-phenoxybenzyl alcohol is made from 3-phenoxybenzaldehyde obtained by oxidation of corresponding alcohol [8] or by the Ullman reaction [9]. The reaction of aldehyde with HCN gave the racemic cyanohydrin (31) from which *R*-isomer (32) was separated using the enzyme D-oxynitrilase. The compound (31) on treatment with acid chloride gave a diastereomeric mixture of esters which on crystallisation gave cypermethrin (3) or decamethrin (4) (Fig. 3.11).

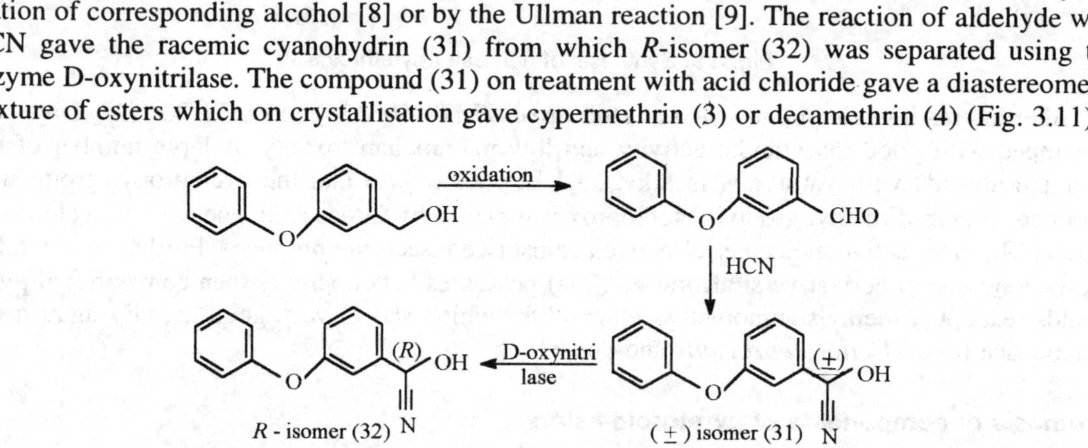

Fig. 3.11 Preparation of R-isomers of cypermethrin/decamethrin.

Acid components of the pyrethroid esters

B. Chrysanthemic acid and analogues

The original synthesis of (±)-(Z)-chrysanthemic acid (34) was given by Staudinger *et al.* [10a, 10b] by reacting 2,5-dimethyl hexa-2,4-diene (33) and ethyl diazo acetate followed by hydrolysis (Fig. 3.12).

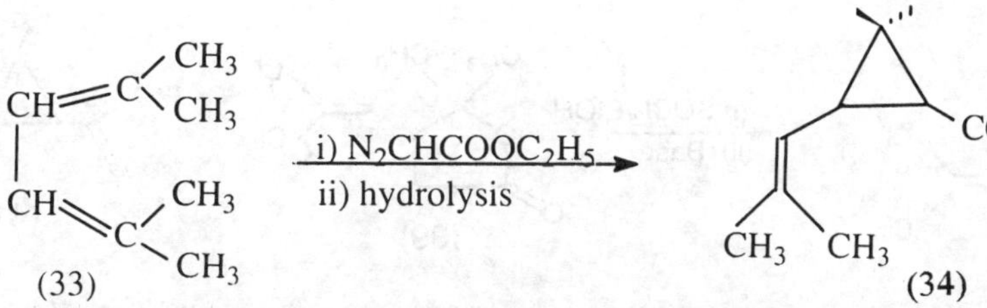

Fig. 3.12. Preparation of (±)-Z-chrysanthemic acid.

Matsumoto *et al.* [11] selectively synthesised (*E*)-chrysanthemate (35) by reacting (±) phenyl 3-methyl-2-butenyl sulfone with ethyl-3-methacrylate in the presence of Cu-dust (Fig. 3.13). This ester was later synthesised by Martel and Huynh [12].

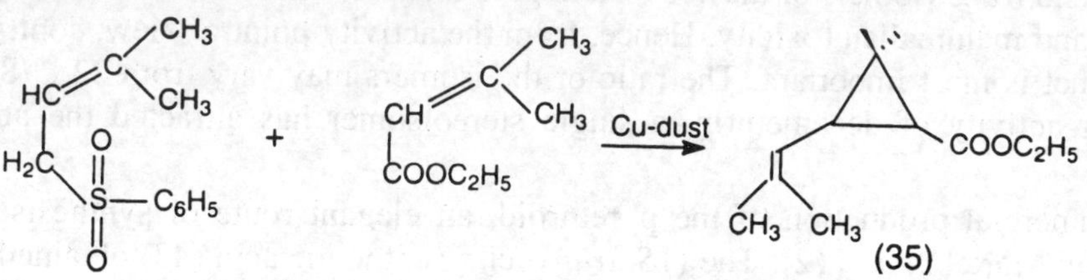

Fig. 3.13. Preparation of (E)-ethyl-chrysanthemate.

The (±)-(*E*)-pyrethric acid (36) was obtained from (D)-(*E*) chrysanthemic acid by oxidation followed by optical resolution to yield pure (+)-isomer, forming the acid component of pyrethrin-II, cinerin-II and jasmolin-II [Matsui and Yamada, 13].

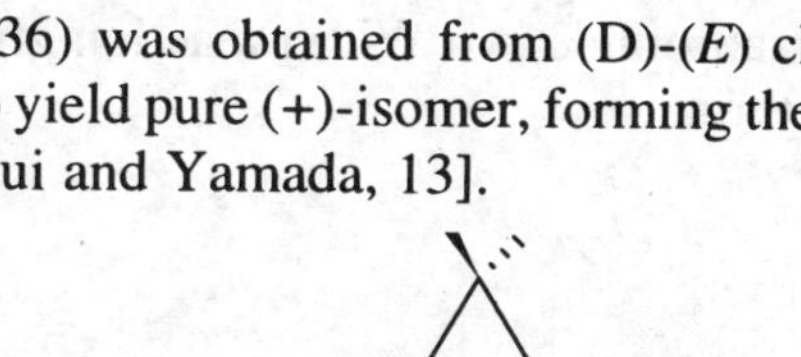

C. Dihalovinyl acids

The (±) *cis, trans*-dichlorovinyl analogue of chrysanthemic acid (40), used for the preparation of permethrin and cypermethrin was synthesised by Farkas *et al.* [14a] and Collonge *et al.* [14b]. Dichlorodiene (38) was made by electrolytic reduction of (37) through acetate either by dehydrocoupling of *iso*butylene and vinylidenechloride with palladous acetate or by dehydration of hydroxy intermediate (Fig. 3.14).

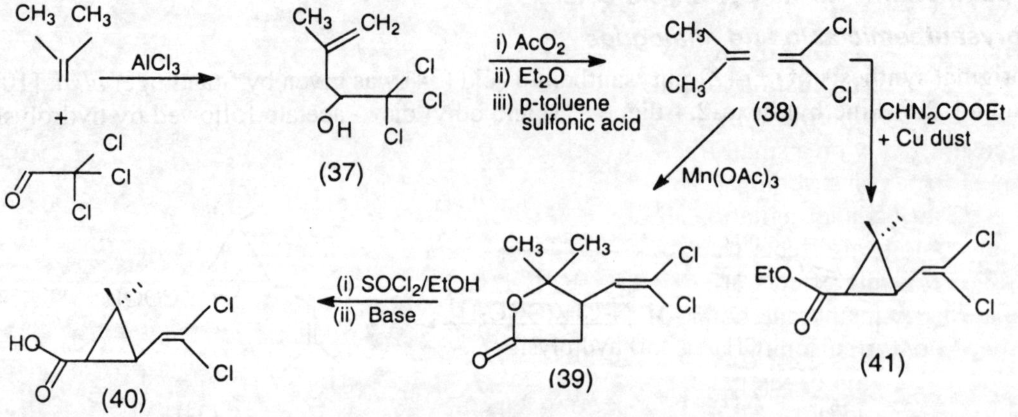

Fig. 3.14. Preparation of (±)cis-trans dichlorovinyl chrysanthemic acid.

Alternatively, the 2-carbon unit is added using manganic acetate and the lactonic product (39) formed was converted using thionyl chloride into the required acid derivative (40) or by treatment of (38) with ethyl diazoacetate in the presence of Cu-dust (41).

The *cis* and *trans* isomers of the above acid give esters of different insecticidal potency, species specificity and mammalian toxicity. Hence, from the activity point of view, controlling their ratio in the product is most important. The ratio of the isomers may vary from 22 : 78 to 45 : 55. The outstanding activity of decamethrin, a single stereoisomer has attracted the attention of most chemists.

For commercial production of the pyrethroid, an elegant route of synthesis (Fig. 3.15) was developed by Martel *et al.* [12]. The (1S, *trans*) chrysanthemic acid (43) obtained after resolution of the (±) *trans* form provides the 1R acidic component for bioallethrin, S-bioallethrin and bioresmethrin which is used as the starting material for the desired (1R, *cis*) caronaldehyde (42). This on treatment with carbon tetrabromide in a Wittig reaction gives (1R, *cis*) dibromovinyl acid, used in the sysnthesis of decamethrin.

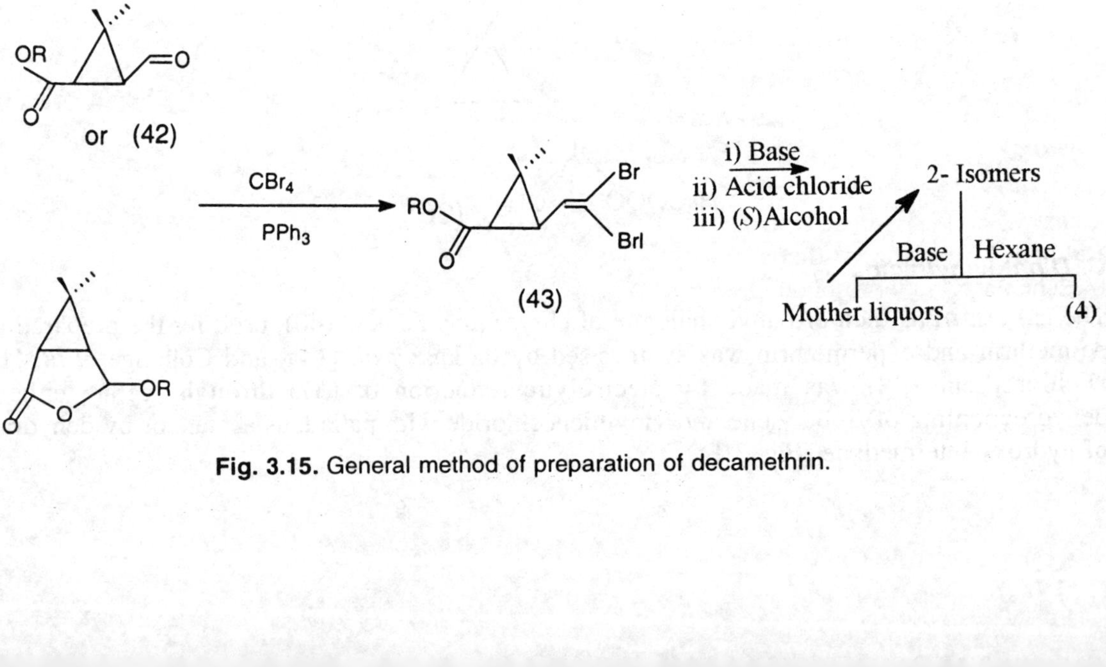

Fig. 3.15. General method of preparation of decamethrin.

Mode of action

The primary site of action by the pyrethroids is the central nervous system symptomised by hyper activity, lack of co-ordination, tremors and convulsions followed by paralysis and eventual death. Control of Na^+ ion permeability through nerve membrane is vital to nerve function. Pyrethroids act stereoselectively on a small fraction of the first voltage dependent Na^+ ion channel in excitable nerve membranes on several parts of insect nervous system. The major effect is to delay the closing of the Na^+ ion channel agitation gate. During few h. after exposure, the initial symptoms and electro-physiological effects were observed on the nervous system but death occurred only as a result of irreversible effects after prolonged action on the insect nervous system. Although the voltage sensitive membrane bound Na^+ ion channel is the major site of action, it is not the only site responsible for poisoning [15]. The insect eventually dies due to complex series of secondary effects like disruption of the balance of neurohormones within insect body and loss of body fluid leading to eventual death.

Metabolism

The metabolism and environmental fate of pyrethroids have been studied in detail [16] and the degradation pathways are found to be mainly photochemical and biological. The metabolism of photostable pyrethroids has been studied in presence of both organic and aqueous solvents and also on plant and inert surfaces. Introduction of the α-cyano group into 3-phenoxy benzyl esters tends to reduce the photostability to some extent but in non-ester pyrethroids such as etofenprox, the photostability depends on nature of the central linkage. Although the pyrethroids are lipophilic, they do not get accumulated in mammal tissues. Metabolism of pyrethroids along with photodecomposition by plants has been studied in detail. They are metabolised by ester cleavage, oxidation and hydroxylation of both alkyl groups and aromatic rings. The resulting metabolites such as carboxylic acids, alcohols and phenols are conjugated with sugars like glucoronic acid, or amino acids such as glycine and tyrosine and excreted along the urine and faeces.

Metabolism of pyrethroids in insects is rather complex in nature. The metabolic pathways include oxidation at one or more sites in the molecule along with ester hydrolysis followed by secondary oxidation, to yield a large number of polar and non-polar metabolites.

Pyrethroids bind strongly to soil and are rapidly degraded to carbon dioxide in most soil types under both aerobic and anerobic conditions.

Photochemistry

(i) Non-cyclopropanoid pyrethroids

The synthetic pyrethroids constitute a large group of pesticides whose photochemistry has been extensively investigated. Unlike natural pyrethroids whose photolabile nature limit their agricultural application, the synthetic pyrethroids are designed to be photostable. Besides being photostable, they still undergo photochemical degradation in well defined manner sometimes giving rise to products of toxicological significance. The photochemistry of a few important synthetic pyrethroids are discussed in this section.

Of the early four commercially important synthetic pyrethroids, permethrin, fenvalerate, cypermethrin and decamethrin, fenvalerate is unique in not having a cyclopropane group and, therefore, having somewhat different photochemical behaviour. Fenvalerate (9) undergoes fairly rapid degradation, when exposed to UV light of longer wave-length (> 290 nm), but more slowly in sunlight specially on a solid deposit impregnated on glass surface (~ 40 days). The absorption maxima (204 and 276 nm) of fenvalerate indicated that the excited state can be both singlet and triplet to give a variety of products. However, these are formed primarily by cleavage of the ester group both at the acyl and alkyl oxygen bond. The cleavage products being radicals, undergo further changes to give final stable products. The principal photoproduct, formed to the extent of 70% in solution is mainly the CO_2 extrusion product, which can lose further a chlorine atom to give a corresponding dechlorinated product in trace amount (Fig. 3.16). Another important dimeric compound is formed by decarboxylation of the p-chlorophenylacyl radical followed by dimerization. Besides the dimeric compound, a large number of other minor products have also been isolated. Fenvalerate is particularly a good substrate for photoelimination of CO_2, since the benzyl radicals formed from both acid and alcohol part of the molecule are stable and produce dimers in high yield, than from any other pyrethroid molecule.

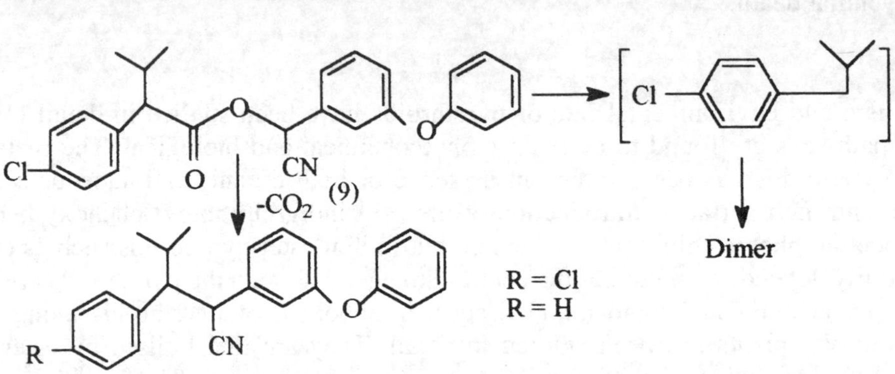

Fig. 3.16. Photolysis of fenvalerate.

It may be mentioned that the principal dimeric photoproduct is as stable as DDT under sunlight and UV. It will constitute an environmental load but fortunately, it is not much toxic. The most toxic products formed from fenvalerate are 3-phenoxybenzoyl cyanide and, the corresponding benzyl cyanide but are fortunately formed in trace quantities.

Recent work [17, 18] with flucythrinate (10) which contains a more easily degradable $OCHF_2$ group in place of the 4-Cl group of fenvalerate, has also shown a similar pattern of photolysis. Besides a dimer from the α-cyano benzyl alcohol part, the major photolytic products include the CO_2 extrusion product and the dimer (Fig. 3.17).

Fluvalinate (11), another pyrethroid with aromatic ring in the acid part in which the isopropyl group is separated from the benzene ring, shows interesting variation from the above two. Here the CO_2 extrusion product is formed in extremely small quantity showing that the radical formed by CO_2 elimination from acid part is highly unstable and degrades rapidly into the corresponding haloaniline derivative which is the major product of photolysis (Fig. 3.18). It is accompanied by small quantities of formanilide formed presumably by further oxidation of the isopropyl group.

(ii) Cyclopropanoid pyrethroids

The general pattern of photolysis of cyclopropanoid pyrethroids, although similar to the fenvalerate type, is quantitatively very different. Permethrin (2), the first synthetic cyclopropanoid molecule,

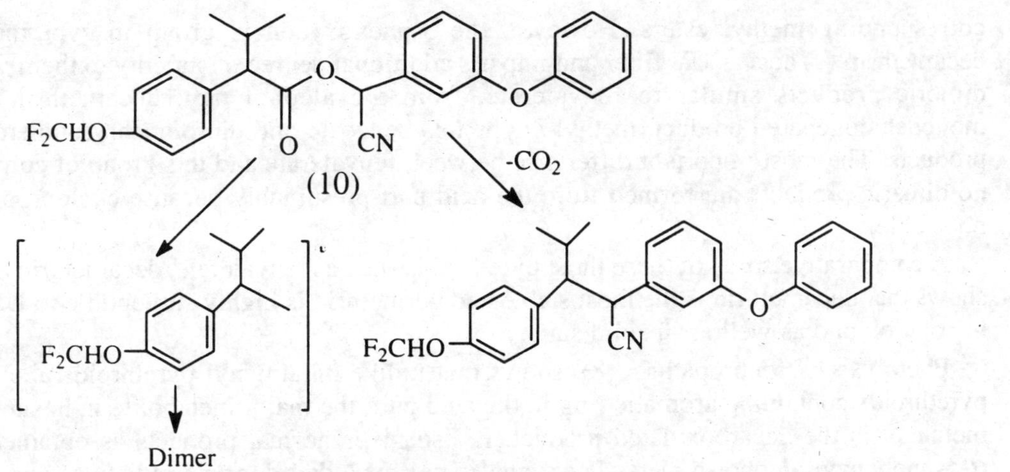

Fig. 3.17. Photolysis of flucythrinate.

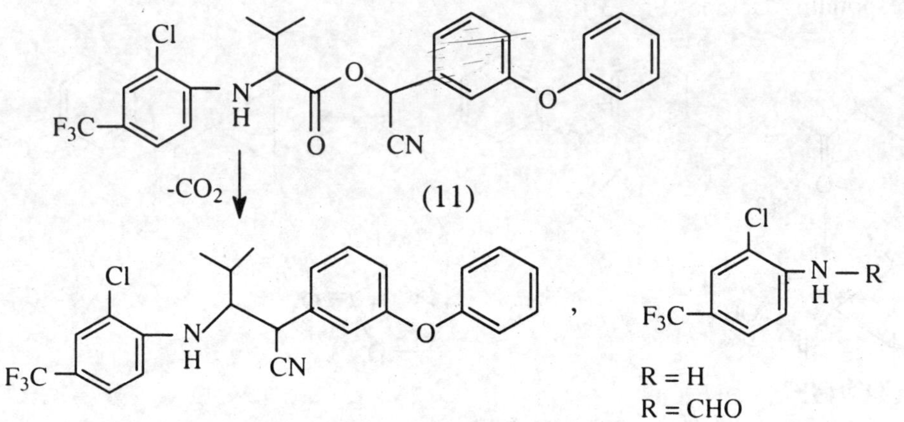

Fig. 3.18. Photolysis of fluvalinate.

gave major products by ester cleavage. The product mixture consisted of *cis*- and *trans*- isomers of dichlorovinyl chrysanthemic acids and *m*-phenoxybenzyl alcohol, when the photolysis was done in hexane or acetonitrile-water solution. In methanol, the corresponding esters and ethers were formed. The formation of *cis-trans* mixtures from *cis*-cyclopropanoid pyrethroids showed that they have an additional site of photoreaction. The isomeric mixture of DV acids and esters are formed by opening of the cyclopropane group and recombination. Thus *cis*-permethrin, the active isomer first changes to *trans*-permethrin and finally gives the cleavage products as *cis- trans-* mixture of DV acids. No dimeric products have however been isolated by recombination of acid or alcohol part or by loss of CO_2 during photolysis of permethrin as the radicals formed are very unstable.

The α-cyanobenzyl esters of dihalovinyl acids like cypermethrin (3) and decamethrin (4) have very similar photolysis patterns. The principle sites of photoreaction in these molecules are ester cleavage like in other pyrethroids but giving more stable cyanobenzyl radicals and the cyclopropyl bond whose cleavage and recombination yields *cis*- and *trans*- isomers. Contrary to earlier belief, the dihalovinyl carboxylic acid is not very photostable as seen in separate experiments with the corresponding methyl esters. However, the 3-phenoxybenzyl group in cypermethrin (3) and decamethrin (4) acts as UV filter and imparts additional degree of stability to them. Besides giving dimeric products similar to fenvalerate, from the alcohol part, decamethrin (4) gives the monodehalogenated product, methyl-m-phenoxybenzoate and dibromochrysanthemic acid as by-products. The most important difference between fenvalerate and this group of compounds is that no dimeric products are formed from the acid part presumably because cyclopropyl radicals are not stable.

A comparative study of these three pyrethroids such as fenvalerate, decamethrin and permethrin shows that decamethrin is the most stable and permethrin is highly susceptible to UV and sunlight in solution and as well as in solid state.

Photolysis of fenpropathrin (8), shows that unlike dihalovinyl pyrethroids above but like the pyrethroids containing aromatic ring in the acid part, the major metabolite in hexane as well as in methanol is the decarboxylated product. It also gives normal products as obtained earlier from α-cyanobenzyl alcohol moiety. If oxygen is not excluded during irradiation, one of the methyl groups of the cyclopropane ring undergoes oxidation giving rise to cyclopropane dicarboxylic acid and corresponding lactone (Fig. 3.19).

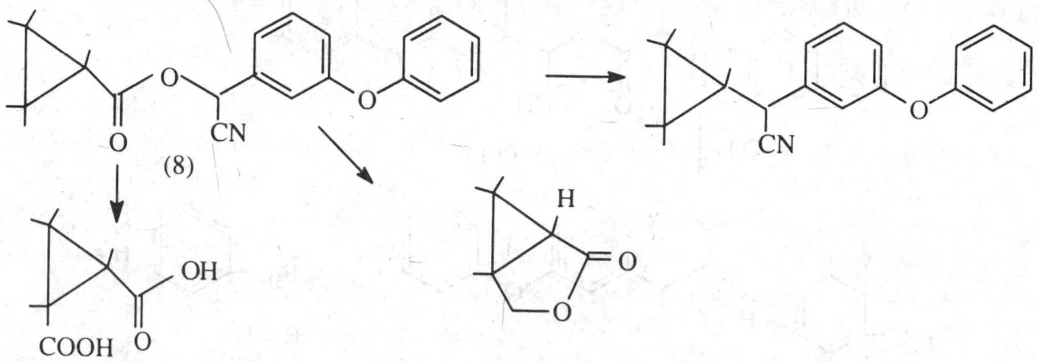

Fig. 3.19. Photolysis of fenpropathrin.

REFERENCES

1. Schechter, M.S., Green, N. and LaForge, F.B. Constitution of pyrethrum flowers XXII, *J. Am. Chem. Soc* **71**, 3165 (1949).

2. Elliott, M., Fairham, A.W., Janes, N.F., Needham, P.H., Pulman, D.A. and Stevenson, J.H. A photostable pyrethroid. *Nature* (London) **246**, 169 (1973).

3. Elliott, M., Farnham, A.W., Janes, N.F. and Soderlund, D.M. Insecticidal activity of the pyrethrins and related compounds. Part XI. Relative potencies of isomeric cyano-substituted m-phenoxy benzyl esters *Pestic. Sci.*, **9**, 1121-16 (1978).

4a. Ohno, N., Fujimoto, K., Okuno, Y., Mizutani, T., Hirano, M., Haya, N., Honda, T. and Yoshioka, H. *Third Internatn. Congr. Pestic. Chem.* Helsinki, abs. 346 (1974).

4b. *Ibid.* A new class of pyrethroid insecticides; α-substituted phenyl acetic acid esters. *Agr. Biol. Chem.* **38**, 881 (1974).

4c. *Ibid.* 2-Arylalkanoates, a new group of synthetic pyrethroid esters not containing cyclopropane carboxylates. *Pestic. Sci.* **7**, 241-246 (1976).

5. Henrick, C.A., Garcia, B.A, Staal, G.B., Cerf, D.C., Anderson, R.J,. Gill, K., Chinn, H.R., Labovitz, J.N., Leippe, M.M., Woo, L.R., Carney, R.L., Gordon, D.C. and Kohn, G.K. 2-Anilino-3-methyl butyrates and

2-(isoindolin-2-yl)-3-methyl butyrates, two novel groups of synthetic pyrethroid esters not containing a cyclopropane ring. *Pestic. Sci.* **11**, 224 (1980).

6. Plummer, E.L. In. *Pesticide synthesis through Rational Approaches* (Ph.S. Magee, G.K. Kohn and J.J. Menn, eds.) *Symp. Sr.* 255, American Chemical Society, Washington, D.C., 297 (1984).

7. Jap. Kokai 73.61450; B. pat. 1489325/1977.

8. Elliott, M., Farnham, A.W., Janes, N.F., Needham, P.H. and Pulman, D.A. Synthetic insecticide with a new order of activity. *Nature* **248**, 710 (1974).

9. Badar, A. *Aldrichimica Acta*, **9**, 49 (1976)/Belg. Pat. 842 177/1976.

10a. Staudinger, H. and Ruzicka, L. Insektentotende Stoffe I. Uber *iso*lierung und Konstitution des wirk samen teiles des dalmatinischen Insekten pulvers. *Helv. Chem. Acta*, **7**, 177 (1924).

10b. *Ibid.* Insektentotende Stoffe X. Uber die synthese vonpyrethrinen. *Helv. Chem. Acta*, **7**, 448 (1924).

11. Matsumoto, T., Nagai, A. and Takahashi, Y. The stereoselective synthesis of trans chrysanthemum mono-carboxylic acid *Bull. Chem. Soc. Japan*, **36**, 481 (1963).

12. Martel, J. and Huynh, C. Synthese de l' acide chrysanthemique (Note preliminaire) 11. Access stereoselectif au trans chrysanthemate d ethyle. *Bull. Soc. Chem. France* **985** (1967).

13. Matsui, M. and Yamada, Y. Studies on chrysanthemic acid Part VIII, Synthesis of pyrethric acid. *Agr. Biol. Chem.* **27**, 373 (1963).

14a. Farkas, J., Kourim, P. and Sorm, F. Relation between chemical structure and insecticidal activity in pyrethroid compounds. 1. An analogue of chrysanthemic acid containing chlorine in the side chain. *Collection Czechoslov. Chem. Commun.* **24**, 2230 (1959);

14b. Collonge, J. and Perrot, A. Etude sur les trichloromethyl carbinols secondairies ethyleniques et satures 1. Preparation. *Bull. Soc. Chem.* France, 204 (1957).

15. Naumann, K. Chemistry of Plant Protection, 4. *Synthetic pyrethroid insecticides : Structures and properties* (W.S. Bowers, W.Ebing, D. Martin, R. Wegler, eds.) Springer-Verlag, Berlin (1990).

16. J.P. Leahey (ed.). *The pyrethroid insecticides*. Tayler and Francis, London (1985).

17. Chattopadhyaya, S. and Dureja, P. Photolysis of flucythrinate. *Pestic. Sci.* **31**, 163-173 (1991).

18. Holmstead, R.L., Casida, J.E. and Ruzo, L.O. Pyrethroid photodecomposition : Permethrin. *J. Agric. Food Chem.*, **26** (3), 590-595 (1978).

Section III.   Chemistry of Miscellaneous Insecticides

- Nitromethylene heterocycles
- Benzophenone hydrazones
- Bicyclo orthocarboxylates
- Pyrimidin amines
- Pyrido [3,2-f] indoline -2,4,9 triones
- Pyrazoles derivatives

- Chloronicotinyl derivatives
- Indole derivatives
- Sulfonyl pyrroles derivatives
- 1,3,5-Thiadiazine derivatives
- *References*

A few promising insecticides reported recently are discussed in the text.

Nitromethylene heterocycles

Soloway *et al.* [1] screened a series of nitromethylene heterocyclics and found that 2-(nitromethylene) tetrahydro 1,3-thiazin analogues (1, 2) exhibited good insecticidal activity with low mammalian toxicity and less persistence. But these compounds are photolabile. To make a photostable product, a formyl derivative of nitro methylene was synthesised with reasonably photostability and wide spectrum of insecticidal activity. These compounds also exhibited activity against insects resistant to other pesticides.

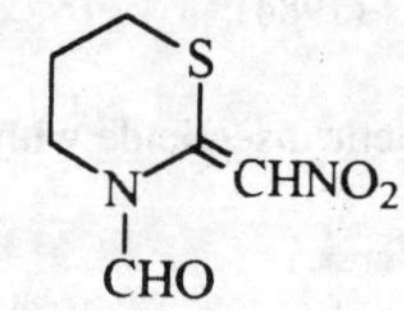

(1) 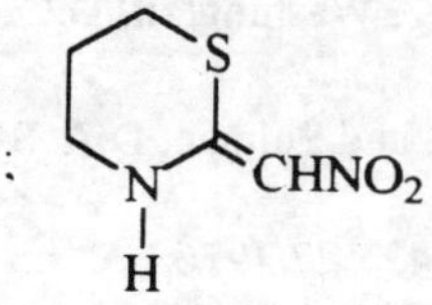 : (2) 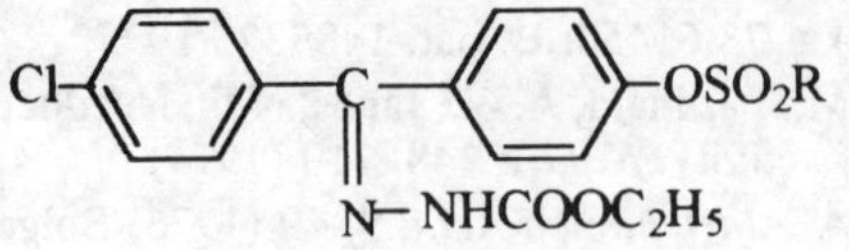 : (3) R = CH₃ or CH₂Br

$$R = CH_3 \text{ or } CH_2Br$$

Benzophenone hydrazones

A series of ethoxy carbonyl derivatives were synthesized which effectively controlled the population of caterpillars, flies, termites and locusts [2]. The most active derivative amongst them is (3)

Bicyclo orthocarboxylates [3, 4, 5]

These compounds affect the neuromuscular function in cockroaches and inhibit GABA mediated synaptic transmission probably by blocking the chloride channels. Addition of cyano group into t-butyl bicyclo orthobenzoates (4) enhances the activity by many fold against houseflies or mice.

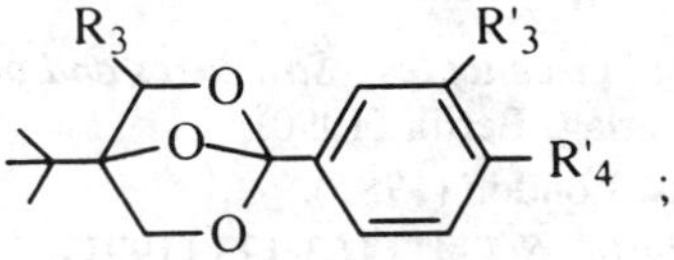

(4) 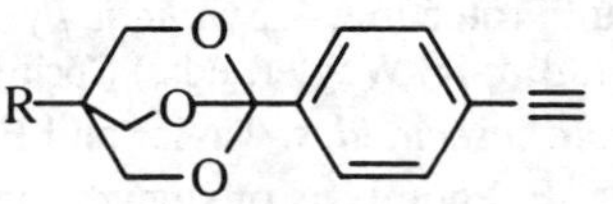 ; (5) 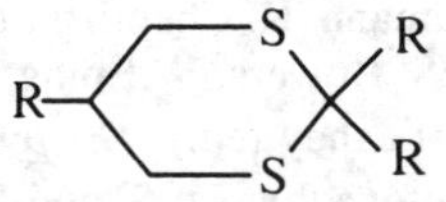 ≡ ; (6)

Similarly 1-(4-ethynyl phenyl)-2,6,7-trioxabicyclo [2,2,2] octanes (5) and 1,3-dithianes (6) with appropriate substituents exhibits insecticidal activity at the GABA gated chloride channel.

Thiourea derivatives

A number of thiourea derivatives such as 1,3,5-thiadiazines (7) and 5-phenyl perhydro 1,3,5-thiadiazines have been found to possess good larvicidal activity. Addition of para phenoxy group (8) enhances their activity against insects, mites and whiteflies in cotton, and *Plutella* in vegetables.

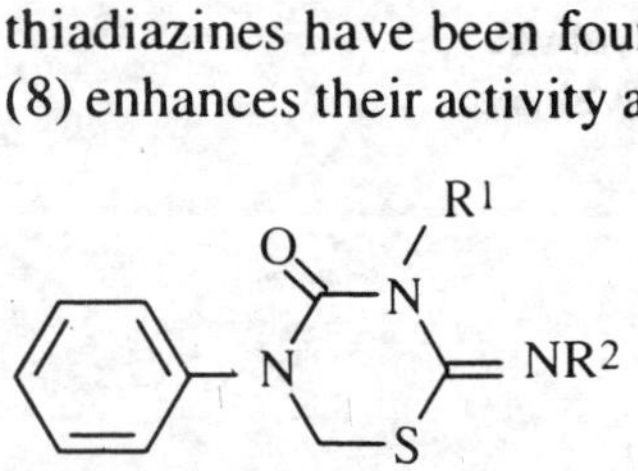

R^1 & R^2 = alkyl or phenyl groups

(7)   (8)

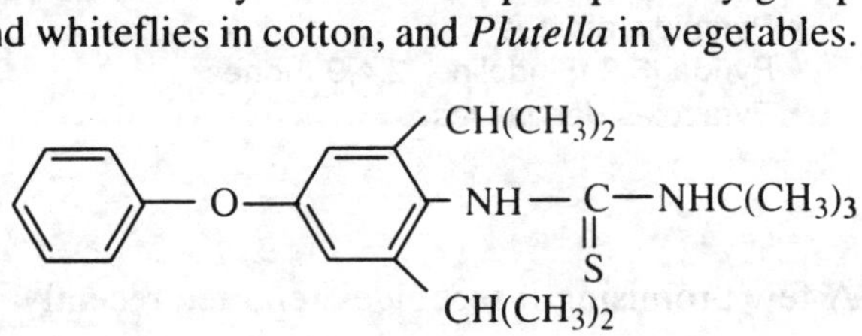

Synthesis of 1,3,5-thiadiazine

The synthesis and reactivity of 1,3,5-thiadiazine (9) has been investigated by several authors. The N-substituted N-chloromethyl carbamoyl chlorides were reacted with 1,1-disubstituted thioureas in presence of a base to give dihydro-1,3,4-thiadizines (Fig. 3.20). 5-Phenyl-1,3,5-thiadiazine is a potent larvicide against *Hemiptera*.

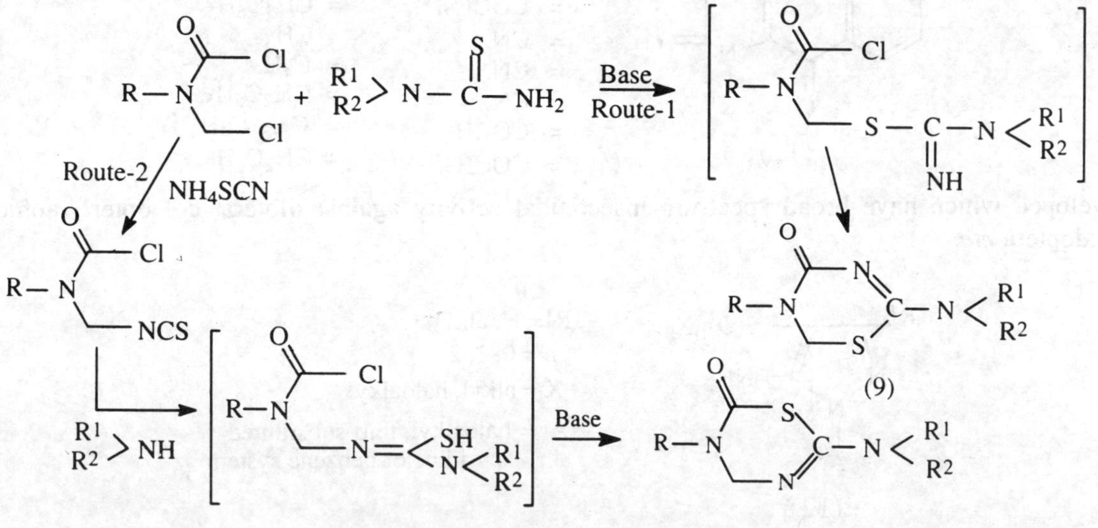

Fig. 3.20. Synthesis of 1,3,5-thiadiazine.

Pyrimidine amines

A series of N-substituted pyrimidin-4-amines (10) as core structure have been prepared and screened for their toxicity against insects such as brown rice plant hopper, diamond back moth and two spotted spider mite.

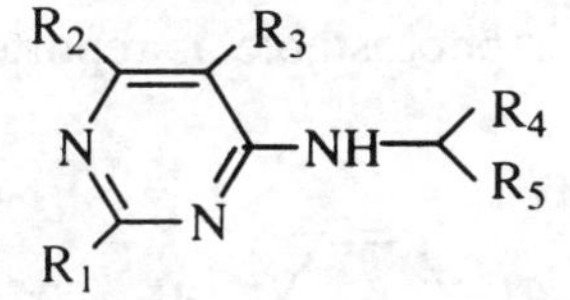

R_1 = H, alkyl
R_2 = alkyl
R_3 = halogen
R_4 = H, alkyl, cycloalkyl
R_5 = 3 or 4 phenoxyphenyl or alkyl

(10)

SAR studies indicate that methyl, fluoromethyl and ethyl group at 6-position were most suited for activity and substitution at 2-position had little contribution. The geometry and stereochemistry around 1 position of the benzyl or alkylamine also play an important role in determining the activity.

Pyrido [3,2-f] indoline-2,4,9-triones

A series of pyrido indolinetrions (11) have been synthesised in order to examine their bio-activity. These compounds exhibit both bactericidal and insecticidal activity.

Pyrazole derivatives

A series of 5-amino-4-halogen alkyl thio (sulfoxy-, sulfonyl-)-1-aryl pyrazoles (12) have been

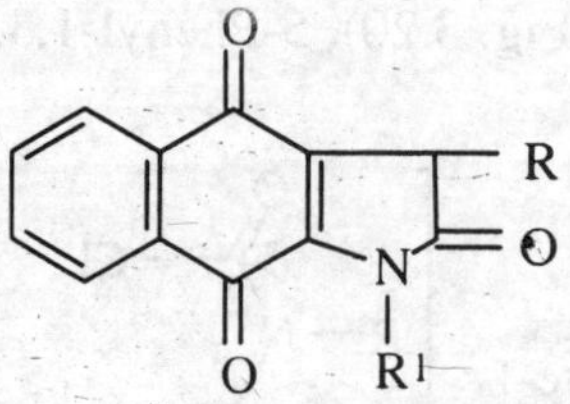

$$R = COOC_2H_5, \quad R^1 = CH_3$$
$$= COOC_2H_5, \quad = C_4H_9$$
$$= COOC_2H_5, \quad = CH_2C_6H_5$$
$$= CN, \quad = CH_3$$
$$= CN, \quad = C_4H_9$$
$$= CN, \quad = CH_2C_6H_5$$
$$= COCH_3, \quad = CH_3, C_4H_9$$
$$= COCH_3, \quad = CH_2C_6H_5$$

(11)

developed which have broad spectrum insecticidal activity against diptera, coleoptera, aphids, lepidoptera *etc.*

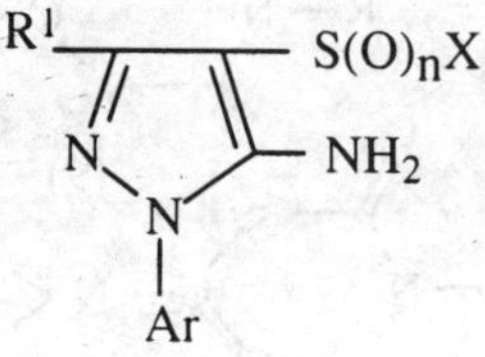

$R^1 = H$, alkyl
$n = 0, 1, 2$
X = alkyl, haloalkyl

Ar = haloalkyl, thio-substituted
pyridine or benzene system

(12)

The 4-thioaryl pyrazoles act on the GABA-regulated chlorine ion channel and their is no cross resistance against phosphatic, carbamate and pyrethroid groups of compounds.

Chloronicotinyl derivative [6]

Amongst chloronicotinyl group of compounds, **imidacloprid** 6-chloro-5-pyridyl methyl-2-nitro iminoimidazolidine (13), is found to be systemic in action and effective against sucking pests such as aphids. Unlike nitromethylene derivatives (14), it is photostable as reported by Soloway [1].

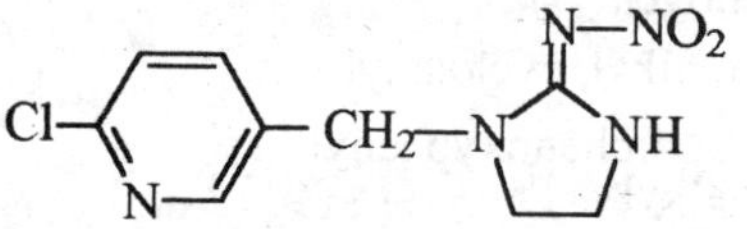

(13) 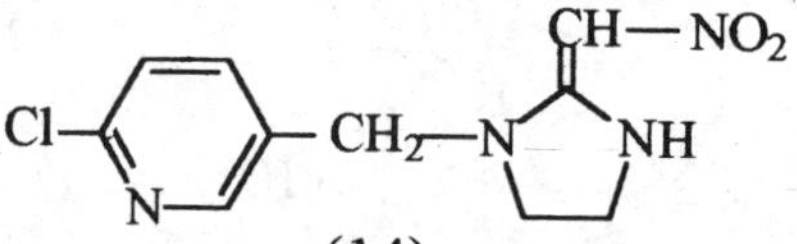   ;   (14)

Crotonamide derivatives

A series of 2 cyano-3-amino-4-substituted-N-(substituted phenyl or 4-pyridyl)-crotonamides (15) have been prepared and found to be highly feed-retardant and lethal to certain *lepidoptera* insects. N-phenyl (16) and 4-pyridyl analogues (17) were found to be the most active derivatives.

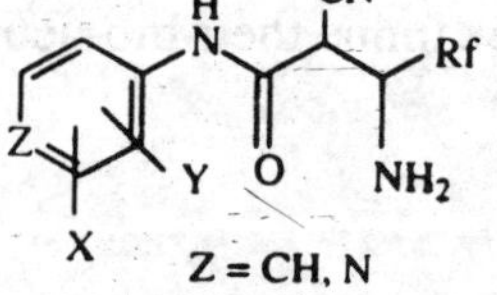

$Z = CH, N$
Rf = Perfluoroalkyl

(15)   ;   (16) 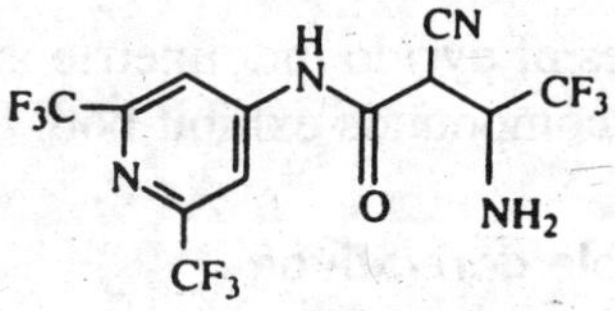   ;   (17)

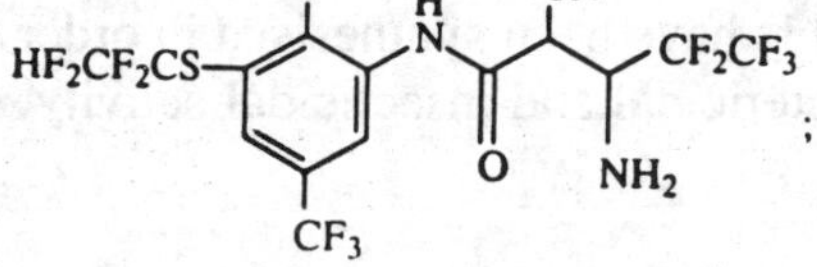

Indole derivatives

The insecticidal activity of a series of synthesised indole derivatives has revealed that these compounds (18, 19) exert good insecticidal activity against *Spodoptera litura* and *Heliothis armigera*.

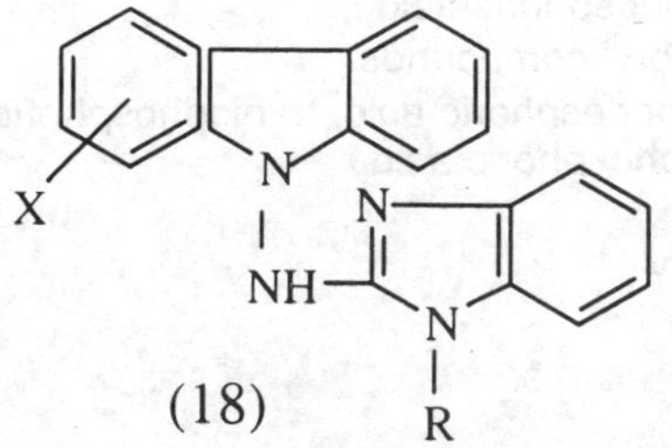

(18) 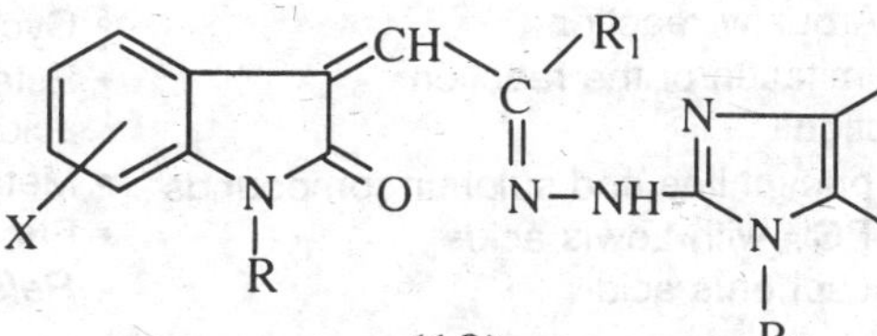   ;   (19)

X= H, F; R= H,CH$_3$,COCH$_3$; R$_1$= C$_6$H$_5$,C$_6$H$_4$Br,C$_6$H$_4$F,3-CH$_3$, 4-F.C$_6$H$_4$; R$_2$= H,C$_6$H$_5$

Sulfonyl pyrrole derivatives

A series of 2-aryl-3-trifluoro methyl sulfonyl pyrroles (20) have been developed based on AC-303,630 (21), a broad spectrum insecticide derived from natural product. These pyrrole derivatives possess good insecticidal activity. The isomeric 3-aryl trifluoromethyl, sulfonyl pyrroles (22) have also been prepared to compare their insecticidal activity to that of 2-aryl pyrrole derivative.

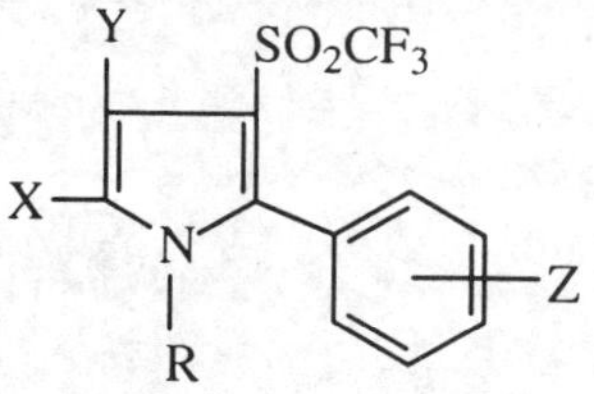

(20) 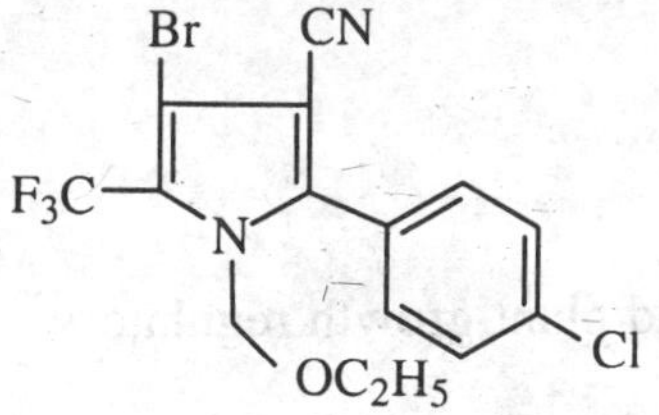   (21) 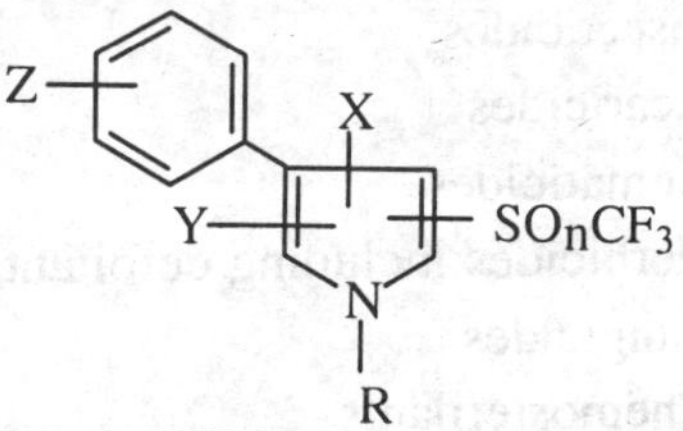   (22)

REFERENCES

1. Soloway, S.B., Henry, A.C., Kellmeyer, W.D., Padgett, W.M., Powell, J.E., Roman, S.A., Tieman, C.H., Corey, R.A. and Horne, C.A. Nitromethylene insecticide. In : *Advances in Pesticide Science* (H. Geissbuhler ed.) Part-2, Pergamon Press, New York, 206-217 (1979).

2. Giles, D.P., Copping, L.G. and Willis, R.J. Benzophenone hydrozones, a group of novel insecticides. *Proc. of British Crop Prot. Conf.*, Brighton, **2**, 405 (1984).

3. Crombie, L. (ed.), *Recent advances in the chemistry of insect control, II*, Royal Society of Chemistry, Cambridge, 212 (1990).

4. Palmer, C.J., Cole, L.M., Smith, I.H., Moss, M.D.V. and Casida, J.E. Structural aspects and mechanism of the selective proinsecticidal action of silylated 1-(4-ethynylphenyl)-2,6,7-trioxabicyclo [2,2,2] octanes. *Seventh Internatn. Congr. Pestic. Chem.*, Hamburg. abs. 0IC-12, 152(1990).

5. Wacher, V., Toia, R.R., Casida, J.E. 2,5-disubstituted-1,3-dithane S-oxides : Preparation, structural and stereochemical assignment, and mammalian GABA receptor potency. *Seventh Internatn. Congr. Pestic. Chem.* Hamburg, abs. 01A-18, 30 (1990).

6. Shiokawa, K., Tsuboi, S. and Moriya, K. and Kagabu, S. *Chloronicotinyl Insecticides : Development of Imidacloprid*. In : *Proc. Sr. Eighth Internatn. Congr. Pestic. Chem. Options 2000* (N.N. Ragsdale, P.C. Kearney and J.R. Plimmer, eds.) American Chemical Society, Washington D.C., 49-58 (1995).

Section IV.   Chemistry of Organophosphorus Insecticides

- Introduction
- History
- Merits and Demerits
- Mode of action
- Michaelis – Arbuzov reaction
- Scope and limitation of the reaction
- Perkow reaction
- Reaction of phosphites and sulphur compounds
- Reaction of PCl$_3$ with Lewis acids
- Esters of phosphorus acid

- Esters of phosphinic acid
- Esters of phosphonic acid
- Esters of phosphoric acid
- Esters of thiophosphoric acid
- Cyclic phosphoryl compounds
- Esters of dithiophosphoric acid, trithiophosphoric acid and pyrophosphoric acid
- Metabolism
- Photochemistry
- *References*

Among pesticides of synthetic origin, organophosphorus pesticides play an important role in controlling insect pests both in agriculture as well as in public health. Their large scale application is due to their easy biodegradability, low persistence and high effectivity. Currently 150 organophosphorus compounds are being used as pesticides in the world. More than 1,00,000 metric tonnes (MT) organophosphorus pesticides are produced annually in U.S.A. alone. In India the production is around 20,000 MT per year comprising 24 compounds.

Organophoshorus compounds show activity against variety of crop pests [1,2,3]. These can·be broadly classified as :

1. Insecticides
2. Acaricides
3. Nematicides
4. Herbicides including defoliants and plant growth regulators
5. Fungicides
6. Chemosterilants

History of organophosphorus pesticides

Investigations on organophosphorus pesticides were first started by Lassaigne in 1820 who prepared phosphatic esters. The subject was put on sound footing by Michaelis in Germany in late 19th century. He along with the Russian chemist A.E. Arbusov conducted extensive research on the chemistry of trivalent phosphorus compounds.

During the Second World War, Saunders in England and Schrader in Germany worked on toxic phosphorus compounds. Saunders synthesised nerve poison compound including DFP, diisopropyl fluorophosphate (1) and its analogues such as tetramethyl phosphorodiamidic fluoride, dimefox (2). In 1937, Schrader discovered a systemic insecticide called OMPA, octamethyl phosphoramidate which was later named as **Schradan** (3). He also synthesised a number of organophosphorus insecticides including TEPP.

$$(i\text{-}C_3H_7O)_2P(O)F \quad ; \qquad \begin{matrix} (CH_3)_2\,N \\ \\ (CH_3)_2\,N \end{matrix}\!\!\!>\!\!P(O)F \quad ; \qquad \begin{matrix} (CH_3)_2\,N \\ \\ (CH_3)_2\,N \end{matrix}\!\!\!>\!\!P(O)\text{-}O(O)P\!\!\!<\!\!\begin{matrix} N\,(CH_3)_2 \\ \\ N\,(CH_3)_2 \end{matrix}$$

$$(1) \qquad\qquad\qquad (2) \qquad\qquad\qquad\qquad (3)$$

Advancement in the knowledge of organophosphorus chemistry, mode of action and SAR studies led to the development of parathion, a powerful insecticide. Although it is extremely toxic to mammals, it effectively controls a number of agricultural insects which are resistant to organo-chlorine insecticides. Common use of this compound has been banned because of the high mammalian toxicity but encapsulated form (Pancap) is still used in extreme cases to control insects.

Minor structural modification of parathion molecule led to the development of a number of toxic pesticides like fenitrothion, malathion and DDVP. Recently many of these phosphatic compounds have been shown to be effective fungicides, herbicides, PGR, defoliants and chemosterilants.

A few toxic compounds used during war time were **sarin** (4), **soman** (5) and **tabun** (6). The abnormal side effect of phosphatic compounds was first observed for dialkyl phosphorofluoridate (DFP), [(RO)$_2$-P-(O)-F] by Lange and Kueger in 1932 [4].

$$
\begin{array}{ccc}
\underset{\text{i}C_3H_7O}{\overset{CH_3}{\diagdown}} P(O)F \; ; &
\underset{(CH_3)_3CCHO}{\overset{CH_3}{\diagdown}} P(O)F \; ; &
\underset{C_2H_5O}{\overset{(CH_3)_2N}{\diagdown}} P(O)CN \\
& \underset{CH_3}{|} & \\
(4) & (5) & (6)
\end{array}
$$

Merits of phosphatic compounds as pesticides over chlorinated pesticides

   i) Phosphatic compounds show high insecticidal and acaricidal activity but now many of them show fungicidal, nematicidal and herbicidal activity.

  ii) They possess broad spectrum of activity against number of insect pests.

 iii) They easily breakdown to non-toxic metabolites causing no environmental pollution.

  iv) Many of them are systemic in nature which ensures their effectivity even at lower concentrations with little or no left over residues.

   v) Due to high insect toxicity these compounds are required in low quantity during application.

  vi) Unlike chlorinated hydrocarbons, these compounds have very low persistence and do not accumulate in the ecosystem.

 vii) They have low chronic toxicity in general.

Demerits

   i) Due to their high mammalian toxicity these compounds are preferred for quick action, but they also require special care during handling, storage and application.

  ii) These compounds have a foul smell.

Mode of action

Organophosphorus compounds have higher anticholinesterase activity as they are structurally similar to acetylcholine. The function of cholinesterases is to hydrolyse acetylcholine. Phosphatic compounds phosphorylate the vitally important enzymes esterases inhibiting normal functions in mammals and insects with the accumulation of acetylcholine which disrupts the normal life and finally leads to death.

The inhibition of cholinesterases is irreversible whereas with carbamate the inhibition is reversble. For maximum inhibition efficiency, the phosphatic compounds should fit into the active centres of the esterase in a 'key in a lock' fashion. Hence the activity of the organophosphorus compounds depends greatly on the structure of the ester groups (Fig. 3.21).

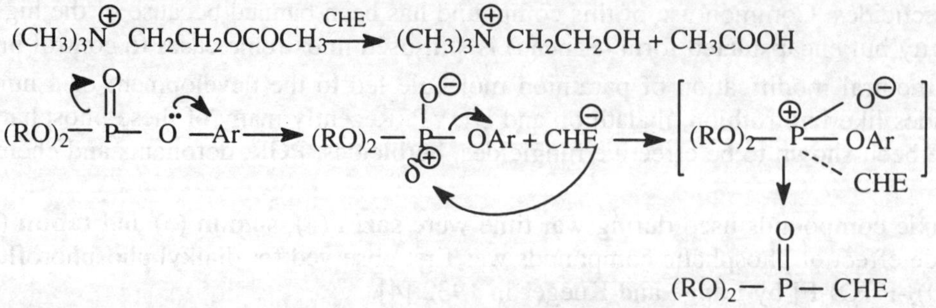

$$(CH_3)_3\overset{\oplus}{N} \ CH_2CH_2OCOCH_3 \xrightarrow{\text{CHE}} (CH_3)_3\overset{\oplus}{N} \ CH_2CH_2OH + CH_3COOH$$

Fig. 3.21. Inhibition of cholinestrase.

Michaelis-Arbuzov reaction

The Michaelis-Arbuzov rearrangement variously called as Arbuzov rearrangement or Arbuzov reaction resulting in Arbuzov transformation is one of the most important reactions for the formation of carbon-phosphorus bonds [5]. It involves the reaction of a trivalent phosphorus ester with alkyl halides. It was originally discovered by Michaelis, a German Chemist in 1898 and explored in detail by Arbuzov a Russian Chemist. This reaction is employed for the synthesis of phosphonates, phosphinic acid esters and phosphine oxides. In its simplest form, the Arbuzov reaction is an interaction between an alkyl halide with and a trialkyl phosphite yielding a dialkyl alkyl phosphonate (Fig. 3.22).

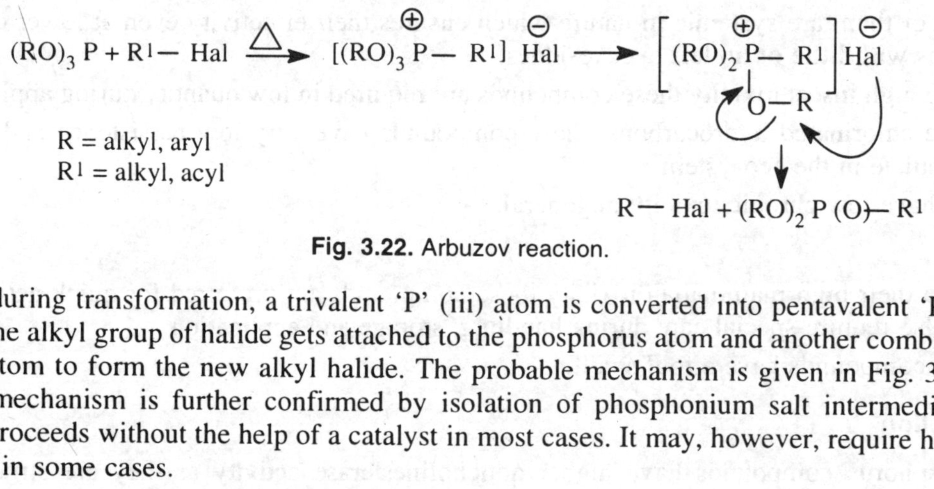

R = alkyl, aryl
R¹ = alkyl, acyl

Fig. 3.22. Arbuzov reaction.

Thus during transformation, a trivalent 'P' (iii) atom is converted into pentavalent 'P' (v). In general, the alkyl group of halide gets attached to the phosphorus atom and another combines with halogen atom to form the new alkyl halide. The probable mechanism is given in Fig. 3.22. The reaction mechanism is further confirmed by isolation of phosphonium salt intermediate. The reaction proceeds without the help of a catalyst in most cases. It may, however, require heating or a catalyst in some cases.

Scope and limitation of the reaction

In general, the Arbuzov reaction comprises a reaction between alkyl halide and trialkyl phosphite as shown below (Fig. 3.23), where A and B may be same or different.

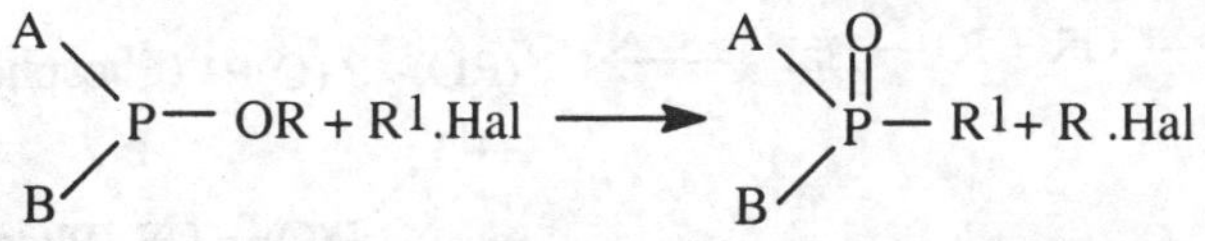

$$A \diagdown P - OR + R^1.Hal \longrightarrow A \diagdown \overset{O}{\underset{\parallel}{P}} - R^1 + R.Hal$$

Fig. 3.23. Symbolic representation of Arbuzov reaction.

This leads to a wide choice of phosphites as well as alkyl halides to provide a large number of phosphonates.

The reactivity of halides follows the sequence R CO-hal > R CH_2-hal > R_2 CH-hal > $RR'R''C$-hal; RI > R Br > RCl. The primary alkyl halides react normally but the t-alkyl halides and simple aryl and vinyl halides do not react with the trialkyl phosphites. Benzyl, diphenylmethyl and triphenylmethylhalides give expected phosphonates as do halogen methyl derivatives of condensed aromatic hydrocarbons (Fig. 3.24).

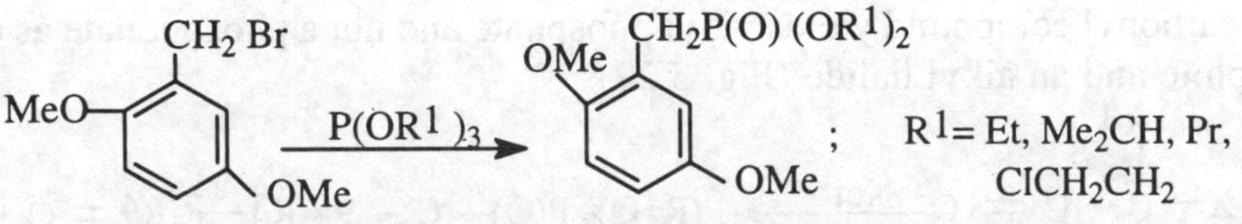

Fig. 3.24. Reaction of bromomethyl derivative with phosphite.

$R^1 = $ Et, Me_2CH, Pr, $ClCH_2CH_2$

Mostly the primary aliphatic halides are used. Although a few secondary halides also react satisfactorily. Other organic halides such as esters of monohalo monocarboxylic acids react to give α-carboxyalkyl phosphonates (Fig. 3.25).

$$R^1OOC— CH_2 — Hal + P(OR)_3 \xrightarrow{-RHal} R^1OOC— CH_2P(O)(OR)_2$$

Fig. 3.25. Reaction of monohalocarboxylate with phosphite.

Various other halides such as isocyanide dichloride, imidochloride, haloalkyl amide, propargyl halide and acylchlorides also react to give phosphonates.

Saturated α-chloro/bromoketones and aldehydes give normal Arbusov product but with poor yield. However, α-iodo ketone [5] gives the normal phosphonates. Simple aryl halides are not good for the Arbuzov reaction, however, iodobenzene reacts with phosphite at 60°C to produce Arbuzov product through a free radical mechanism [6].

The general structure of the phosphorus ester may be written as ABP-OR where A & B may be primary alkoxy, secondary alkoxy, aryloxy alkyl, aryl or dialkylamino group. The reactivity of A,B increases in the order aryloxy < alkoxy < aryl < arkyl < dialkylamino. The R should be aliphatic for the reaction to proceed smoothly. If A and B in the phosphite ABP-OR are alkoxy groups, a phosphonate is obtained. When one is alkoxy and the other is alkyl or aryl group, then phosphinate is produced. If both A & B are alkyl or aryl groups, then a t-phosphine oxide is obtained as shown in Fig. 3.26.

Use of catalyst

Normally Arbuzov reaction proceeds without the help of a catalyst but in some cases, catalysts such as nickel halides, cobalt salts and copper powder or simply heating is required.

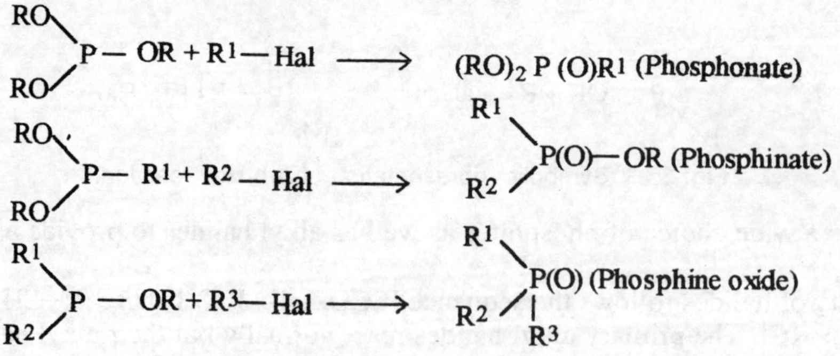

$$RO \diagdown P - OR + R^1 - Hal \longrightarrow (RO)_2 P (O)R^1 \text{ (Phosphonate)}$$

$$RO \diagdown P - R^1 + R^2 - Hal \longrightarrow \begin{matrix} R^1 \diagdown \\ R^2 \diagup \end{matrix} P(O) - OR \text{ (Phosphinate)}$$

$$\begin{matrix} R^1 \diagdown \\ R^2 \diagup \end{matrix} P - OR + R^3 - Hal \longrightarrow \begin{matrix} R^1 \diagdown \\ R^2 \diagup \end{matrix} P(O) \underset{R^3}{|} \text{ (Phosphine oxide)}$$

Fig. 3.26. Reaction of alkylhalide with phosphorus ester.

Perkow Reaction

In 1952 Perkow [6] observed for the first time that a trialkyl phosphite on reaction with a α-halogenated carbonyl compound yields vinyl phosphate and not a phosphonate as expected from a trialkyl phosphite and an alkyl halide (Fig. 3.27).

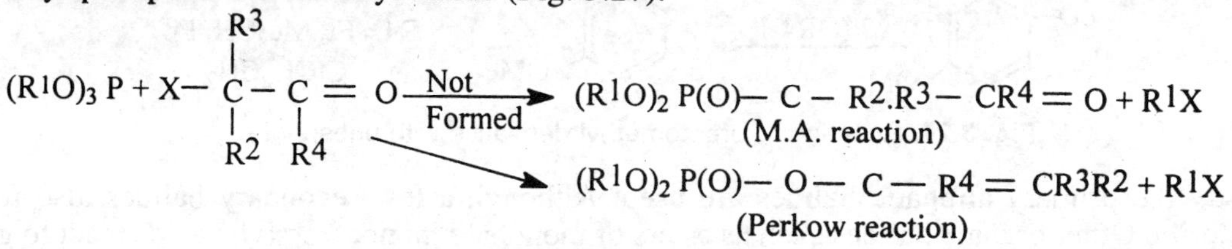

$$(R^1O)_3 P + X - \underset{\underset{R^2}{|}}{\overset{\overset{R^3}{|}}{C}} - \underset{\underset{R^4}{|}}{C} = O \xrightarrow[\text{Formed}]{\text{Not}} (R^1O)_2 P(O) - C - R^2.R^3 - CR^4 = O + R^1X$$
$$\text{(M.A. reaction)}$$
$$\searrow (R^1O)_2 P(O) - O - C - R^4 = CR^3R^2 + R^1X$$
$$\text{(Perkow reaction)}$$

Fig. 3.27. Perkow reaction.

The reactivity of α-halocarboxyl compounds decreases in the order α-haloaldehydes > α-haloketones > α-haloesters. But with last two compounds both the reactions occur competitively. The Perkow reaction takes place smoothly with α-haloaldehyde. The nature of halogen also affects the composition and the reaction rate which is of the order Cl > Br > I. The Perkow reaction has been employed for the preparation of vinyl phosphates such as chlorofenvinphos, dichlorvos, mevinphos, monocrotophos and phosphamidon. Many of these compounds are used as insecticides.

Four possible mecahnisms [7, 8] have been proposed for the Perkow reaction which depend on the site where phosphite attacks at the initial step.

  i)   halogen atom
 ii)   α-carbon atom
iii)   carbonyl carbon
iv)   carbonyl oxygen

  i)   When the phosphite attacks at halogen atom, a halophosphonium salt is produced as an intermediate which finally gives vinyl phosphate. But the formation of halophosphonium salt as an intermediate product appears unlikely in the Perkow reaction as confirmed by other reactions (Fig. 3.28).

 ii)   If the reaction is initiated by the attack of phosphite on α-carbon, a β-ketophosphonium salt should be produced as an intermediate. But the formation of enol-phosphonium salt from β-ketophosphonium salt is an extremely difficult process (Fig. 3.29).

iii)   The initial nucleophilic attack of the phosphite compound on the carbonyl carbon, an elec-

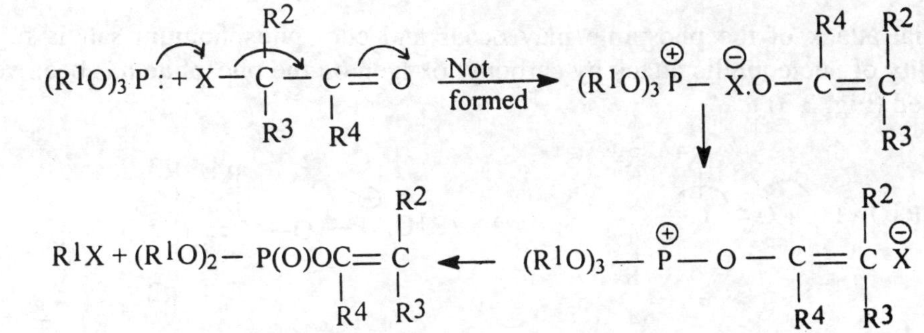

Fig. 3.28. Attack of phosphite on halogen atom.

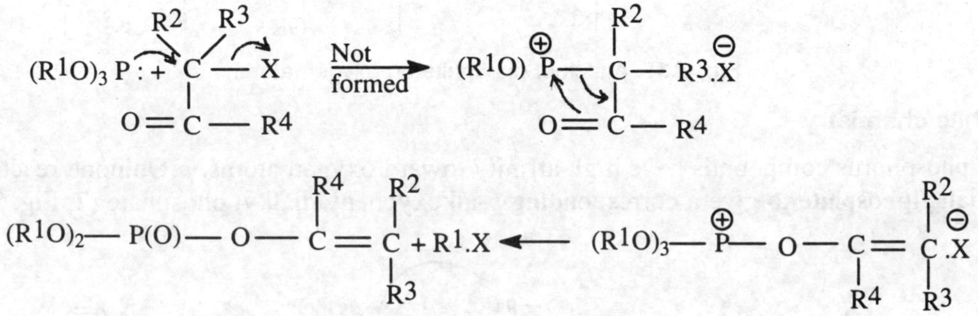

Fig. 3.29. Attack of phosphite on a carbon atom.

trophilic centre in the molecule, appears to be reasonable. This is further supported by the formation of α-hydroxyalkyl phosphonate during Perkow reaction carried out in alcohols. This mechanism involves a rearrangement process similar to the base catalysed transformation of trichlorfon into dichlorvos (Fig. 3.30).

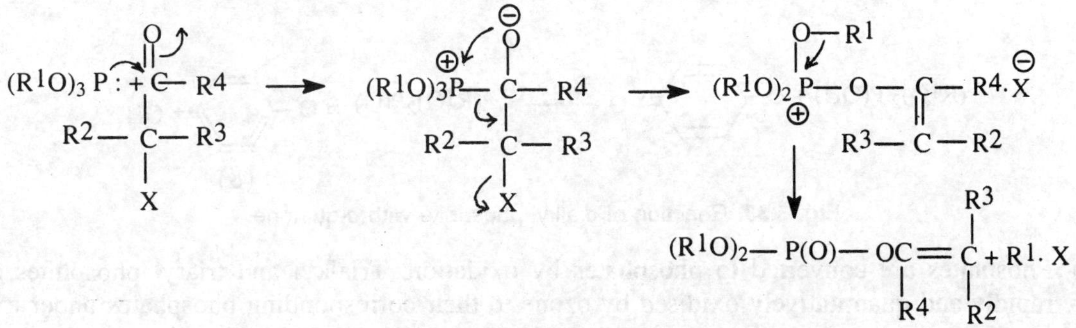

Fig. 3.30. Attack of phosphite on carbonyl carbon atom.

However, the Perkow reaction is stereospecific and this mechanism cannot explain the formation of some vinyl phosphates.

iv) Trialkyl phosphites may attack the oxygen atoms of carbonyl groups of certain compounds such as quinones. Hence the carbonyl oxygen appears to be one of the probable centres where the initial attack of the phosphite may occur and enol phosphonium salt is formed. The possibility of nucleophilic attack by carbonyl oxygen on the phosphite has been vehemently supported (Fig. 3.31).

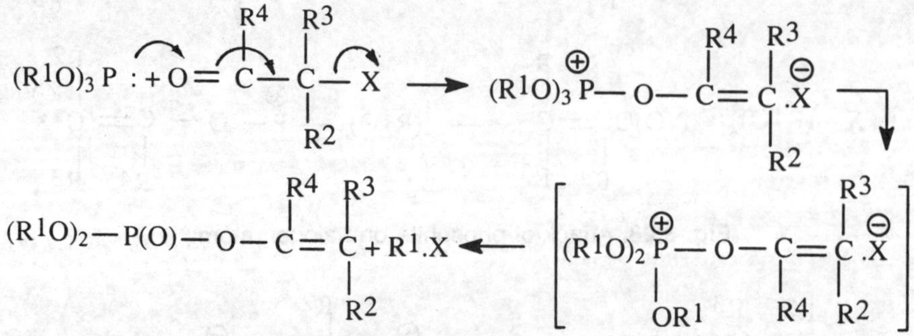

Fig. 3.31. Attack of phosphite on oxygen atom.

Phosphite chemistry

i) T-phosphorus compounds have high affinity toward oxygen atoms. p-Quinone reacts with a trialkylphosphite to give a corresponding p-alkoxyphenyldialkyl phosphate (7, Fig. 3.32).

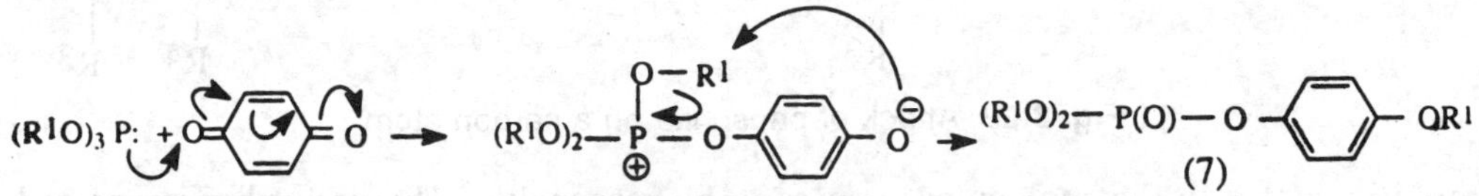

Fig. 3.32. Reaction of phosphite with p-quinone.

ii) The reaction of dialkyl phosphite with p-quinone in presence of a base gives p-hydroxyphenylphosphate (8, Fig. 3.33).

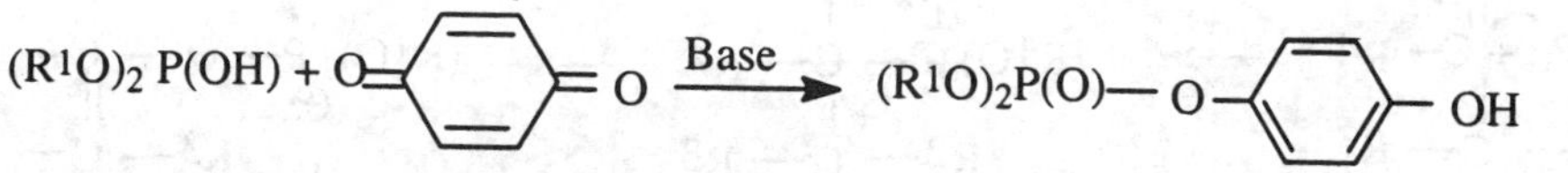

Fig. 3.33. Reaction of dialkyl phosphite with p-quinone.

iii) Phosphites are converted to phosphates by oxidation. Trialkyl and triaryl phosphites are rapidly and quantitatively oxidised by ozone to their corresponding phosphates under mild conditions.

Preparation of some important phosphite intermediates

These reactions can be expanded further to prepare phosphate triesters directly from dialkyl phosphites. Thus the anthelmintic agent **haloxon** is prepared by employing the reaction of bis-β-chloroethylphosphite with 3-chloro-4-methyl-7-hydroxy coumarin and CCl_4 in presence of Et_3N (9, Fig. 3.35).

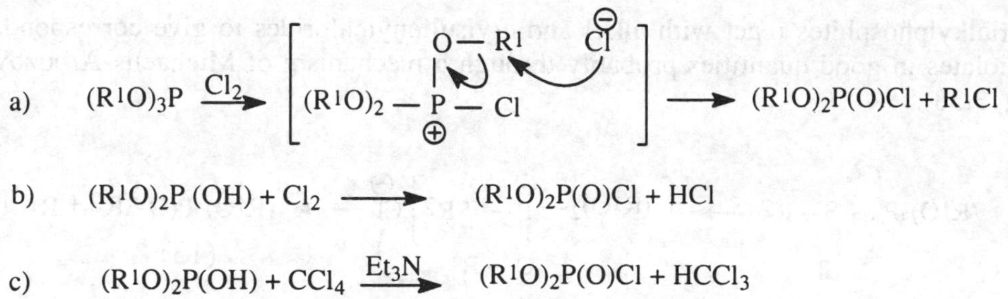

a) $(R^1O)_3P \xrightarrow{Cl_2}$ $\left[(R^1O)_2 - \overset{\oplus}{\underset{}{P}} - Cl \right]$ $\longrightarrow (R^1O)_2P(O)Cl + R^1Cl$

b) $(R^1O)_2P(OH) + Cl_2 \longrightarrow (R^1O)_2P(O)Cl + HCl$

c) $(R^1O)_2P(OH) + CCl_4 \xrightarrow{Et_3N} (R^1O)_2P(O)Cl + HCCl_3$

Fig. 3.34. Preparation of phosphoryl chloride.

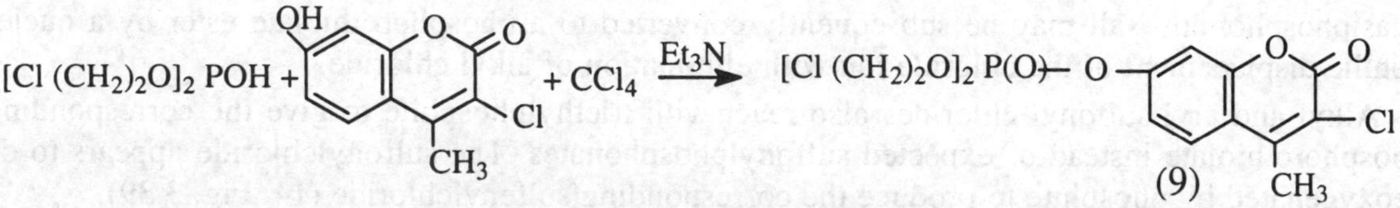

$[Cl(CH_2)_2O]_2 POH +$ $+ CCl_4 \xrightarrow{Et_3N} [Cl(CH_2)_2O]_2 P(O) - O -$

(9)

Fig. 3.35. Preparation of haloxon.

d) Alkyl phosphites are produced by reaction of phosphorus trichloride and alcohols. In the presence of a *t*-amine the reaction proceeds to form trialkyl phosphites (10). But in absence of base however the reaction stops to give dialkyl phosphites (11, Fig. 3.36).

$$PCl_3 + 3R^1OH + 3NR_3^2 \longrightarrow (R^1O)_3P + 3R_3^2 \, N.HCl$$

(10)

$$PCl_3 + 3R^1OH \longrightarrow (R^1O)_2POH + R^1Cl + 2HCl$$

(11)

Fig. 3.36. Reaction of phosphorus trichloride with alcohols.

e) Mixed dialkyl phosphites may be prepared by a transesterification reaction using phosphoric acid as catalyst (12, Fig. 3.37).

$$(R^1O)_2POH + R^2OH \xrightarrow{H_3PO_4} \genfrac{}{}{0pt}{}{R^1O}{R^2O} {>} P.OH + R^1OH$$

(12)

Fig. 3.37. Reaction of mixed dialkyl phosphite with alcohols.

Few reactions of phosphites with sulphur compounds [9]

Trialkylphosphites are active nucleophilic agents and bring substitution at a bivalent sulphur atom linked to a good leaving group. Dialkylphosphites are much less active, however, their salts behave as trivalent phosphorus compounds which are active enough to react with certain sulphur compounds.

a) Trialkylphosphites react with alkyl and arylsulfenylchlorides to give corresponding phosphorothiolates in good quantities probably through a mechanism of Michaelis-Arbuzov reaction (13, Fig. 3.38).

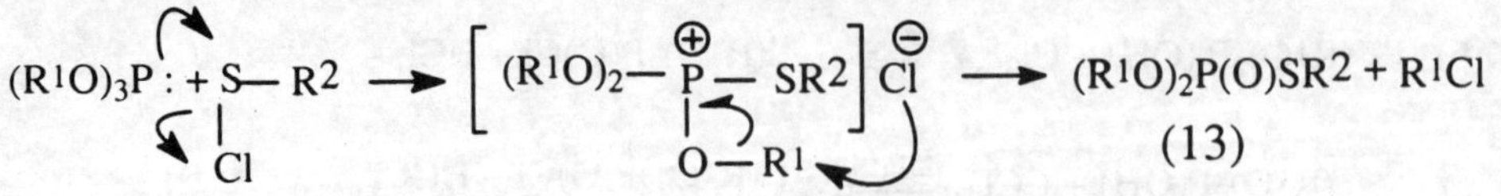

$$(R^1O)_3P : + S— R^2 \longrightarrow \left[(R^1O)_2— \overset{\oplus}{\underset{\underset{O—R^1}{|}}{P}}— SR^2 \right] \overset{\ominus}{Cl} \longrightarrow (R^1O)_2P(O)SR^2 + R^1Cl$$

$$\qquad\qquad\qquad\qquad\qquad\qquad\qquad\qquad\qquad\qquad\qquad\qquad (13)$$

Fig. 3.38. Reaction of trialkyl phosphite with sulfenyl chloride.

In phosphite the phosphorus atom attacks the S-atom to displace halide. The intermediate quasiphosphonium salt may be subsequently converted to a phosphorothiolate ester by a nucleophilic displacement of the chloride ion with elimination of alkyl chloride.

Alkyl and aryl-sulfonyl chlorides also react with triethylphosphite to give the corresponding phosphorothiolate instead of expected sulfonylphosphonates. The sulfonylchloride appears to be deoxygenated by phosphite to produce the corresponding sulfenylchloride (14, Fig. 3.39).

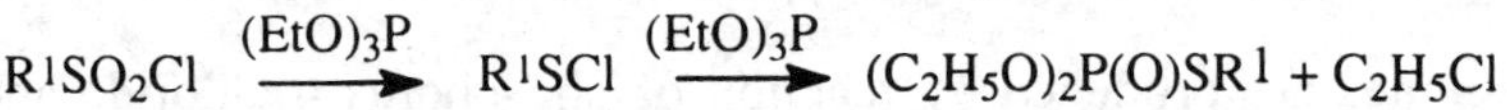

$$R^1SO_2Cl \xrightarrow{(EtO)_3P} R^1SCl \xrightarrow{(EtO)_3P} (C_2H_5O)_2P(O)SR^1 + C_2H_5Cl$$

$$\qquad\qquad\qquad\qquad\qquad\qquad\qquad\qquad\qquad (14)$$

Fig. 3.39. Reaction of phosphite with sulfonylchloride.

Phosphorothiolates react similarly with phenyl sulfenylchlorides to give phosphorodithiolates such as **edifenphos** (15, Fig. 3.40).

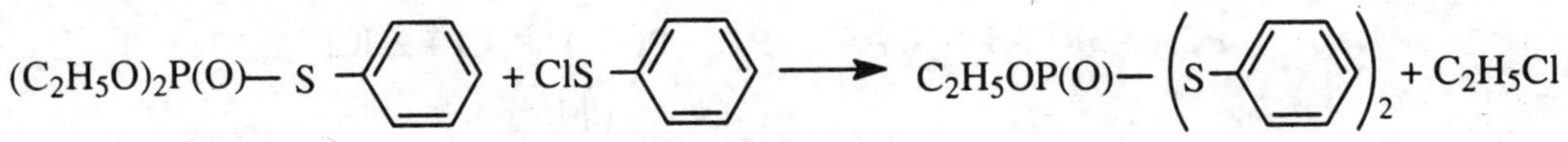

$$(C_2H_5O)_2P(O)— S \text{—⬡—} + ClS \text{—⬡—} \longrightarrow C_2H_5OP(O)— \left(S\text{—⬡—} \right)_2 + C_2H_5Cl$$

$$\qquad\qquad\qquad\qquad\qquad\qquad\qquad\qquad\qquad\qquad\qquad\qquad (15)$$

Fig. 3.40. Reaction of phosphorothiolate with sulfenylchloride.

A similar reaction occurs between sulfenyl compounds and dialkyl phosphites to give compounds like **cerezin** (16, Fig. 3.41).

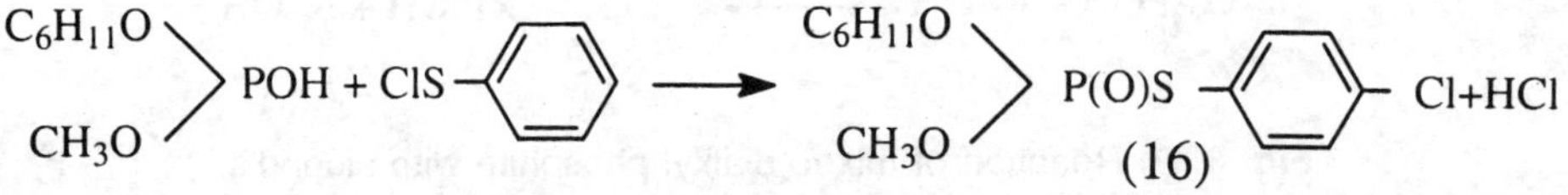

$$\begin{matrix} C_6H_{11}O \\ \\ CH_3O \end{matrix} \Big\rangle POH + ClS\text{—⬡—} \longrightarrow \begin{matrix} C_6H_{11}O \\ \\ CH_3O \end{matrix} \Big\rangle P(O)S \text{—⬡—} Cl + HCl$$

$$\qquad\qquad\qquad\qquad\qquad\qquad\qquad\qquad\qquad\qquad (16)$$

Fig. 3.41. Reaction of dialkyl phosphite with sulfenylchloride.

Thiocyanates react with trialkylphosphites to give phosphorothiolates and nitriles like demetons (17, Fig. 3.42).

Disulfides also react with phosphites to cleave S-S bond and the formation of a P-S bond. The **trialkylphosphite** may react with aliphatic disulfides at elevated temperature (18, Fig. 3.43).

$$(C_2H_5O)_3 P + NCS (CH_2)_2SC_2H_5 \longrightarrow (C_2H_5O)_2 P(O)S(CH_2)_2SC_2H_5 + C_2H_5CN$$

$$(17)$$

Fig. 3.42. Reaction of thiocyanate with phosphite.

$$(R^1O)_3P + R^2SSR^3 \longrightarrow \left[(R^1O)_3\overset{\oplus}{P}SR^2 \right] \overset{\overset{\ominus}{S}R^3}{\longrightarrow} (R^1O)_2P(O)SR^2 + R^1SR^3$$

$$(18)$$

Fig. 3.43. Reaction of disulfide with phosphite.

An interesting reaction of disulfides with cyclic phosphoramidates occurs to give phosphoroamidothiolate. In presence of diaryl or alkylaryldisulfides, the cyclic structure opens up to form O-(ω-substituted thioalkyl) phosphoramidothiolates (19, Fig. 3.44).

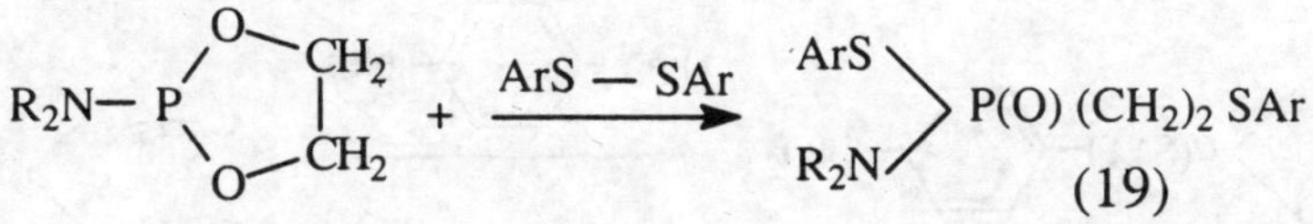

$$(19)$$

Fig. 3.44. Reaction of disulfide with cyclic phosphoramidate.

Trialkylphosphite react with chlorourea to give dialkoxy phosphinyl urea (20) having herbicidal property (Fig. 3.45).

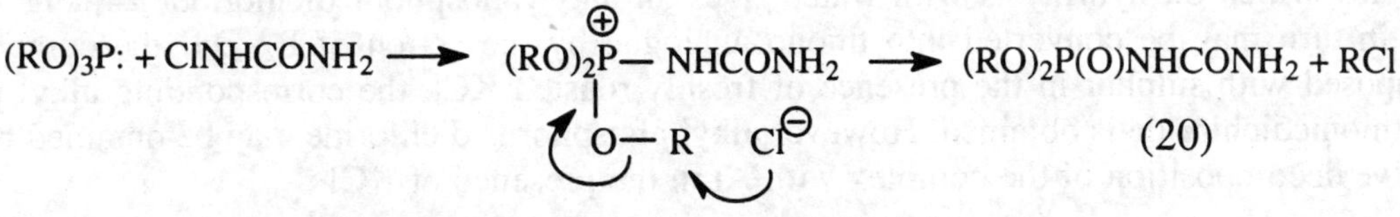

$$(RO)_3P: + ClNHCONH_2 \longrightarrow (RO)_2\overset{\oplus}{P} - NHCONH_2 \longrightarrow (RO)_2P(O)NHCONH_2 + RCl$$

$$(20)$$

Fig. 3.45. Reaction of chlorourea with phosphite.

Trialkylphosphites undergo an interesting reaction with azides. The reaction with hydrazoic acid appears to proceed through a trialkyl phosphoramidate intermediate which then rearranges to form N-alkyl phosphoramidate, (21, Fig. 3.46).

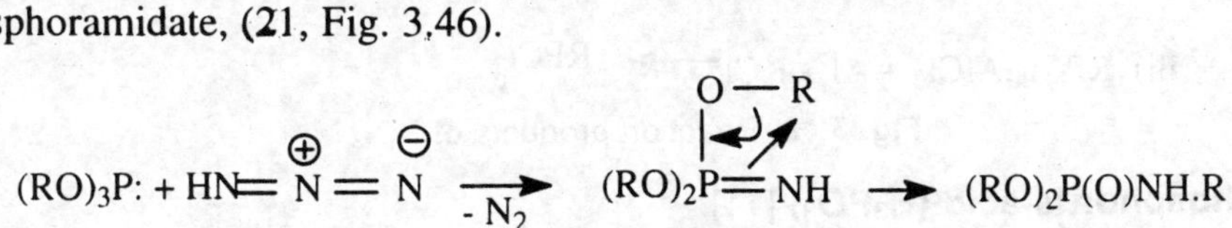

$$(RO)_3P: + H\overset{\oplus}{N} = \overset{\ominus}{N} = N \xrightarrow{-N_2} (RO)_2P = NH \longrightarrow (RO)_2P(O)NH.R$$

$$(21)$$

Fig. 3.46. Reaction of azide with phosphite.

Dialkyl phosphites react with O,O-dialkyl S-morpholinodithiophosphate to give dithiopyrophosphate esters (22, Fig. 3.47).

Reaction of PCl₃ with Lewis acids [10]

a) On heating aromatic hydrocarbons with excess amount of PCl_3 for several h. in presence of

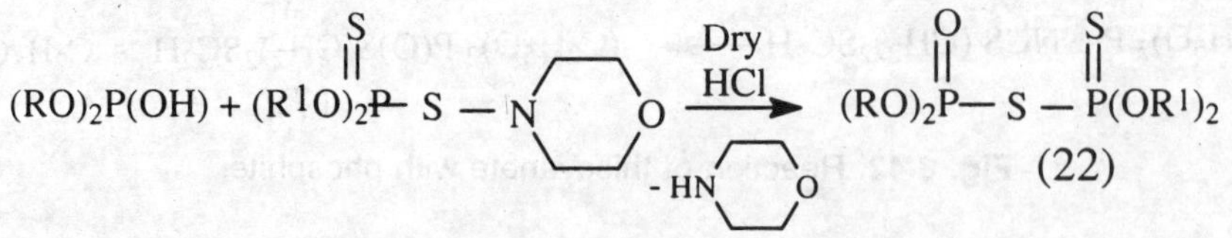

$$(RO)_2P(OH) + (R^1O)_2\overset{\overset{S}{\|}}{P}-S-N\underset{-HN}{\bigcirc}O \xrightarrow{\text{Dry} \atop \text{HCl}} (RO)_2\overset{\overset{O}{\|}}{P}-S-\overset{\overset{S}{\|}}{P}(OR^1)_2$$

$$(22)$$

Fig. 3.47. Reaction of phosphite with morpholinodithiophosphate.

aluminum chloride, the Friedel Crafts reaction takes place. The complex formed is decomposed by $POCl_3$ or pyridine to give aryl phosphorousdichloride. Phosphorous dichloride reacts with thiophosphoryl chloride or elemental sulphur to give an aryl phosphonothionicdichloride which can be converted into **EPN**, an insecticide (23, Fig. 3.48).

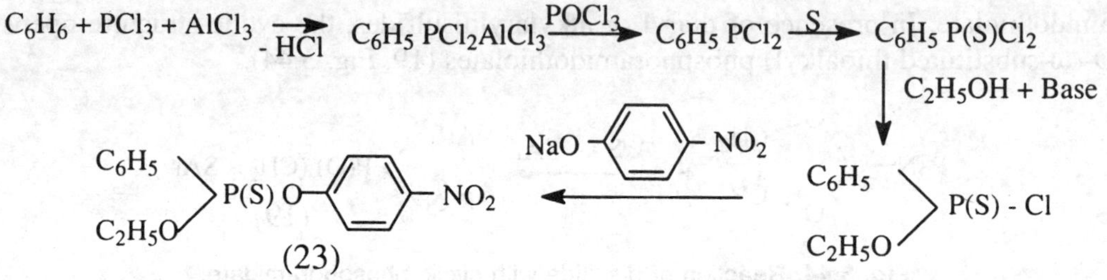

$$C_6H_6 + PCl_3 + AlCl_3 \xrightarrow{-HCl} C_6H_5 PCl_2AlCl_3 \xrightarrow{POCl_3} C_6H_5 PCl_2 \xrightarrow{S} C_6H_5 P(S)Cl_2$$

$$\downarrow C_2H_5OH + Base$$

$$\underset{C_2H_5O}{\overset{C_6H_5}{>}}P(S)\,O-\bigcirc-NO_2 \xleftarrow{\quad} NaO-\bigcirc-NO_2 \qquad \underset{C_2H_5O}{\overset{C_6H_5}{>}}P(S)-Cl$$

$$(23)$$

Fig. 3.48. Preparation of EPN.

b) PCl_3 also reacts with alkyl halides in presence of a catalyst like aluminum chloride to yield a complex which on hydrolysis with water gives an alkyl phosphonicdichloride (24, Fig. 3.49) which in turn may be converted into thiono analogue by the action of P_2S_5. If the complex is decomposed with sulphur in the presence of freshly roasted KCl, the corresponding alkyl phosphonothionicdichloride is obtained. However alkyl phosphorus dichloride may be obtained by the reductive decomposition of the complex with Al in the presence of KCl.

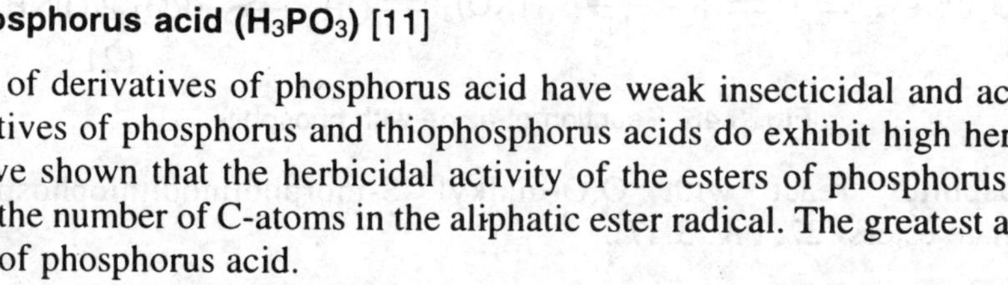

i) $\quad PCl_3 + RCl + AlCl_3 \longrightarrow RPCl_4.AlCl_3 \xrightarrow{H_2O} RP(O)Cl_2$

$$(24)$$

ii) $\quad RPCl_4.AlCl_3 + S + KCl \xrightarrow{\triangle} RP(S)Cl_2 \xleftarrow{P_2S_5}$

iii) $\quad RPCl_4.AlCl_3 + Al + KCl \longrightarrow RPCl_2$

Fig. 3.49. Reaction products of PCl_3.

i) Esters of phosphorus acid (H_3PO_3) [11]

A large number of derivatives of phosphorus acid have weak insecticidal and acaricidal activity but some derivatives of phosphorus and thiophosphorus acids do exhibit high herbicidal activity. SAR studies have shown that the herbicidal activity of the esters of phosphorus acids increases with increase in the number of C-atoms in the aliphatic ester radical. The greatest activity is shown by the tri esters of phosphorus acid.

2,4-DEP, falone®, tris(2,4-dichlorophenoxy ethyl) phosphite is produced by the reaction of phosphorus trichloride with 2,4-dichlorophenoxy ethanol in presence of pyridine or t-amine (25, Fig. 3.50).

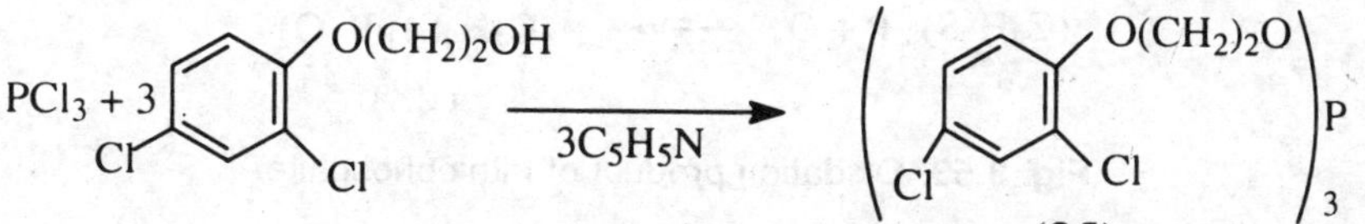

Fig. 3.50. Preparation of 2,4-DEP.

The reaction product after removal of amine hydrochloride is used without further purification for the preparation of emulsive concentrates. It is a thick oily liquid with a weak odour which decomposes if distilled under normal pressure. It is practically insoluble in water but soluble in organic solvents with LD_{50} (rat) 850 mg/kg. Oxidation converts it to corresponding phosphates. When falone reacts with water it is first converted to the corresponding dialkylphosphite which on hydrolysis and oxidation yields 2,4-chlorophenoxy ethanol and phosphoric acid (26, Fig. 3.51).

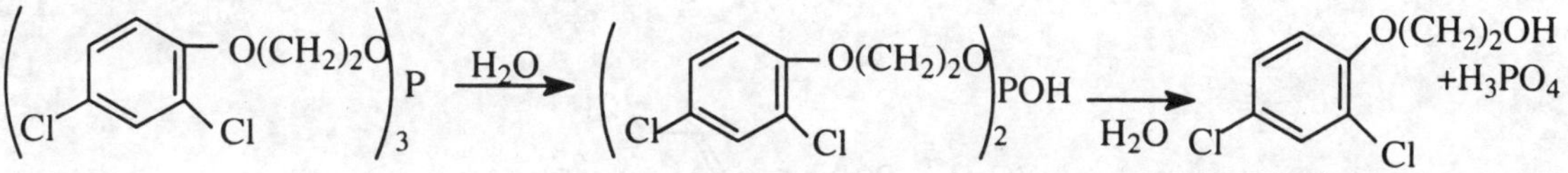

Fig. 3.51. Reaction products of falone.

The action of falone in the soil apparently is based on its conversion to 2,4-D. Falone is recommended for the control of weeds in corn, potatoes, strawberries and some other crops at 4-8 kg/ha. For treatment of soil it is used usually in the form of aqueous emulsion.

Bis (2,4-dichlorophenoxy ethyl) phosphite also has herbicidal properties but is inferior to falone in activity.

2. Tributyltrithiophosphite, merphos®

Tributyltrithiophosphite (27) is obtained by the reaction of butyl mercaptan with phosphorus trichloride at an elevated temperature (Fig. 3.52).

$$PCl_3 + 3\ C_4H_9SH \longrightarrow (C_4H_9S)_3\ P + 3HCl$$
$$(27)$$

Fig. 3.52. Preparation of tributyltrithiophosphite.

It is light oily liquid having bp 150-152° at 2 mm of Hg, almost insoluble in water but highly soluble in organic solvents. The LD_{50} (rat) is 350 mg/kg. This compound is used in the form of aqueous emulsion at dosages of 1-2 kg/ha for the defoliation of cotton crop.

Merphos is one of the best defoliants because of high activity. Merphos is slowly oxidised by the oxygen of the air to tributyltrithiophosphate. This reaction is accelerated by heating and can serve as a preparatory method for tributyl trithiophosphate (28, Fig. 3.53). Water slowly hydrolyzes tributyltrithio-phosphite with the formation of butyl mercaptan (Fig. 3.54).

High defoliant activity of some homologues of tributyltrithiophosphite was observed but these compounds are not much in use.

$$(C_4H_9S)_3 \ P + O \longrightarrow (C_4H_9S)_3 \ P \ (O)$$

$$(28)$$

Fig. 3.53. Oxidation product of trithiophosphite.

$$(C_4H_9S)_3 \ P + H_2O \xrightarrow[\text{Slow}]{\triangle} (C_4H_9S)_2 \ POH \xrightarrow[+ \ O]{H_2O} C_4H_9SH + H_3PO_4$$

Fig. 3.54. Hydrolysis product of trithiophosphite.

ii) Esters of phosphinic acid [12]

In general the phosphinothionates do not possess much insecticidal activity. A few phosphinothionates found active are discussed in this chapter

2,4,5-Trichlorophenyl diethylphosphinothionate, Agvitor (29)

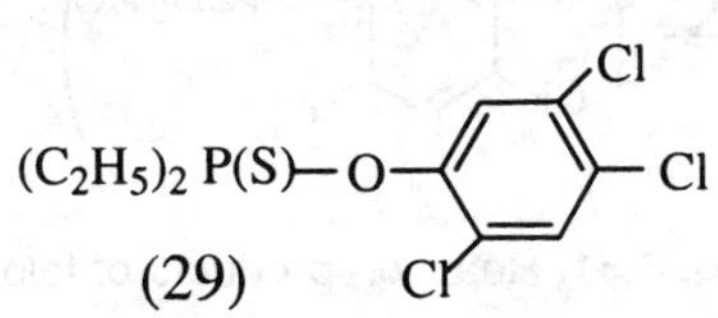

Agvitor is the phosphinate analogue of phosphorothionate insecticide ronnel and the phosphonothionate insecticide trichloronate. It effectively controls carrot and onion flies with LD_{50} (rat) 100 mg/kg. The mammalian toxicity is somewhat less than the phosphonate analogue trichloronate and the phosphinate analogues such as **fenthion** (30) and **azinphosmethyl** (31) seem to be quite active as insecticides.

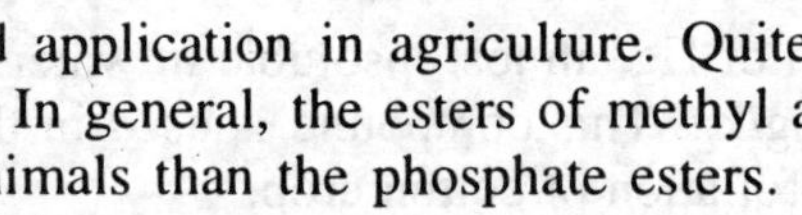

iii) Esters of phosphonic acids [13]

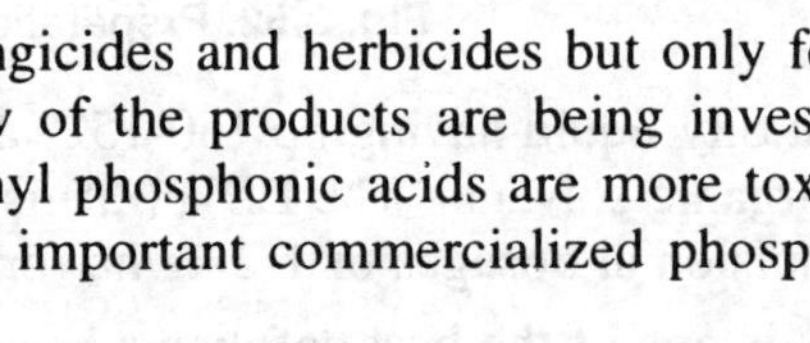

Phosphonates also are effective as insecticides, fungicides and herbicides but only few of them have found application in agriculture. Quite a few of the products are being investigated and developed. In general, the esters of methyl and ethyl phosphonic acids are more toxic to warm blooded animals than the phosphate esters. A few important commercialized phosphonates are discussed below.

Trichlorfon, dipterex® O,O-dimethyl 1-hydroxy 2,2,2,-trichloroethyl phosphonate is prepared by the reaction of dimethyl phosphite with chloral (32, Fig. 3.55).

It is solid with mp 83-84°C and dissolves readily in water (15.4 g in 100 ml), ether, chloroform and benzene. It is relatively stable in acid but gets converted to dichlorvos in alkali (33, Fig. 3.56). It gives dimethylhydrogenphosphate and dichloroacetaldehyde upon hydrolysis. It has a high

$$(CH_3O)_2 \, POH + CCl_3CHO \xrightarrow[\text{at r.t.}]{\text{Kept}} (CH_3O)_2 \, P(O) - CH(OH) - CCl_3 \tag{32}$$

Fig. 3.55. Preparation of trichlorfon.

$$(CH_3O)_2 \, P(O) - CH \overset{\displaystyle |}{\underset{\displaystyle OH}{|}} C \overset{\displaystyle Cl}{\underset{\displaystyle Cl}{-}} Cl \longrightarrow (CH_3O)_2 \, P(O) - OCH = CCl_2 \tag{33}$$

Fig. 3.56. Conversion of trichlorfon to dichlorvos.

insecticidal activity particularly against diptera. It is useful for the control of both sucking and chewing insect pests of field crops, vegetables and seed crops with LD_{50} (rat) 630 mg/kg. Trichlorfon is a poor inhibitor of acetyl cholinesterase but *in situ* is readily converted into a strong inhibitor dichlorvos (32) at physiological pH conditions. This reaction goes faster in presence of a base.

EPN, O-ethyl O-4 (nitrophenyl) phenyl phosphonothionate (23, Fig. 3.48) is solid substance with mp 36°C. This is relatively stable under neutral and acidic conditions but gets hydrolysed in alkali solutions to phenyl phosphonothioic acid, ethanol and nitrophenol. It is an insecticide and acaricide and useful to control many phytophagus mites and insect species including rice stem borer and boll weevil with LD_{50} (rat) 40 mg/kg.

Cyanofenphos, surecide®, p-cyanophenyl O-ethyl phenyl phosphonothionate (34, Fig. 3.57) can be prepared as follows.

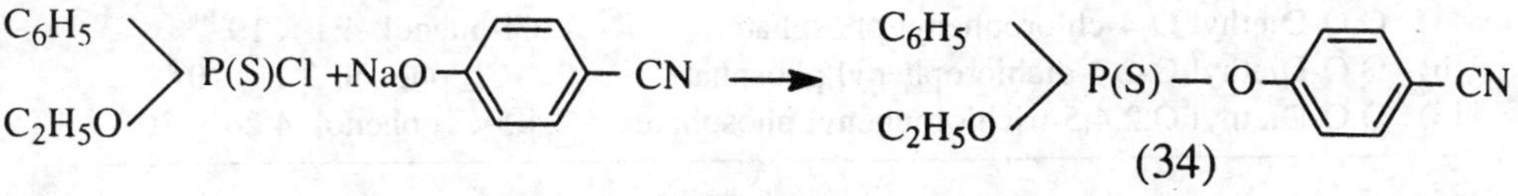

$$\begin{array}{c} C_6H_5 \\ C_2H_5O \end{array}\!\!\!> P(S)Cl + NaO\!\!-\!\!\langle \bigcirc \rangle\!\!-\!\!CN \longrightarrow \begin{array}{c} C_6H_5 \\ C_2H_5O \end{array}\!\!\!> P(S) - O\!\!-\!\!\langle \bigcirc \rangle\!\!-\!\!CN \tag{34}$$

Fig. 3.57. Preparation of cyanofenphos.

It is white solid having mp 83°C with LD_{50} (rat) 30 mg/kg. It is an insecticide found effective against rice stem borer, cotton boll worm, cabbage worm and aphids. The methyl ester homologue is more toxic to both mammals and insects and a second substitution in the cyanophenyl ring does not decrease the mammalian toxicity.

Fonofos, dyfonate®, ethyl S-phenyl O-ethyl phosphonothiolothionate (35, Fig. 3.58) is prepared as follows.

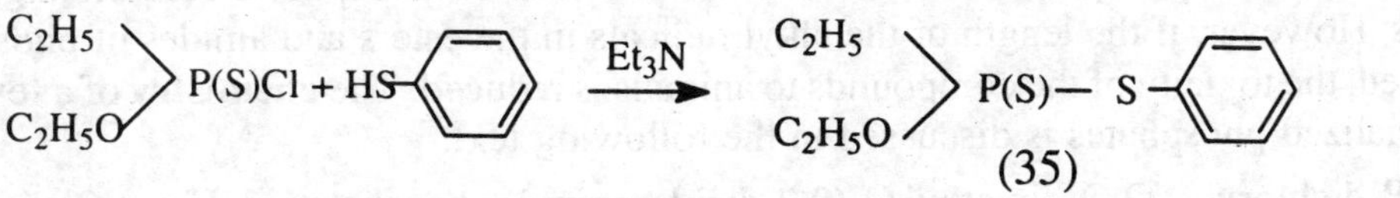

$$\begin{array}{c} C_2H_5 \\ C_2H_5O \end{array}\!\!\!> P(S)Cl + HS\!\!-\!\!\langle \bigcirc \rangle \xrightarrow{Et_3N} \begin{array}{c} C_2H_5 \\ C_2H_5O \end{array}\!\!\!> P(S) - S\!\!-\!\!\langle \bigcirc \rangle \tag{35}$$

Fig. 3.58. Preparation of fonofos.

It is a pale yellow liquid with mercaptan like odour having bp 130°/0.1 mm/Hg. It is highly stable in soil and controls soil insects like corn root worms, wire warms, cut worms and maggots.

The LD_{50} (rat) is 16.5 mg/kg. The main route of cleavage is through P-S bonding. Both the alkyl groups of fonofos can be changed without any appreciable reduction in insecticidal activity.

Trichloronate, phytosol®, O-ethyl 2,4,5-trichlorophenylethylphosphono-thionate (36, Fig. 3.59) is prepared as follows.

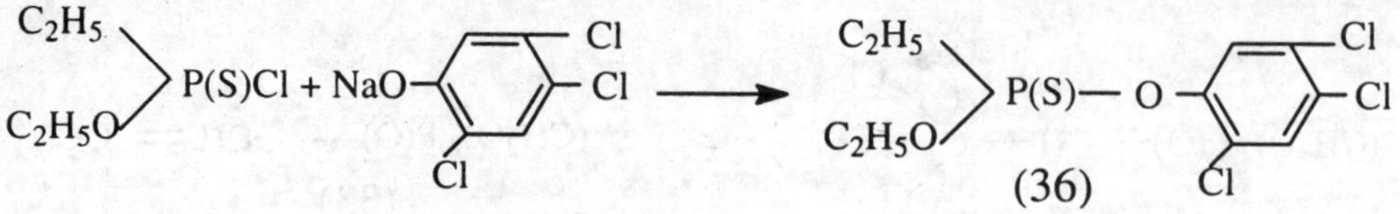

Fig. 3.59. Preparation of trichloronate.

It is a liquid having bp 108°C/0.01 mm Hg with LD_{50} (rat) 50 mg/kg. It is a non-systemic insecticide with prolonged residual activity, effective to control soil pests like root maggots and wireworms.

iv) Esters of phosphoric acid [14]

The insecticidal and acaricidal activity of the phosphatic compounds is found to be more than phosphite, as one of the ester radicals is acidic in nature. The higher the dissociation constant of such an alcohol or phenol (acid), the more toxic is the compound towards insects and animals. In the following phosphate series, the activity depends on the dissociation constants of different phenols such as

Phosphates	Dissociation constant
i) O,O-Diethyl O-4-chlorophenyl phosphate	4-Cl-phenol, 4.1×10^{-10}
ii) O,O-Diethyl O-2,4-dichlorophenyl phosphate	2,4-Cl_2 phenol, 3.1×10^{-8}
iii) O,O-Diethyl O,2,4,5-trichlorophenyl phosphate	2,4,5-Cl_3 phenol, 4.26×10^{-8}

Thus the 2,4,6-trichlorophenol has the highest activity. Further, the insecticidal property can be enhanced by introducing nitro and methyl mercapto groups into the aromatic radicals. In general, the activity of the aryl phosphates with substitution in position 4 is higher than in position 2 and 3. But, inspite of their high insecticidal activity, the mixed alkyl aryl esters of phosphoric acid have not been recommended in agriculture because of high mammalian toxicity.

However, certain aliphatic halogen containing esters of phosphoric acids and enol phosphates are found useful in agriculture and public health programme as these are less toxic to mammals.

The dialkylfluorophosphates and also the amides of fluorophosphoric acid are highly toxic to mammals. However, if the length of the alkyl radicals in the esters and amides of phosphoric acid is increased, the toxicity of the compounds to animals is reduced. The **chemistry** of a few important commercialized phosphates is discussed in the following text.

DDVP, dichlorvos, O,O-dimethyl O-(2,2-dichlorovinyl) phosphate (33, Fig. 3.60) (cf. Fig. 3.56) is synthesized as follows.

It is a colourless liquid having bp 74°C/mm Hg with LD_{50} (rat) 80 mg/kg. In water it hydrolyses to diethyl phosphoric acid. On bromination it gives dibrom, a powerful insecticide. Analogues of

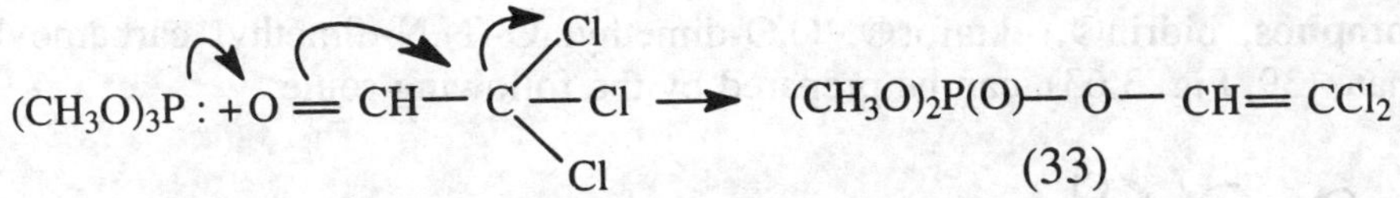

$$(CH_3O)_3P: +O = CH - C - Cl \longrightarrow (CH_3O)_2P(O) - O - CH = CCl_2$$

with Cl and Cl substituents on the carbon.

(33)

Fig. 3.60. Preparation of dichlorvos.

DDVP such as dipropyl and dibutyl derivatives are less toxic to mammals in comparison to the parent molecule.

Phosdrin, mevinphos®, O,O-dimethyl O-(1-methyl-2-carbomethoxy) vinyl phosphate (37, Fig. 3.61) is prepared as shown below.

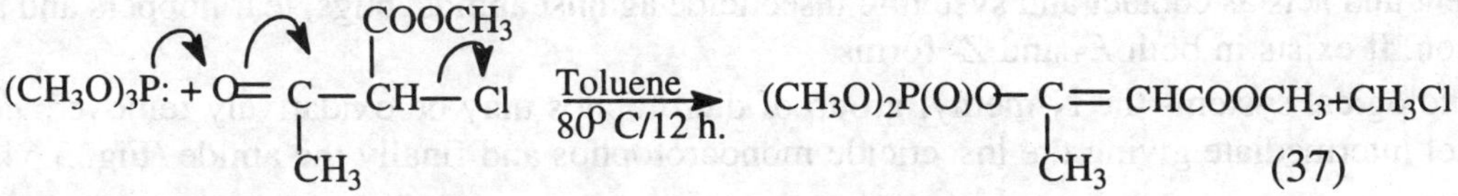

$$(CH_3O)_3P: +O = \underset{CH_3}{\overset{COOCH_3}{C}} - CH - Cl \xrightarrow[80^\circ C/12\,h.]{Toluene} (CH_3O)_2P(O)O - \underset{CH_3}{\overset{}{C}} = CHCOOCH_3 + CH_3Cl$$

(37)

$$\underset{CH_3}{\overset{(CH_3O)_2P(O)O}{}} C = C \overset{H}{\underset{COOCH_3}{}} \quad ; \quad \underset{CH_3}{\overset{(CH_3O)_2P(O)O}{}} C = C \overset{COOCH_3}{\underset{H}{}}$$

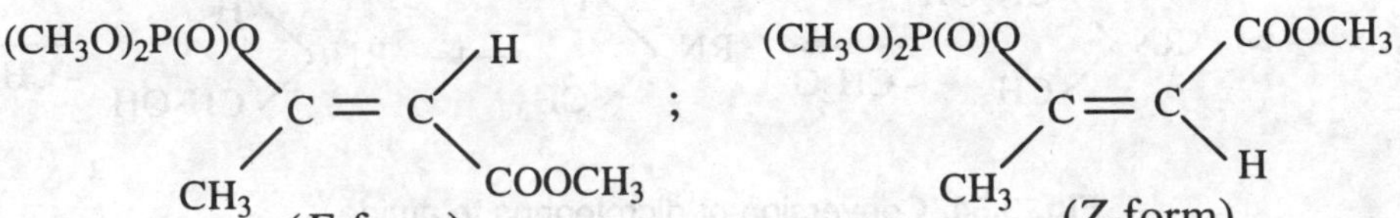

(*E*-form)   (*Z*-form)

Fig. 3.61. Preparation of phosdrin.

Phosdrin was developed by Shell research in 1953. It is a mixture of both *E*-isomer (60%) and *Z*-isomer (40%), however the *E*-isomer is more active than the *Z*-isomer. It is a liquid having bp 106-107°C at 1 mm Hg with LD_{50} (rat) 4-7 mg/kg. It has both contact and systemic activity and is effective to control both sucking and chewing insects.

iii) **Monocrotophos**, azodrin®, O,O-dimethyl O-(N-methylcarbamoyl 1-methyl) vinyl phosphate (38, Fig. 3.62) is as follows.

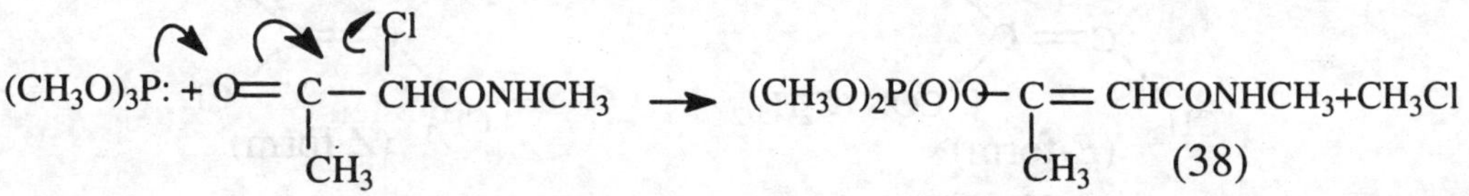

$$(CH_3O)_3P: +O = \underset{CH_3}{\overset{Cl}{C}} - CHCONHCH_3 \longrightarrow (CH_3O)_2P(O)O - \underset{CH_3}{\overset{}{C}} = CHCONHCH_3 + CH_3Cl$$

(38)

$$\underset{CH_3}{\overset{(CH_3O)_2P(O)O}{}} C = C \overset{H}{\underset{CONHCH_3}{}} \quad ; \quad \underset{CH_3}{\overset{(CH_3O)_2P(O)O}{}} C = C \overset{CONHCH_3}{\underset{H}{}}$$

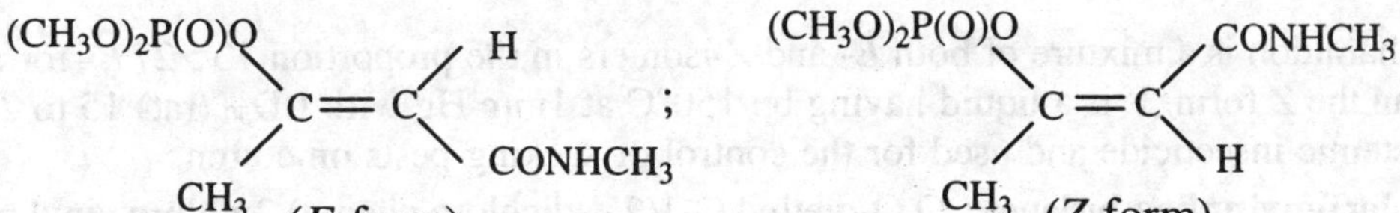

(*E*-form)   (*Z*-form)

Fig. 3.62. Preparation of monocrotophos.

The pure compound is solid having mp 53-55°C with LD_{50} (rat) 21 mg/kg. It is a powerful contact and systemic insecticide and is effective against a variety of foliage insects particularly boll worms on cotton. It exists in both *E*- and *Z*- forms.

iv) **Dicrotophos**, bidrin®, ektafos®, O,O-dimethyl O-(N,N-dimethyl carbamoyl 1-methyl) vinyl phosphate (39, Fig. 3.63), can be prepared by the following route.

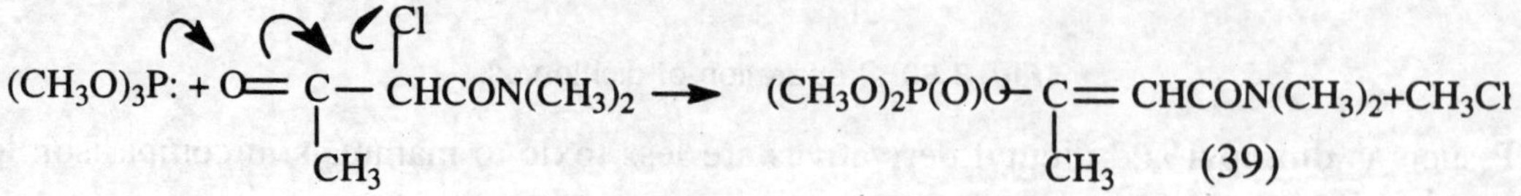

Fig. 3.63. Preparation of dicrotophos.

It is a yellow liquid having bp 90-95°C at 10^{-3} mm Hg with LD_{50} (rat) 22 mg/kg. It is moderately persistent and acts as contact and systemic insecticide against aphids, bugs, leaf hoppers and mites on cotton. It exists in both *E*- and *Z*- forms.

In biological systems the N-methyl group of dicrotophos may be oxidatively removed *via* the methylol intermediate giving the insecticide monocrotophos and finally the amide (Fig. 3.64).

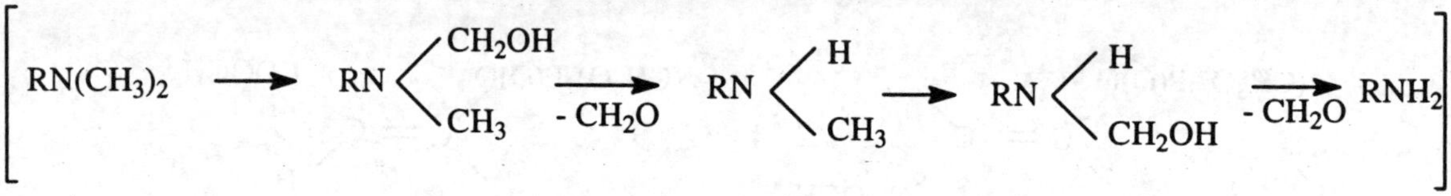

Fig. 3.64. Conversion of dicrotophos to amide.

v) **Phosphamidon**, dimecron®, O,O-dimethyl O-(1-methyl 2-chloro-2(N,N-dimethyl carbamoyl vinyl) phosphate (40, Fig. 3.65), is prepared as follows.

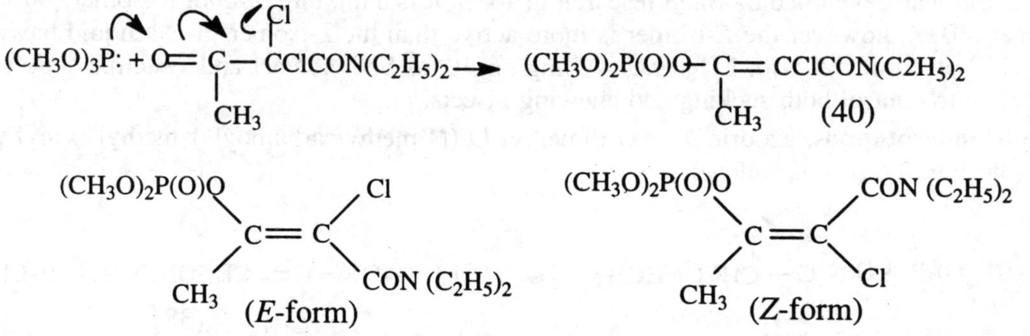

Fig. 3.65. Preparation of phosphamidon.

Phosphamidon is a mixture of both *E*- and *Z*-isomers in the proportion 73 : 27 *E*-form is more active than the *Z* form. It is a liquid having bp 150°C at 1mm Hg with LD_{50} (rat) 15 to 27 mg/kg. It is a systemic insecticide and used for the control of sucking pests on cotton.

vi) **Chlorfenvinphos**, birlane®, O,O-diethyl O-1(2,4-dichloro phenyl) 2-chloro vinyl phosphate (41, Fig. 3.66), can be prepared by the following method.

Birlane is a mixture of both *E*- and *Z*-isomers but the bulk is *Z*-form. It is an amber liquid having bp 168-170°/0.5 mm Hg. The LD_{50} (mice) is 117-200 mg/kg. Since it is stable against hydrolysis, it persists for a longer time in soil and can be used for the control of soil insects.

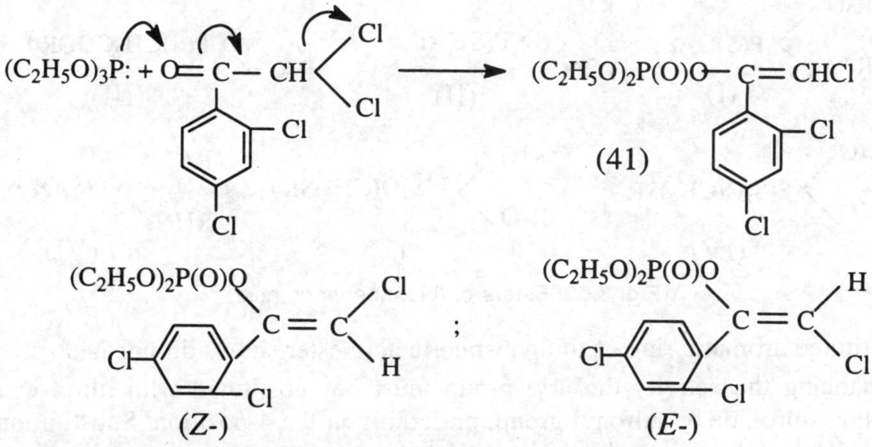

Fig. 3.66. Preparation of chlorfenvinphos.

vii) **Tetrachlorvinphos**, gardona® O,O-dimethyl O-1 (2,4,5-trichlorophenyl) 2-chlorovinyl phosphate (42, Fig. 3.67), is prepared as follows :

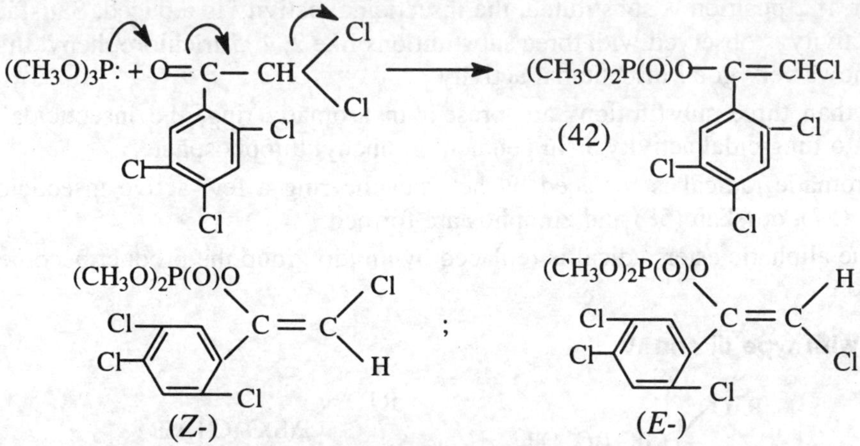

Fig. 3.67. Preparation of tetrachlorvinphos.

Tetrachlorvinphos (technical) having mp 97-98°C contains 98% E-isomer with LD_{50} (rat) 4000 mg/kg. It is effective against lepidoptera, diptera and coleoptera and is used against pests of vegetable, fruit crops and stored products.

Esters of thiophosphoric acid

When an oxygen atom is replaced by sulphur in phosphoric acid, the products show considerable decrease in mammalian toxicity without any loss in insecticidal and acaricidal activity (Fig. 3.68).

S.A.R. studies with type (I)

i) Maximum activity is observed when R^1 and R^2 are methyl, ethyl or mixed.

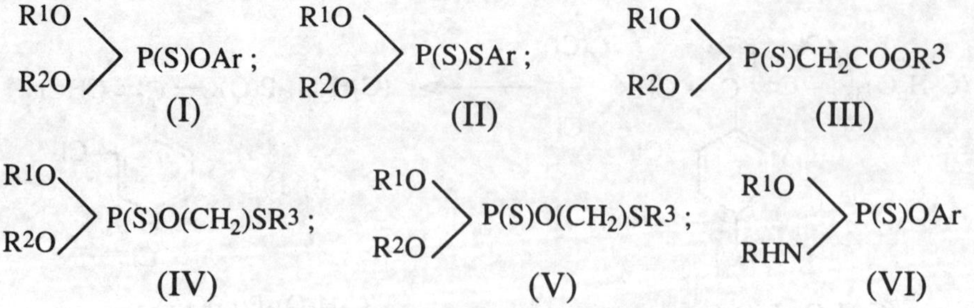

$$\begin{matrix} R^1O \\ R^2O \end{matrix} \Big> P(S)OAr \; ; \qquad \begin{matrix} R^1O \\ R^2O \end{matrix} \Big> P(S)SAr \; ; \qquad \begin{matrix} R^1O \\ R^2O \end{matrix} \Big> P(S)CH_2COOR^3$$

(I)   (II)   (III)

$$\begin{matrix} R^1O \\ R^2O \end{matrix} \Big> P(S)O(CH_2)SR^3 \; ; \qquad \begin{matrix} R^1O \\ R^2O \end{matrix} \Big> P(S)O(CH_2)SR^3 \; ; \qquad \begin{matrix} R^1O \\ RHN \end{matrix} \Big> P(S)OAr$$

(IV)   (V)   (VI)

Fig. 3.68. Esters of thiophosphoric acid.

ii) Unsubstituted aromatic ring of thiophosphoric acid ester shows diminished activity.

iii) For enhancing the activity the aryl group must be substituted with nitro, cyano, methyl mercaptan, sulfoxide or sulfonyl group, preferably in the 4-position. Substitution in 2 and 3 position shows reduced activity.

iv) If second substituent is added, the mammalian toxicity is reduced without lowering the insecticidal activity. If the addition of methyl or halogen group is in position 3 as in fenitrothion, the product shows low mammalian toxicity without losing the insecticidal activity. However, if 2 position is substituted, the insecticidal activity is reduced. Satisfactory insecticidal activity is observed with three substitutions like 2, 4, 5-trichlorophenyl thiophosphate which shows very good insecticidal activity.

v)   If more than three substitutions are present in aromatic ring, the insecticidal activity is changed to fungicidal activity as in pentachlorophenyl thiophosphate.

vi) When aromatic radical is replaced by heterocyclic ring a few active insecticides such as diazinon (57), dursban (58) and zinophos are formed.

vii) When one aliphatic ester radical is replaced by amido group the product becomes herbicide like zytron

SAR studies with type III and IV

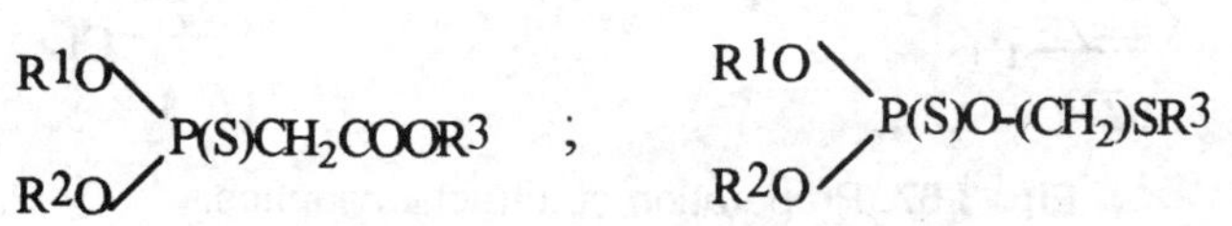

$$\begin{matrix} R^1O \\ R^2O \end{matrix} \diagup P(S)CH_2COOR^3 \; ; \qquad \begin{matrix} R^1O \\ R^2O \end{matrix} \diagup P(S)O-(CH_2)SR^3$$

i)   If the total number of carbon atoms in R^1 and R^2 is more than four, the insecticidal activity is reduced.

ii) If the number of methylene groups between the phosphorus and sulphur atoms is increased, the activity of the compound is reduced. The maximum activity is observed when the number of methylene groups is 1 or 2.

iii) If sulphur atom is oxidised to sulfoxyl or leads to the formation of sulfonium ion, the toxicity of the compound is increased.

iv) When R^3 is substituted by various substituents, sometimes active molecules are formed.

v)   All compounds of structure III and IV have systemic action. The contact effect of the thiono isomers is comparatively lower than that of thiolo isomers.

vi) When sulphur atom is replaced by nitrogen atom the compound becomes toxic to insects and mites.

vii) If the number of carbon atoms exceeds three, the insecticidal activity is decreased.

viii) Compounds containing aromatic radical as in R^3 show a lower insecticidal activity than the compounds of aliphatic series.

Chemistry of intermediate phosphorus compounds used for the preparation of phosphorothioates (Fig. 3.69)

i) $PSCl_3 + R^1OH \xrightarrow{\text{Base}} R^1OP(S)Cl_2 + HCl$

ii) $R^1OP(S)Cl_2 + R^2OH \xrightarrow{\text{Base}} \begin{matrix} R^1O \\ \diagup \\ R^2O \end{matrix} P(S)Cl + HCl$

iii) $R^1OP(S)Cl_2 + NaOR^2 \longrightarrow \begin{matrix} R^1O \\ \diagup \\ R^2O \end{matrix} P(S)Cl + NaCl$

iv) $R^1OPCl_2 + S \text{ (or } PSCl_3) \longrightarrow (R^1O)P(S)Cl_2$

v) $(R^1O)_2 PCl + S \text{ (or } PSCl_3) \longrightarrow (R^1O)_2 P(S)Cl$

vi) $P_2S_5 + 4R^1OH \longrightarrow 2(R^1O)_2 P(S)SH + H_2S$

vii) $2(R^1O)_2 P(S)SH + 3Cl_2 \longrightarrow 2(R^1O)_2 P(S)Cl + 2HCl + S_2Cl_2$

viii) $P(O)Cl_3 \xrightarrow[-HCl]{+R^1OH} R^1O\,P(O)Cl_2 \xrightarrow[-HCl]{+R^1OH} (R^1O)_2 P(O)Cl$

Fig. 3.69. Preparation of phosphorothioates.

Isomerization of phosphorothionates (Thiono-Thiolo rearrangement) [15]

The reaction shown in Fig. 3.70 is called Pichchemuka (Pistochimuka) rearrangement. The possible mechanism is as follows. A phosphorothionate anion is first formed by the dealkylation of phosphorothionate ester. This is a strong nucleophilic agent which can react again with the alkylated product. Dialkyl phosphorothioic acids are preferentially alkylated at S-atom. Thus the realkylated product is called phosphorothiolate and in the course of dealkylation, phosphorothionate esters are often isomerized to phosphorothiolates (Fig. 3.71).

$$(R^1O)_2P(S)\,OR^2 \rightleftharpoons (R^1O)_2P(O)\,SR^2$$

Fig. 3.70. Thiono-thiolo rearrangement.

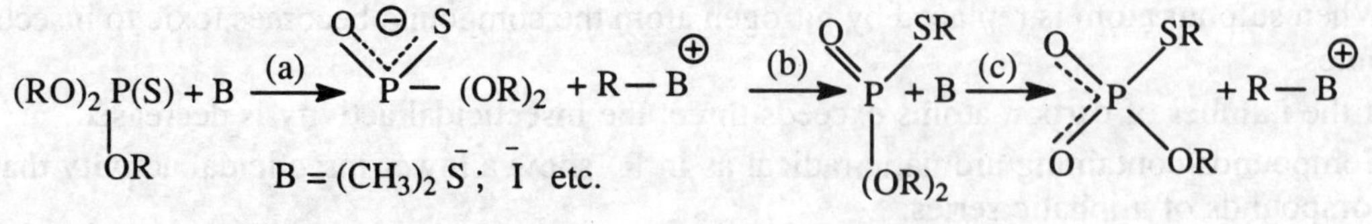

Fig. 3.71. Isomerisation of phosphorothionate to phosphorothiolate.

Oxidative desulfuration of phosphorothioates

Phosphorothioates are oxidised easily to phosphate esters. The oxidative desulfuration reaction is also conveniently applied to the enzymatic analysis of organophosphorus pesticides.

For preparative purpose, HNO_3 is often used as an oxidant. Thus coroxon (43, Fig. 3.72a) can be prepared in 87% yield by the action of HNO_3 (d= 1.5) from its thiono analogue, coumaphos. Similarly a phosphonothionate such as EPN is also converted to a corresponding phosphonate oxon (44, Fig. 3.72b).

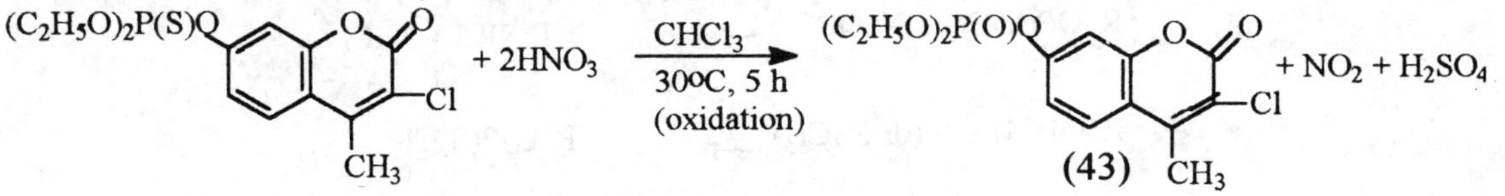

Fig. 3.72a. Oxidation of coumaphos to coroxon.

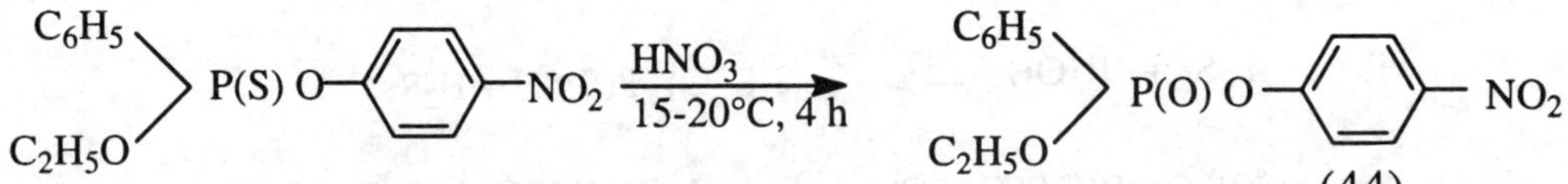

Fig. 3.72b. Conversion of EPN to oxon analogue.

Dinitrogen tetroxide is also used as an oxidant. Parathion is oxidised to paraoxon (90%) at room temperature using N_2O_4 (45) and this can be extended to malathion (46, Fig. 3.73) and other dithioates.

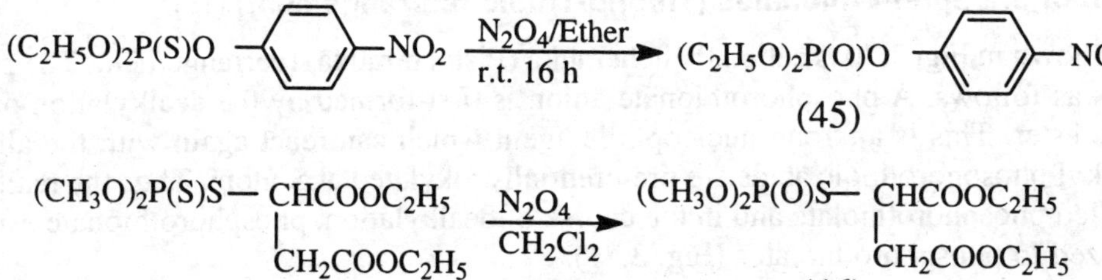

Fig. 3.73. Conversion of thioates to oxon analogue.

Chemistry of some important phosphorothionates (phosphorothioates)

Parathion, folidol® and thiophos®, O,O-diethyl 4-nitrophenyl phosphorothioate (47, Fig.

3.74), was discovered by Schrader in 1944. Ethyl and methyl parathion have been widely used against number of insects. Currently the use of parathion is banned in many countries because of **high** mammalian toxicity. It is prepared from

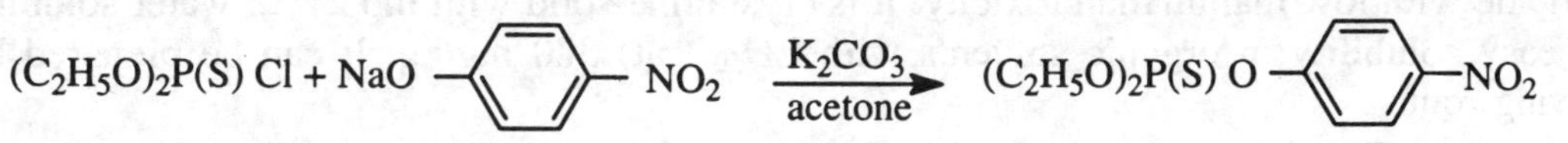

$(C_2H_5O)_2P(S)$ Cl + NaO —⬡— NO_2 $\xrightarrow[\text{acetone}]{K_2CO_3}$ $(C_2H_5O)_2P(S)$ O —⬡—NO_2

(47)

Fig. 3.74. Preparation of parathion.

Pure parathion is a yellow liquid having bp 113°C/0.05 mm Hg, v.p. at 20°, 0.57×10^{-5} mm Hg, d^{25} 1.265; η_D^{25} 1.5368. It is highly soluble in most organic solvents and hydrolysed in alkali with solubility in water of about 24 ppm. Thermal isomerisation gives the thioloderivative (48).

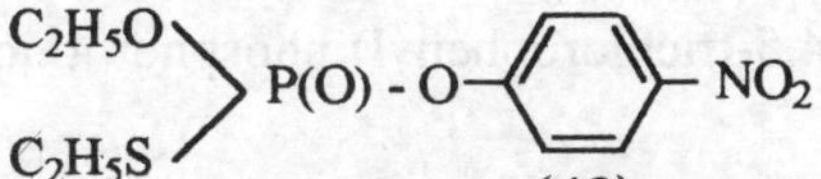

$\begin{matrix} C_2H_5O \\ C_2H_5S \end{matrix}$> P(O) - O —⬡— NO_2

(48)

This thiolo isomer is more reactive and potent as anticholinesterase agent but it is less effective as insecticide than the thionoanalogue. Half-life of parathion when sprayed on grape vines as emulsion concentrate at 2 lb/acre, was 2.2 days. However, the residue was stabilized in the bark at the level of 25 ppm, being maintained for at least 70 days. The LD_{50} (rat) is 7 mg/kg.

Methyl parathion is prepared in the same way as parathion. It hydrolyses 4.3 times faster in alkali than parathion and has LD_{50} (rat) 25-50 mg/kg. It is a good methylating agent and it is assumed that the high methylating ability of methyl parathion is the cause of its lower mammalian toxicity without reducing insect toxicity. The *iso*propyl analogue possesses less mammalian toxicity than methyl/ethyl parathion.

Fenitrothion, sumithion®, O,O-dimethyl 3-methyl 4-nitrophenyl phosphorothioate (43, Fig. 3.75), was commercialized by Sumitomo Chemical Company Ltd., Japan as an experimental insecticide. It has high insecticidal activity with low mammalian toxicity and is prepared as follows.

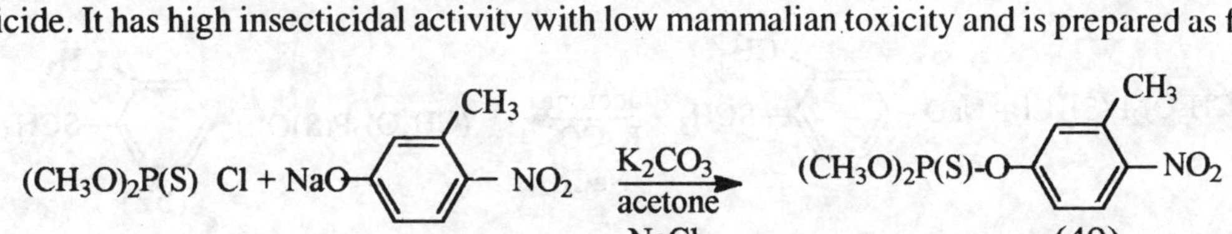

$(CH_3O)_2P(S)$ Cl + NaO—⬡—NO_2 (with CH_3) $\xrightarrow[\substack{\text{acetone} \\ \text{-NaCl}}]{K_2CO_3}$ $(CH_3O)_2P(S)$-O—⬡—NO_2 (with CH_3)

(49)

Fig. 3.75. Preparation of fenitrothion.

It is a liquid having bp 95°C/0.01 mm Hg, d^{25} 1.5528 with LD_{50} (rat) 142-1000 mg/kg. It is soluble in alcohols, ethers and aromatic hydrocarbons. It is more stable than methyl parathion while the half life in 0.01 N NaOH at 30°C is 272 min. It is isomerized partially into S-methyl isomer during distillation under reduced pressure.

For agricultural use, a mixture of both O,O-dimethyl 3-methyl 4-nitrophenyl-phosphorothioate with O,O-dimethyl 3-methyl 6-nitrophenylphosphorothioate has been recommended. The 6-nitro isomer is about 200 times less effective than the 4-nitro isomer as insecticide. The rate of demethylation in mammals is however, greater with fenitrothion than methyl parathion.

Chlorthion, O,O-dimethyl 3-chloro 4-nitrophenyl phosphorothioate (50, Fig. 3.76), is a contact insecticide with low mammalian toxicity. It is crystalline solid with mp 21°C, water solubility 40 ppm, easy solubility in organic solvents with LD_{50} (rat) 380 mg/kg. It can be prepared by the following route :

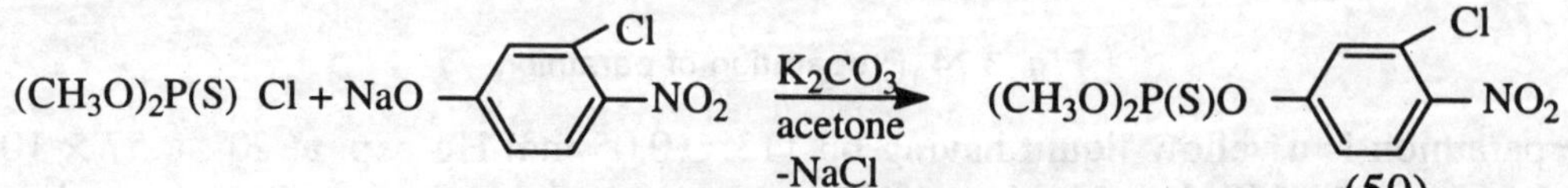

$$(CH_3O)_2P(S)\ Cl + NaO \underset{\text{-NaCl}}{\overset{K_2CO_3}{\underset{\text{acetone}}{\longrightarrow}}} (CH_3O)_2P(S)O$$

Cl

NO$_2$ (50)

Fig. 3.76. Preparation of chlorthion.

The detoxification is due to ester hydrolysis and methyl ether bond cleavage in mammals.

Ronnel, O,O-dimethyl O-(2,4,5-trichlorophenyl) phosphorothioate (51, Fig. 3.77), is prepared as follows :

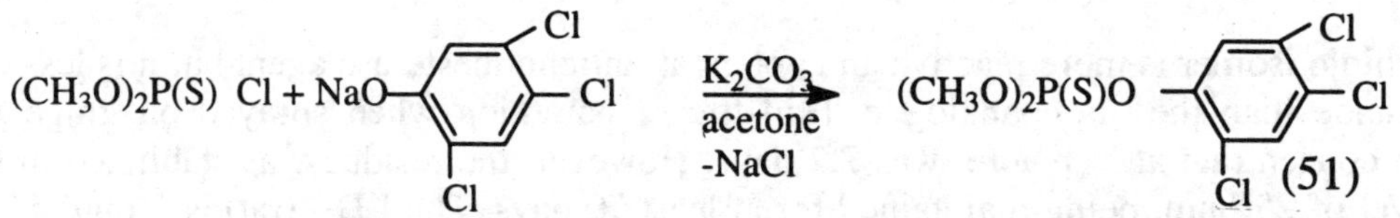

$$(CH_3O)_2P(S)\ Cl + NaO \underset{\text{-NaCl}}{\overset{K_2CO_3}{\underset{\text{acetone}}{\longrightarrow}}} (CH_3O)_2P(S)O$$

Cl

Cl (51)

Fig. 3.77. Preparation of ronnel.

Ronnel is crystalline solid having mp 41°C. In weakly alkaline medium it is hydrolysed with the formation of O-methyl O-2,4,5-trichlorophenyl thiophosphoric acid but in strong alkaline medium the main product is O,O-dimethyl thiophosphoric acid.

Fenthion, baytex® O,O-dimethyl 3-methyl 4-methylthiophenyl phosphorothioate (52, Fig. 3.78) is prepared as follows :

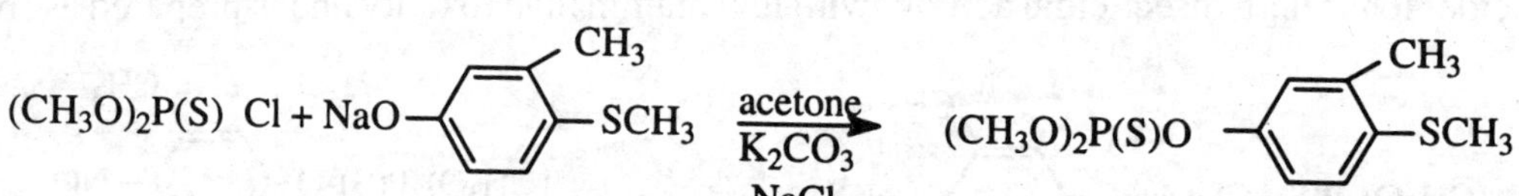

$$(CH_3O)_2P(S)\ Cl + NaO \underset{\text{-NaCl}}{\overset{\text{acetone}}{\underset{K_2CO_3}{\longrightarrow}}} (CH_3O)_2P(S)O$$

CH$_3$

SCH$_3$ (52)

Fig. 3.78. Preparation of fenthion.

Fenthion is a liquid having bp 87/0.01 mm Hg, water solubility 54 mg/litre with LD_{50} 215-245 mg/kg. It is soluble in most organic solvents and gets oxidised to sulfoxide, sulfone and oxon derivatives in plants and animals (53, Fig. 3.79).

It is an insecticide with systemic action and is effective against flies and mosquitoes. The O-ethyl analogue of fenthion is an effective systemic insecticide while its sulfoxide derivative acts as nematicide and soil insecticide.

Quinalphos, ekalux, O,O-dimethyl O-quinoxalin-2yl-phosphorothioate (54, Fig. 3.80), devel-

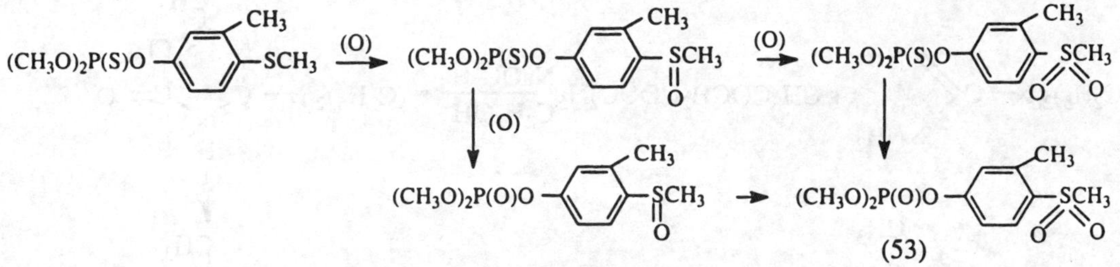

Fig. 3.79. Formation of sulfoxide and sulfone.

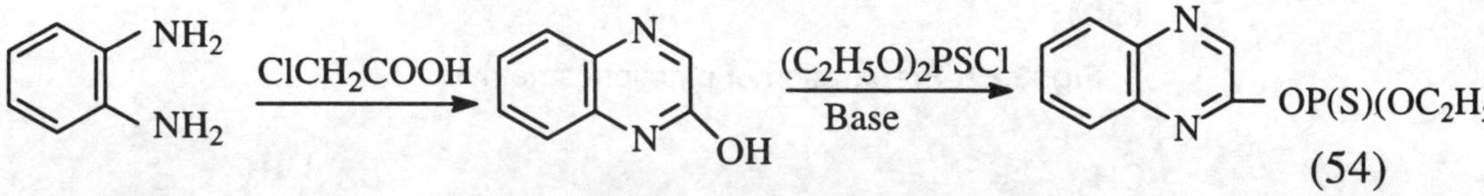

Fig. 3.80. Preparation of quinalphos.

oped by Bayer A.G. (1969) is prepared by the reaction between *o*-phenylenediamine and chloro-acetic acid followed by condensation of the product with O,O-diethyl phosphorochlorothioate.

It is a broad spectrum contact and systemic insecticide having LD_{50} (rat) 70 mg/kg and is used to control insects in cereals, vegetables and many other crops. It is quite safe as it decomposes in plants within a few days of application.

Triazophos, hostathion®, O,O-diethyl O(1-phenyl-1-H-1,2,4-triazol-3yl) phosphorothioate (**55**, Fig. 3.81) was introduced by Hoechst A.G. (1970) and is derived from 1,2,4-triazole.

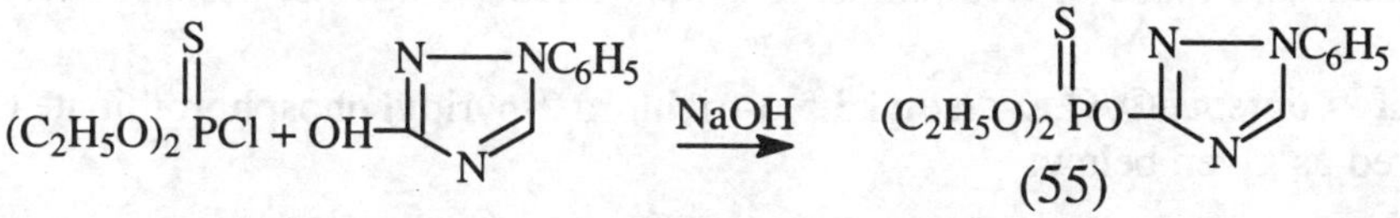

Fig. 3.81. Preparation of trizophos.

Triazophos is a broad spectrum contact insecticide and acaricide and is used to control number of insects in cereals and vegetables.

Pirimiphosmethyl, O,O-dimethyl 2-dimethylamino 4-methyl pyrimidin-6-yl phosphorothioate (**56**, Fig. 3.82) introduced by ICI (1970) is derived from diethylguanidine, ethylacetoacetate and dimethyl phosphorochloridothioate.

It is a broad spectrum insecticide effective against number of pests in stored products and public health. The LD_{50} (rat) is 2000 mg/kg. It is very effective against insects resistant to organochlorine insecticides. Pirimiphosethyl, prepared in the same way having LD_{50} (rat) 140 mg/kg, is a broad spectrum insecticide, effective against diptera and coleoptera. It also possesses fungicidal activity.

Diazinon, O,O-diethyl-2 *iso*propyl 6-methyl 4-pyrimidinyl phosphorothioate (**57**, Fig. 3.83), introduced by the Geigy company in 1952, is prepared as follows.

It is a colourless oily liquid having bp 89/0.1 mm Hg with LD_{50} (rat) 108-250 mg/kg and η_D^{20}

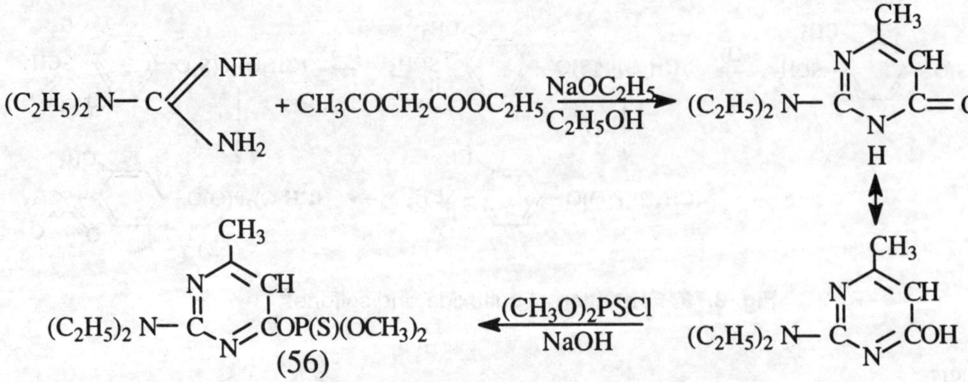

Fig. 3.82. Preparation of pirimiphosmethyl.

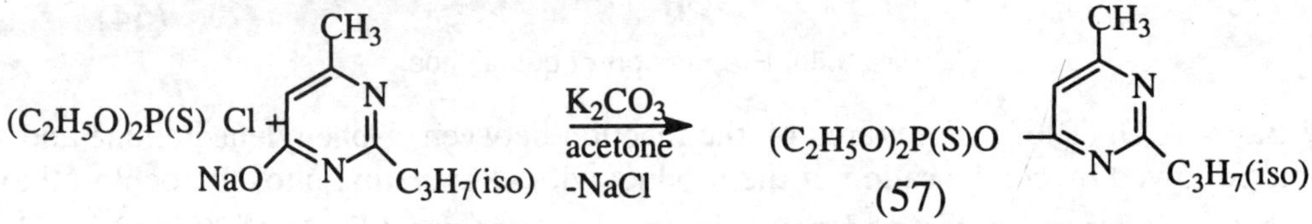

Fig. 3.83. Preparation of diazinon.

1.4978 to 1.4981. It is readily soluble in most organic solvents but sparingly soluble (40 ppm) in water. It is a contact insecticide used for soil application and foliar spray.

Diazinon shows a relatively long residual action and is quite effective for soil, fruit and vegetable crops and insect pests of rice. When applied to paddy field it is absorbed and translocated in the leaf sheath and blade of rice plants. It is also effective against pests in the household and live stock.

Chlorpyrifos, dursban®, O,O-diethyl 3,5,6-trichloro 2-pyridyl phosphorothioate (58, Fig. 3.84) can be prepared as given below.

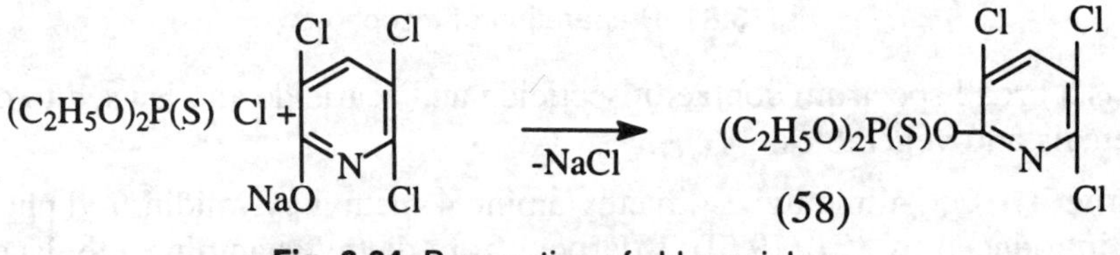

Fig. 3.84. Preparation of chlorpyriphos.

Chlorpyrifos is solid having mp 42-43°C with LD_{50} (rat) 163 mg/kg. It is readily soluble in organic solvents but insoluble in water. It is stable except under strong acid and alkaline conditions. It is moderately persistent and retains activity in soil for quite sometime. Now it is recommended to replace aldrin for the control of white ants. It also effectively controls mosquito and fly larvae, sucking and chewing insects and soil borne plant pests. The methyl homologue is much less toxic to mammals with LD_{50} (rat) 1500 mg/kg and more effectively controls adult mosquitoes than the larvae.

Coumaphos, O,O-diethyl O-(3-chloro-4-methylcoumarinyl-7) phosphorothioate (59, Fig. 3.85) is prepared as follows :

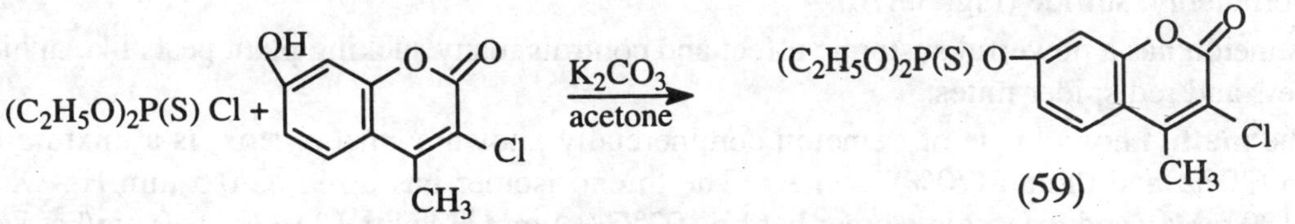

Fig. 3.85. Preparation of coumaphos.

Coumaphos is solid having mp 95°C with LD_{50} (rat) 55-200 mg/kg. It is a good insecticide for the control of ectoparasites of domestic animals, as it has low mammalian toxicity. Its chemical reaction with dilute and concentrated alkali follows different routes (Fig. 3.86). The pyrone ring is opened up by the action of dilute KOH, while on prolonged action of dilute alkali, cleavage of an ethoxy group occurs. But on heating with concentrated alkali it is completely decomposed.

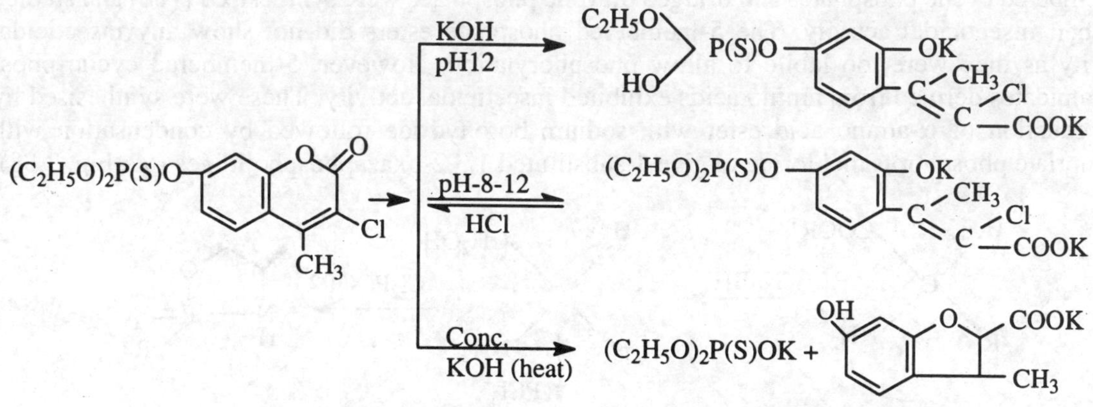

Fig. 3.86. Reaction products of coumaphos at different pH.

Demeton, systox®, O,O-diethyl 2-ethyl thioethyl phosphorothioate was first synthesised by Schrader in 1950. The technical grade is a mixture of both thiono (70%) (60, Fig. 3.87a) and thiolo (30%) isomers (61, Fig. 3.87b). This thiono isomer demeton-O is an oily liquid having bp 106°/0.4 mm with LD_{50} (rat) 30 mg/kg. The thiolo isomer demeton-S is also an oily liquid having bp 100°/0.25 mm Hg with LD_{50} (rat) 1.5 mg/kg. The compound demeton-O (60) is prepared as follows.

$$(C_2H_5O)_2P(S) Cl + HO(CH_2)_2SC_2H_5 \xrightarrow{NaOH} (C_2H_5O)_2P(S)\text{-}O(CH_2)_2SC_2H_5$$
$$(60)$$

Fig. 3.87a. Preparation of demeton-O.

$$(C_2H_5O)_2P(O) SK + Cl (CH_2)_2 SC_2H_5 \longrightarrow (C_2H_5O)_2P(O)S\text{—} (CH_2)_2SC_2H_5$$
$$(61)$$

Fig. 3.87b. Preparation of demeton-S.

The compound demeton-O is isomerized partially to thiolo derivative during its preparation. The pure thiolo isomer of demeton (61) is obtained from salts of diethyl thiophosphoric acid and 2-chlorodiethyl sulfide (Fig. 3.87b).

Demeton has a powerful systemic effect and controls many sucking plant pests like aphids, leaf hoppers and red spider mites.

The **methyl** homologue of demeton commercially known as metasystox, is a mixture of both thiolo (30%) and thiono (70%) isomers. The thiono isomer has bp 93°C/0.5 mm Hg with LD_{50} (rat) 180 mg/kg and the thiolo isomer has bp 102°C/0.4 mm Hg with LD_{50} (rat) 40 mg/kg. Demeton is more stable than methyl demeton.

Cyclic phosphoryl compounds

Eto [16a] synthesized several cyclic phosphoryl compounds utilizing the neurotoxic effect of tri-O-cresyl phosphates (TOCP).

Three types of cyclic phosphoryl compounds such as 5-membered cyclic phosphoramidates, 6-membered cyclic phosphates and bridged bicyclic phosphates were synthesized [16b] and studied for their insecticidal activity. The 5-membered phosphate esters did not show any insecticidal activity as they were too labile to allow phosphorylation. However, 5-membered cyclic phosphoramidates derived from amino acids exhibited insecticidal activity. These were synthesised by the reduction of α-amino acid ester with sodium borohydride followed by condensation with appropriate phosphorus dichloride to give 4-substituted 1,3,2-oxazaphospholidines (62, Fig. 3.88).

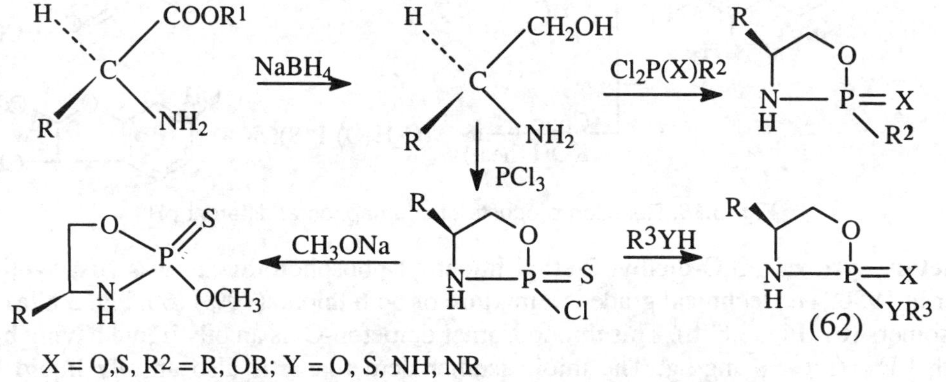

X = O,S, R2 = R, OR; Y = O, S, NH, NR

Fig. 3.88. Preparation of 1,3,2-oxazaphospholidines.

The sulphur analogues such as 2-ethoxy-1,3,2-oxazaphospholidine 2-sulfides derived from various amino acids exhibit good insecticidal activity. SAR study shows that the presence of hydrophobic alkyl group at 4-position is essential for insecticidal activity. The most active insecticide in this series is (4S)-4-isobutyl-2-methoxy-1,3,2-oxazaphospholidine 2-sulfide derived from L-leucine. The conversion of TOCP into cyclic phosphate involves a number of steps (cf. metabolism).

SAR study shows that biocidal activity of saligenin cyclic phosphate is influenced by the exocyclic substituent on P-atom. The aryl esters derived from TOCP metabolite is neurotoxic with no insecticidal activity. The presence of alkyl group instead of aryl group on phosphorus atom imparts high insecticidal activity which makes methyl saligenin cyclic phosphorothionate or 2-me-thoxy-4H-1,3,2-benzodioxa phosphorin 2-sulfide, commercially known as salithion, as the most promising insecticide. It is prepared by reaction of saligenin and methyl phosphoro-dichloridothionate in presence of aqueous NaOH solution (63, 64, Fig. 3.89).

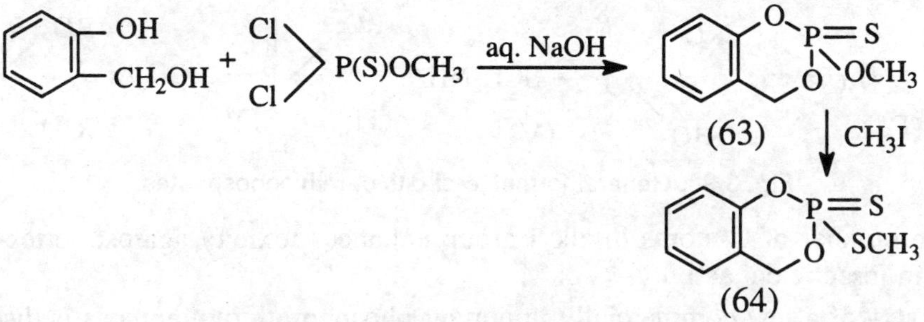

Fig. 3.89. Preparation of salithion and MTBO.

SAR studies indicate that the insecticidal activity of salithion decreases with the introduction of electron withdrawing or releasing group. The S-methyl thiolate isomer of salithion, MTBO, 2-methylthio-4H-1,3,2-benzodioxaphosphorin 2-oxide (64), showed high insecticidal activity. It also exhibited fungicidal activity and controlled blast disease of rice caused by *P. oryzae*.

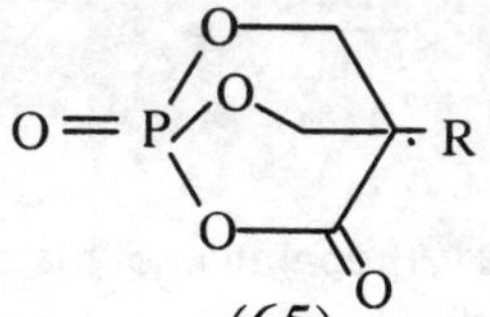

Bridged bi-cyclic phosphates [(2,6,7-trioxa-1-phosphabicyclic [2,2,2] octanes] (65), were shown to possess high mammalian toxicity, [17]. The bulky branched alkyl groups such as *t*-butyl and *iso*-propyl group impart high toxicity. These compounds show lack of anticholinesterase activity probably due to size and shape of the molecule. QSAR study showed that toxicity was correlated with the hydrophobicity and steric factors of the bridgehead substituents and not with the electronic factor. Electrophysiological experiment using the longitudinal muscle of earthworm indicated that they acted as antagonist of γ-aminobutyric acid (GABA) to inhibit neurotransmission.

v) Esters of dithiophosphoric acid, trithiophosphoric acid and pyrophosphoric acid [18]

The general formulae of the pesticides derived from dithio and trithio-phosphoric acid are shown in Fig. 3.90.

In general the toxicity of phosphorothiolothionates/trithiolothionates towards mammals is less than that of phosphorothioates but chemical stability and duration of their action under field conditions are more. Many derivatives of dithiophosphoric acid especially those containing het-erocyclic radicals have high activity against sucking and chewing pests. SAR studies indicate that mixed esters in which R, R^1 & R^2 are methyl groups, have least toxicity towards vertebrates. An

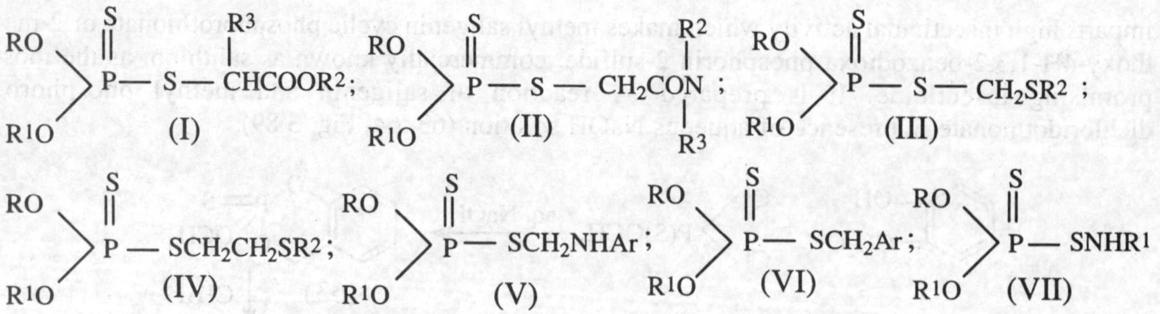

Fig. 3.90. General formulae of dithio, trithiophosphates.

increase in the number of C-atoms in alkyl group enhances toxicity against vertebrates without much change in insecticidal activity.

The **chemistry** of a few commercially important phosphorothiolothionates is discussed in the following text.

Malathion, cythion® S[1,2-di(ethoxy-carbonyl) ethyl] O,O-dimethyl phosphorothiolothionate or O,O-dimethyl S(1,2-dicarbethoxy) ethyl phosphorodithioate (66, Fig. 3.91), was first introduced by American Cyanamid company and has been acknowledged as the first organophosphate insecticide with high selective toxicity. It is prepared as follows.

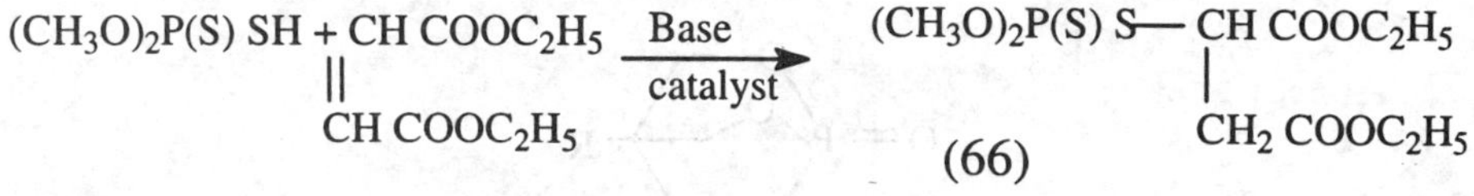

$$(CH_3O)_2P(S)\,SH + \underset{\underset{CH\,COOC_2H_5}{\|}}{CH\,COOC_2H_5} \xrightarrow[\text{catalyst}]{\text{Base}} \underset{\underset{CH_2\,COOC_2H_5}{|}}{(CH_3O)_2P(S)\,S\!\!-\!\!CH\,COOC_2H_5}$$

(66)

Fig. 3.91. Preparation of malathion.

It is a yellow liquid having bp 120°C/2 mm Hg with LD_{50} (rat) 1375 mg/kg. Its solubility in water is 45 ppm, but readily dissolves in most organic solvents. It is hydrolysed in aqueous solution above pH 7.0 and below pH 5.0. Heavy metals particularly iron catalyze the decomposition.

Chemical reaction. Heating converts malathion to *iso*malathion (67, Fig. 3.92) a highly toxic compound.

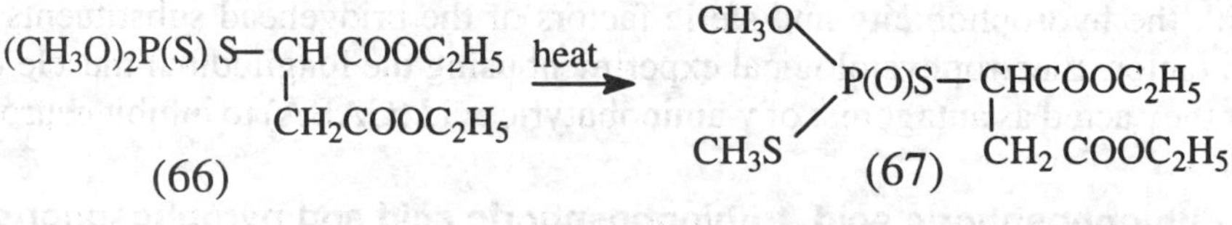

$$\underset{\underset{CH_2COOC_2H_5}{|}}{(CH_3O)_2P(S)\,S\!\!-\!\!CH\,COOC_2H_5} \xrightarrow{\text{heat}} \underset{CH_3S}{\overset{CH_3O}{>}}P(O)S\!\!-\!\!\underset{\underset{CH_2\,COOC_2H_5}{|}}{CHCOOC_2H_5}$$

(66)   (67)

Fig. 3.92. Conversion to *iso*malathion.

Reaction with acid and alkali yield mercaptosuccinic ester and ethylfumarate respectively (Fig. 3.93).

It is considered as one of the safest insecticide suitable for the control of sucking and chewing pests of fruits and vegetables and mosquitoes and flies. Owing to low mammalian toxicity it is used on large scale by WHO for eradication of *Anopheles*. Detoxification of malathion is due to carboxyesterase activity.

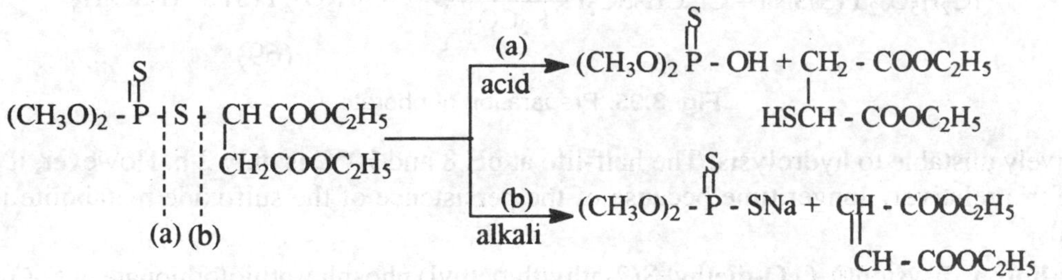

Fig. 3.93. Reaction of malathion with acid and alkali.

Based on malathion, a series of new organophosphorodithioates have been developed [19] having high insecticidal activity against susceptible strain of *Musca domestica* with LD_{50} (rat) 0.4-47 μg/insect. These are also found effective against malathion resistant G strain of *Musca domestica*. These dithioates are prepared by condensation of phosphorodithioic acid with maleamic ester which in turn can be prepared by reaction of aliphatic amines with maleic anhydride.

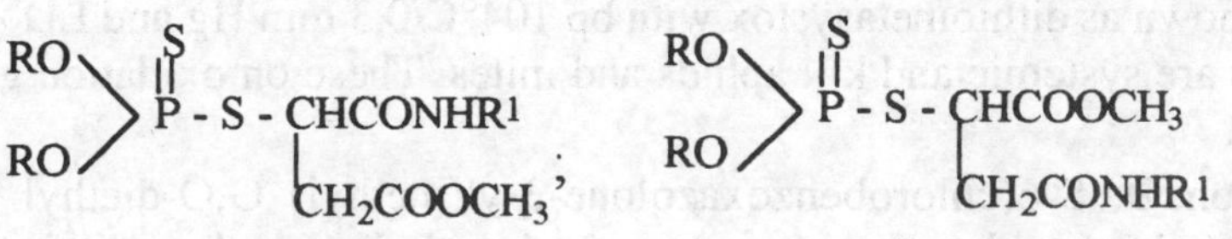

Phenthoate, cidial®, O,O-dimethyl S(1-carboethoxybenzyl) phosphorodithioate or O,O-dimethyl S (α-ethoxycarbonylbenzyl) phosphorothiolothionate (68, Fig. 3.94), is prepared as follows :

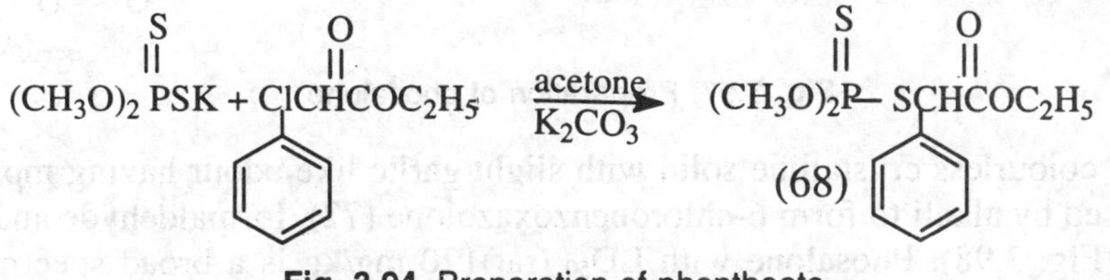

Fig. 3.94. Preparation of phenthoate.

Phenthoate is liquid having bp $70°C/2 \times 10^{-5}$ mm Hg with LD_{50} (rat) 4700 mg/kg. It is soluble in most organic solvents. It is relatively susceptible to heat, with 50% decomposition at 120°C for 110 h. Phenthoate has a broad spectrum of insecticidal and acaricidal activities and is effective against codling moth and scale insects. It controls the insect pests on vegetables, citrus, tea and rice. In several homologues prepared and tested for pesticidal activity it was found that mammalian toxicity was remarkably influenced by the presence of alkyl group of the carboxyester. Out of these, the fluoroethyl ester appeared to be most toxic.

Phorate, thimet®, O,O-diethyl S-(ethyl thiomethyl) phosphorothiolothionate or O,O-diethyl S-(ethylthiomethyl) phosphorodithioate (69, Fig. 3.95), can be prepared as shown below.

Phorate is liquid having bp 100°C/0.4 mm Hg with LD_{50} (rat) 1.1-2.3 mg/kg. It is effective against sucking plant pests and is systemic in nature with good contact and vapour action. Phorate

$$(C_2H_5O)_2 \; P(S)SNa + Cl \; CH_2SC_2H_5 \xrightarrow[K_2CO_3]{acetone} (C_2H_5O)_2 \; P(S) \; SCH_2SC_2H_5$$

$$(69)$$

Fig. 3.95. Preparation of phorate.

is relatively unstable to hydrolysis. The half-life at pH 8 and 70°C is only 2 h. However, it protects plants for a relatively longer time because of the persistence of the sulfoxide metabolite in plants and soils.

Disulfoton, disyston®, O,O-diethyl S(2-ethylthioethyl) phosphorothiolothionate or O,O-diethyl S-(2-ethyl thioethyl) phosphorodithioate (70, Fig. 3.96), is prepared as given below.

$$(C_2H_5O)_2 \; P(S)SK + ClCH_2CH_2SC_2H_5 \xrightarrow[K_2CO_3]{acetone} (C_2H_5O)_2 \; P(S) \; SCH_2CH_2SC_2H_5$$

$$(70)$$

Fig. 3.96. Preparation of disulfoton.

Disulfoton is liquid having bp 113°C/0.4 mm with LD_{50} (rat) 12.5 mg/kg. Thiometon the methyl derivative is also known as dithiometasystox with bp 104°C/0.3 mm Hg and LD_{50} (rat) 70-120 mg. Both the analogues are systemic and kill aphids and mites. These on oxidation give sulfoxide and sulfone derivatives.

Phosalone, zolone®, S-6(chlorobenzoxazolone-3-yl(methyl) O,O-diethyl phosphorothiolothionate or O,O-diethyl S-6 (chlorobenzoxazolone-3-yl methyl) phosphorodithioate (71, Fig. 3.97), can be prepared as follows.

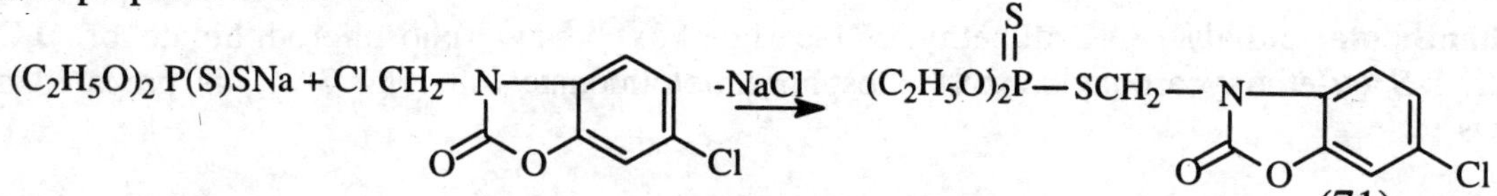

Fig. 3.97. Preparation of phosalone.

Phosalone is colourless crystalline solid with slight garlic like odour having mp 47-48°C. It is readily hydrolysed by alkali to form 6-chlorobenzoxazolone (72), formaldehyde and diethyl phosphorodithioate (Fig. 3.98). Phosalone with LD_{50} (rat)120 mg/kg is a broad spectrum insecticide and acaricide, effective against caterpillars, aphids and mites.

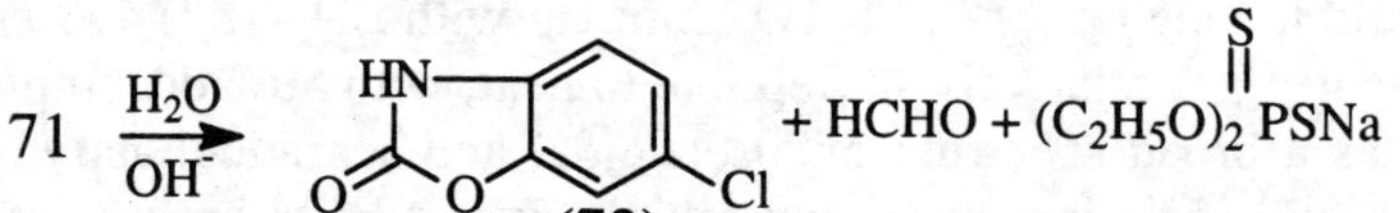

Fig. 3.98. Reaction of phosalone with alkali.

Azinphosmethyl, guthion®, O,O-dimethyl S-(4-oxo-1,2,3-benzotriazin-3(4H)-yl methyl) phosphorodithioate or O,O-dimethyl S-(4-oxobenzotriazino-3methyl) phosphorothiolothionate (73, Fig. 3.99), is prepared as follows.

Azinphosmethyl is white crystalline substance with mp 73-74°C. Its hydrolysis takes place under alkali and acid conditions. Under natural conditions however both homologues of azinphos

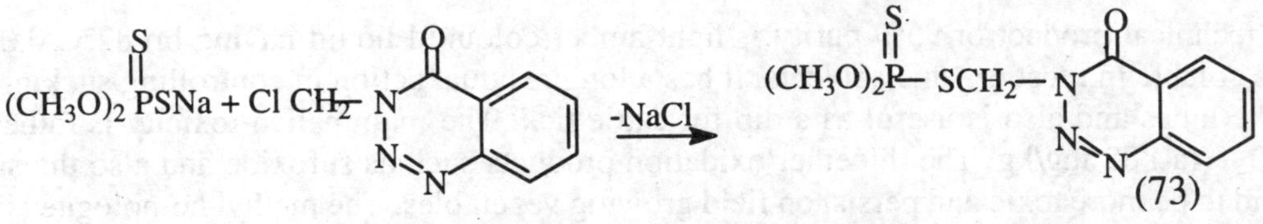

Fig. 3.99. Preparation of azinphosmethyl.

have a long residual activity. Azinphosethyl forms colourless needles having mp 53°C. Azinphosmethyl/ethyl are non-systemic insecticides and acaricides for use on field crops, fruits and cotton. The LD$_{50}$ (rat) for methyl and ethyl derivatives are 15 mg/kg and 17.5 mg/kg respectively.

Dimethoate, rogar®, O,O-dimethyl S-(N-methylcarbamoylmethyl) phosphorothiolothionate or O,O-dimethyl S-(N-methylcarbamoylmethyl) phosphorodithioate (74a, Fig. 3.100), can be prepared as follows.

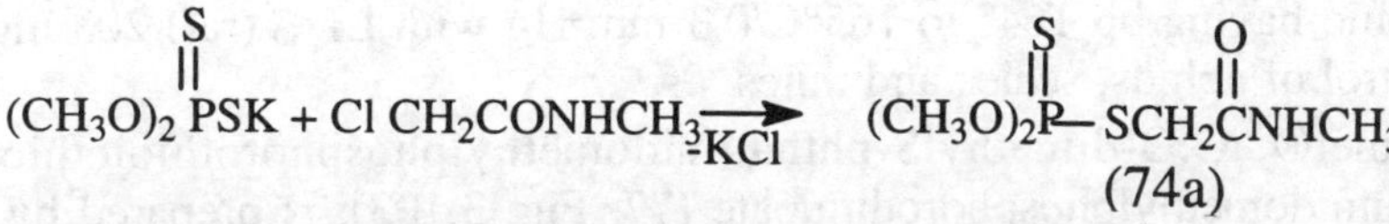

Fig. 3.100. Preparation of dimethoate.

Dimethoate is a colourless crystalline compound having mp 51° to 52°C with LD$_{50}$ (rat) 600 mg/kg. On heating it gives *iso*dimethoate which is more toxic. The technical product of dimethoate may contain several by-products like O,O,S-trimethyl phosphorothiolothionate and dimethoate acid methyl ester with the oxygen analogue of dimethoate also present in traces. These impurities increase the toxicity of dimethoate. Dimethoate is relatively unstable and decomposes on storage. The main degradation mechanism is the alkylation reaction resulting in the synthesis of dimethyl phosphorodithioic acid. In animals hydrolysis of dimethoate by amidase is important while the cleavage of P-O, P-S and S-C linkages occurs to a considerable degree. It is used for the control of sucking insects and mites on ornamentals, vegetables, cotton and fruits.

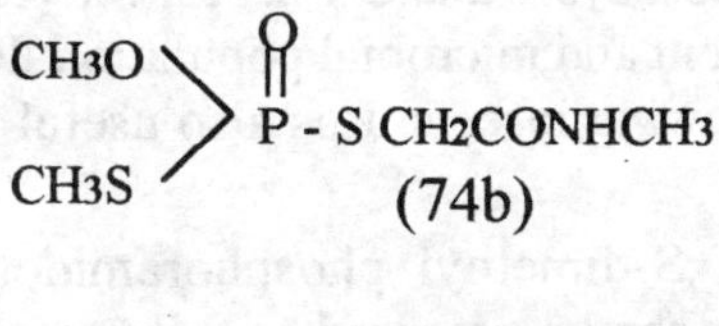

(74b)

Carbophinothion, trithion®, S-(p-chlorophenylthiomethyl) diethyl phosphorothiolothionate or S-(p-chlorophenyl thiomethyl) diethyl phosphorodithioate (75, Fig. 3.101), is prepared as shown below.

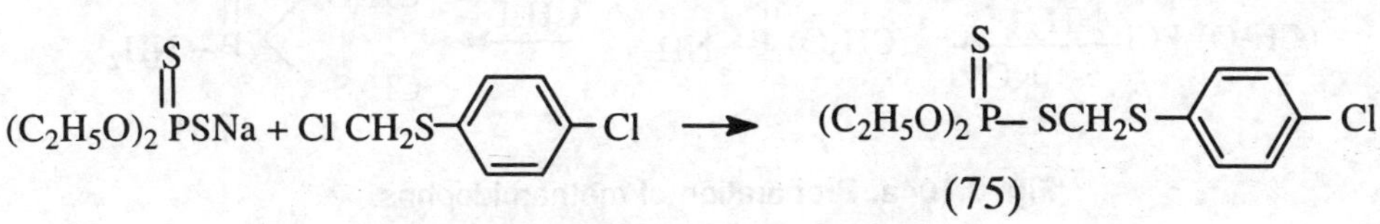

Fig. 3.101. Preparation of carbophinothion.

The technical product of 95% purity is light amber-coloured liquid having bp 82°C/0.01 mm Hg. It is soluble in most organic solvents. It has a long residual action in controlling sucking plant pests like mites and also is useful as a dip for cattle tick. The mammalian toxicity is rather high with LD_{50} (rat) 32 mg/kg. The thioether oxidation products such as sufoxide and also the sulfone are found to be more toxic and persist on field growing vegetables. The methyl homologue is found to be much less toxic to mammals than the ethyl homologue.

Ethion, O,O,O,O-tetraethyl S,S-methylenebisphosphorothiolothionate or O,O,O,O-tetra-ethylS,S-methylenebisphosphorodithioate (76, Fig. 3.102), can be prepared as follows.

$$(C_2H_5O)_2P(S)SH + BrCH_2Br \xrightarrow{NaOH} (C_2H_5O)_2P(S)SCH_2SP(S)(OC_2H_5)_2$$

(76)

Fig. 3.102. Preparation of ethion.

It is yellow liquid having bp 164° to 165°C/0.3 mm Hg with LD_{50} (rat) 208 mg/kg. Ethion is useful for the control of aphids, scales and mites.

Imidan, phosmet®, O,O-dimethylS-phthalimidomethylphosphorothiolothionate or O,O-dimethyl S-phthalimidomethylphosphorodithioate (77, Fig. 3.103), is prepared by the following route.

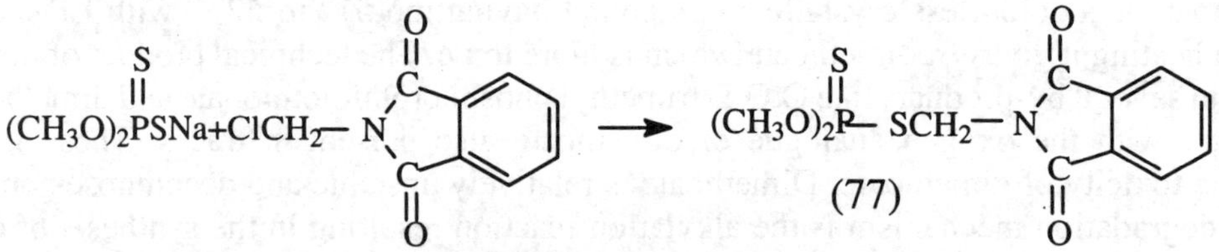

(77)

Fig. 3.103. Preparation of imidan.

Imidan is white crystalline solid with offensive odour having mp 72°C. The half-life in aqueous solutions at pH 4.5, 7.0 and 8.3 is 13 days, 12 and 4 h. respectively. It persists for 3 to 19 days in soil depending on the moisture content and microbial population. It is a broad spectrum insecticide, effective to control sucking and chewing insects. It is also useful to control cattle grub with LD_{50} (rat) 230 mg/kg.

Methamidophos, monitor®, O,S-dimethyl phosphoramidothiolate or O,S-dimethyl phos-phoramidothioate (78, Fig. 3.104a), can be prepared as follows.

(78)

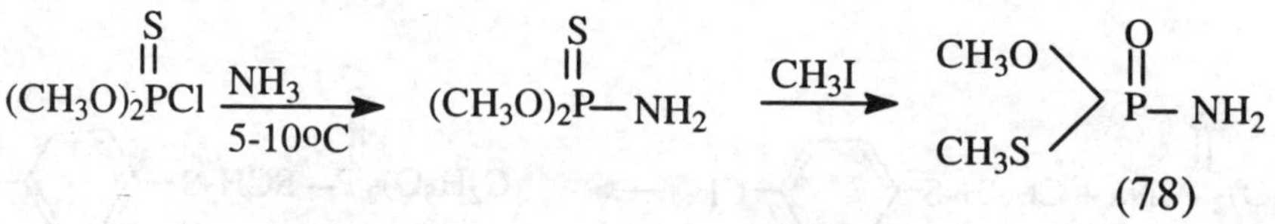

Fig. 3.104a. Preparation of methamidophos.

Methamidophos is colourless solid having mp 44.5°C, highly soluble in water but slightly soluble in ether. It is a broad spectrum insecticide particularly active against caterpillars and aphids and also has acaricidal activity. It has contact and systemic action against both biting and sucking pests. The mammalian toxicity is very high with LD_{50} (rat) 30 mg/kg. The higher alkyl (ethyl, n-propyl ester) homologues show high insecticidal activity. O-Ethyl S-methylphosphoromidothio-ate is more effective than the methyl homologue against some insects like tobacco bud worms. N-alkylation of methamidophos imparts insecticidal property whereas N-acetylation does not cause any change in the insecticidal activity except reducing the mammalian toxicity. The thionothioate analogue of methamidophos is also much less toxic to mammals than the thiolate with LD_{50} 250-500 mg/kg but the thionoisomer is completely inactive.

Acephate, orthene®, O,S-dimethyl N-acetyl phosphoroamidothiolate or O,S-dimethyl N-acetylphosphoramidothioate (79, Fig. 3.104b), is prepared as given below.

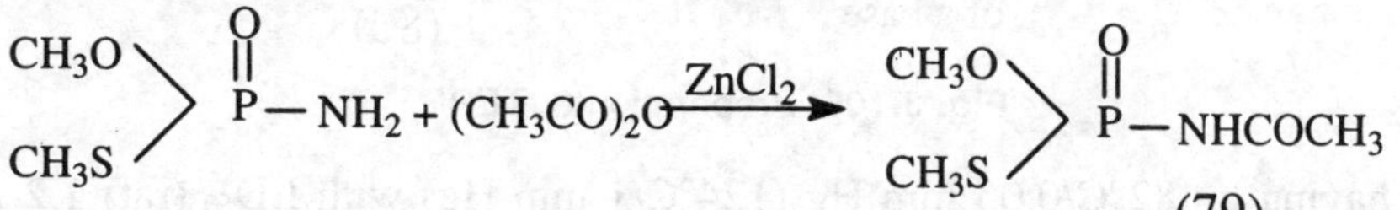

Fig. 3.104b. Preparation of acephate.

Acephate, the N-acetyl derivative of the methamidophos was commercialized by the Chevron Chemical Co., USA in the year 1971. It is prepared by standard procedure using acetic anhydride in presence of Lewis acid such as $ZnCl_2$, $FeCl_3$ or BF_3-etherate. It is solid having mp 91° to 92°C, highly soluble in water with LD_{50} (rat) 945 mg/kg. It is a systemic insecticide with moderate persistence and effective not only against sucking pests but also chewing insects. It is a poor inhibitor of cholinesterase but is activated metabolically in plant or animal system.

The systemic insecticide N-malonyl ethyl ester derivatives like acephate, was designed which can be hydrolysed by esterases in plants yielding a carboxyl group. The acid gets transported to other sites and is oxidised to produce methamidophos which may impart the insecticidal activity (Fig. 3.105).

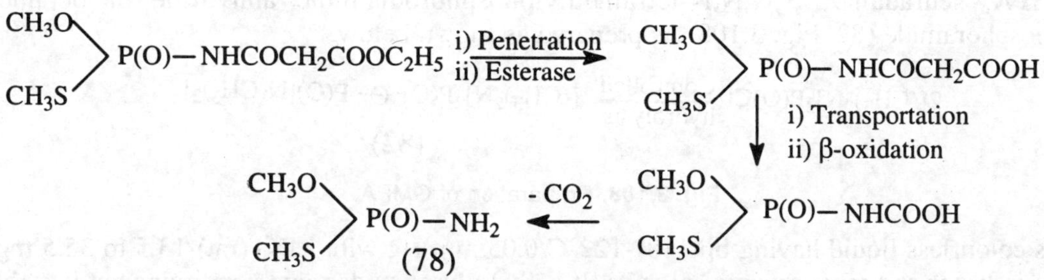

Fig. 3.105. Transformation of N-malonylethylester.

Esters of pyrophosphoric acid [20]

The insecticidal property of pyrophosphate, TEPP was discovered by Schrader in Germany in the year 1943 which was followed by detailed studies on other TEPP related compounds. SAR studies on pyrophosphates indicate

  i) that the insecticidal activity of TEPP is reduced with an increase in size of alkyl group.

ii) tetraalkyl monothiopyrophosphates possess greater insecticidal activity than the TEPP but the toxicity to mammals also is increased. Introduction of a second sulphur atom on the 'P'-atom lowers the toxicity of the compound to mammals without reducing the insecticidal activity.

iii) Introduction of a third S-atom on the phosphorus moiety decreases the insecticidal activity of the compound.

TEPP, bladan®, O,O,O,O-tetraethyl pyrophosphate (80, Fig. 3.106), is prepared as follows.

$$(C_2H_5O)_2 P(O)Cl \xrightarrow[\text{in presence of a base}]{\text{Controlled hydrolysis}} (C_2H_5O)_2 P(O)-O-P(O)(OC_2H_5)_2$$

(80)

Fig. 3.106. Preparation of TEPP.

It is liquid having bp 82°C/0.05 mm Hg (124°C/1 mm Hg) with LD_{50} (rat) 1.2-2 mg/kg. It is soluble in water and most organic solvents. TEPP decomposes rapidly in presence of moisture and vanishes within 48 h. of application, forming harmless non-toxic compounds. It acts as a contact insecticide and acaricide, effective in the control of aphids and mites in active stages.

Sulfotep, O,O,O,O-tetraethyldithiopyrophosphate (81, Fig. 3.107), can be prepared as follows.

It is light yellow liquid having bp 136° to 139°C at 2 mm Hg with LD_{50} (rat) 5 mg/kg. It is sparingly soluble in water and is highly stable against hydrolysis. It is an acaricide and insecticide with contact and fumigant action.

$$TEPP \xrightarrow{S} (C_2H_5O)_2P(S)-O--P(S) P(S)(OC_2H_5)_2$$

(81)

Fig. 3.107. Preparation of sulfotep.

OMPA, schradan®, N,N,N,N-tetramethylphosphorodiamidic anhydride or octamethyl pyrophosphoramide (82, Fig. 3.108), is prepared as shown below.

$$2[(CH_3)_2N]_2P(O)Cl \xrightarrow{\text{Controlled hydrolysis}} [(CH_3)_2N]_2P(O)-O-P(O)[N(CH_3)_2]_2$$

(82)

Fig. 3.108. Preparation of OMPA.

It is colourless liquid having bp 118°-122°C/0.03 mm Hg with LD_{50} (rat) 13.5 to 35.5 mg/kg, soluble in water and most organic solvents. It is hydrolysed under acid conditions but is stable in water and alkali. It is used for the control of aphids and mites on citrus and apple.

Metabolism [21]

Metabolism may be divided into two parts : a) activative, and b) degradative.

a) **Activative metabolism** is a process where the molecule is converted from a poorer to a stronger anticholinesterase effect.

b) **Degradative metabolism** is the reverse process. The fact is that in the majority of cases, the activating step is absolutely necessary for the conversion of a latent molecule to a potent molecule and the degradative step is a conversion of toxic molecule to a harmless compound.

a) Activative (Oxidative)

The activative reactions may be classified as follows :

a) P (S) to P (O) (Desulfuration i.e. oxidation)

b) Hydroxylation

c) Thioether oxidation

d) Hydroxylation cum cyclisation

e) Isomerisation

g) Side chain oxidation

The first four reactions occur in vertebrates and insect body while the first three occur only in plants. In mammals the catalytic system is present in the microsomes of liver.

a) Activation of P(S) to P(O)

This activation is a prerequisite to toxicity. The increase in anticholinesterase potency is about 10,000 fold for parathion and malathion. In vertebrates the oxidation reaction takes place in liver microsomes in presence of NADPH plus oxygen while in insects it occurs in the fat bodies (Fig. 3.109).

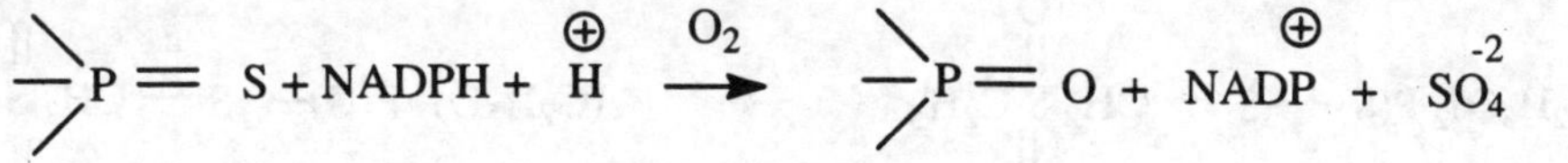

$$\overset{\diagup}{\underset{\diagup}{}}\!\!\!-\!P = S + NADPH + \overset{\oplus}{H} \xrightarrow{\;\;O_2\;\;} \overset{\diagup}{\underset{\diagup}{}}\!\!\!-\!P = O + \overset{\oplus}{NADP} + SO_4^{-2}$$

Fig. 3.109. Oxidation in liver microsomes.

b) Hydroxylation

OMPA, schradan, is stable towards basic hydrolysis and shows no anticholinesterase activity but is toxic to certain insects probably due to oxidative metabolism to a toxic molecule (Fig. 3.110).

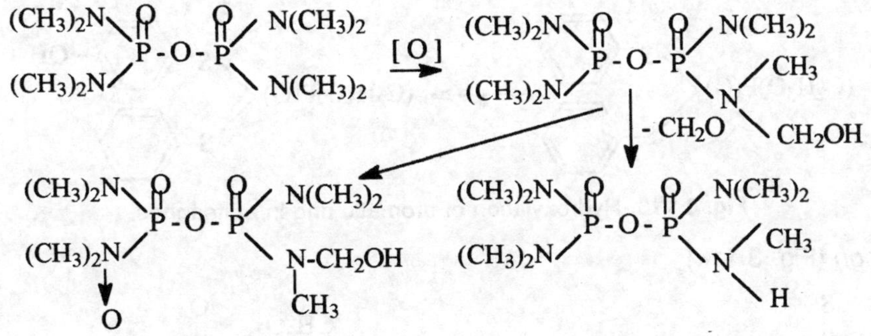

Fig. 3.110. Oxidative metabolism of OMPA.

c) Oxidation and cyclisation

TOCP, (tri-*o*-cresyl phosphate) is not an insecticide but it produces a neurotoxic molecule *in situ* (Fig. 3.111) responsible for its insecticidal properties.

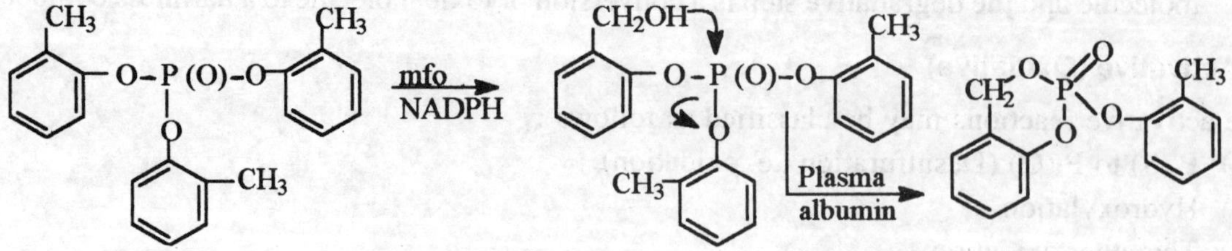

Fig. 3.111. Transformation of TOCP.

d) Thioether oxidation

In plant system the activation process is prominent as observed in case of phorate. The oxidation of thioether to sulfoxide is rapid in all cases followed by sulfone formation (Fig. 3.112).

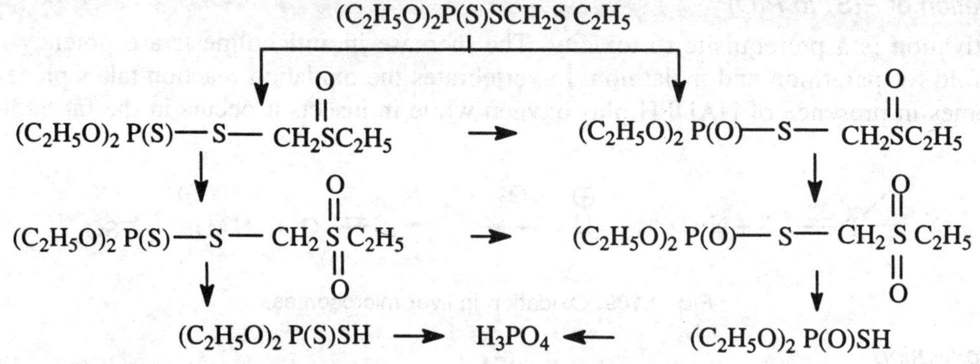

Fig. 3.112. Oxidation of thioether in phorate.

f) Hydroxylation of an aromatic ring (Fig. 3.113)

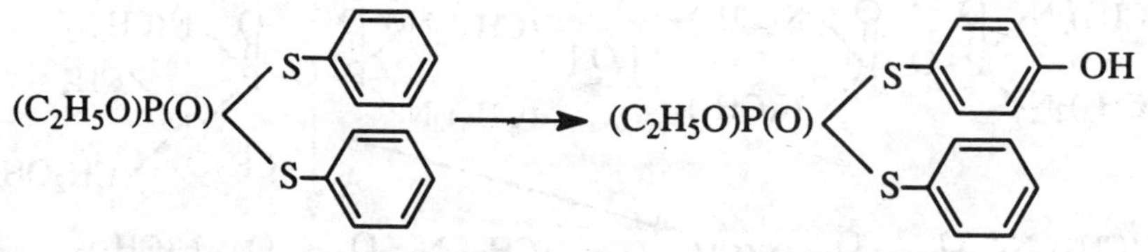

Fig. 3.113. Hydroxylation of aromatic ring in edifenphos.

g) Isomerisation (Fig. 3.114)

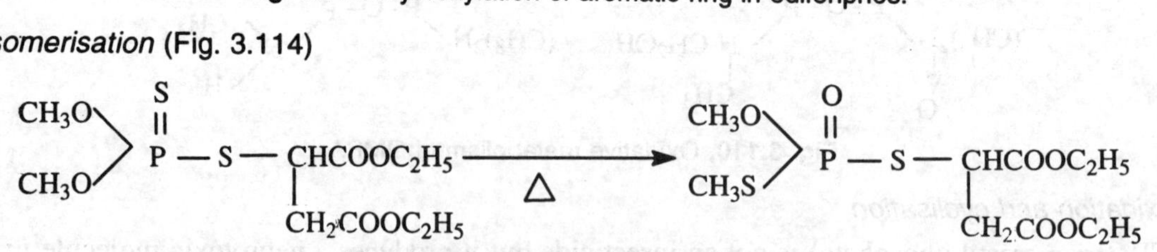

Fig. 3.114. Isomerisation of malathion.

h) Side chain oxidation (Fig. 3.115)

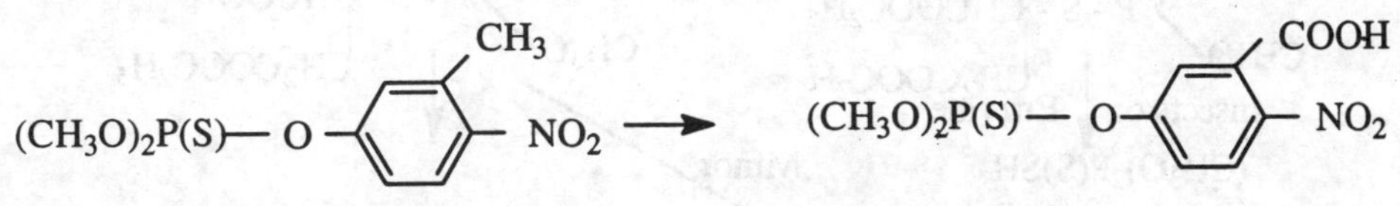

Fig. 3.115. Oxidation of side chain in fenitrothion.

b. Degradation [22]

Degradation occurs primarily by a hydrolytic route which creates an anionic site attached to the phosphorus atom thereby reducing its positive character. The other minor routes of degradation are reductive process and dealkylation.

Phosphatases. The commonest hydrolysis is by phosphatase enzymes which hydrolyse any phosphorus ester, at the anhydride bond including P-O-C, P-S-C, P-F and others (Fig. 3.116).

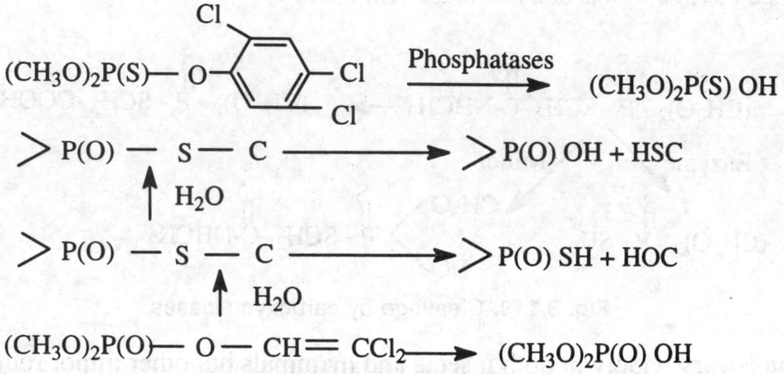

Fig. 3.116. Hydrolysis by phosphatases.

a) **De-alkylation** is a minor route of degradation. Demethylation is observed in some cases but de-ethylation is difficult (Fig. 3.117).

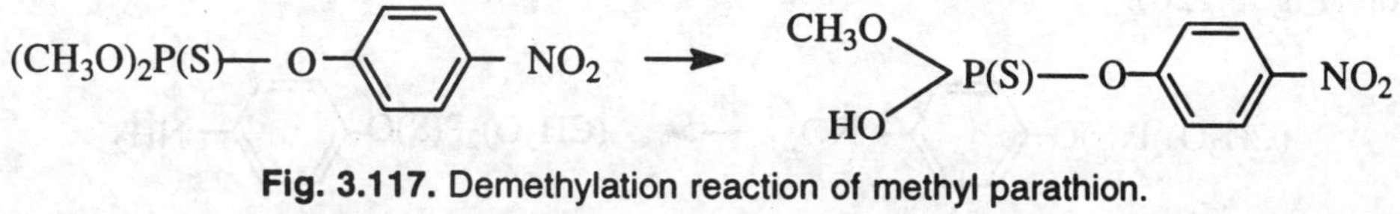

Fig. 3.117. Demethylation reaction of methyl parathion.

Degradation also depends on the chain length and branching of the alkyl substituents in the alkoxy group, the type of ester linkage such as P(S), P(O) in phosphates, phosphonates and the presence of substitution in the aromatic ring.

b) Carboxyesterases (specific enzymes)

Cleavage of compounds containing a carboxyester group COOR occurs and in some cases may constitute the predominant degradative pathway. In case of malathion carboxyesterase activity predominates in mammals but not in insects. Thus malathion has a selective toxicity (Fig. 3.118).

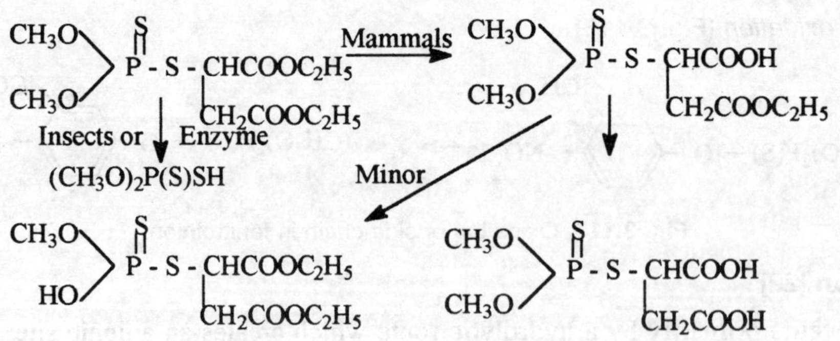

Fig. 3.118. Cleavage by carboxyesterases.

c) Amidases

Whenever a carboxyamide group $CONR_2$ is present in an organophosphate molecule there is a possibility of its cleavage by carboxyamidase (Fig. 3.119).

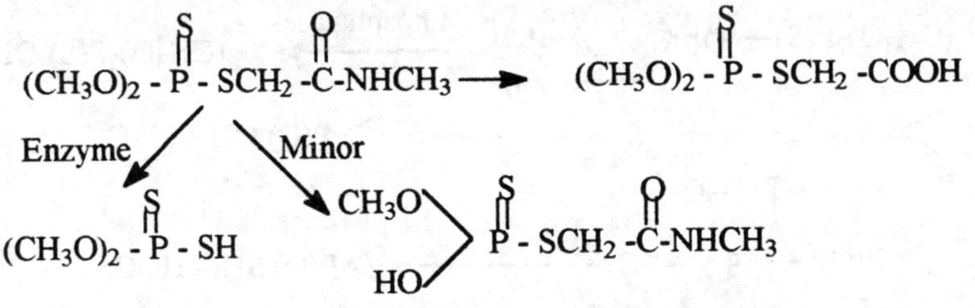

Fig. 3.119. Cleavage by carboxyamidases.

Amidase routes vary widely in both insects and mammals but other minor routes of degradation are de-alkylation and by the enzyme phosphatase are family common.

d) Reductive degradation

Reduction of a nitro group to amino group occurs as observed in case of methyl parathion and fenitrothion (Fig. 3.120).

Fig. 3.120. Reduction to amine in methyl parathion.

Photochemistry [23]

Most of the organophosphorus compounds except a few which contain aryl groups, do not have a proper chromophoric group. They undergo photochemical reactions mainly by ester cleavage as well as the conversion of P-S bond to P-O bond. Unlike chemical ester cleavage where the major product is formed by de-esterification of the P-O bond connecting the leaving group, photochemical cleavage is rather nonspecific. The products are formed by the loss of not only the leaving group but also by the loss of alkyl substituents. Some organophosphorus compounds containing a photoreactive leaving group can undergo further changes after de-esterification to yield secondary products. Phototransformation of few compounds which yielded interesting results are described here.

Quinalphos [24] being used in India and elsewhere possesses low persistence due to its high susceptibility to UV and sunlight. Besides yielding the normal cleavage product quinoxaline-2-ol, it gives several other abnormal products like quinoxaline-2-thiol, sulfide and disulfide (Fig. 3.121). The ease of formation of these products from quinalphos must involve a facile photochemical thiono-thiolo rearrangement in the molecule. The formation of these products also takes place during storage of technical grade quinalphos which needs to be stabilised to prevent this undue rearrangement and consequent loss of activity. It is believed that both the sulfide and the disulfide formed by photochemical degradation have toxicological significance.

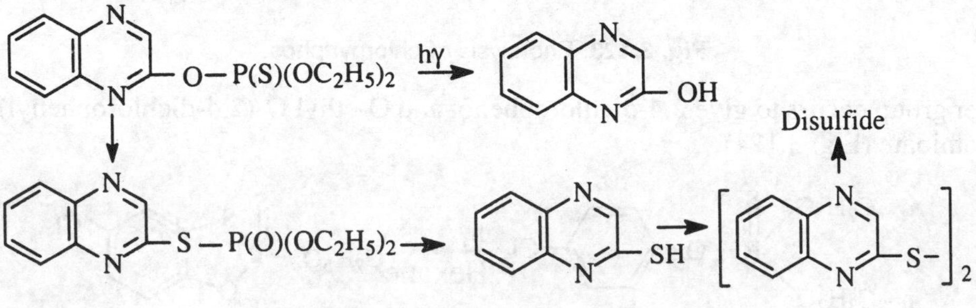

Fig. 3.121. Photoproducts of quinalphos.

Phosalone

Phosalone is an important broad-spectrum insecticide commonly used in India and elsewhere. On photolysis it gives 6-chloro-3-mercaptomethyl 2-oxobenzoxazole, as an intermediate which can be further oxidised to disulfide, a major photoproduct (Fig. 3.122). Phosalone can also undergo-N-CH_2-or S-CH_2 cleavage to yield a variety of minor products. Formation of dechlorophosalone and phosalone oxon suggested that phosalone is degraded by both dehalogenation and oxidation processes.

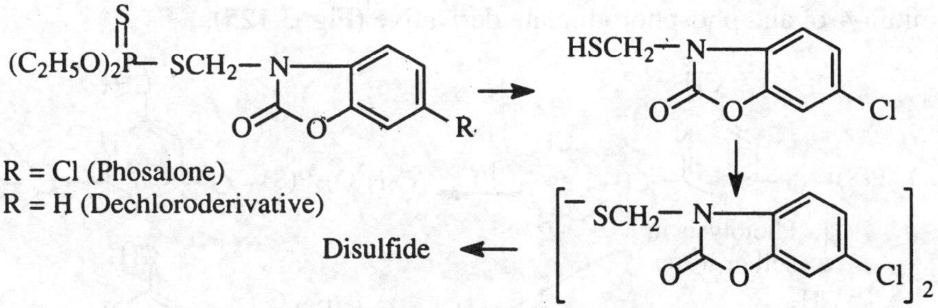

R = Cl (Phosalone)
R = H (Dechloroderivative)

Fig. 3.122. Photolysis of phosalone.

Similarly **chlorpyriphos** on irradiation in methanol or hexane under UV light gives mono and didechlorinated products along with phosphorothiolate (Fig. 3.123).

Prothiophos on photolysis in hexane solution produces cyclised phospholone, a product formed via cleavage of P-S-alkyl linkage followed by rearrangement (25). But in aqueous solution, cleav-

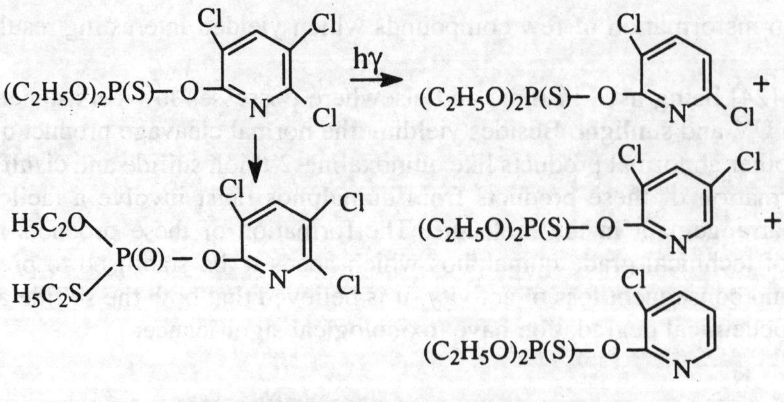

Fig. 3.123. Photolysis of chlorpyriphos.

age of ester group occurs to give 2,4-dichlorophenol and O-ethyl O-(2,4-dichlorophenyl) hydrogen phosphorothioate (Fig. 3.124).

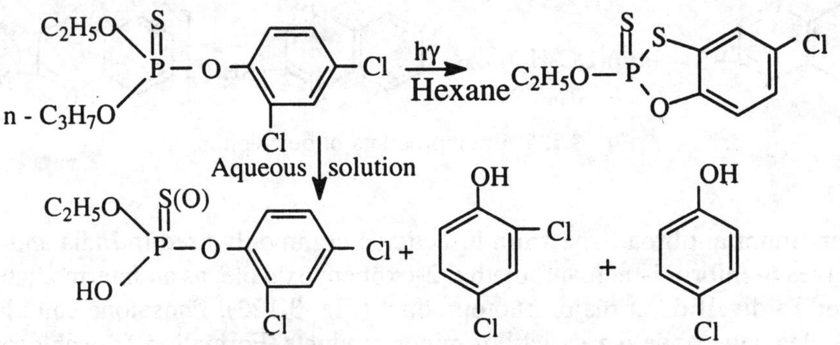

Fig. 3.124. Photolysis of prothiophos.

Diazinon a rice insecticide on photolysis gives hydroxydiazinon [26] along with ester hydrolysis to yield pyrimidin-4-ol and phosphorothiolate derivative (Fig. 3.125).

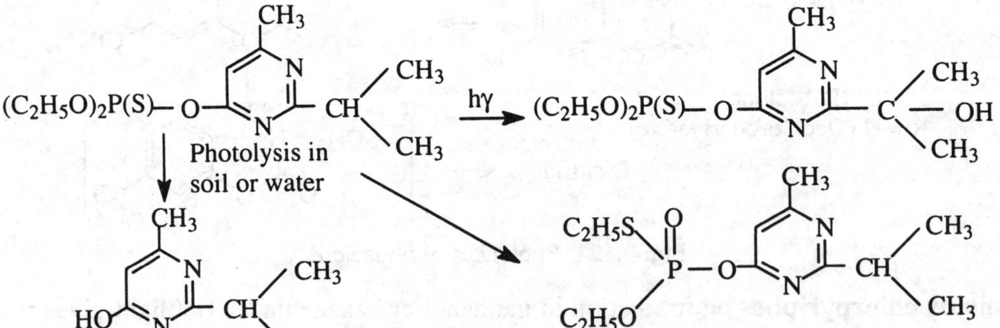

Fig. 3.125. Photolysis of diazinon.

O,O,S-Triethyl thiophosphate and O,O-diethyl O-phenylthiophosphate were identified as the major product of the photolysis of **parathion** in aqueous THF or methanol along with some minor products like nitrophenol.

Glyphosate on photolysis in water yielded aminomethyl phosphoric acid as the major product. **Edifenphos** on irradiation with UV light gave mainly ethyl S-phenyl hydrogen phosphorothiolate and ethyl dihydrogen phosphate.

REFERENCES

1. Eto, M.(ed.). *Organophosphorus Pesticides : Organic and Biological Chemistry*, C.R.C. Cleaveland, Ohio (1974).
2. Fest, C. and Schmidt, K.J. (eds.). *The Chemistry of Organophosphorus Pesticides*. Springer, Berlin (1973).
3. Matolcsy, G., Nadasy, M. and Andriska, V. (eds.). *Pesticide Chemistry*. Elsevier, Amsterdam, 108-161 (1988).
4. Longe, W. and Krueger, B. Uber Ester der Monofluorophosphorsaure. *Chem. Ber.* **65**, 1958 (1932).
5. Bhattacharya, A.K. and Thygarajan, G. The Michaelis-Arbusov rearrangement. *Chem. Rev.* **81**, 415-430 (1981).
6. Perkow, W. Umsetzungen mit alkylphosphiten. I. Mitteil : Umlagerungen bei der Reaktion mit chloral und bromal. *Chem. Ber.* **8**, 755 (1954).
7. Kirby, A.J. and Warren, S.G. (eds). *The Organic Chemistry of Phosphorus*. Elsevier, Amsterdam (1967).
8. Hudson, R.F. (ed.). *Structure and Mechanism in Organophosphorus Chemistry*. Academic Press, London (1965).
9. Morrison, D.C. The reaction of sulfenyl chloride, with trialkyl phosphites. *J. Am. Chem. Soc.* **77**, 181 (1955).
10. Kinnear, A.M. and Perren, E.A. Formation of organophosphorus compounds by the reaction of alkyl chlorides with phosphorus trichloride in the presence of aluminium chloride. *J. Chem. Soc.* 3437 (1952).
11. Gier, D.W. Dialkyl and diphenyl esters of aryloxyacetyl phosphoric acids as herbicides, US. Pat. 3, 382, 060; CA 70, 4273 (1969).
12. Fukuto, T.R. The chemistry of organic insecticides. *Ann. Rev. Entomol.* **6**, 313 (1961).
13. Schrader, G. Die Entwicklung never insektizider phosphorasaure-Ester. Verlag Chemie, Weinhein 281 (1963).
14. Scherer, O. and Stachler, G. Insecticidal phosphates. Ger. Pat. 1,283,592, CA, 70, 56712 (1965).
15. Burn, A.J. and Cadogan, J.I.G. The reactivity of organophosphorus compounds IX. The reaction of thionates with alkyl iodides. *J. Chem. Soc.* 5532 (1961).
16a. Eto, M., Casida, J.E. and Eto, T. Hydroxylation and cyclization reactions involved in the metabolism of tri-O-cresyl phosphate. *Biochem. Pharmacol.* **11**, 337 (1962).
16b. Eto, M. Development of insecticidal cyclic phosphoryl compounds through chemical and biochemical approaches. *J. Environ. Sci. Health*, B **18** (1), 119-145 (1983).
17. Bellet, E.M. and Casida, J.E. Bicyclic phosphorus esters : High toxicity without cholinesterase inhibitors. *Science.* **182**, 1135-1136 (1973).
18. Melnikov, N.N. and Schetovska, S., K.D. *Dokl Akad. Nauk.* SSSR 86, 543 (1952).
19. Zerba, E., deLicastro, S.A., Secceacini, E., Wallace, Y.G. Synthesis and insecticidal activity of dialkyl phosphorodithioates derivatives of methyl esters of maleamic acids. *Eighth Internatn. Congr. Pestic. Chem.*, Washington D.C. (1994).
20. Toy, ADF. The preparation of tetraethyl pyrophosphate and other tetraalkyl pyrophosphates. *J. Am. Chem. Soc.* **70**, 3882 (1948).
21. O'Brien, R.D. (ed.). *Insecticides Action and Metabolism*. Academic Press, New York, (1967).

22. Dauterman, W.C. Biological and non-biological modifications of organo-phosphorus compounds. *Bull Wld. Hlth Org.* **44**, 133 (1971).

23. Crosby, D.G. Experimental approaches to pesticide photodecomposition. *Residue Rev.* **5**, 91 (1963).

24. Dureja, P., Walia, S. and Kumar, Jitendra. Photochemical reactions of organophosphorus pesticides. *Pestic. Res. J.* **2** (1), 1 (1990).

25. Takase, I., Oyama, H. and Ueyama, I. Photodecomposition of prothiophos (O-2,4-dichlorophenyl O-ethyl S-propylphosphorodithioate). *J. Pestic. Sci.* **7**, 463 (1982).

26. Pardue, J.R., Hausen, E.A., Barron, R.P. and Chem, J.T. Diazinon residues on field sprayed kale. Hydroxydiazinon—a new alteration product of diazinon. *J. Agric. Food Chem.* **18**, 405 (1970).

Section V.   Chemistry of Carbamate Insecticides

- Introduction
- Carbamate compounds
- Uses
- Mode of action

- Metabolism
- Photochemistry
- *References*

Carbamates are important group of synthetic compounds with high insecticidal activity and a reasonable rate of biodegradation. The idea of developing carbamate insecticides came from the alkaloid physostigmine (1), an active ingredient of calaban beans. Based on the structure of this alkaloid, attempts were made to synthesize a series of N-substituted carbamate molecules for screening as insecticides :

(1)   ;   (2)

In 1951, the Geigy company [1] introduced 1-*iso*propyl-3-methyl pyrazolyl 5-dimethylcarba-mate (2), but due to its high toxicity towards mammals, it was not commercialized. Subsequently several carbamates were prepared and screened for insecticidal activity [2].

Carbamates [3, 4, 5]

The base for all carbamates is carbamic acid (3), the monoamide of carbon dioxide which is unstable in free form but decomposes to CO_2 and NH_3. On the other hand salts of carbamic acid, called carbamates are very stable. One such stable salt is ammonium carbamate (4), used as commercial insecticide and rodenticide along with aluminium phosphide. Carbamic acid can be stabilised by formation of a simple alkyl ester such as ethyl carbamate called as urethane (5). This is used as bactericide and cosolvent for pesticides, but is carcinogenic in nature.

(3)   ;   (4)   ;   (5)

In addition to alkyl esters, aryl esters of carbamic acid are also prepared. One such compound is phenylcarbamate (6).

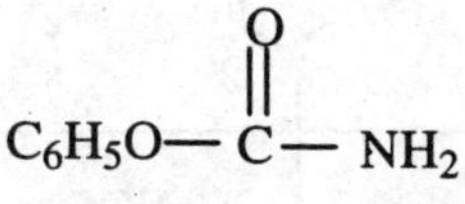

$$C_6H_5O-\overset{\overset{\textstyle O}{\|}}{C}-NH_2$$

(6)

If one of the protons attached to the nitrogen atom is replaced by a methyl group, the resulting compound phenyl N-methylcarbamate was found to be mildly toxic against few insect species. The majority of commercial and experimental methylcarbamate insecticides are prepared by substitution of the phenyl ring.

Other aryl methylcarbamates with insecticidal activity can be prepared from aromatics such as 1-naphthol or 2,3-dihydro-2, 2-dimethyl-7-hydroxybenzofuran and 4-hydroxy benzothiophene. The resulting insecticides are **carbaryl**, 1-naphthyl N-methylcarbamate (7), **carbofuran**, 2,3-dihydro-2, 2-dimethyl-7-benzofuranyl N-methylcarbamate (8) and **propoxur**, 2-*iso*proxyphenyl N-methylcarbamate (9).

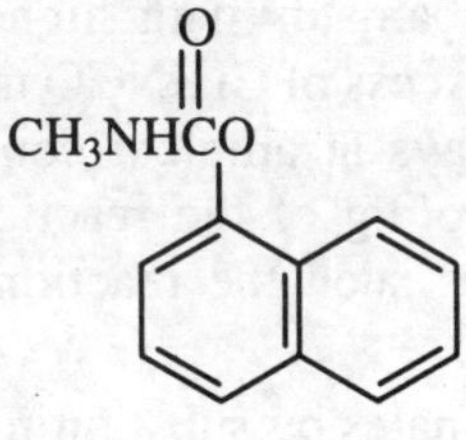

(7) 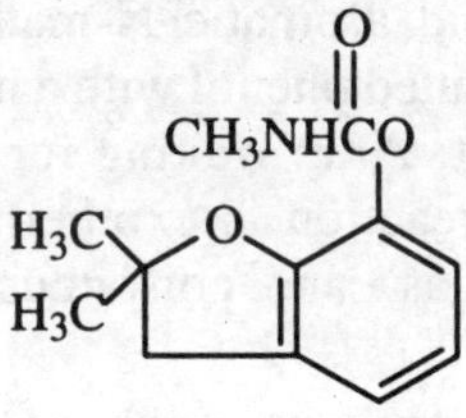   (8) 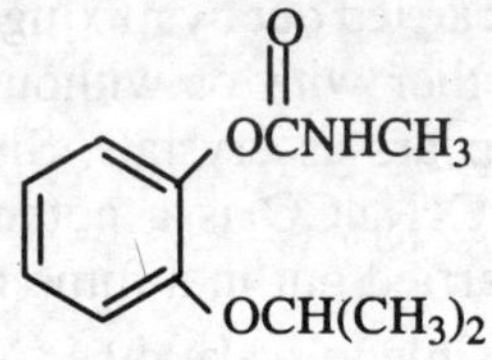   (9)

The latest group of successful methylcarbamate insecticides which are derived from aliphatic oximes, commercialized are **aldicarb**, 2-methyl-2-(methylthio) propionaldehyde O-(methyl carbamoyl) oxime (10), **methomyl**, S-methyl-N-[methyl carbamoyloxy] thioacetimidate (11) and **thiodicarb** (12).

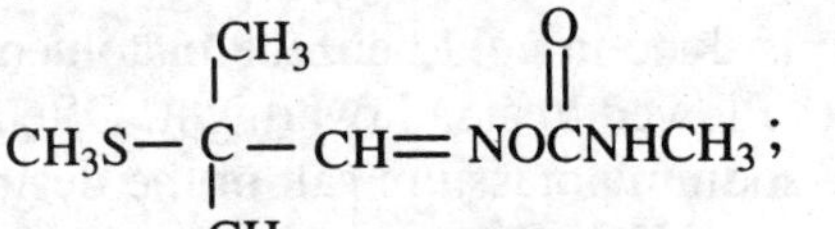

(10) 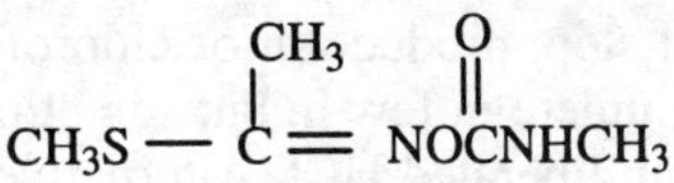   (11)

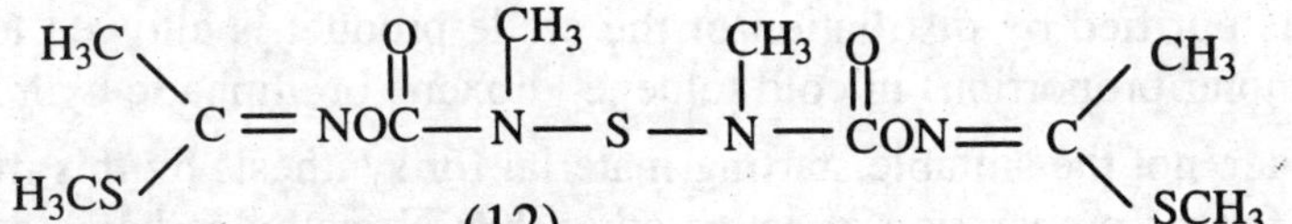

(12)

SYNTHESIS

N-Methylcarbamates

All substituted phenyl N-methylcarbamates have been synthesized by the following three methods (Fig. 3.126) :

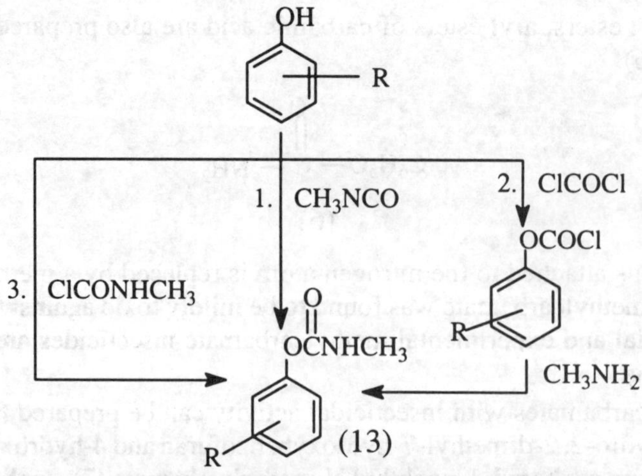

Fig. 3.126. General methods for preparation of carbamate.

Method 1. Esterification of phenol with methyl *iso*cyanate is the easiest and most convenient route for synthesising both aliphatic and aromatic N-methyl-carbamates with high yield. The reaction is carried out by mixing a substituted phenol with a molar excess of CH₃NCO in anhydrous *iso*propyl ether with or without catalyst. After waiting for few days at ambient conditions, the product appears as crystals. Since the reaction is exothermic, cooling of the reaction vessel is necessary. CH₃NCO is a potent "tear gas" and poisonous and as such the reaction should be carefully carried out in a fume hood.

It is possible to make other N-alkylcarbamates and N-arylcarbamates by substituting the *iso*cyanate or instead of starting with the phenol, one could mix CH₃NCO with benzyl alcohol or aliphatic hydroxy compounds including oximes.

Method 2. Phenol is mixed with an excess of phosgene to yield phenyl chloroformate, which then reacts with methylamine to produce N-methylcarbamate. Since it is a two steps reaction, yields are lower than that of method 1.

In the first step, production of chloroformate is done in cold benzene or toluene. Usually the reaction is complete in few h. but sometimes it is allowed to stand overnight. Often triethylamine is added to trap liberated HCl. Alternatively, the sodium/potassium salt of the desired phenol can be placed together with phosgene, avoiding release of HCl. The reaction should be carried out in a fume hood with great care.

The chloroformate is purified by distillation or the crude product is allowed to react directly with methylamine (in molar proportion) in cold toluene, dioxane or aliphatic hydrocarbons.

Substituted phenols are not the suitable starting material for synthesis by this method. Method 2 is particularly useful for the preparation of esters other than N-methylcarbamates.

Method 3. It is rarely employed for the synthesis of N-methylcarbamate. The reaction comprises of addition of a substituted phenol or salt of phenol to methyl carbamoylchloride in solvents such as toluene in presence of weak base for absorption of released acid. The reaction may be completed in a short period without heat or it may be necessary to reflux for several hours. Again using other carbamoylchlorides, a variety of mono or dialkylcarbamates can be produced.

N,N-Dimethylcarbamates

Synthesis of N,N-dimethylcarbamates of heterocyclic enols is accomplished primarily by method 3. For example, **pyramat** (16), crystals appear after refluxing the sodium salt of 2-propyl-4-methyl-6-hydroxypyrimidine (14) with dimethylcarbamoyl chloride (15) in benzene for 12 h. (route 1, Fig. 3.127). The same reaction is used for industrial synthesis of **isolan** from 1-*iso*propyl-3-methyl-5-pyrazolone-5-enolate. Numerous other Na or K salts of cyclic enols have been converted to dimethylcarbamates via this route using other solvents such as THF, methyl ethyl ketone· or acetonitrile with different reflux times. If enol is allowed to react in place of salt, then a HCl scavenger, such as pyridine/triethylamine is added along with the solvent.

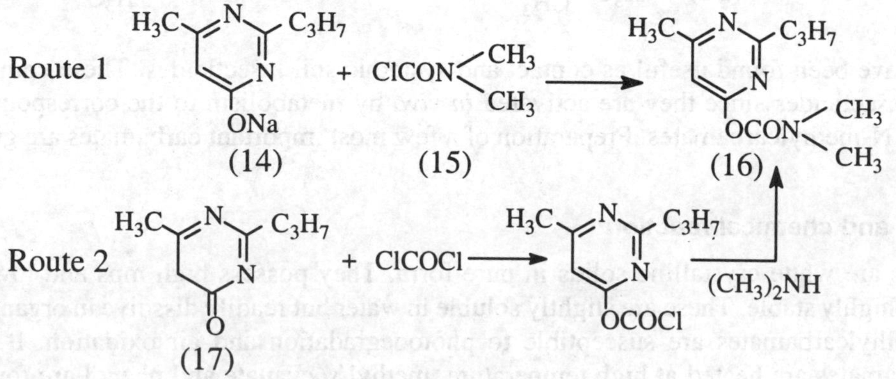

Fig. 3.127. Preparation of pyramat.

Interestingly, dimethylcarbamates can also be prepared by direct action of phosgene on ketone analogues depicted in reaction route-2 (Fig. 3.127). Thus 2-propyl-4-methyl-6-pyrimidone (17) plus phosgene in cold benzene forms the chloroformic acid ester which is separated out. This on subsequent treatment with dimethylamine at ambient temperature yields pyramat after few h. Similarly, **dimetilan**, 2-dimethyl carbamoyl-3-methyl-5-pyrazolyl dimethylcarbamate (18) was prepared in good yield from 3-methylpyrazolone-5(14) mixed with at least two equivalents of dimethylcarbamoylchloride in benzene containing K_2CO_3 as acid scavenger. **Pirimicarb** (19), a pyrimidinyl carbamate having similar biological properties was also prepared by this method.

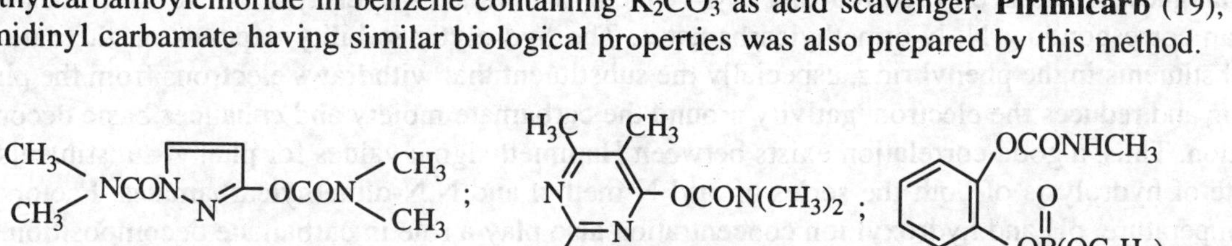

Hetnarski and O'Brien [6] developed an insecticidal compound (20), which contains both carbamyl and phosphoryl group with the hope that the presence of two bioactive groups might lead to very effective and selective insecticides after carbamylation or phosphorylation of the insect enzyme. But in practice, there was no enhancement in the activity as compared to parent carbamates.

Two new highly effective sulphenylated and sulphinylated carbamate insecticides with low mammalian toxicity and phytotoxicity namely carbosulfan (21) and furathiocarb (22) have been recently developed.

(21) ; (22)

These have been found useful as contact and systemic soil insecticides. These compounds are called proinsecticides since they are activated *in vivo* by metabolism to the corresponding active component N-methylcarbamates. Preparation of a few most important carbamates are given in Fig. 3.128.

Properties and chemical reaction

Carbamates are white crystalline solids in pure form. They possess high mps and low vp which make them highly stable. These are slightly soluble in water but readily dissolve in organic solvents. Certain methylcarbamates are susceptible to photodegradation and air oxidation. If phenyl N-methylcarbamates are heated at high temperature, methyl*iso*cyanate and phenol are formed.

$$PhOCONHCH_3 \xrightarrow{\Delta} PhOH + CH_3NCO$$

The most important reaction of carbamates is hydrolysis. It has been observed that N-alkyl groups stabilize carbamate esters. The order of stability is N-*iso*proyl < N-ethyl < N-n-propyl < N-methyl, and the corresponding N,N-disubstituted carbamates are found to be more resistant to hydrolysis than their mono analogues. However, the rate of hydrolysis vary with respect to dimethyl analogues. It has been reported that aryl N-methylcarbamates hydrolyse 10^3 to 10^7 times faster than corresponding N, N-dimethylcarbamates. The hydrolytic stability depends on the nature of substituents in the phenyl ring, especially the substituent that withdraws electrons from the phenyl ring and reduces the electronegativity around the carbamate moiety and enhances basic decomposition. Thus, a good correlation exists between Hammett sigma values for phenyl substituents and rate of hydrolysis of both the series of aryl N-methyl and N,N-dimethylcarbamates. Factors like temperature, pH and hydroxyl ion concentration also play a role in carbamate decomposition. The rate of hydrolysis varies directly with increase in temperature while half-life decreases with the increase of pH value. With variation of concentration of hydroxyl ion or carbamate, indicates that the cleavage of ester bond is of first order with respect to both OH ion and substrate [3]. Fig. 3.129 shows the mechanism of hydrolysis of aromatic unsubstituted carbamates and aromatic N-methylcarbamates. However, in acid solution, the hydrolysis of carbamates generally occurs at very slow rates.

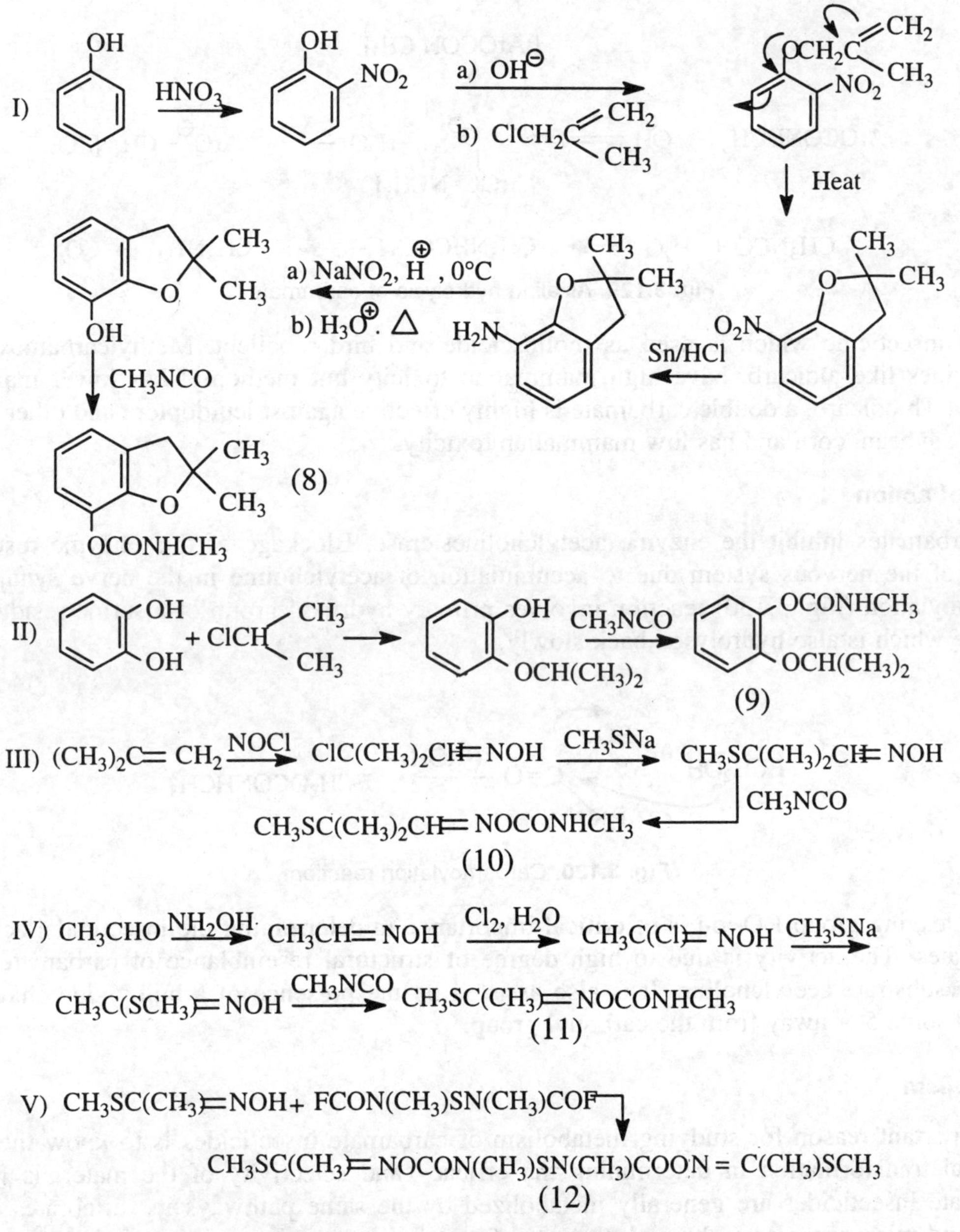

I) Phenol $\xrightarrow{HNO_3}$ o-nitrophenol $\xrightarrow[\text{b) } ClCH_2C(CH_3)=CH_2]{\text{a) } OH^\ominus}$ (o-OCH_2C(CH_3)=CH_2, NO_2) $\xrightarrow{\text{Heat}}$

$\xrightarrow{\text{Sn/HCl}}$ (O_2N-benzofuran, 2,2-dimethyl) $\to$ (H_2N-benzofuran) $\xrightarrow[\text{b) } H_3O^{\oplus}, \Delta]{\text{a) } NaNO_2, H^{\oplus}, 0°C}$ (HO-2,2-dimethyl-2,3-dihydrobenzofuran) $\xrightarrow{CH_3NCO}$ (OCONHCH_3-2,2-dimethylbenzofuran) (8)

II) Catechol $+ ClCH(CH_3)_2 \to$ (o-OH, OCH(CH_3)_2) $\xrightarrow{CH_3NCO}$ (o-OCONHCH_3, OCH(CH_3)_2) (9)

III) $(CH_3)_2C=CH_2 \xrightarrow{NOCl} ClC(CH_3)_2CH=NOH \xrightarrow{CH_3SNa} CH_3SC(CH_3)_2CH=NOH \xrightarrow{CH_3NCO}$

$CH_3SC(CH_3)_2CH=NOCONHCH_3$ (10)

IV) $CH_3CHO \xrightarrow{NH_2OH} CH_3CH=NOH \xrightarrow{Cl_2, H_2O} CH_3C(Cl)=NOH \xrightarrow{CH_3SNa}$

$CH_3C(SCH_3)=NOH \xrightarrow{CH_3NCO} CH_3SC(CH_3)=NOCONHCH_3$ (11)

V) $CH_3SC(CH_3)=NOH + FCON(CH_3)SN(CH_3)COF \to$

$CH_3SC(CH_3)=NOCON(CH_3)SN(CH_3)COON=C(CH_3)SCH_3$ (12)

Fig. 3.128. Preparation of carbamates.

Uses

Carbaryl is a contact insecticide effective against many pests of fruits, vegetables, and cotton while carbofuran is systemic in nature and can control a number of rice insects, mites and nematodes. Propoxur is used as a fumigant in green houses against whiteflies and aphids. Methiocarb is a

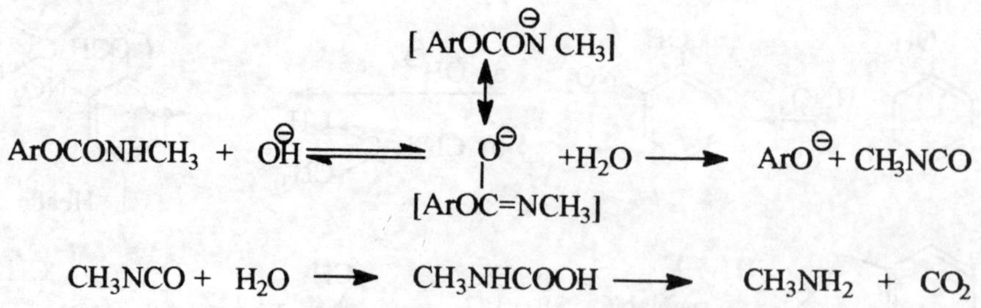

$$[\overset{\ominus}{ArOCON} CH_3]$$

$$ArOCONHCH_3 + \overset{\ominus}{OH} \rightleftharpoons \quad \underset{[ArOC=NCH_3]}{\overset{\ominus}{O}} +H_2O \longrightarrow ArO^{\ominus} + CH_3NCO$$

$$CH_3NCO + H_2O \longrightarrow CH_3NHCOOH \longrightarrow CH_3NH_2 + CO_2$$

Fig. 3.129. Alkaline hydrolysis of carbamates.

contact insecticide which is used as molluscicide and bird repellent. Methylcarbamoyl oxime insecticides like aldicarb, have high mammalian toxicity but methomyl has lower mammalian toxicity. Thiodicarb, a double carbamate is highly effective against lepidoptera and other pests·on cotton, soybean, corn and has low mammalian toxicity.

Mode of action

The carbamates inhibit the enzyme acetylcholinesterase. Blockage of this enzyme results in a failure of the nervous system due to accumulation of acetylcholine in the nerve synapse. The carbamoylation (Fig. 3.130) reaction involves primary hydroxyl group of a serine residue of the enzyme which is also hydrolysed back slowly.

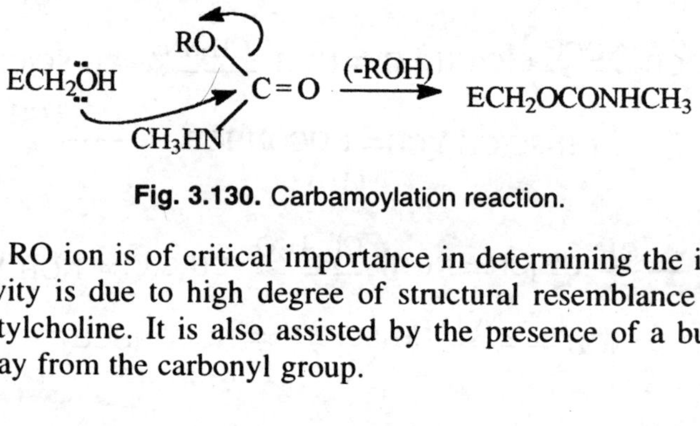

$$ECH_2\overset{..}{O}H \quad \underset{CH_3HN}{\overset{RO}{\diagdown}}C=O \quad \xrightarrow{(-ROH)} \quad ECH_2OCONHCH_3$$

Fig. 3.130. Carbamoylation reaction.

The leaving group RO ion is of critical importance in determining the insecticidal activity of carbamates. The activity is due to high degree of structural resemblance of carbamates to the enzyme substrate acetylcholine. It is also assisted by the presence of a bulky side chain group situated some 5 Å away from the carbonyl group.

Metabolism

The important reason for studying metabolism of carbamate insecticides is to know the role of chemical transformation in determining the efficacy and selectivity of the materials [7]. The carbamate insecticides are generally metabolized by the same pathways in vertebrates, insects, plants and microorganisms although the rate of metabolic degradation is species dependent. The rate difference definitely affects the toxicity of a compound. It is desirable that non-target species including man, are capable of rapidly detoxifying the insecticide. On the other hand, rapid metabolism by the target organism would be undesirable as it leads to resistance generation. The metabolism of carbaryl has been the subject of most investigations among carbamate insecticides [8]. The hydrolysis of the carbamate ester linkage was the major metabolic pathway yielding 1-naphthol, carbon dioxide and methylamine although it was recognised that methyl hydroxylation, ring hydroxylation and conjugation before or after hydrolysis of the carbamate moiety were possible for certain biologically active carbamates. Carbaryl was shown to be metabolized by a rat liver microsome system, requiring NADPH and oxygen to a formaldehyde yielding derivative without loss of anticholinesterase activity.

$$\text{Carbaryl} \xrightarrow{\text{Hydrolysis}} \text{1-Naphthol} + CO_2 + CH_3NH_2$$

Carbaryl was oxidatively metabolised to N-hydroxymethyl derivative (23). The other non-hydrolytic metabolites of carbaryl identified were 4-hydroxy-1-naphthyl N-methylcarbamate (24), 5-hydroxy-1-naphthyl N-methylcarbamate (25) and 5,6-dihydro-5,6-dihydroxy-1-naphthyl N-methylcarbamate (26).

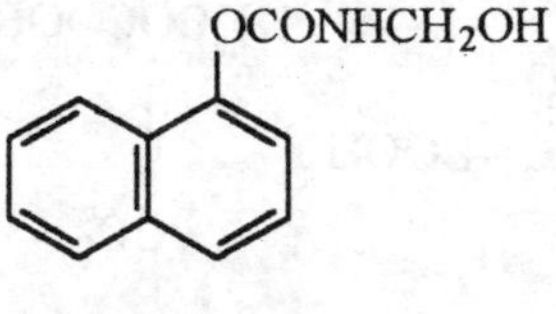

OCONHCH₂OH   OCONHCH₃   OCONHCH₃   OCONHCH₃

(23) 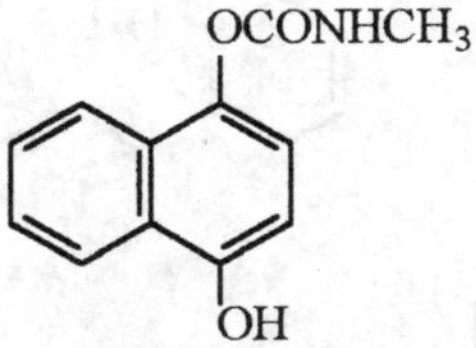   (24) 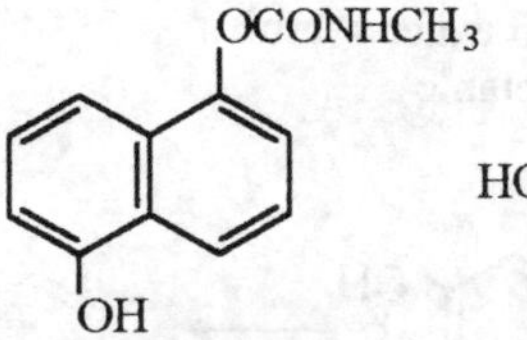   (25) 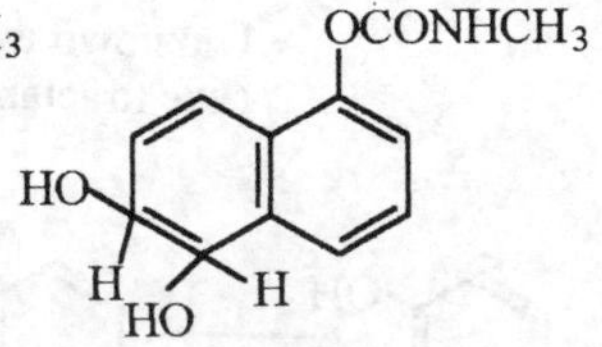   (26)

Water soluble metabolites of carbaryl consist largely of conjugates of the hydroxylated carbamate metabolites shown above. In mammals they appear to be present primarily as glucuronide and sulfate conjugates.

Partial hydrolysis of carbaryl was reported to occur in cotton plants but it is not a major metabolic pathway in plants. In fact, the carbamate ester linkage appears to be quite stable. In insect pests, the degree of hydrolysis of the carbamate ester depends on the species involved. However, oxidative metabolism is usually the major pathway of metabolism.

Earlier it was proposed that the critical step in carbaryl metabolism in insects was its hydrolysis to 1-naphthol and methylamine. Carbaryl and baygon were metabolised by American cockroaches and by houseflies through non-hydrolytic pathways. Non-conjugated metabolites were tentatively identified as 5, 6-dihydro-5, 6-dihydroxy-1-naphthyl N-methylcarbamate, 1-naphthyl N-hydroxy methylcarbamate and the 4- and 5-hydroxy analogues of carbaryl. It was assumed that the water soluble metabolites were conjugates of these hydroxylated metabolites of carbaryl.

Carbaryl metabolism by microorganisms isolated from soil may be predominantly hydrolytic or oxidative. Seldom there is a mixture of both hydrolytic and oxidative products as found in animals.

A strain of *Achromobacter* and an unidentified bacterium degraded carbaryl to 1-naphthol and other products which appeared to be identical to those of *Pseudomonas* sp. produced from naphthalene.

The soil fungi studied, metabolise carbaryl by non hydrolytic mechanisms and produce the same kind of metabolites formed in plants (Fig. 3.131). A series of soil fungi like *Aspergillus terreus, Fusarium oxysporium* and *Penicillium* spp. are capable of metabolising carbaryl to 1-naphthyl N-hydroxy methylcarbamate, 4-hydroxy-1-naphthyl-N-methylcarbamate and 5-hydroxy-1-naphthyl-N-ethylcarbamate.

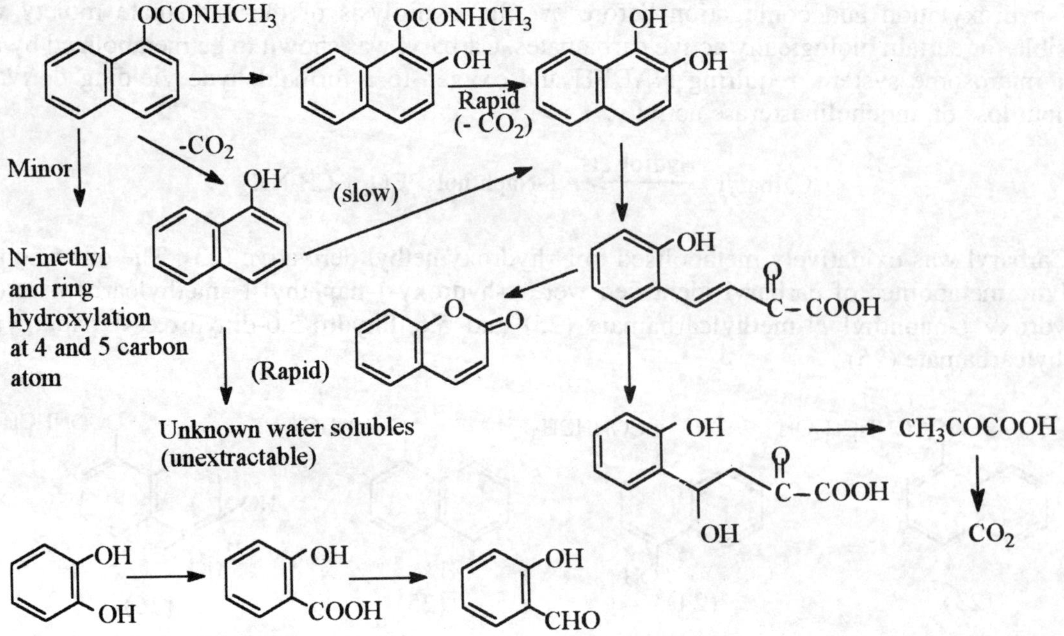

Fig. 3.131. Proposed metabolic pathway of carbaryl in soil.

Carbofuran (Furadan)

Metabolism of carbofuran has been extensively investigated in mammals, insects and plants (Fig. 3.132) and there is an excellent understanding of the degradative pathway of carbofuran in these organisms. Number of review articles and books on various aspects of carbofuran metabolism have been published.

Carbofuran metabolites are of water soluble nature. Only a small percentage extracted into ether comprised of 2,3-dihydro-2, 2-dimethyl-3-hydroxybenzofuranyl-7-N-methylcarbamate (27), 2,3-dihydro-2,2-dimethyl-3-hydroxybenzofuranyl-7-N-hydroxy-methylcarbamate (28) and trace amounts of 2, 3-dihydro-2,2-dimethyl-3-keto benzofuranyl-7 N-methylcarbamate (29).

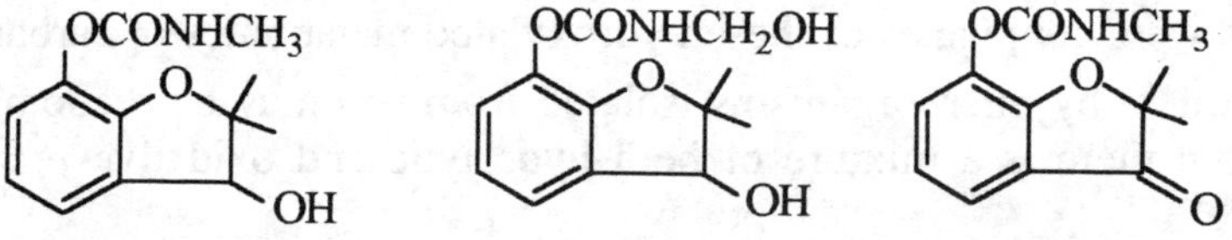

Acid treatment of the water soluble metabolites yielded 50% of the radiocarbon as the 3-ketophenol, the hydrolysis products of metabolite (29); 21% as carbofuran phenol. 1.5% as 3-hydroxy phenol, the hydrolysis products of metabolite (27), 15% as 3-hydroxy carbofuran (27) and 4% as the 3-hydroxy-N-hydroxymethyl derivative (28).

Aldicarb. Because of its high mammalian toxicity, aldicarb is applied commercially in the form of granules. Since the compound is commonly used on cotton crop, the plant metabolism has been

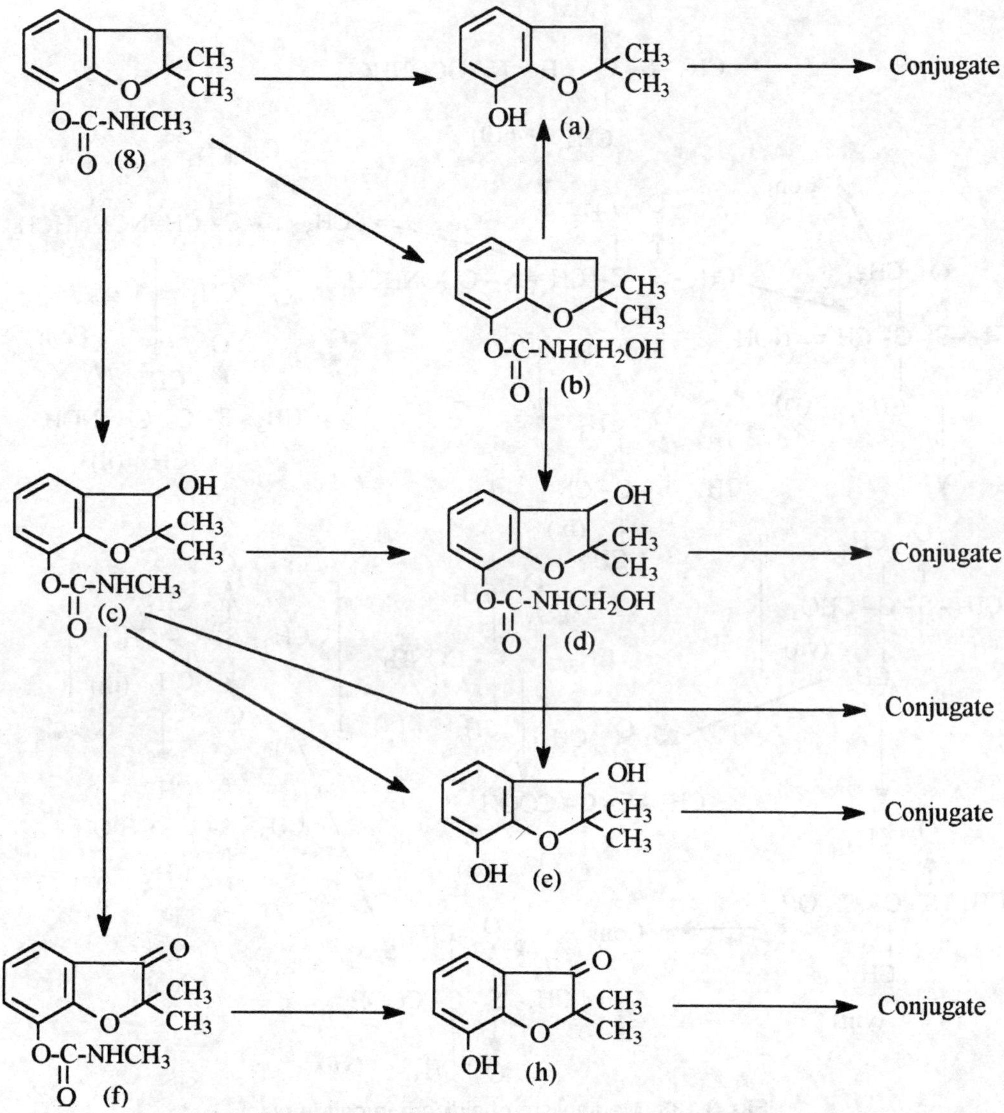

Fig. 3.132. Metabolism of carbofuran in animals.

studied mostly with cotton plants. Studies with rat liver enzymes showed that it is converted to sulfoxide, 2-methyl-2 (methylsulfinyl) propionaldehyde O-(methylcarbamoyl) oxime and to water soluble unknown metabolites (Fig. 3.133). The formation of sulfone (30) from sulfoxide (31) was in very small quantity.

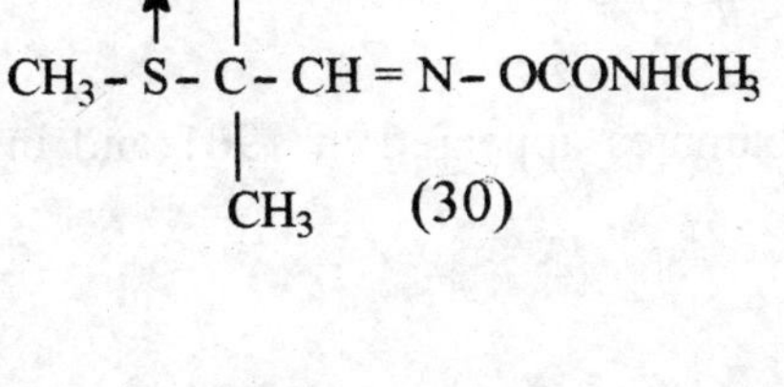

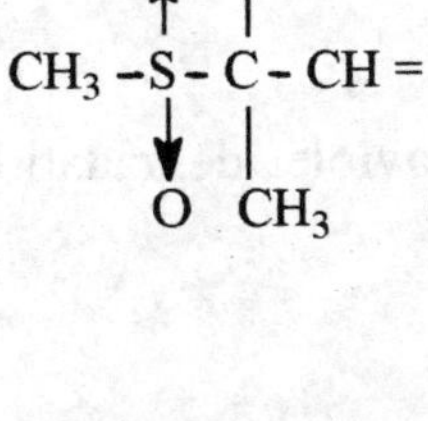

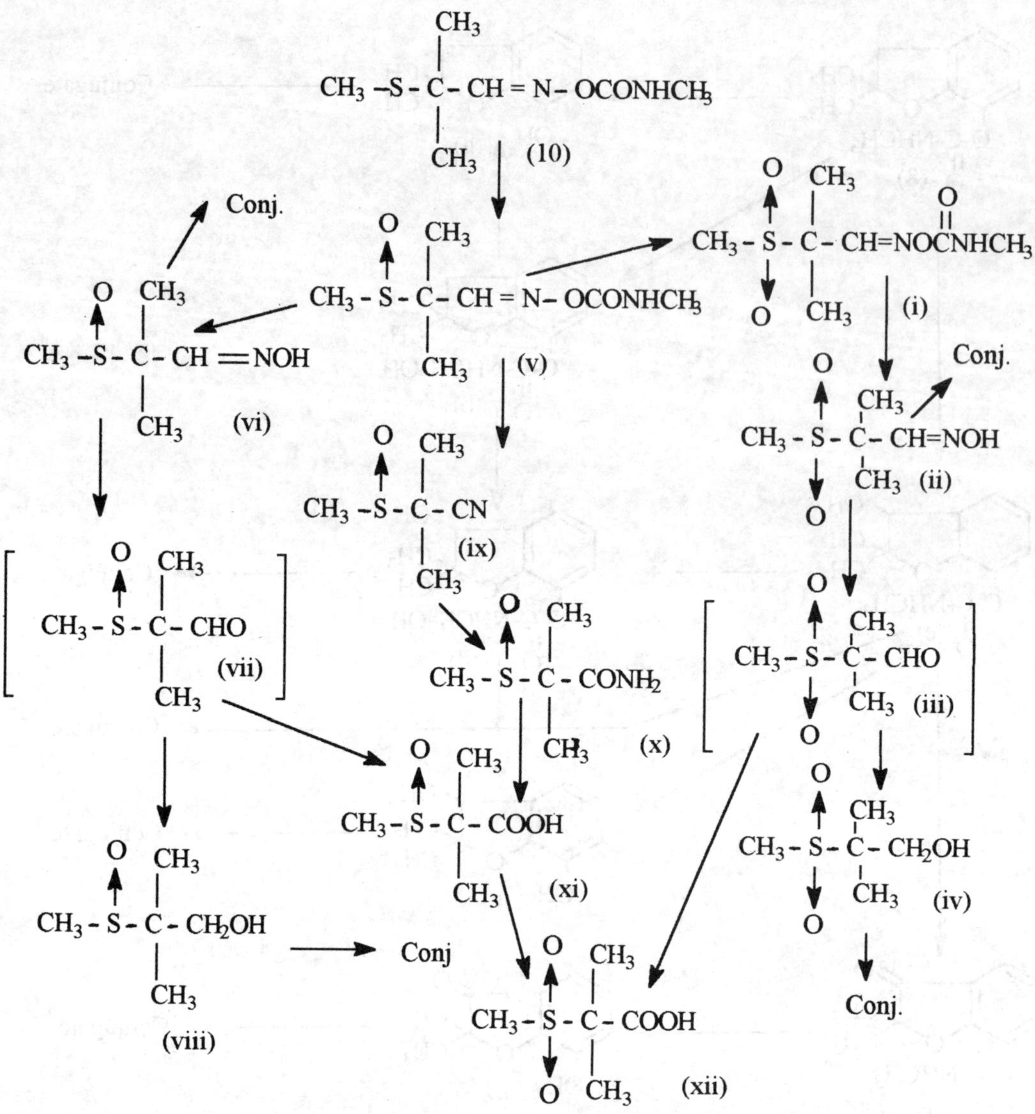

Fig. 3.133. Metabolism of aldicarb in cotton plant.

Metabolism of aldicarb in cotton plants involves oxidation to sulfoxide and sulfone analogues and their hydrolysis leads to the formation of water soluble metabolites. A few water soluble metabolites have also been identified [9].

Methomyl. The metabolic pathway of methomyl in rats and plants involves degradation to carbon dioxide and acetonitrile. There is no formation of S-methyl-N-hydroxythioacetimidate nor the sulfoxide or sulfone of methomyl [10]. In soils, only carbon dioxide is formed.

Photochemistry [3]

The first report about ultraviolet degradation of carbamates appeared in 1961 and in 1965 by

Crosby *et al.* [11]. In general, the photolysis of carbamates involves hydrolysis to phenols or rearrangement to other carbamates and oxides. Silk and Unger [12] have published several papers on the mechanism of photoreaction of carbamates where the product formation depended mainly on solvents and wavelength. The photochemical dissociation proceeds via i) photo-Fries rearrangement which yields phenol and benzamide products and ii) the other results in C-N bond cleavage product followed by hydrogen abstraction to produce monomethyl aminocarbamate. A more interesting photoproduct was obtained in case of 4-methylphenyl N-methylcarbamate. A solution of carbamate (0-21 M) in *t*-butanol resulted in 25% hydrolysis to *p*-cresol and 57% conversion plus rearrangement to N-methyl 2-hydroxy-5-methylbenzamide (Fig. 3.134).

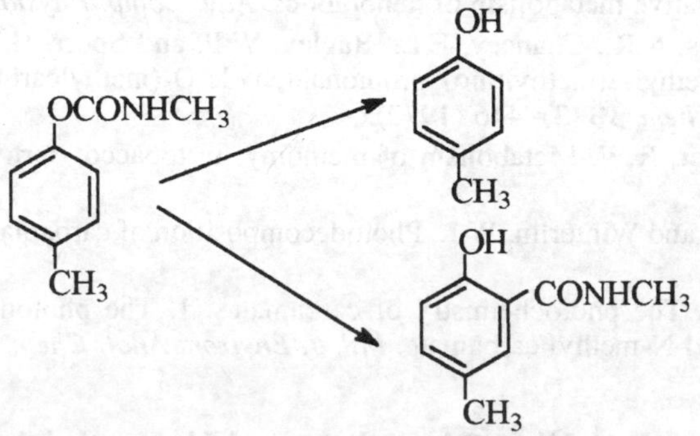

Fig. 3.134. Photolysis of 4-methylphenyl N-methylcarbamate in t-butanol.

Carbaryl formed only one minor inhibitory product after exposure to sunlight or weak UV light, but with intense UV light and using different solvents, several products were formed and identified. Two more carbamates **zectran**, 4-dimethylamino-3, 5-xylyl-N-methylcarbamate and **metacil**, 4-dimethyl amino-3-methyl phenyl N-methylcarbamate were found to be highly prone to photolysis by UV light (254 nm) and produced several photoproducts (Fig. 3.135) [12].

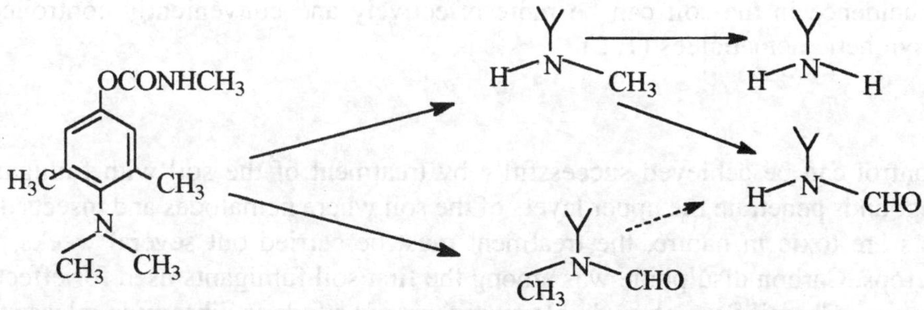

Fig. 3.135. Photolysis of dimethylamino group of zectran and metacil.

Photolysis of carbamates like baygon and carbaryl at different pH, duration of exposure and light intensity was studied in detail. The major degradative pathway was found to be the ester hydrolysis yielding phenol and 1-naphthol respectively.

REFERENCES

1. Geigy, J.R. Swiss Pat. 279, 553, March 1, and 282, 655, August 16 (1952).

2. Gysin, H. *Uber einige nene Insektizide, Chemica,* **8**, 208 (1954).

3. Kuhr, R.J., Dorough, H.W. and Zweig, G. (eds.). *Carbamate insecticides. Chemistry, Biochemistry and Toxicology.* CRC Press, USA (1976).

4. Matolcsy, G., Carbamates. In : *Pesticide Chemistry,* (G. Matolcsy M. Nadasy and V. Andriska, eds.), Elsevier, New York, 90-107 (1988).

5. Cremlyn, R.J. (ed) *Agrochemicals.* John Wiley & Sons, New York, 140-156.

6. Hetnarski, B. and O'Brien, R.D. Preparation and properties of phenylcarbamates with phosphoryl and carbamyl substituents. *J. Agric. Food Chem.* **20**, 543-546 (1972).

7. O'Brien, R.D. (ed.). *Insecticides, Action and Metabolism,* A.P., New York (1967).

8. Smith, J.N. The comparative metabolism of xenobiotics. *Adv. Comp. Physiol. Biochem.,* **3**, 173 (1968).

9. Bartley, W.J., Andrawes, N.R., Chancey, E.L., Bagley, W.P. and Spurr, H.W. The metabolism of temik aldicarb pesticide [2-methyl-2(methylthio) propionaldehyde-O-(methylcarbamoyl) oxime] in the cotton plants. *J. Agric. Food Chem.* **18** (3), 446 (1970).

10. Harvey, J. Jr. and Reiser, R.W. Metabolism of methomyl in tobacco, corn and cabbage. *J. Agric. Food Chem.,* **21**, 775 (1973).

11. Crosby, D.G., Leitix, F. and Winterlin, W.L. Photodecomposition of carbamate insecticides. *J. Agric. Food Chem.* **13**, 204 (1965).

12. Silk, P.J. and Unger, I. The photochemistry of carbamates, 1. The photodecomposition of Zectran : 4-dimethylamino-3,5-xylyl-N-methyl carbamate. *Intl. J. Environ. Anal. Chem.* **2,** 213 (1973).

Section VI.   Chemistry of Nematicides

- Introduction
- Fumigants
- Organophosphorus compounds
- Carbamates
- Bionematicides
- Mode of action
- *References*

Nematodes are soil inhabitants which parasitize the roots of crop plants causing considerable yield loss. Their incidence in the soil can be more effectively and conveniently controlled by using natural and synthetic nematicides (1, 2).

Fumigants

Chemical control can be achieved successfully by treatment of the soil with fumigants. Highly volatile compounds penetrate the upper layers of the soil where nematodes and insects dwell. Since the fumigants are toxic in nature, the treatment must be carried out several weeks prior to the planting of crops. Carbon disulphide was among the first soil fumigants used for effective control of sugar eelworm. Chemicals such as dichloropropene and ethylene dibromide release toxic fumes which kill the nematodes. Another fumigant chloropicrin, was used extensively in Britain and USA to control nematodes. D-D, 1,3-dichloropropene both *cis* (Z-) and *trans* (E-) isomers are effective against eelworms and their cysts. D-D can be prepared by distillation of petroleum cracking product followed by chlorination

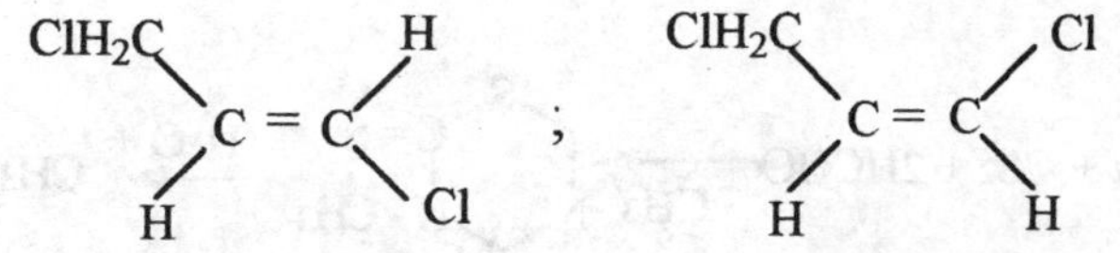

$$\underset{H}{\overset{ClH_2C}{\diagdown}}C=\underset{Cl}{\overset{H}{\diagup}} \quad ; \quad \underset{H}{\overset{ClH_2C}{\diagdown}}C=\underset{H}{\overset{Cl}{\diagup}}$$

(E-isomer)   **(Z-isomer)**

It has the added advantage of increasing the availability of nutrient nitrogen through triggering the microbial activity in the soil. D-D is not persistent in soil as it is easily hydrolysed to 3-chloroallyl alcohols.

Ethylene dibromide, 1,2-dibromoethane, and **methyl bromide** are also used as effective nematicides (soil sterilization in glass houses). But methyl bromide is too volatile for soil treatment and also an ozone depleting agent. Recently a new fumigant carbonyl sulfide (CS) has been developed and registered for use by CSIRO (Australia) which might be a good substitute for methyl bromide.

Nemagon, DBCP, 1,2-dibromo 3-chloropropane, prepared by the treatment of bromine with allyl chloride (Fig. 3.136) is used as soil fumigant. It is less phytotoxic than D-D.

$$H_2C = CH-CH_2Cl + Br_2 \rightarrow Br-CH_2\,CHBrCH_2Cl$$

Fig. 3.136. Preparation of nemagon.

Methyl isothiocyanate was first introduced in 1959 as soil fumigant for the control of soil fungi, insects, nematodes and weed seeds. Being phytotoxic its treatment has to be given sufficiently in advance to allow the chemical to degrade before the crop is planted. MIC is applied in the form of metham sodium which slowly decomposes to methyl isocyanate (MIC) in presence of soil moisture (Fig. 3.137).

$$H_3C-\underset{\overset{|}{H}}{N}-\underset{\overset{|}{S^\ominus}}{C}=S \xrightarrow[H_2O]{} H_3C-N=C=S+S\overset{\ominus}{H}$$

Fig. 3.137. Formation of methyl isocyanate.

Metham sodium can be obtained by the reaction of CS_2 with methylamine in presence of NaOH (Fig. 3.138).

$$CH_3NH_2 + CS_2 + NaOH \longrightarrow CH_3NH.\underset{\overset{|}{SNa}}{C}=S + H_2O$$

Fig. 3.138. Preparation of metham sodium.

Thiadazine-2-thione, dazomet (1), is obtained by the reaction of CS_2, methyl amine and formaldehyde (Fig. 3.139).

The compound acts by decomposition in the soil to give MIC. It is used for the control of soil fungi, nematodes and insects.

Soil fumigation is not cost effective for field crops because of high volatility. It is sometimes used for glass house crops but the technique is not ecofriendly.

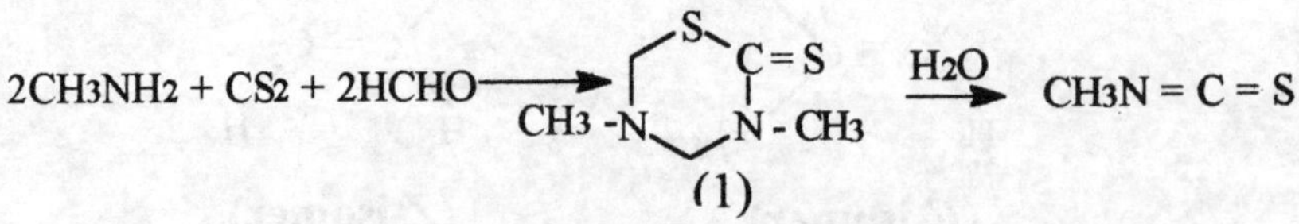

$$2CH_3NH_2 + CS_2 + 2HCHO \longrightarrow \underset{(1)}{[\text{ring structure}]} \xrightarrow{H_2O} CH_3N=C=S$$

Fig. 3.139. Preparation of dazomet.

Recently non-volatile compounds like organophosphorus compounds [3, 4] have been developed which are gaining popularity as nematicides.

Organophosphorus compounds

Organophosphorus compounds are applied to the soil by spraying their aqueous suspensions. They quickly disperse and persist relatively for longer time in the soil. Some of them are water soluble, exhibit systemic properties, and control nematodes which have entered root zone. A few effective nematicides are :

Diamidafos known as Dowco 169, Nellite, phenyl N, N$'$ dimethyl phosphorodiamidate (2) is a hydrophilic molecule with low adsorption on to soil organic matter and very effective against root knot nematodes. A related amidate **fenamiphos** or Nemacur (3), is a good nematicide but possesses high mammalian toxicity. Similarly phosphorothioate containing heterocyclic moiety thionazin, **zinophos** (4), is an effective soil nematicide but exhibits high mammalian toxicity.

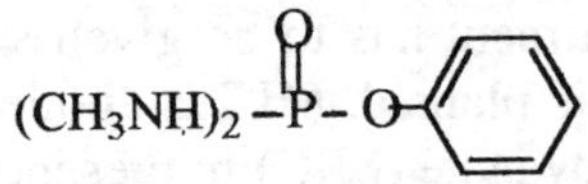

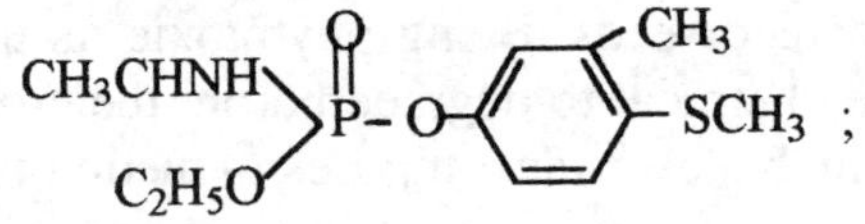

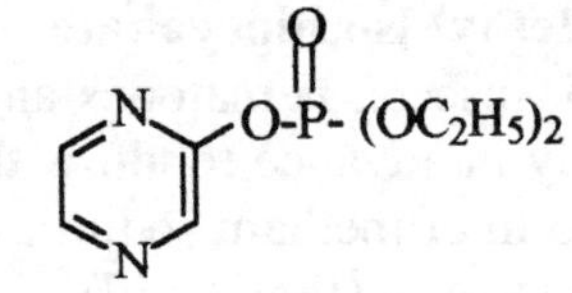

$$(2) \qquad\qquad (3) \qquad\qquad (4)$$

Prophos, Mocap (5), an effective nematicide against corn root worm can be prepared by reaction of propylmercaptan and phosphorodichloridate (Fig. 3.140).

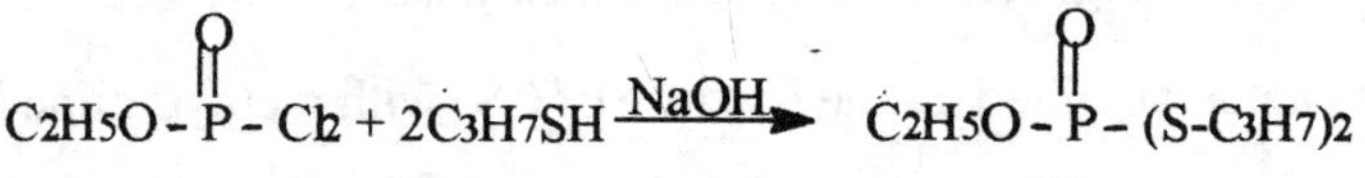

$$C_2H_5O-\overset{O}{\underset{\|}{P}}-Cl_2 + 2C_3H_7SH \xrightarrow{NaOH} C_2H_5O-\overset{O}{\underset{\|}{P}}-(S-C_3H_7)_2$$
$$(5)$$

Fig. 3.140. Preparation of prophos.

Dichlorfenthion, VC-13 nematicide, 2,4-dichlorophenyl diethyl phosphorothioate, (6a), is obtained by the reaction of sodium salt of 2,4-dichlorophenol with diethyl thiophosphorochloridate (Fig. 3.141). It is highly persistent and controls nematodes and soil insects.

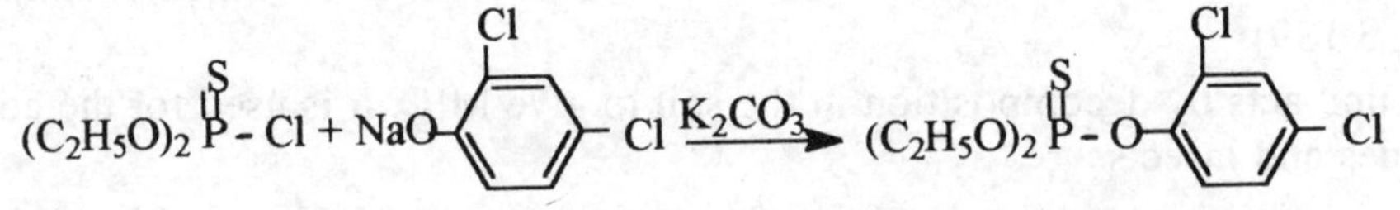

$$(C_2H_5O)_2\overset{S}{\underset{\|}{P}}-Cl + NaO\!-\!\!\langle\text{ring}\rangle\!-\!Cl \xrightarrow{K_2CO_3} (C_2H_5O)_2\overset{S}{\underset{\|}{P}}-O\!-\!\!\langle\text{ring}\rangle\!-\!Cl$$
$$(6a)$$

Fig. 3.141. Preparation of dichlorfenthion.

FMC 67825, Rugby, O-ethyl, S,S-di-*sec*.butyl phosphorodithioate (6b), is a broad spectrum nematicide.

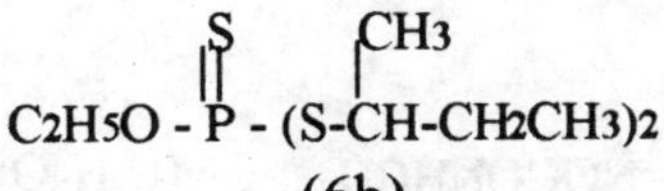

$$C_2H_5O - \overset{\overset{S}{\|}}{P} - (S-\overset{\overset{CH_3}{|}}{CH}-CH_2CH_3)_2$$

(6b)

A large number of phosphorothioates and phosphonates were synthesised at Indian Agricultural Research Institute, New Delhi, India [5, 6, 7]. Some of the compounds like 7a, 7b and 7c exhibited good activity against root-knot nematode *Meloidogyne incognita*. These compounds were prepared by reacting phosphorochloridates with substituted phenols.

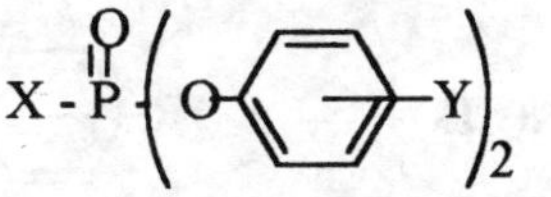

$$X - \overset{\overset{O}{\|}}{P} \left(O - \overset{}{\bigcirc} - Y \right)_2$$

(7a)

X = CCl$_3$, CH$_3$S, Cl$_2$CH,
    CH$_3$, CH$_3$CCl$_2$, ClCH$_2$CH$_2$
Y = 2,4-Cl$_2$, 2,4,5-Cl$_3$, Cl$_5$

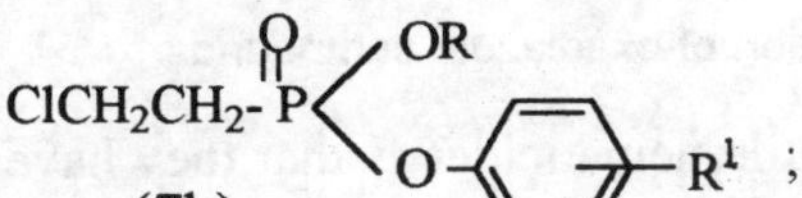

$$ClCH_2CH_2 - \overset{\overset{O}{\|}}{P} \overset{OR}{\underset{O - \bigcirc - R^1}{<}}$$

(7b)

R = OCH$_3$, OC$_2$H$_5$
Rl= 2, 4,5-Cl$_3$, Cl$_5$

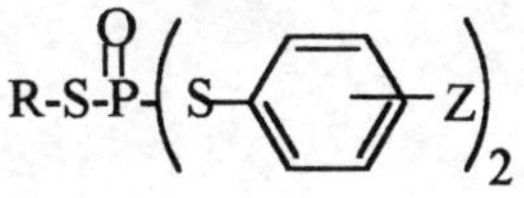

$$R-S-\overset{\overset{O}{\|}}{P} \left(S - \bigcirc - Z \right)_2$$

(7c)

R = CH$_3$, C$_2$H$_5$, i-C$_3$H$_7$
Z = H, CH$_3$, Cl

Carbamate compounds

The next important non-volatile nematicides are carbamates [3, 8]. The commonly used carbamates are carbofuran, carbosulfan, aldicarb and oxamyl.

Carbofuran is a systemic nematicide used for corn, sugar beet, sugarcane, peanut, tobacco and vegetables. **Carbosulfan** (8) is another nematicide (also referred as pro-insecticide) used in vegetables like *Brassica* spp. and carrots.

The oxime carbamates like aldicarb (9) and oxamyl (10) are highly active systemic nematicides used on wide variety of crops vegetables (potatoes, onions etc.), strawberries and ornamental plants.

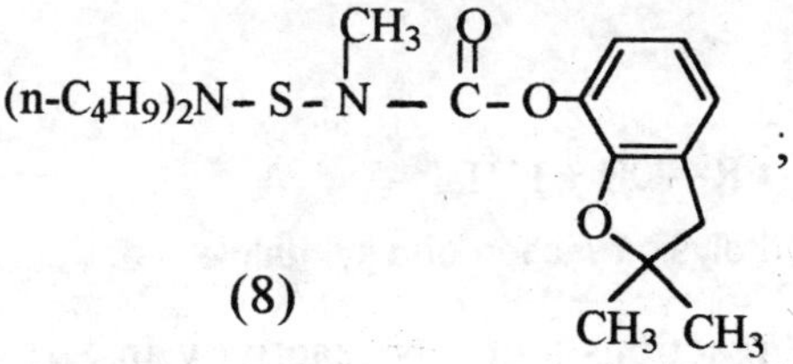

$$(n\text{-}C_4H_9)_2N - S - \overset{\overset{CH_3}{|}}{N} - \overset{\overset{O}{\|}}{C} - O - \bigcirc$$

(8)

$$CH_3S-\overset{\overset{CH_3}{|}}{\underset{CH_3}{C}}-CH = N\overset{\overset{O}{\|}}{OC}-NHCH_3$$

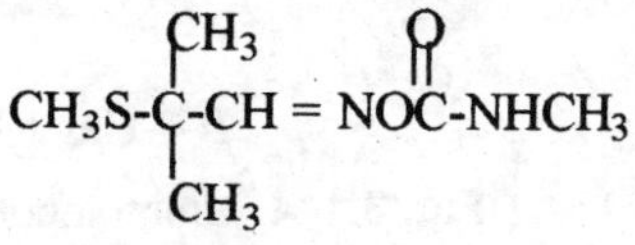

(9)

AC-64, 475, cyclic methylene (diethoxy phosphonyl) dithioiminocarbonate (11) introduced by American Cyanamide Company has been found to be an effective broad spectrum nematicide and insecticide which probably decomposes to the toxic thiocyanate ion [9a].

$$(CH_3)_2NCOC(SCH_3)=NOCONHCH_3 \quad ; \quad (C_2H_5O)_2 - \overset{\overset{O}{\|}}{P} - N = C\overset{S}{\underset{S}{\diagdown}}CH_2$$

$$(10) \qquad\qquad\qquad\qquad (11)$$

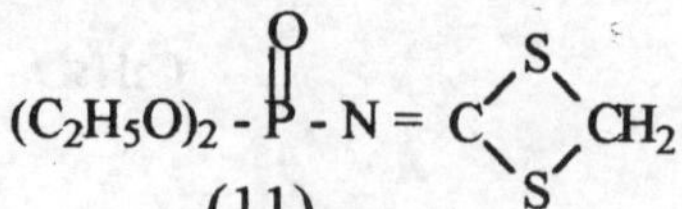

A series of aryl substituted 2-mercapto-1,3,4-oxadiazole compounds (12) were synthesised [10] by the reaction of aromatic acid hydrazides with tetramethyl thiuramdisulfide (Fig. 3.142). A few of these compounds were found to be quite active against *M. incognita*.

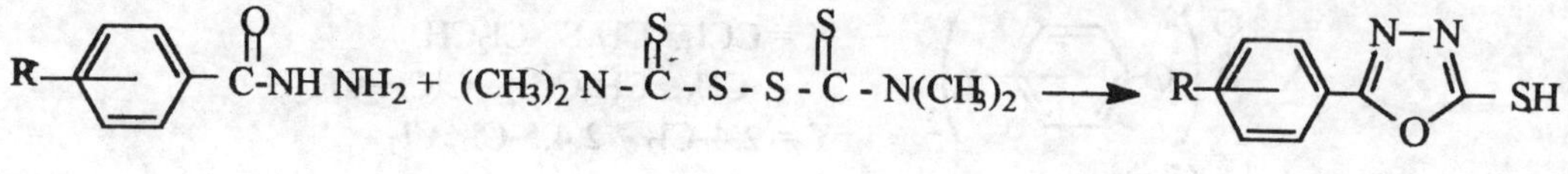

Fig. 3.142. Preparation of oxadiazole compounds.

The only disadvantage of these non-volatile nematicides is that they have a long residual life and many possess high mammalian toxicity. Hence bionematicides would play, in near future, a major role in the nematode management programme.

Bionematicides are discussed in Chapter 2.

Mode of action

Volatile nematicides like methyl bromide, EBD and D-D act on the nucleophilic sites like OH, SH and NH_2 groups of the vital enzymes in the nematodes (Fig. 3.143).

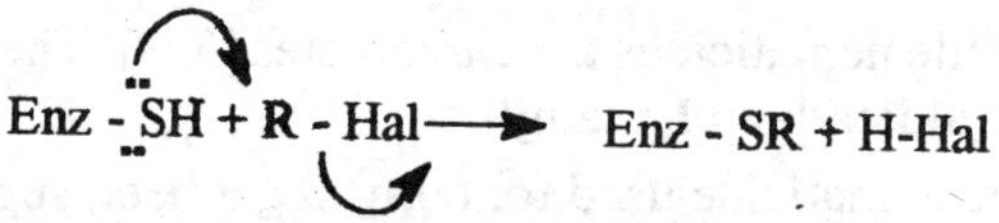

$$Enz - \overset{..}{S}H + R - Hal \longrightarrow Enz - SR + H\text{-}Hal$$

Enz = Remainder of the enzyme molecule

Fig. 3.143. Reaction of enzyme with alkyl halide.

This reaction is a bimolecular nucleophilic substitution reaction (S_N2) which again competes with unimolecular hydrolysis (S_N1) reaction carried out by excess water present in the soil for deactivation (Fig. 3.144).

$$R\text{---}Hal + H_2O \rightarrow R\text{---}OH + H\,Hal$$

Fig. 3.144. Unimolecular hydrolysis reaction of alkyl halide.

Thus molecules with high reactivity in S_N2 reactions and low reactivity in S_N1 would be the most active nematicides. Alkyl isothiocyanates also exert their nematicidal activity in a similar way (Fig. 3.145) (S_N2 type reaction).

$$R-N = C = S + Enz\text{-}SH \longrightarrow R\text{-}NHC = S$$
$$\underset{S\ Enz}{|}$$

Fig. 3.145. Reaction of enzyme with alkyl *iso*thiocyanate.

The movement of non-volatile nematicides depends on the degree of adsorption on soil colloids. The quantity of adsorption depends on the nature of the compound and soil characters like particle size, temperature, moisture and organic matter content. It is well established that soils rich in organic matter adsorb nematicides more readily and remove the chemicals from the soil solution. As a result, the chemical becomes unavailable for the control of nematodes present in soils. The partition co-efficient (Q) of the chemical in soil is given by the equation.

Q = Chemical in soil organic matter/Chemical in soil solution

For a chemical to be a good nematicide the value of Q should be very low [9] and for carbamate compounds the value of Q is approximately 10 while for organophosphorus compounds it is 150.

The organophosphorus and carbamate nematicides also react with cholinesterase enzyme by phosphorylation or carbamoylation similar to their respective insecticides.

REFERENCES

1. Woods, A. (ed.). *Pest Control—A Survey*. McGraw Hill, London, 102 (1974).

2. Ware, G.W. (ed.) *Fundamentals of Pesticides*. 2nd Ed., Thomson Publication, Calif., USA, 169 (1986).

3. Matolcsy G. Nematocides. In : *Pesticide Chemistry*. (G. Matolcsy, M. Nadasy and V. Andriska eds.) Elsevier, New York, 256-260 (1988).

4. Eto, M. (ed.) *Organophosphorus Pesticides—Organic and Biological Chemistry*, CRC Press, Cleveland, Ohio, 301-304 (1974).

5. Gupta, R.L. and Roy, N.K. Recent developments in the field of organophosphorus nematicides. *Pesticides* **21** (10), 15-22 (1987).

6. Panda, A.K., Prasad, D. and Roy, N.K. Nematicidal activity of O-aryl O-ethyl 2-chloroethyl phosphonates on soybean infected with *M. incognita. Ann. Pl. Prot. Sci.*, **6** (1), 56-62 (1998).

7. Gupta,R.L., Roy, N.K. and Prasad, D. Evaluation and quantitative structure activity relationship of S-alkyl S,S-diaryl phosphorotrithioates for the nematicidal activity against root knot nematode (*M. incognita*). *Fundam. appl. Nematol.* **16** (2), 97-102 (1993).

8. Tomlin, C. (ed.). *The pesticide manual*. 10th Ed., British Crop Prot. Council Publications, U.K. (1995).

9a. Whitney and Aston, J.L. AC64, 475—A new systemic nematicide and insecticide. *Proc. Eighth British Insecticide and Fungicide Conf.* Brighton. **2**, 625-632 (1975).

9b. Hague, N.G.M. Nematicides—past and present. *Proc. Eighth British Insecticide and Fungicide Conf.*, Brighton. **3**, 837-851 (1975).

10. Rusu, G.G., Barba, N.A., Gutzu, J.E., Okopnyi, N.S. and Krimer, M.Z. Synthesis and nematocidal activity of some new 5-aryl-2-aryl-2-mercapto 1,3,4-oxadiazolederivatives. *Ninth Internatn. Congr. Pestic. Chem.*, London, **1** (1-4), abs. 1A-015 (1998).

Section VII.   Chemistry of Rodenticides

- Introduction
- Methods for the control of rodents
- Characteristics of an ideal rodenticide
- Classification of rodenticides
- Inorganic compounds
- Fumigants
- Botanical origin
- Synthetic origin
- SAR studies of 4-OH-coumarin derivatives

- Formulations
- Advantages and disadvantages
- Alternative ways of control
- Chemosterilants
- Mode of action
- Applications
- Resistance to rodenticide
- Future outlook in rodent management
- *References*

Rodents such as rats, mice, gophers and ground squirrels spread diseases like plague, ratbite fever and leptospiral jaundice in human beings. They damage the standing crops and cause substantial loss during storage of the produce. Valuable installations like lead pipes, metal sheathed cables, insulated electrical and telephone wirings, plastics and hardwoods and household items like clothes, books and furniture are cut and damaged by these rodents. It is estimated that about 33 million tonnes of food is lost due to rodents every year in the world. Control of the rodent menace can save as much as 42% of our annual food grain productivity. So there is an urgent need to control and eliminate the ever increasing rodent population in country.

Methods for the control of rodents [1-4]

i) **Mechanical** method of eliminating rodents involves the use of different types of traps. The method is ideally suited for domestic application where rodent population is not large.

ii) **Biological** control of rodent population involves the use of predators like cats or using baits, carrying bacteria like *Salmonella enteretidis*. Maintenance of the predators and non-specificity of the disease causing organisms have greatly limited the use of this method.

iii) **Fumigation.** Volatile compounds such as CS_2 are commonly used to control rats.

iv) **Chemical methods.** Quick and effective control of rodents on large scale has been possible only through the use of chemical rodenticides.

v) **Alternative methods of control.** These involve the use of attractants, repellents and chemosterilants.

Characteristics of an ideal rodenticide

- The toxic action should be slow so as to allow the animal to consume a lethal dose.
- It should be palatable and odourless.
- Symptoms of acute poisoning should be absent. Bait shyness must be avoided.
- The poison should be specific to the species to be controlled.
- The manner of death should not be cruel and make the surviving population suspicious.
- The susceptibility should be age, sex or strain dependent
- There should be no danger of secondary poisoning through animals eating the poisoned rodents.
- Consumption of the chemical should not lead to development of resistance.

- The chemical when mixed with the bait must be stable under varied environmental conditions.
- For easy removal of corpses, the animal should preferably die in the open space.

CLASSIFICATION OF RODENTICIDES

Inorganic compounds

In general these compounds are non-selective in action and for this reason, they are not frequently used. Barium carbonate is one of the oldest compounds is still used as rodenticide. Zinc phosphide, however, has proved to be valuable for the extermination of rats.

Zinc phosphide (Zn_3P_2), a greyish black powder has LD_{50}, 40 mg/kg : It has a strong disagreeable odour which surprisingly does not deter rats and is sometimes said to possess certain attractive properties. The instability of Zn_3P_2 in presence of moisture reduces the effectivity of the baits. For lasting effects, baits are sometimes wrapped in waxed or waterproof paper or mixed with mineral oil rather than water. In U.K., cereal baits such as soaked wheat or medium oat meal containing 2.5% zinc phosphide have been used successfully. The advantages of low cost and secondary toxicity together with fairly good safety record have made it a commonly used rodenticide. Aluminium phosphide is another rat poison having attributes similar to zinc phosphide.

Arsenious oxide has a toxicity depending upon the particle size of the preparation, for white rats LD_{50} is 60 mg/kg at < 5 μ diameter but 148 mg/kg are required if the granule size is > 100 μ. It is one of the oldest rodenticides which is not in use now due to general restriction on the sale of arsenic compounds and inconsistent results. Thallous sulphate (Tl_2SO_4), a cumulative poison, is one of the most effective rodenticide having LD_{50}, 16 mg/kg which is not generally used due to health hazards. Yellow phosphorus is also used as rodenticide.

Fumigants [5]

Rodent menace in warehouses, food stores and granaries is not always effectively controlled by poison baiting, trapping and other direct control methods. It may often be beneficial to use fumigation technique for better control. The penetrating attribute of fumigants make them reach inaccessible areas ensuring complete rodent extermination.

Most frequently employed fumigants are hydrogen cyanide (HCN) and methyl bromide (MeBr) but a number of other gases and volatile liquids such as SO_2, solid CO_2, CS_2 and CO are also in vogue.

The extermination of outdoor colonies of rodents is normally carried out with hydrogen cyanide gas. It is customary to blow or inject calcium cyanide $Ca(CN)_2$ in granular or dust form into a burrow so that when it comes into contact with moist air or soil the HCN gas is liberated and kills rodents. Methyl bromide has also been used in burrow fumigation. Ampoules containing the volatile liquid, MeBr are carefully broken deep in the burrow system to kill the rodents.

Botanical origin [6]

Red squill is the oldest known rodenticide of botanical origin. It is extracted from the bulb of the lily like plant *Urginea maritima* belonging to family *Liliaceae*. The toxic principle is known as **scilliroside** (1) [7]. It is crystalline, yellow in colour, mp 168°C, soluble in alcohols and dioxane but insoluble in water, LD_{50} to rats 0.43-0.7 mg/kg.

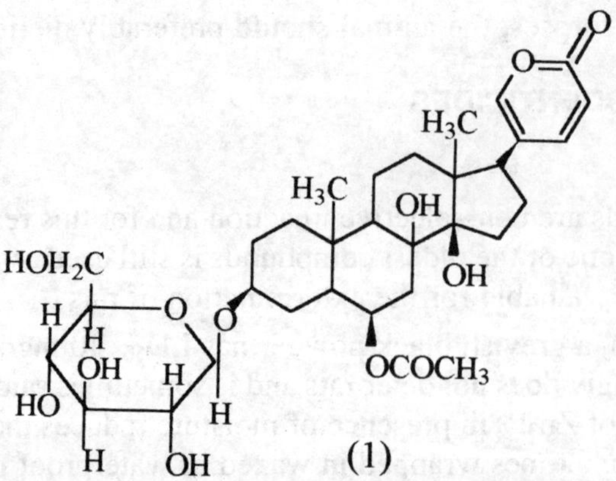

(1)

The main disadvantage of red squill as a rodenticide is the variable potency of the extract which sometimes cause violent death, and is one of the reasons for banning it in many countries.

Strychnine [8], is obtained from the seeds of *Strychnos nux-vomica* and other related species of family *Logamiaceae*. It has a mp of 268-290°C and forms salts with strong acids. Strychnine (2) is effective against house mice and other rodents with LD_{50} 50 mg/kg. Since it is toxic to warm blooded animals, it is banned in the U.K.

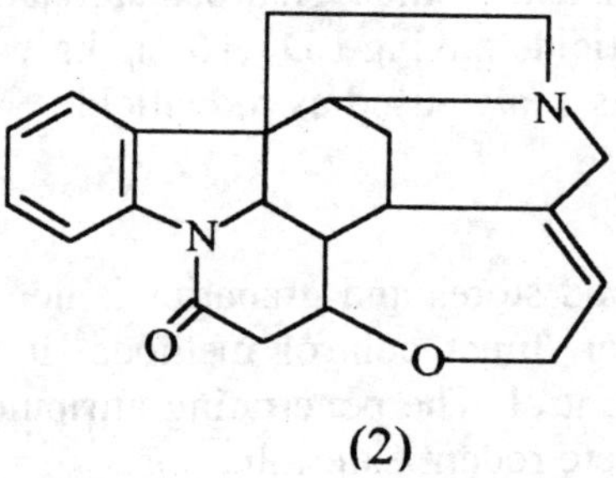

(2)

Reserpine is an alkaloid, extracted from the roots of various species of *Rauwolfia* specially *R. serpentina* and *R. vomitoria* [9]. At present, it is produced synthetically for the treatment of hypertension. A number of analogues of reduced potency like Yohimbine and Desperidine have been extracted from *Rauwolfia* species. Reserpine (3) is generally much more effective in controlling mice than rats with LD_{50} to mice 200 mg/kg.

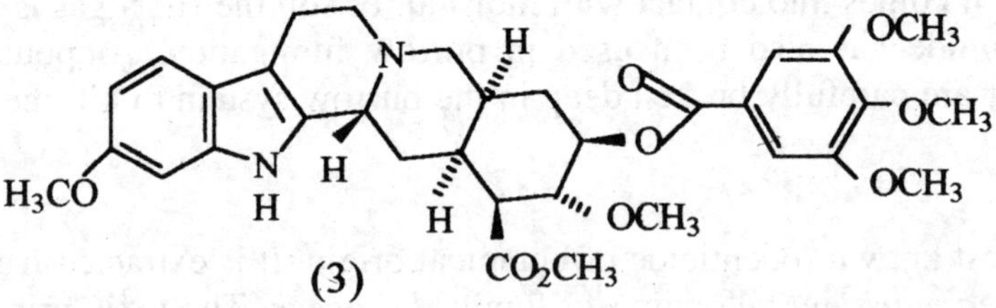

(3)

Ergocalciferol and cholecalciferol [10] have been introduced recently as rodenticides. Ergocalciferol (4) known as Vit. D_2 and cholecalciferol (5) as Vit. D_3 are generally used in small quantities for maintaining good health. However, large quantities of these vitamins can cause toxic symptoms leading to an increase in calcium levels in the blood. The resulting hypercalcemia gives rise to calcification and degeneration of various soft tissues, particularly kidneys, lungs and heart. These can be recommended as alternative to warfarin which is developing resistance to rodents.

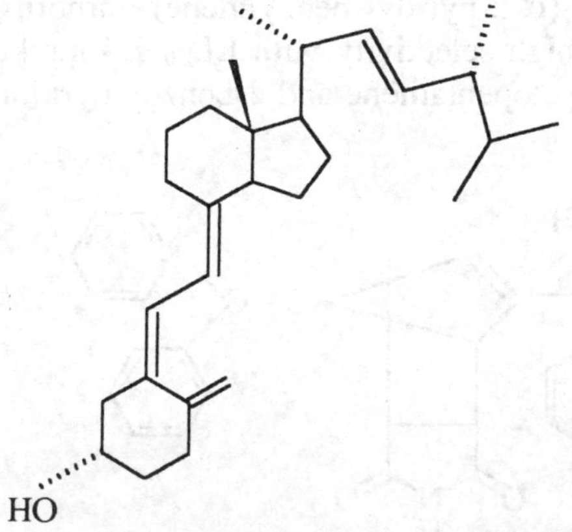

HO   HO

(4) 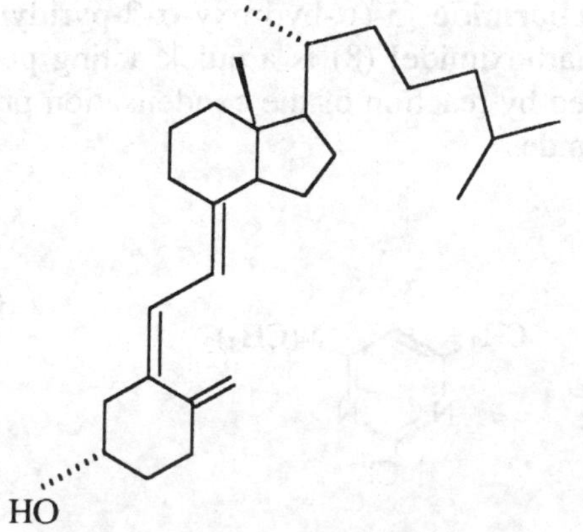   (5)

Synthetic origin [11, 12]

A large number of synthetic organic compounds have been found effective as rodenticides. Some important rodenticides amongst those are described below.

ANTU, Alpha naphthylthiourea (6), is probably the first synthetic organic rodenticide developed and is prepared by treating α-naphthylamine with ammonium thiocyanate (Fig. 3.146), mp 198°C and is almost insoluble in water (0.06%), but soluble both in triethylene glycol (9%), and acetone (2.4%) with LD_{50} 7 mg/kg.

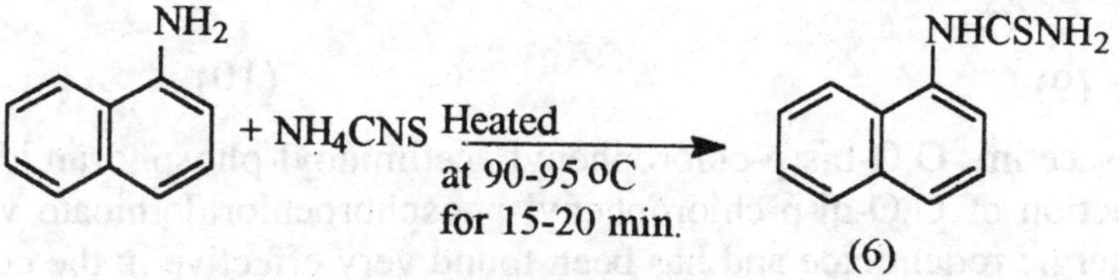

NH_2 $NHCSNH_2$

+ NH_4CNS $\xrightarrow[\text{for 15-20 min.}]{\text{Heated at 90-95 °C}}$

(6)

Fig. 3.146. Preparation of naphthylthiourea.

Crimidine, Castrix, 2-chloro-4-dimethylamino-6 methyl pyrimidine (7), was developed in Germany during the Second World War. This is a powerful convulsive agent with LD_{50} 1 mg/kg. It is prepared by chlorination of the condensation product of acetoaceticethyl ester and urea and subsequent reaction of the intermediate formed with dimethyl amine.

Sodium fluoroacetate, FCH_2COONa, is used as rodenticide under designation "1080" with LD_{50} 3 mg/kg, bp 167°C, mp 31°C. It occurs naturally in the South African plant and is used for the control of rats, mice, ground squirrels and other predators but due to its high mammalian toxicity, its use is limited to professional pest control operations only. Many derivatives of monofluroacetic acid which possess similar toxicity appear in the literature. These are :

(a) FCH_2-C(O)R   where R = NH NH Ph and its derivatives

(b) FCH_2-CH(OH)-R   where R = H, CH_2F or CH_2Cl

(c) FCH_2CONH_2

Norbormide [5-(α-hydroxy-α-2-pyridyl benzyl)-7-(α-2-pyridyl benzylidene) norborn-5-ene-2,3-dicarboximide] (8) is a quick acting poison with high selectivity with LD_{50} 7.4 mg/kg, It is prepared by reaction of the condensation product of cyclopentadiene and 2-benzoylpyridine with maleimide:

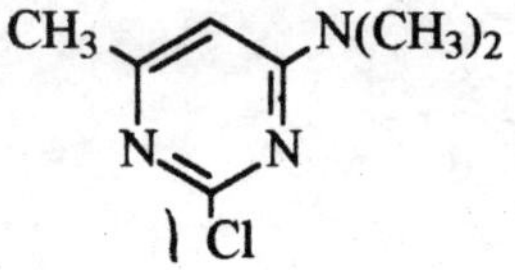

(7)

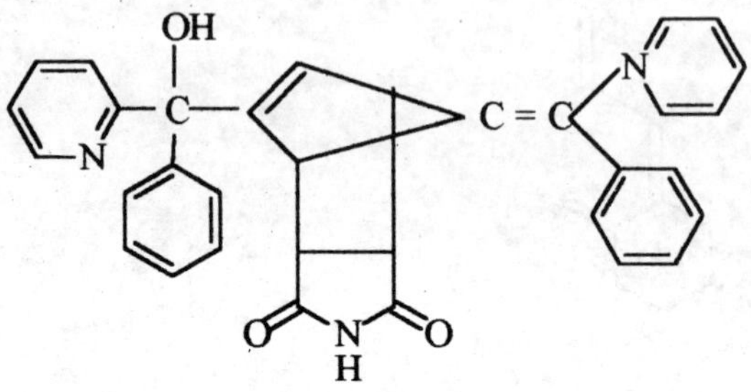

(8)

Alphachloralose (9), is specific to mice and acts by slowing the metabolic processes, so that the rodent dies by hypothermia. The LD_{50} is 300 mg/kg. It is prepared by reaction of chloral with glucose and exists in two stereoisomers (and). It is primarily a sleep inducing narcotic drug and also used to control mice and birds.

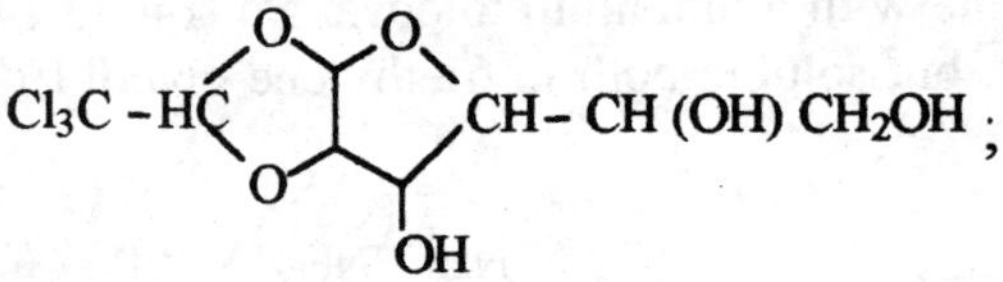

(9)

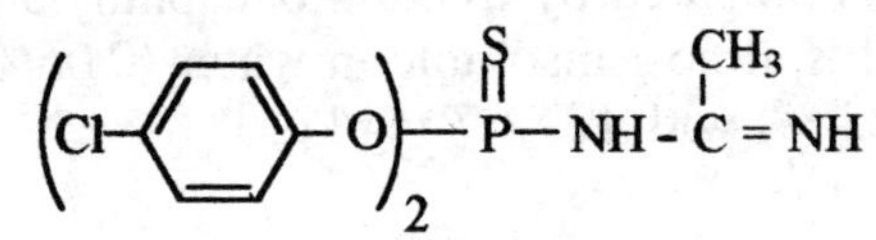

(10)

Gophacide, phosacetim, O,O-bis(p-chlorophenyl)acetimidoyl-phosphoramidothioate (10), is prepared by the reaction of O,O-di-p-chlorophenyl phosphorochloridothioate with acetamidine. This is a new cholinergic rodenticide and has been found very effective in the control of deemice and pocket gophens.

Chlorosilatran (11), is claimed to be an effective fast-acting control agent with LD_{50} 1-4 mg/ka. The **1,1'-methylene bis thiosemicarbaze** (12), is also a fast-acting rodenticide.

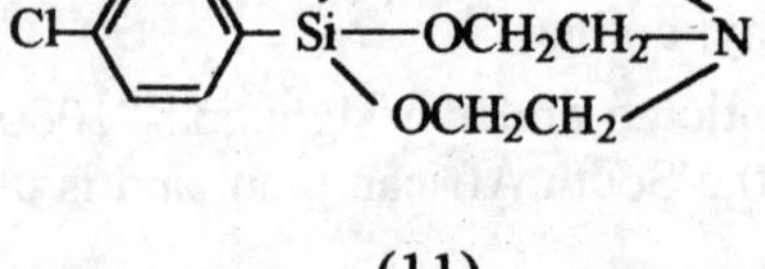

 , NH_2-C(S) NH $NHCH_2NHNH$-C(S)NH_2

(11)   (12)

Flupropadine (13), is very effective against rats and mice. The urea pyriminyl (14) is a very potent broad spectrum rodenticide, effective as single dose against rats resistant to warfarin.

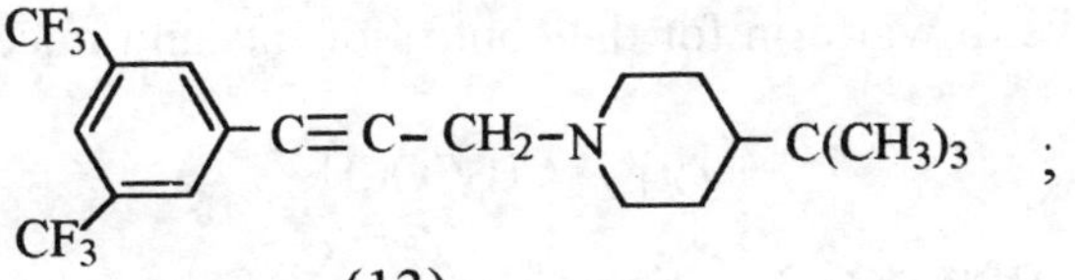

(13) 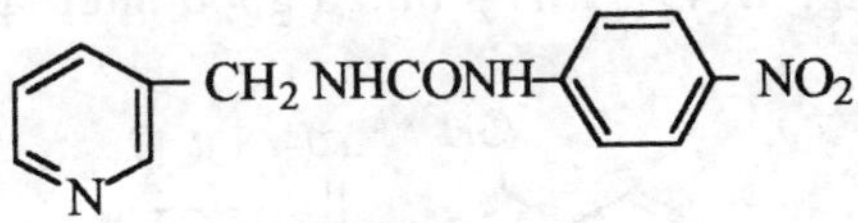   ;   (14)

Hydroxycoumarins

A large number of compounds containing 4-hydroxy coumarin group [13] have been found very effective. These compounds were developed as a result of efforts made by Campbell and Link, and have anticoagulant action [13]. These are known as chronic rodenticides. The most important characteristic of this class of compounds is that the symptoms of poisoning are so much delayed that the animal is unable to associate the discomfort with bait consumption and continues to feed until a lethal dose has been ingested. The cumulative slow acting nature of these materials is characteristic of this type of poison.

Dicoumarin, 3,3^1-methylene bis (4-hydroxycoumarin) (16), is prepared by condensation of 4-hydroxycoumarin (15) with formaldehyde (Fig. 3.147).

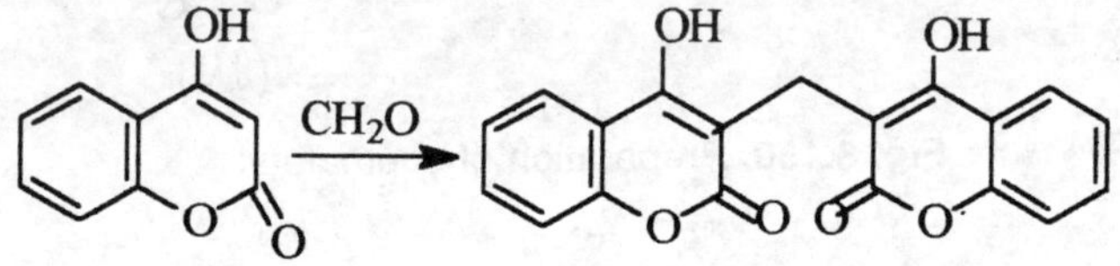

(15)   (16)

Fig. 3.147. Preparation of dicoumarin.

Continued ingestion of this compound at 2 mg level per day produces fatal hemorrhage, due to interference with the action of vitamin K. It has low toxicity in acute dosages with LD_{50} (oral) to rat 540 mg/kg.

Warfarin, 3-(1-phenyl-2 acetylethyl)-4 hydroxycoumarin (17), can be prepared as follows (Fig. 3.148).

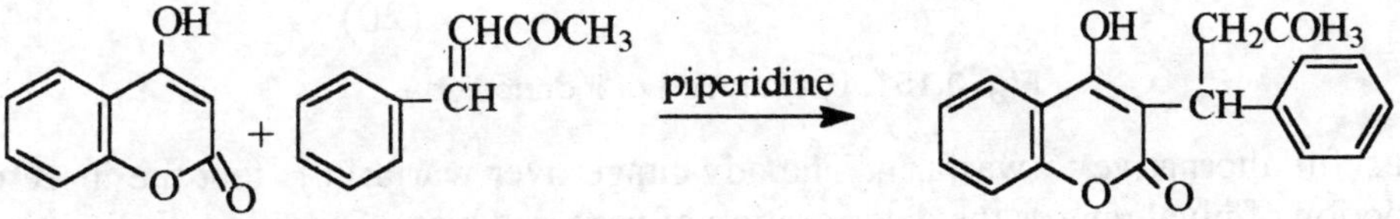

(17)

Fig. 3.148. Preparation of warfarin.

It is a widely used anticoagulant, effective against rats and house mice. It is odourless and tasteless and is readily accepted because rodents do not tend to show bait shyness after testing the material once. They continue to consume it till its anticlotting properties have caused death through internal hemorrhage.

Coumachlor, 3-(2-acetyl-1-p-chlorophenylethyl) 4-hydroxycoumarin (18), is prepared as follows (Fig. 3.149). It is found a good alternative to warfarin for the control of rats and mice.

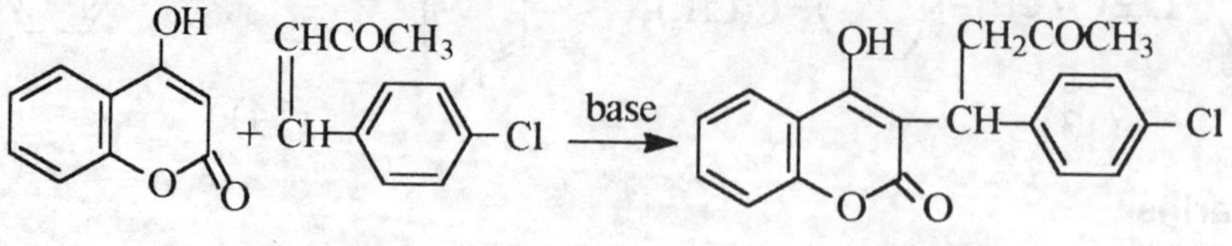

Fig. 3.149. Preparation of coumachlor.

Coumafuryl, fumarin, 3-(2-acetyl-1-furylethyl) 4-hydroxy coumarin (19) is prepared as follows (Fig. 3.150).

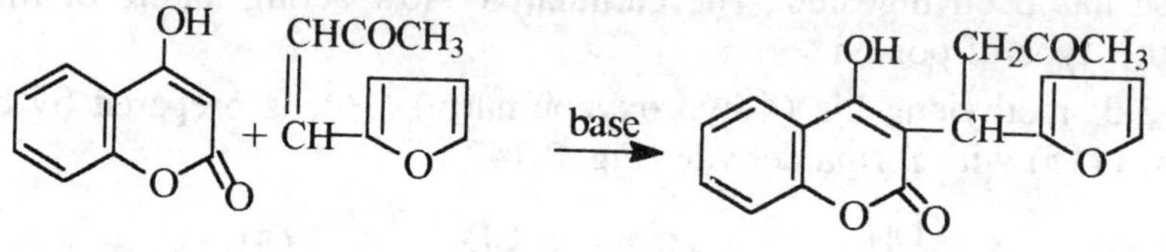

Fig. 3.150. Preparation of coumafuryl.

Indanediones

The rodenticidal action of 1,3-indanedione is based on its anticoagulant activity similar to coumarins. The most active compound is **Pival**, pindone, 2-pivalyl-1, 3-indanedione (20), and is prepared as shown below (Fig. 3.151).

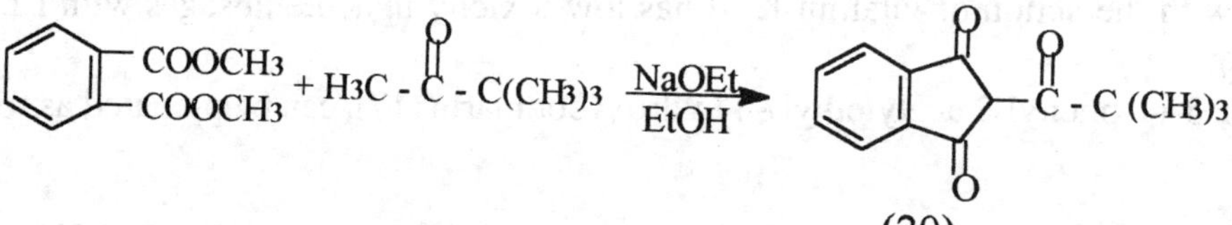

Fig. 3.151. Preparation of indanedione.

It is an useful alternative to warfarin. The advantage over warfarin is that the insecticidal and fungistatic action of pival retards the deterioration of prepared baits. Other indanedione derivatives found useful as rodenticides are :

Diphacinone, 2-diphenylacetyl-1,3-indanedione (21), can be prepared as follows (Fig. 3.152) :

Chlorophacinone, 2-[1-(p-chlorophenyl)-1-phenyl]acetyl-1,3-indanedione (22), is prepared as follows (Fig. 3.153) :

Brodifacoum, 3-[3-(4-bromobiphenyl-4-yl)-1,2,3,4-tetrahydronaphth-1-yl)-4-hydroxy-

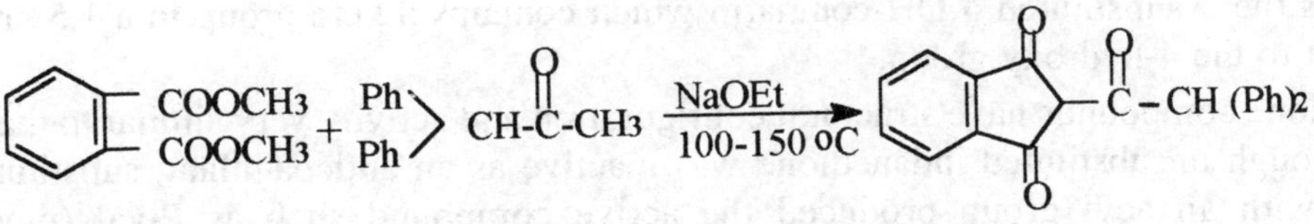

Fig. 3.152. Preparation of diphacinone.

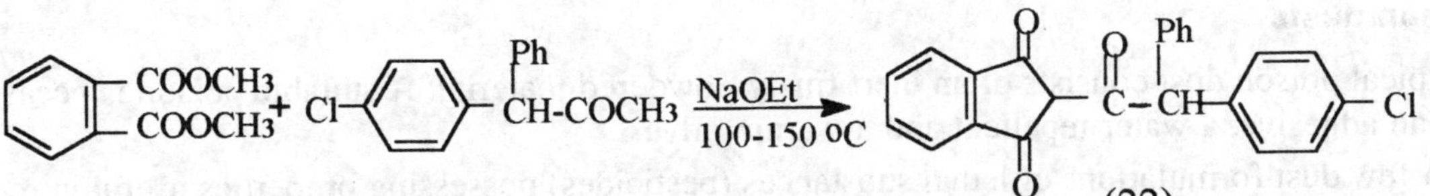

Fig. 3.153. Preparation of chlorophacinone.

coumarin (23a), has the capability to control resistant rodents as well as several noncommensal species at very low doses. It is very effective against Norway rats, roof rats and house mice and no bait shyness is observed.

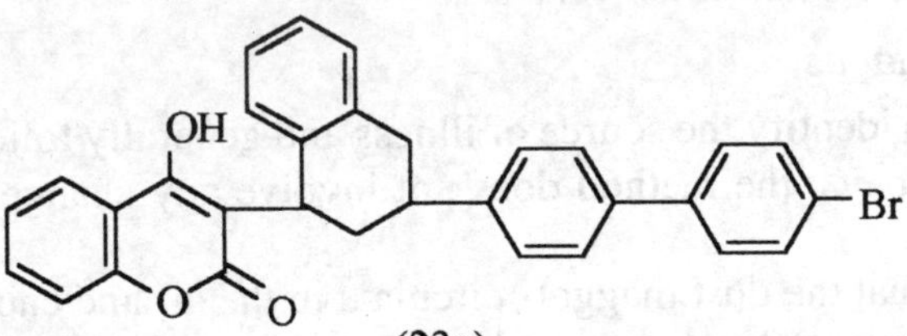

(23a)

Another most potent analogue is **bromadiolone** (23b), capable of achieving high mortality in Norway rats and many resistant species.

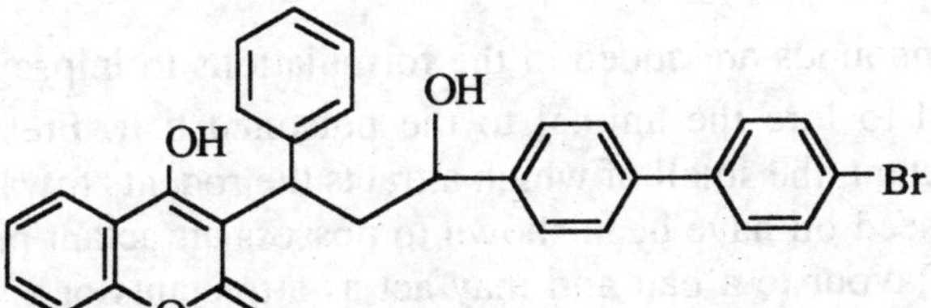

(23b)

Structure-activity relationships of 4-OH-coumarin derivatives

In the dicoumarin series, activity was decreased on substitution of the methylene group between the two coumarin moieties with short aliphatic chains or phenyl group. Replacement of CH_2-group by sulfur atom led to retention of only about 0.05% of the activity. The most interesting synthetic derivative is the 3-substituted 4-OH-coumarin which contains a keto group in a 1,5 arrangement with respect to the 4-hydroxy group.

Indanedione compounds have structural configuration and activity very similar to that of dicoumarin. Although unsubstituted indanedione was inactive as an anticoagulant, substitution in the 2-position with an acyl group produced the active compound such as Pival (pindone) and diphacinone. The relationship of the C=O groups of these compounds suggests that they are competitive inhibitors for vitamin K which is prosthetic group in the formation of prothrombin.

FORMULATIONS

Poison dusts

A typical poison dust consists of an inert finely powdered material. A suitable poison is occasionally an adhesive, a water repellent and a warning dye.

A few dust formulations of lethal substances (pesticides) possessing properties useful in rodent control programme, are :

DDT & BHC (micronised form 20-25% a.i.)

Primarily these preparations act as adhering contact dusts. It is possible that their lethal dose can be eventually ingested by a rodent as the chemical smeared on the feet and fur gets transferred to the mouth during normal cleaning and grooming activity. This method, therefore, requires far higher concentration of the active compound than that used in baits as the amount ingested during cleaning and grooming is expected to be very less.

Advantages and disadvantages

- Rodents are unable to identify the source of illness and generally follow the same travel route.
- Unlike the poisoned baits, the method does not involve any change in the feeding habits of rodents.
- The disadvantage is that the dust maggots circulate in the air and cause respiratory problems. They may also contaminate food, feed and water proving to be potential health hazards.
- Locating the routes frequented by rodents is a cumbersome exercise.
- Need for the use of bulk quantities make them uneconomical.

Alternative ways of control

Additives : These compounds are added in the formulations to improve their properties.

Attractants are added to lure the animal to the poisoned bait. Fresh raw linseed oil is an example of a simple attractant, the smell of which attracts the rodents towards the bait. Many other essences and oils like aniseed oil have been shown to possess attractant properties. Additives like sugars impart their own flavour to a bait and may act as attractants or secondary foods. Maltose at a 2-30% concentration is considered to improve palatability of various baits.

Repellents. Control can be carried out with repellents but the effect is short lived because these compounds only keep the rodents away and do not kill them. Thiram, OMPA and aniline complex of TNB are generally recommended for this purpose. The repellent technique is very useful in the fields where rodents damage crops, seedlings and trees. The essential attributes of the repellents are (i) retain effectivity throughout the season, (ii) safe to handle and easy to apply, (iii) do not damage other plants and trees, and (iv) no non-target toxicity. Rodent repellents commonly used on packaging materials like boxes, sacks and stored articles are : i) actidione (cycloheximide), ii) malathion, and iii) pentachlorobenzylmercaptan.

Vital cables and electrical wiring are sometimes damaged by rodents. Repellents which can be used for coating them are i) NN-dimethyl-S-t-butyl sulfenyl dithiocarbamate, ii) tributyltin salts, and iii) dodecylamine and its salts.

e) Chemosterilants

Another alternative approach to eradicate rodents is through biogenetic control. The extent of infestation can be appreciably reduced to the point of virtual extinction with the introduction of infertile animals in the community. Although infertility can be induced in a number of ways, but the most convenient way is through the use of chemosterilants. Introduction of chemosterilants without any preliminary control measures generally proves to be unsatisfactory on account of their delayed action. A chemosterilant in conjunction with a selective rodenticide ensures effective control.

Female rat sterility

The following steroidal compounds (24), are effective in sterilisation of female species :

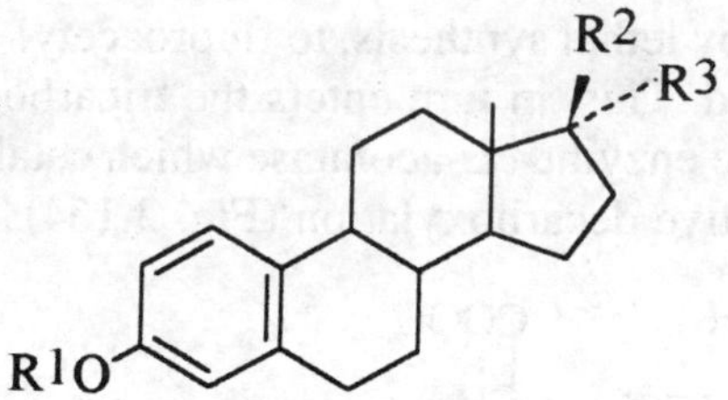

a) R^1 = Me, R^2 = OH; R^3 = C ≡ CH
      (mestranol)

b) R^1 = CH_2C ≡ CH, R^2 = $OCOCMe_3$, R^3 = H

c) R^1 = —◁ , R^2 = OH, R^3 = C ≡ CH
      (quinestrol)

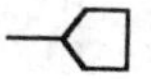

Some non-steroidal compounds are able to terminate pregnancy. These are derivatives of **clomiphene** like MER-25 (25).

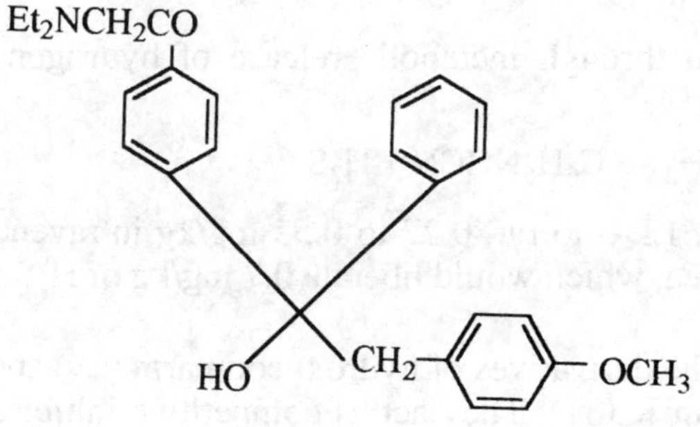

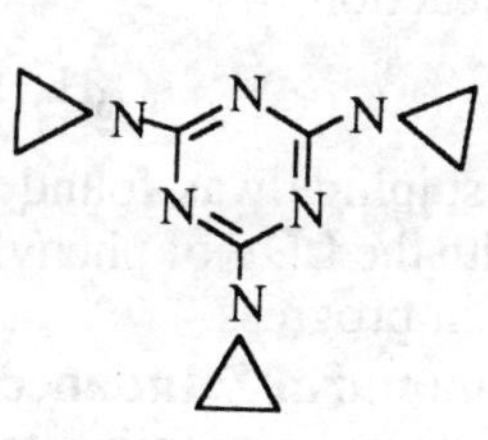

Male rat sterility

It is more convenient to induce sterility in the male animal as it involves the use of only the antispermatogenic agents. These are **triethylene melamine** (26), **furadantin** (27) and **chlorohydrins** (28).

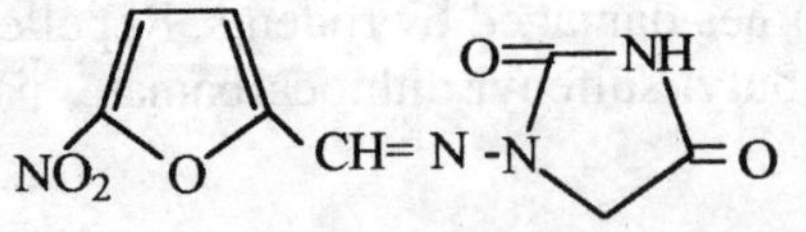

$$NO_2 \quad O \quad CH= N -N \quad O=NH \quad =O \quad ; \quad ClCH_2 - CH(OH) - CH_2(OH)$$

(27)   (28)

Mode of action

Strychnine is well known to produce toxic convulsions and lowers synaptic resistance in the nerve cord by a not well understood inhibition of the inhibitory cells.

Red squill contains scillaren and other cardiac glycosides which produce convulsions and respiratory failure.

Zinc phosphide reacts with hydrochloric acid in the rodent stomach to release phosphine gas PH_3 which is highly reactive and poisonous.

Thallium sulfate is a general cellular poison and resembles arsenicals. Presumably it inactivates -SH groups or other reactive centres in enzymes, producing a variety of neurological, circulatory and gastrointestinal disorders.

Fluoroacetate is rapidly converted in the body through by lethal synthesis, to fluoroacetyl-CoA which condenses with oxaloacetate to form fluorocitric acid. This, in turn enters the tricarboxylic acid cycle and acts as a powerful competitive inhibitor of the enzyme *cis*-aconitase which catalyzes the conversion of citric acid to isocitric acid prior to oxidative decarboxylation (Fig. 3.154).

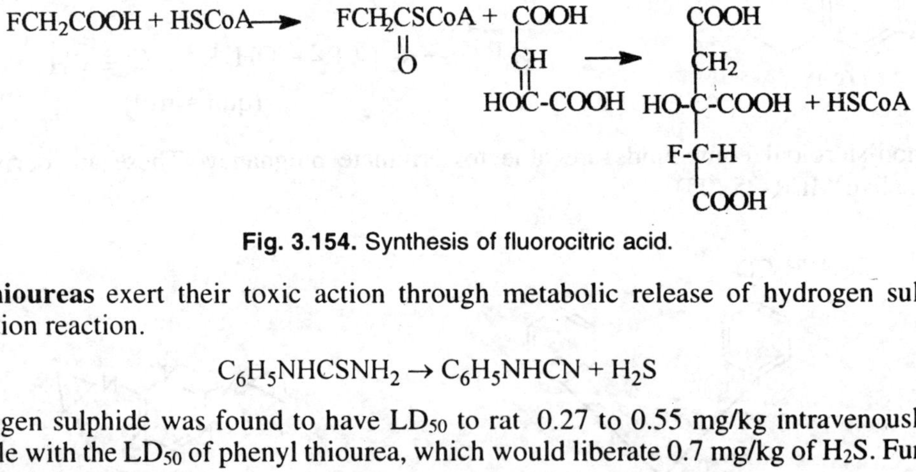

$$FCH_2COOH + HSCoA \longrightarrow FCH_2CSCoA + \overset{COOH}{\underset{O}{\overset{||}{}}} \quad \overset{|}{\underset{HOC-COOH}{\overset{CH}{||}}} \longrightarrow \overset{COOH}{\underset{HO-C-COOH}{\overset{|}{CH_2}}} + HSCoA$$

$$F-C-H$$
$$COOH$$

Fig. 3.154. Synthesis of fluorocitric acid.

Arylthioureas exert their toxic action through metabolic release of hydrogen sulfide by a desulfuration reaction.

$$C_6H_5NHCSNH_2 \rightarrow C_6H_5NHCN + H_2S$$

Hydrogen sulphide was found to have LD_{50} to rat 0.27 to 0.55 mg/kg intravenously, a value compatible with the LD_{50} of phenyl thiourea, which would liberate 0.7 mg/kg of H_2S. Further study in this area is in progress.

Hydroxycoumarin & Indanedione The derivatives of hydroxycoumarin (29) and indanedione (30) have structures resembling to vitamin K (31). They act as competitive inhibitors of vitamin K which acts as the prosthetic group for *in vivo* synthesis of prothrombin. Prothrombin is the precursor of thrombin, which catalyzes the conversion of fibrinogen to fibrin, a skeleton of protein fibres that promote coagulation. Many factors are involved in this complicated process of coagulation.

(29)   (30)   (31)

Application

The use of rodenticides like all other pesticides varies with the habits and habitat of the pest. The rodenticides differ widely in their toxicity to man, domestic animals and wild life. Baits are the most practical means of using rodenticide and involve mixing the compound with food, cut pieces of fruit and meat. Impregnated grains of corn or wheat are especially effective for mice. The concentration of these compounds generally used in baits are : Antu 1-3%, Warfarin, 0.025%, Pindone 0.25%, Strychnine 0.5 to 1% and Zinc phosphide 0.5 to 1%.

In situations where large scale control programmes are carried out, as in plague control, dusting of the burrows and runways of rats with a mixture of rodenticide with talc or other diluent is more effective. Sodium fluoroacetate is used in water or cereal baits for control of ground squirrels and predators.

Resistance to rodenticides (warfarin resistance)

There are two kinds of resistance that may develop from the use of rodenticides.

The first type is an acquired tolerance to a poison that builds up in the rodent pests during treatment and is not passed on to the offspring. This acquired resistance may also arise from the use of acute poisons.

A more recently encountered type of resistance may appear after frequent use of anticoagulant poisons like warfarin. The resistance so developed can pass from one generation to the next and poses a more serious problem. The present situation is not very bad as reported widely although some resistance has already developed in some species of rat. The present conditions are that there is basically no abnormal movement of resistant rats although the area of resistance is increasing. The number of resistant rats in particular area seem to have reached a plateau. The number of rats existing currently in infested areas is of the same order as that existed before the induction of resistance. Consequently there is no increase in the spread of rat borne diseases.

Future outlook of rodent management

- Research and development should aim at discovering new ecofriendly rodenticides with different modes of action to replace the presently used less effective compounds.
- Search for selective chemosterilants has to be stepped up.
- Developing improved formulations procedures like microencapsulation for targetted application.
- Educating the society to keep the surrounding tidy and uninhabitable by the rodents.

REFERENCES

1. Nadasy, M. Rodenticides. In *Pesticide Chemistry* (G. Matolcsy, M. Nadasy and V. Andriska eds.) Elsevier, New York, 261-270 (1988).
2. Bentley, E.W. Review of currently used anticoagulants. Seminar on Rodents and Rodent Ectoparasites, Geneva (W.H.O. Vector Control 66/217, 89 (1966).
3. Gutteridge, N.J.A. 'Chemicals in Rodent Control'. Chemical Society Reviews, **1** (3), 381 (1972).
4. Bean, J.R. and Hudson, R.H. Acute oral toxicity and tissue residues of Thallium sulphate in golden eagles. *Bull. Environ. Toxicol.*, **15**, 118 (1976).
5. Buckle, A.P. and Smith, R.H. Rodent Pests and their control CAB International, Wallingford, (1994).
6. Elliot, A.C. Rodenticides. In *Agrochemicals from Natural Products* (C.R.A. Godfrey, ed.) Marcel Dekkar, Inc. New York, 341-366 (1994).
7. Marsh, R.E. and Howard, W.E. *Proc. Third Internatn. Biodegrad. Symp.* (J.M. Shapley and A.M. Kaptan, eds.). *Applied Science*, London, 317-329 (1975).
8. Robinson, R. The constitution of strychnine, *Experientia*, **2**, 28 (1946).
9. Meehan, A.P. The rodenticide activity of reserpine and related compounds. *Pestic. Sci.* **11**, 555-561 (1980).
10. *Ibid*. Rats and Mice, Rentokil Ltd. East Grinstead (1984).
11. Richter, C.P. Biological factors involved in poisoning rats with alphanaphthyl thiourea (ANTU). *Proc. Soc. Exp. Biol. Med.* **63**, 364 (1945).
12. Apperson, C.S., Sanders, O.T. and Kaukeinen, D.E. Laboratory and field trials with rodenticide, brodifacoum against warfarin resistant Norway rats. *Pestic. Sci.*, **12**, 662-668 (1981).
13. Campbell, H.A. and Link, K.P. *J. Biol. Chem.* **138**, 21 (1941).

Section VIII.   Chemistry of Fungicides

- Introduction
- Natural resistance
- Chemical control measures
- Non systemic fungicides
- Dithiocarbamates
- Organomercurial salts
- Miscellaneous organic compounds
- Sulfenyl compounds
- Organophosphorus compounds
- Systemic fungicides
- Carboxamide and imide compounds
- 2, 5-Dioxo-3-pyridine-1-acetanilide compounds
- Dicarboximide compounds
- Benzimidazole compounds
- Formamide type of compounds
- Alkyl amines and salts
- Alkane carboxylic acid compounds
- Aromatic compounds
- Acylalanine compounds
- Pyridine, and pyrimidine compounds
- Quinaline, and quinoxaline compounds
- Morpholines, and piperidine compounds
- Oxazole, isoxazole and thiazole derivatives
- Azole compounds : Imidazole, Triazole and Oxetane derivatives
- Newer potent fungicides
- Biofungicides
- Mode of action
- Metabolism
- Photochemistry
- *References*

Chemicals used for controlling fungal diseases are called fungicides. The latter part *cide* of this term has been derived from the latin word *caedo* meaning 'to kill'. Some chemicals do not kill fungi but inhibit the growth temporarily. Once the chemical is removed, the fungus revives and such chemicals are termed as 'fungistats'. Other chemicals which inhibit spore production without affecting the vegetative growth are called 'antisporulants'. Compounds like dithiocarbamates, applied prior to fungal infection are called protectants and those like oxathin, which eradicate a fungal infection and cure the plant from fungal attack are called 'therapeutants'. The fungal diseases can be controlled by i) induction of inbuilt resistance, ii) chemical treatment and iii) use of biofungicide.

Natural resistance

Many plants show inbuilt resistance to a variety of fungal diseases. The desired plants can be made disease resistant by transfer of this trait through breeding or using biotechnological contraptions. Considerable success has been achieved in combating soil borne diseases such as panama disease of bananas. The process however is greatly limited by the difficulty in finding a compatible donor and also possibility of curing and reversion, leading to the loss of acquired character.

Chemical control measures [1, 2, 3, 4]

Depending upon the mode of action fungicides can be grouped as non-systemic and systemic fungicides. The non-systemic fungicides either have contact action or act as single dose poison. The systemic fungicides enter the metabolic system of the fungus and accumulate to reach lethal levels.

Non-systemic fungicides

Based on chemical structure fungicides can be grouped into two major classes.

i) Inorganic compounds

Copper compounds include Bordeaux mixture, Burgundy mixture; copper oxychloride and cuprous oxide. *Sulphur* compounds are utilised in various forms such as powdered sulphur, wettable sulphur and lime sulphur. *Mercury compounds* generally used are mercurous and mercuric chloride.

The Bordeaux and Burgundy mixtures got widely accepted because of lack of phytotoxicity and better efficacy and even today copper fungicides constitute the major part of the total consumption. Wettable sulphur in various forms is also being used commonly to combat many fungal diseases. Mercury compounds are also used as seed dressers and are quite effective. However, after the discovery of Minamata disease and pollution effects of mercurial compounds, the use of this group of chemicals has decreased substantially.

ii) Organic compounds

Organotin compounds [5] are classified as contact fungicides. The activity range is similar to that of copper compounds, but because of the phytotoxic nature, these compounds are rarely used in agriculture. The most important fungicide of this group is triphenyltin acetate (1, Fig. 3.155) and its hydrolysed product triphenyltin hydroxide. Triphenyltin acetate (fentin acetate) can be prepared by the following method.

$$(C_6H_5)_3SnCl + CH_3COONa \rightarrow (C_6H_5)_3SnOCOCH_3 + NaCl$$
$$(1)$$

Fig. 3.155. Preparation triphenyltin acetate.

Organoarsenic compounds. Since the compounds of this group are phytotoxic and have high mammalian toxicity, they are not recommended for use in agriculture. An important organic arsenic compound is methylarsinediyl bis(dimethyl) dithiocarbamate (2, Fig. 3.156), used in fruit orchards. It can be prepared as follows.

$$CH_3AsO + 2HCl \longrightarrow CH_3AsCl_2 + H_2O$$

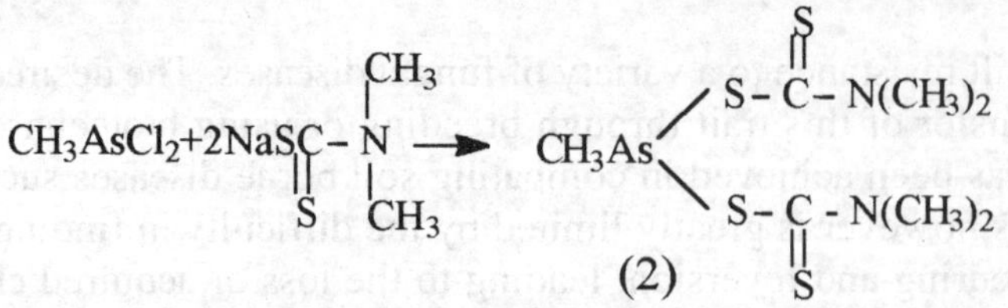

$$CH_3AsCl_2 + 2NaSC\overset{\overset{\displaystyle CH_3}{|}}{\underset{\overset{\displaystyle |}{CH_3}}{\underset{\parallel}{S}}}N \longrightarrow CH_3As\overset{\displaystyle S-\overset{\overset{\displaystyle S}{\parallel}}{C}-N(CH_3)_2}{\underset{\displaystyle S-\underset{\underset{\displaystyle S}{\parallel}}{C}-N(CH_3)_2}{}}$$

(2)

Fig. 3.156. Preparation of methylarsinediyl bis(dimethyl) dithiocarbamate.

Organosulphur compounds. From 1950 onwards, the use of organosulphur fungicides has come into prominence and presently constitute the most important group of fungicides, being used in the country.

Dithiocarbonic acid derivatives [6]. The dithio derivative of carbonic acid is xanthogenic acid which is unstable in the free state. Its anhydride CS_2 possesses good biological properties.

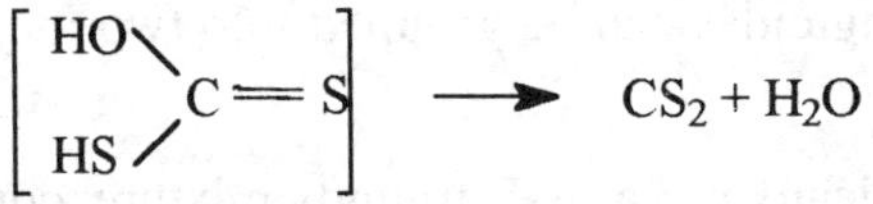

$$\left[\overset{HO}{\underset{HS}{}} C = S \right] \longrightarrow CS_2 + H_2O$$

CS_2 was first used as soil insecticide because it permeates the soil rapidly due to high vapour pressure and diffusibility. Being highly inflammable and explosive it has now been replaced by better alternatives. The amide of dithiocarbonic acid is dithiocarbamic acid which occurs only in the form of its salts and mainly in its ester forms.

Dithiocarbamates are represented by two distinct groups comprising i) N-monoalkyl dithiocarbamates and ii) N,N'-alkylene bisdithiocarbamates. The first group of compounds are synthesized from primary amines while the synthesis of latter group of compounds is accomplished by the reaction of secondary amines with CS_2. The compounds of latter group however cannot be metabolised to *iso*thiocyanates unlike those of the former group.

i) N-monoalkyldithiocarbamates

The simplest one in this category is the sodium salt of methyl dithiocarbamic acid (3, Fig. 3.157), which can be prepared as follows.

$$CS_2 + CH_3NH_2 + NaOH \rightarrow CH_3NH\text{-}C(S)SNa + H_2O$$

(3)

Fig. 3.157. Preparation of methyl dithiocarbamic acid.

The salt slowly hydrolyses in dilute aqueous solution to form acid and finally to methylisothiocyanate (4), the active component.

$$CH_3NH\text{-}C(S)\text{-}SH \rightarrow CH_3N = C = S + H_2S$$
$$(4)$$

The oxidation product of vapam (metham), N,N$'$-dimethylthiuram disulfide (5), is also used as a soil disinfectant

$$CH_3NH\text{-}C(S)\text{-}S\text{-}S\text{-}C(S)\text{-}NHCH_3$$
$$(5)$$

If the reaction of methylamine with CS_2 is carried out in presence of HCHO in acid medium, the solid product formed is called as 3,5-dimethyl-tetrahydro-1,3,5-thiadiazine-2-thione, or dazomet (6, Fig. 3.158) which on hydrolysis forms methyl isothiocyanate.

$$2CH_3NH_2 + CS_2 + 2HCHO + H^+ \longrightarrow$$

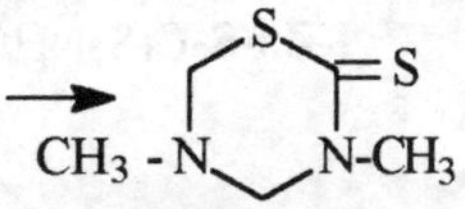

$$(6)$$

Fig. 3.158. Preparation of dazomet.

Of the N-alkyldithiocarbamates, zinc-bis*iso*propyldithiocarbamate (neviram, 7, Fig. 3.159) is a foliage fungicide which is as active as zineb. The synthesis is as follows :

$$(CH_3)_2NH_2 + CS_2 + NH_4OH \rightarrow (CH_3)_2 CH\text{-}NH\text{-}C(S)\text{-}SNH_4$$
$$(CH_3)_2CH\ NH\ C(S)SNH_4 + ZnSO_4 \rightarrow [(CH_3)_2CH\ NH\ C(S)S]_2Zn$$
$$(7)$$

Fig. 3.159. Preparation of zinc-bis *iso*propyldithiocarbamate.

ii) N,N$'$-ethylene-bisdithiocarbamates

These are prepared by the reaction of ethylenediamine with CS_2. Amobam, the diammonium salt (8, Fig. 3.160) is prepared as shown below.

$$(CH3)2NH2 + 2CS2 + 2NH4OH \longrightarrow$$

$$\begin{array}{l} CH_2\text{-}NHC(S)S\ NH_4 \\ | \\ CH_2\text{-}NHC(S)S\ NH_4 \end{array}$$
$$(8)$$

Fig. 3.160. Preparation of amobam.

The Zn-salt of N,N$'$-ethylene bisdithiocarbamic acid, Zineb (9) and Mn-salt Maneb (10) are prepared by reacting aqueous solutions of $ZnSO_4$ and $MnSO_4$ with amobam respectively.

$$\begin{array}{l} CH_2NHC(S)\ S \\ \\ CH_2NHC(S)\ S \end{array} \Big\rangle Zn \qquad \begin{array}{l} CH_2NHC(S)\ S \\ \\ CH_2NHC(S)\ S \end{array} \Big\rangle Mn$$
$$(9) \qquad\qquad\qquad\qquad (10)$$

Both are used as protective foliage fungicides to control number of diseases in different agricultural crops.

Mancozeb, is the managanese-zinc double salt of N,N$^{\prime}$-ethylene bisdithiocarbamic acid (11). This compound is more stable than both Zineb and Maneb. Mancozeb is not phytotoxic because of its polymeric structure. The chemical composition shows that it contains 20% Mn and 2% Zn besides ethylene bisdithiocarbamate ions with a possible structure (11) as

$$[(-S-\underset{\underset{S}{\|}}{C}-HN-CH_2-CH_2-NH-\underset{\underset{S}{\|}}{C}-S-)\,(Mn)_x\,(Zn)_y\,]$$

$$(11)$$

Propineb, is the polymer of zinc propylene 1, 2-bisdithiocarbamate (12)

$$[-Zn-S-C(S)NHCH_2-CH(CH_3)NH-C(S)S-]_x$$

$$(12)$$

Its fungicidal activity considerably surpasses that of Zineb particularly against pathogens of scab. The effect of its manganese zinc complex is even more powerful than others.

Matolcsy [7] prepared a highly effective co-polymer with linear structure (13, Fig. 3.161) by the reaction of the piperazinium salt of ethylene-bisdithiocarbamic acid with formaldehyde (Fig. 3.161).

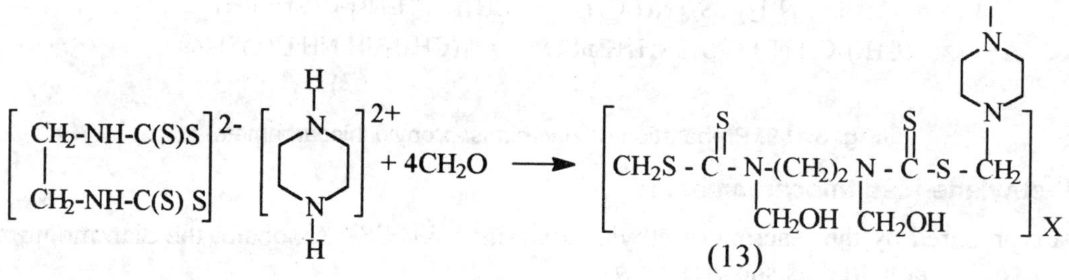

$$\left[\begin{array}{c}CH_2-NH-C(S)S \\ | \\ CH_2-NH-C(S)\ S\end{array}\right]^{2-}\left[\begin{array}{c}H \\ | \\ N \\ | \\ N \\ | \\ H\end{array}\right]^{2+} + 4CH_2O \longrightarrow \left[\begin{array}{c} S\ \ \ \ \ \ \ \ \ \ \ S \\ \| \ \ \ \ \ \ \ \ \ \ \ \| \\ CH_2S-C-N-(CH_2)_2\ N-C-S-CH_2 \\ | \ \ \ \ \ \ \ \ \ \ | \\ CH_2OH\ CH_2OH \end{array}\right]_X$$

$$(13)$$

Fig. 3.161. Preparation of co-polymer from piperazinium salt of dithiocarbamic acid with formaldehyde.

Dialkyldithiocarbamates

Of the dialkyldithiocarbamates belonging to the second group of dithiocarbamates, the dimethyldithiocarbamates have highest activity. With increase of the alkyl group, the fungitoxicity is reduced. These compounds are prepared by the reaction of dimethylamine with CS_2 in presence of alkali hydroxide (Fig. 3.162).

$$(CH_3)_2NH+CS_2+NaOH \longrightarrow (CH_3)_2N-C(S)-SNa$$

$$(14)$$

Fig. 3.162. Preparation of diram.

The Na-salt of dimethyl dithiocarbamate is diram (14) which gives metal salt ferbam (15), ziram (16) and thiram (17) (Fig. 3.163).

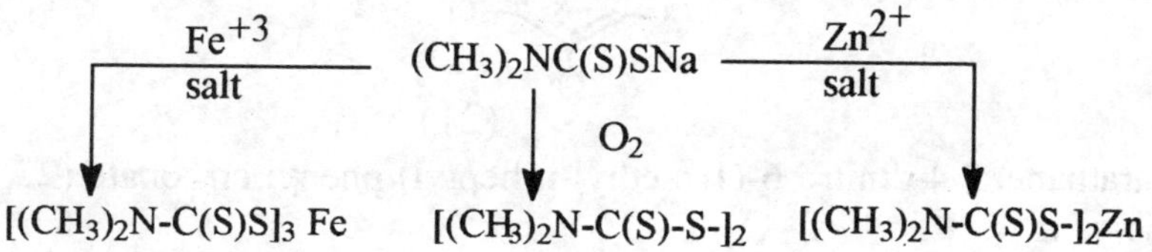

$$Fe^{+3} \text{ salt} \quad\longleftarrow\quad (CH_3)_2NC(S)SNa \quad\xrightarrow{\quad Zn^{2+} \text{ salt}\quad}$$

$$\downarrow O_2$$

$$[(CH_3)_2N\text{-}C(S)S]_3 \, Fe \qquad [(CH_3)_2N\text{-}C(S)\text{-}S\text{-}]_2 \qquad [(CH_3)_2N\text{-}C(S)S\text{-}]_2Zn$$

$$(15) \qquad\qquad\qquad (16) \qquad\qquad\qquad (17)$$

Fig. 3.163. Preparation of ferbam, ziram, thiram.

These metal salts are particularly effective against apple scab, blue mould and give good protection against the seedling diseases of vegetables. Another molecule N,N^{l}-dimethylthiocarbamoylthioacetic acid (18) has properties similar to those of auxins and acts as a systemic fungicide.

A group of compounds having the general formula R-C(S)-S-S-C_2H_5 (19), where, R = dimethylamine, diethylamine, piperidine and 1,2-ethylene amino group, were developed with activity similar to that of TMTD and captan.

$$(CH_3)_2N\text{-}C(S)\text{-}SCH_2COOH \quad ;$$

$$M = Zn, Cd, Cu$$

$$(18)$$

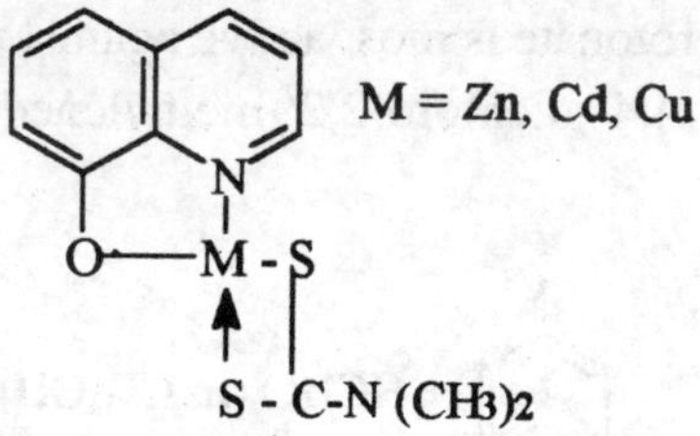

$$(20)$$

Matolcsy *et al.* [8] prepared mixed ligand chelates, having general structure (20) by treating the sodium salt of 8-oxyquinolinate and dimethyldithiocarbamate with the salts of bivalent metals. The fungicidal action of this product is stronger than that of their physical mixture in molar ratio.

b) Organomercuric salt

A number of organomercuric salts are being used as a seed dressings. Some of these are methoxyethylmercuric chloride ($CH_3OC_2H_4HgCl$), phenylmercuric acetate (PhHgOAc) and ethylmercuric chloride (C_2H_5HgCl).

v) Miscellaneous organic compounds [9]

Chlorthalonil (21), dinocap (22) pentachlorophenol, pentachloronitrobenzene, dichlorophen (23), chloranil (24), captan (25), folfet (26), captafol (27), and difolpet (28) are used as effective fungicides to control large number of fungal diseases [8].

The **chemistry** of a few important miscellaneous fungicides is discussed in the following text.

Chlorthalonil, 2,4,5,6-tetrachloroisophthalonitrile, (21), is a colourless crystalline compound effective against *Phytophthora* and rust fungi. The activity is caused by the high reactivity of the Cl-atom with thiol enzymes of the fungi. It is chemically stable in both acid and alkaline media.

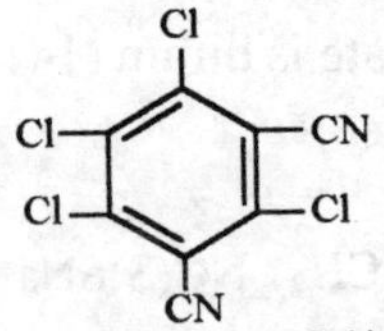

(21)

Dinocap, Karathane, 2,4-dinitro 6-(1-methyl-n-heptyl) phenyl crotonate (22, Fig. 3.164). It is prepared as follows .

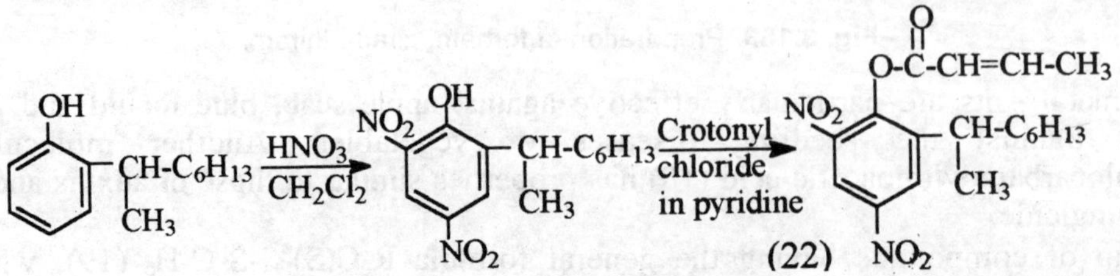

Fig. 3.164. Preparation of dinocap.

It is brown liquid. The technical product, a mixture with its isomeric compound 2,6-dinitro 4-*iso*octylphenyl crotonate is most active against powdery mildew disease having LD_{50} 980 mg/kg.

Dichlorophen, 4,4'-dichloro-2,2'-methylenediphenol) (23, Fig. 3.165) is prepared by the following method

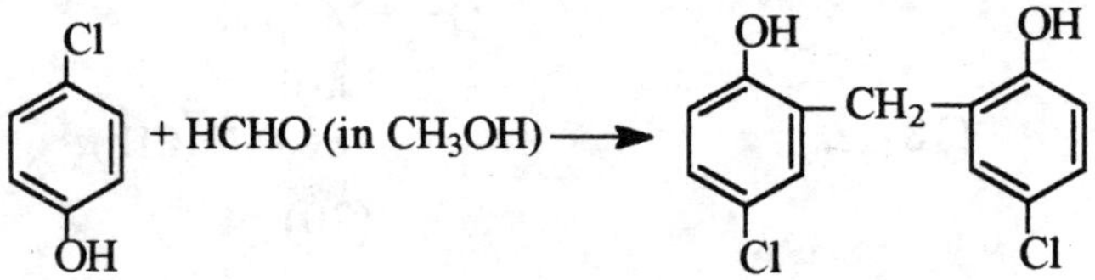

(23)

Fig. 3.165. Preparation of dichlorophen.

The compound is colourless crystalline solid having good fungicidal activity with LD_{50} 2000 mg/kg The activity is due to the presence of a hydroxyl group in ortho position with respect to the bridge connecting the two rings and halogen substitution in para position of the OH-group. The presence of more than two OH-groups or two Cl-atoms in the molecule reduces the activity.

Chloranil, 2,3,5,6-tetrachloro-1,4-benzoquinone (24), is prepared by the oxidation of pentachlorophenol.

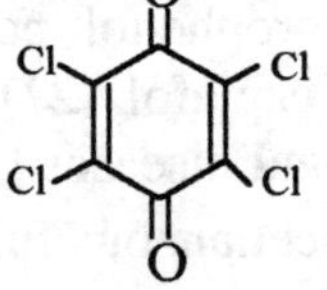

(24)

It is generally used for seed treatment against smut and other diseases in vegetables. It decomposes photochemically or hydrolyses if used as foliar spray.

Sulfenyl group of compounds. The N-trichloromethane sulfenyl derivatives of methane, and ethane sulfonic amides possess fungicidal activity. The first active compound containing the trichloromethane sulfenyl group that gained wide application was **captan**, N-(trichloromethylthio)-1,2,3,6-tetrahydrophthalimide (25). The preparation comprises of four steps (Fig. 3.166).

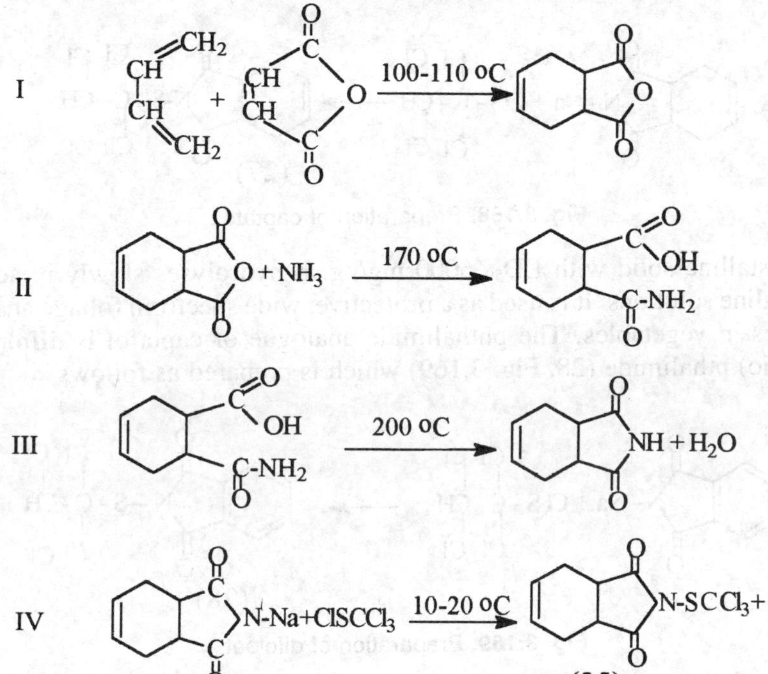

Fig. 3.166. Preparation of captan.

Captan is crystalline solid, effective against almost all the important fungal diseases of plants. It is a foliage fungicide with protective action but is also used for seed treatment and as soil fungicide. Owing to its lipid solubility, it penetrates the cell membrane to react with many thiol containing enzymes essential to the cell. It is hydrolysed in aqueous solution. At pH 7 and 28°C, the half life is 2.5 h. with LD_{50} 9000 mg/kg.

Folpet, N-(trichloromethylthio) phthalimide (26, Fig. 3.167) is prepared as follows.

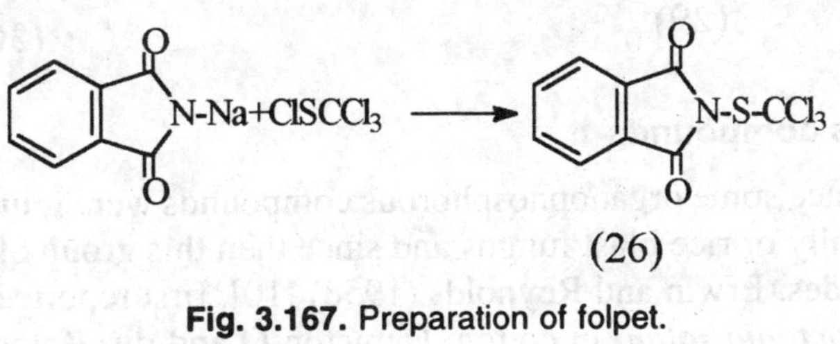

(26)

Fig. 3.167. Preparation of folpet.

Folpet is crystalline solid. It is hydrolysed in aqueous medium at ambient temperature. The rate of hydrolysis increases in hot and alkaline solutions. Its mode of action and field of application is similar to those of captan. It is specially effective against downy mildew of vine and foliage diseases of cereals and can also be used as soil fungicide.

Captafol, N-(1,1,2,2-tetrachloroethylthio)1,2,3,6-tetrahydrophthalimide (27, Fig. 3.168) is prepared as follows.

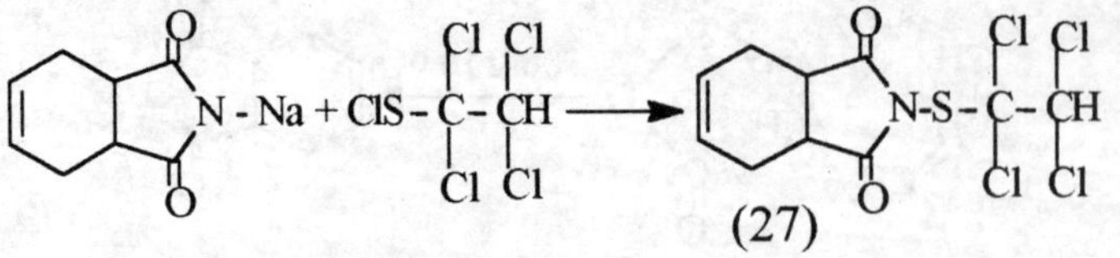

Fig. 3.168. Preparation of captafol.

It is white crystalline solid with LD_{50} 6000 mg/kg. It hydrolyses slowly in aqueous medium and rapidly in alkaline solutions. It is used as a protective, wide spectrum foliage and soil fungicide, to combat diseases in vegetables. The phthalimide analogue of captafol is **difolpet**, N-(1,1,2,2-tetrachloroethylthio) pthalimide (28, Fig. 3.169) which is prepared as follows.

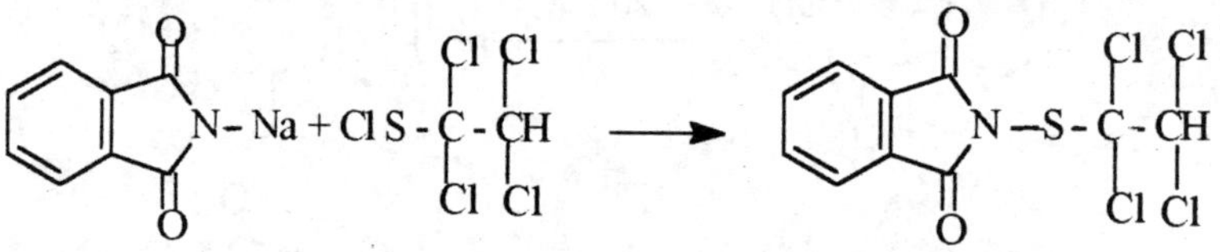

Fig. 3.169. Preparation of difolpet.

The other class of active fungicides containing polyhalogen alkane sulfenyl group is the family of alkyl sulfonic amides. A few of them are **mesulfan** (29), N-trichloromethyl thiochloro-(methanesulfone) anilide and **dichlofluanid** (30), N'-dichlorofluoromethylthio-N,N-dimethyl N'-phenylsulfamide

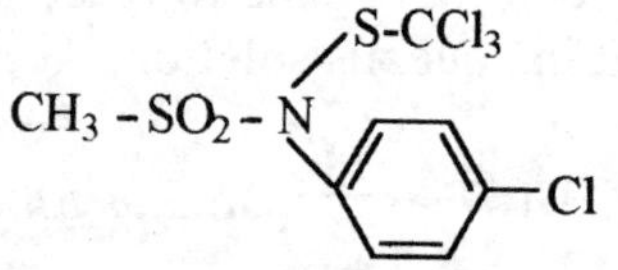

(29)

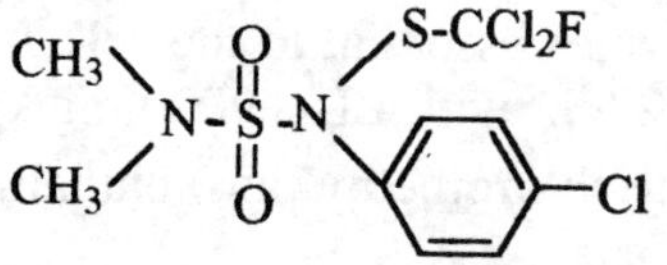

(30)

vi) Organophosphorus compounds

It is only two decades since some organophosphorous compounds were found useful for the control of plant diseases especially of rice blast fungus and since then this group of compounds have been widely used as insecticides. Erwin and Reynolds (1958) [10], first reported the antifungal activity of phorate against *Rhizoctonia solani* in cotton. Demeton-O and disulfotan also exhibit moderate.

antifungal activity. Various groups of organophosphorus compounds which have been developed recently as fungicides can thus be classified as .

Phosphorothiolates

The fungicide **kitazin**, S-benzyldiethylphosphorothiolate (31), used for the control of rice blast was developed in 1965. But this was soon replaced by the more potent *iso*propyl homologue **kitazin**-P, S-benzyl di-isopropylphosphorothiolate (32), having a considerable lower mamalian toxicity than that of kitazin.

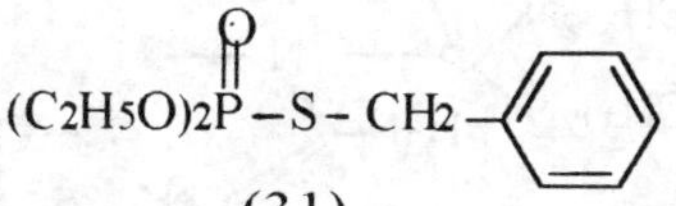

$$(C_2H_5O)_2 \overset{O}{\underset{||}{P}} - S - CH_2 - \bigcirc \quad : \quad [(CH_3)_2CHO]_2 - \overset{O}{\underset{||}{P}} - S - CH_2 - \bigcirc$$

(31) 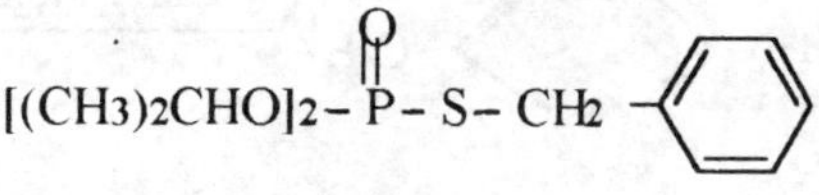　　　　　　　　　　　　　　　(32)

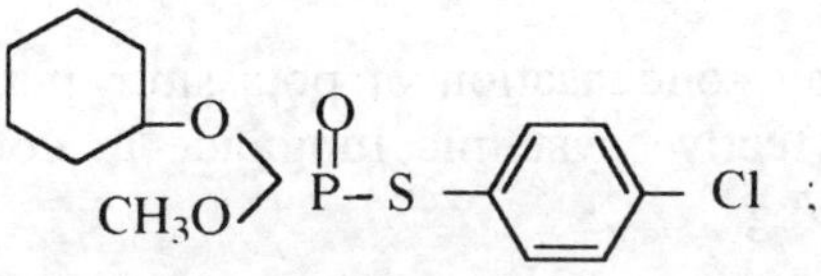

$$\begin{matrix} \bigcirc{-}O \\ CH_3O \end{matrix} \overset{O}{\underset{||}{P}} - S - \bigcirc - Cl \quad : \quad C_2H_5O \overset{O}{\underset{||}{P}} \left(S - \bigcirc \right)_2$$

(33) 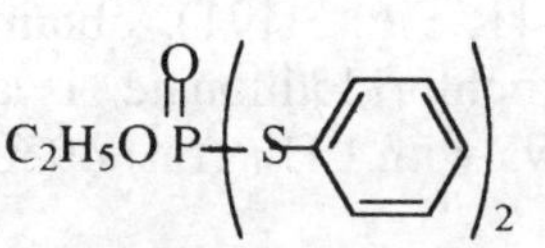　　　　　　　　　　　　　　　(34)

$$\begin{matrix} C_2H_5S \\ C_4H_9O \end{matrix} \overset{O}{\underset{||}{P}} - S - CH_2 - \bigcirc$$

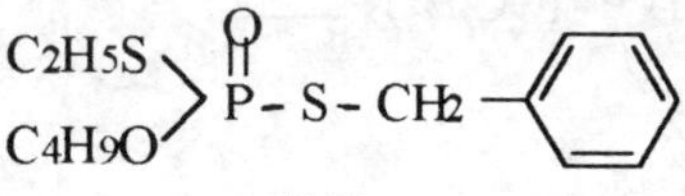

(35)

Cerezin, S-4-chlorophenylcyclohexylmethylphosphorothiolate, (33) and **edifenphos**, hinosan, O-ethyl diphenylphosphorodithiolate, (34) were developed later. Both these compounds are highly effective against rice blast fungus however the former is slightly toxic to fish and is not recommended in the paddy fields used for rearing fish. The latter has been marketed since 1967 and is widely used in India and other rice growing areas. **Connen**, S-benzylO-butyl S-ethylphosphorodithiolate, (35), **HOO34**, ethyl phenyl 2,4, 5-trichlorophenyl phosphate, (36) and **phosdiphen**, ethyl 2, 4-dichlorophenyl phosphate, (37) are also reported to be effective against blast disease of rice. **Fosteyl-Al**, (38), a phosphonate developed by Rhone-Poulene (1977), was found to be effective against downy mildew of vines, tropical crops, vegetable crops and rotting diseases of fruits. The methyl 3-*iso*nonyloxypropylammonium phosphonate (39) has been found useful as seed dressing agent to control seed borne fungi in wheat, maize and rye.

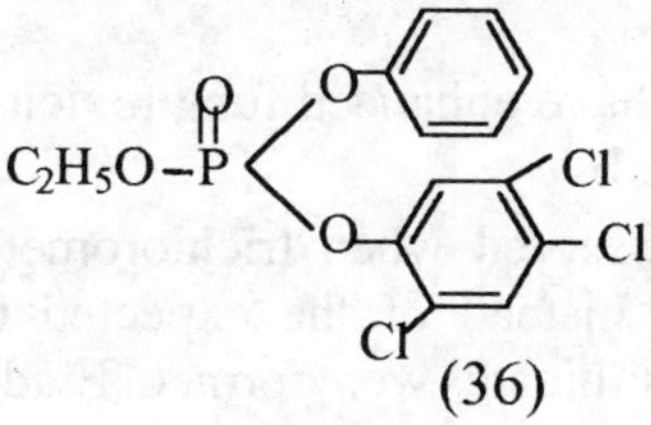

$$C_2H_5O - \overset{O}{\underset{||}{P}} \begin{matrix} O - \bigcirc \\ O - \bigcirc \end{matrix} \overset{Cl}{\underset{Cl}{}} \quad ; \quad C_2H_5O - \overset{O}{\underset{||}{P}} \left(O - \bigcirc - Cl \right)_2$$

(36) 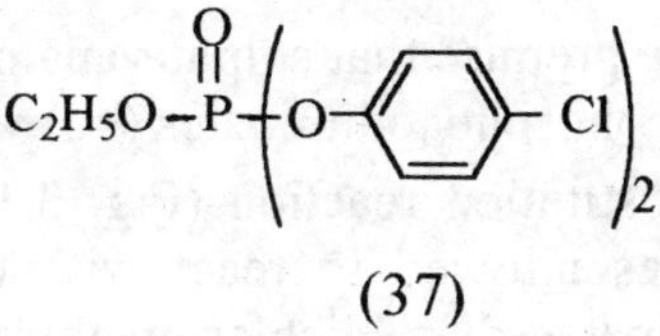　　　　　　　　　　　　　　　(37)

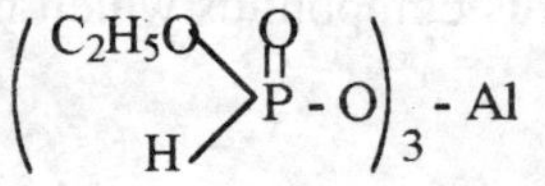

$$\left(\begin{array}{c}C_2H_5O \\ H\end{array}\!\!\!\!>\!\!\!\!\begin{array}{c}O \\ \| \\ P-O\end{array}\right)_3 \!\!-Al \quad ; \quad C_9H_{19}O(CH_2)_3\;\overset{+}{N}H_3\;\overset{-}{O}\!\!\!\!\begin{array}{c}\; \\ \\ CH_3\;O\end{array}\!\!\!\!>\!\!\!\!\begin{array}{c}O \\ \| \\ P-H\end{array}$$

$$(38) \qquad\qquad\qquad (39)$$

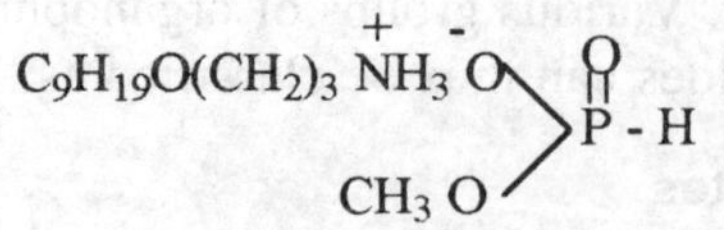

Triamphos (40, Fig. 3.170), a systemic fungicide effective against powdery mildews was commercialized in the year1960 and is prepared as follows.

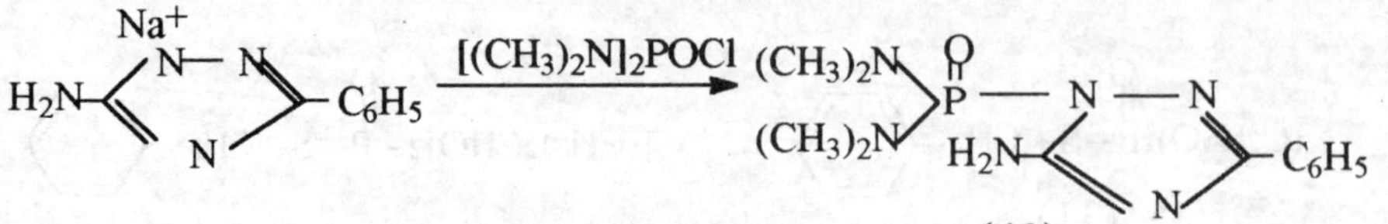

Fig. 3.170. Preparation of triamphos.

Ditalimfos (41, Fig. 3.171), obtained by condensation of potassium phthalimide with O,O-diethylphosphorochloridothionate, is an effective systemic fungicide to control wide range of powdery mildews with LD_{50} (rat) 4930 mg/kg.

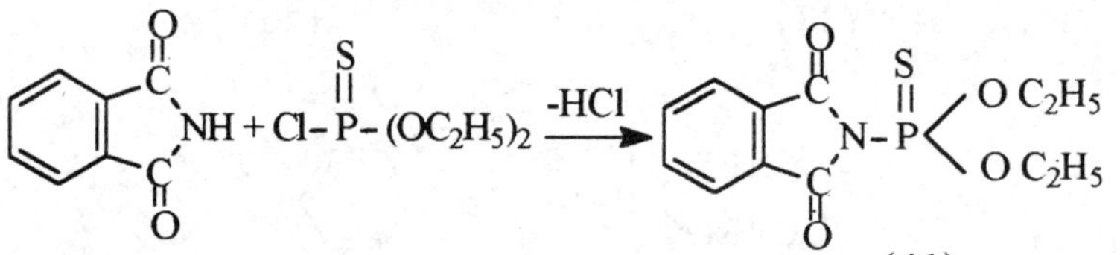

Fig. 3.171. Preparation of ditalimphos.

Recently, a large number of fungicides belonging to phosphonates, phosphates and phosphorothiolates have been reported by Roy [11]. Among these, dichloromethyl diaryl phosphonates; dichloromethyldiphenyl (42), di-p-nitrophenyl (43), 2,4,5-trichlorophenyl (44) and pentachlorophenylphosphonates (45) have exhibited high fungicidal activity against blast disease of rice [12, 13].

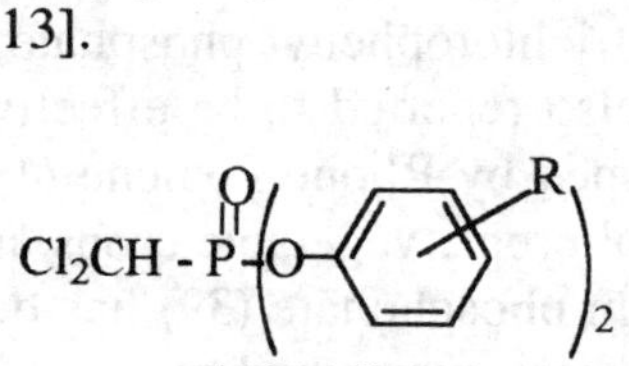

$$Cl_2CH-\overset{\overset{O}{\|}}{P}\!\!\!\left(\!\!O\!\!-\!\!\!\left\langle\;\right\rangle\!\!-\!\!R\right)_2$$

$$(42\text{-}45)$$

R = H, (42)
= 4-NO_2, (43)
= 2,4,5-Cl_3, (44) ;
= Cl_5, (45)

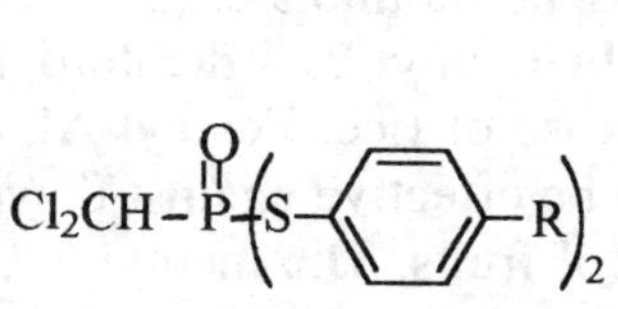

$$Cl_2CH-\overset{\overset{O}{\|}}{P}\!\!\!\left(\!\!S\!\!-\!\!\!\left\langle\;\right\rangle\!\!-\!\!R\right)_2$$

$$(46)$$

Based on the premise that sulphur analogues often have enhanced fungitoxicity, several diaryl dichloromethyl phosphonodithioates (46) were prepared.

A novel substitution reaction (Fig. 3.172) was observed when trichloromethyl phosphonodichloridate was allowed to react with thiophenol. Instead of the expected trichloromethyl-phosphonodithioates, diaryldichloromethylphosphonodithioates were formed. Evidently, during the

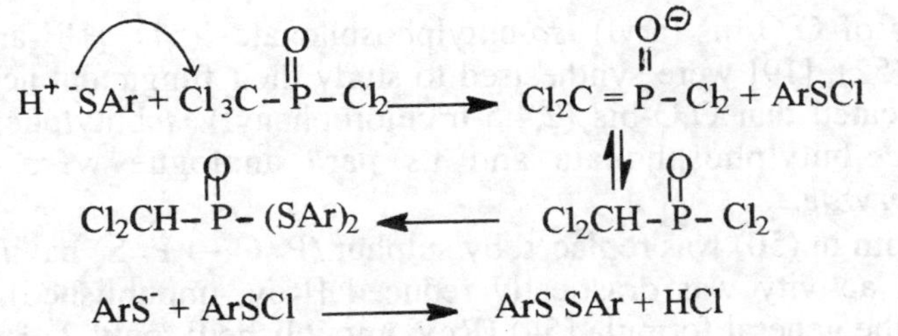

Fig. 3.172. Formation of dichloromethyl phosphonodithioate.

reaction, one of the chlorine atoms of the trichloromethyl group was replaced by hydrogen atom and simultaneous formation of aryl disulfide showed that the thiophenol acted as a reducing agent in this reaction.

Further it was observed that if one of the phenyl group in this molecule is replaced by an alkyl group such as in dichloromethyl phenyl alkyl phosphonates, (47) there was a decline in the activity [Dureja *et al.* 14].

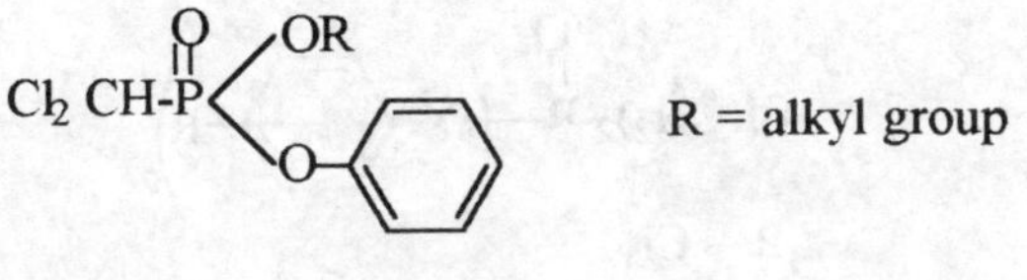

(47) 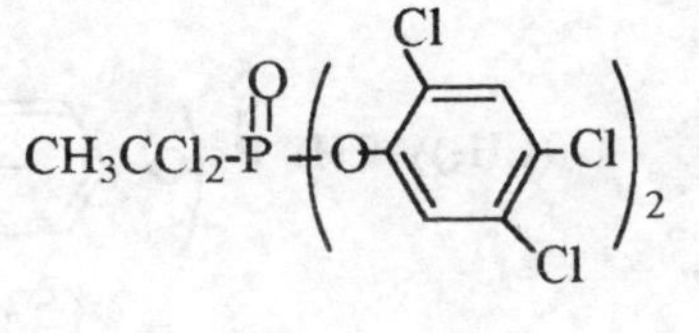   (48)

The effect of lengthening the carbon chain by one carbon such as in diaryl 1,1-dichloroethyl phosphonates, in general, did not enhance the fungicidal activity except in the case of 1,1-dichloroethyl di-(2,4,5-trichlorophenyl) phosphonate (48) [Roy *et al.* 15].

Further, the effect of substitution attached to the alkyl side chain such as in diarylmethylphosphonate, (49) having the general structure was examined [Vasu & Roy 16] to study their activity.

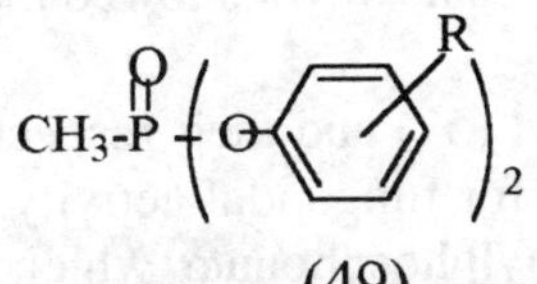

(49) 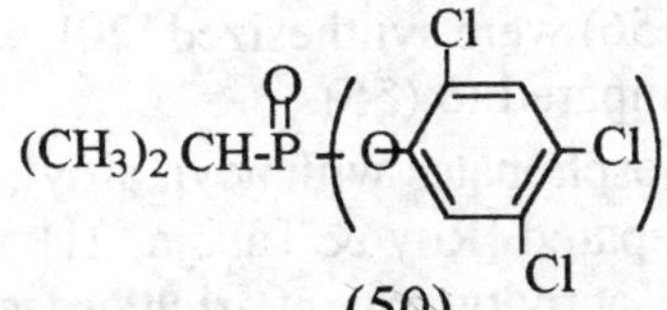   (50)

It was found that the simple phenyl derivative was moderately active but the chlorophenyl analogue exhibited better activity. In this series the most active compound identified was O,O-bis-(2,4,5-trichlorophenyl) methylphosphonate. Further the fungicidal activity was not enhanced by the addition of two more chlorine atoms in the phenyl rings.

The *iso*propyl group in lieu of methyl or ethyl group, sometimes enhanced the fungicidal activity. The fungicidal activity of bisaryl *iso*-propyl phosphonates [Nidiry & Roy 17] was examined and out of the 20 new compounds, the 2,4,5-trichloro analogue (50) exhibited the highest activity against blast disease of rice.

Two more series of O,O-bis (aryl) *iso*-butylphosphonates (51), [18] and O,O-bis (aryl) *sec*-butylphosphonates (52), [19] were synthesised to study their fungicidal activity. SAR studies of both the series indicated that O,O-bis (2,4,5-trichlorophenyl) *iso*butylphosphonate and O,O-bis (2-chlorophenyl) *sec*-butylphosphonate and its para analogue were most potent against *Helminthosporium oryzae*.

When oxygen atom in (50) was replaced by sulphur (P=O → P=S) having the general formula (53), the fungicidal activity was drastically reduced [Roy, unpublished]. In another series of compounds having the general formula (54) [Roy, unpublished], only 2, 4, 5-trichlorophenyl and pentachlorophenyl derivatives were found to be most effective against blast disease of rice.

When one of the phenyl rings was replaced by methyl group having the general formula (55) there was not much variation in fungicidal activity [Roy, unpublished] and the 2,4,5-trichlorophenyl analogue showed the highest activity.

$$(CH_3)_2\text{-}CH\text{-}\overset{\overset{O}{\|}}{P}\left(O\!\!-\!\!\langle\ \rangle\!\!-\!\!R\right)_2 \quad ; \quad CH_3\text{-}CH_2\text{-}\underset{\underset{CH_3}{|}}{CH}\text{-}\overset{\overset{O}{\|}}{P}\left(O\!\!-\!\!\langle\ \rangle\!\!-\!\!R\right)_2$$

$$(51) \hspace{6cm} (52)$$

$$(CH_3)_2\text{-}CH\text{-}\overset{\overset{S}{\|}}{P}\text{-}\left(O\!\!-\!\!\langle\ \rangle\!\!-\!\!R\right)_2 \quad ; \quad Cl(CH_2)_2\text{-}\overset{\overset{O}{\|}}{P}\!\!-\!\!\left(O\!\!-\!\!\langle\ \rangle\!\!-\!\!R\right)_2$$

$$(53) \hspace{5cm} \begin{array}{l} R = Cl_5 \\ R = 2,4,5\text{-}Cl_3 \end{array} \quad (54)$$

$$Cl(CH_2)_2\text{-}\overset{\overset{O}{\|}}{P}\!\!<\!\!\begin{array}{l} O\,CH_3 \\ O\!\!-\!\!\langle\ \rangle\!\!-\!\!R \end{array} \quad ; \quad Cl(CH_2)_2\text{-}\overset{\overset{O}{\|}}{P}\!\!<\!\!\begin{array}{l} O\,C_2H_5 \\ O\!\!-\!\!\langle\ \rangle\!\!-\!\!R \end{array}$$

$$R = 2,4,5\text{-}Cl_3 \hspace{2cm} (55) \hspace{4cm} (56)$$

Similarly on increasing one carbon atom in the alkoxy group, a series of O-aryl-O-ethyl 2-chloro ethylphosphonates (56) were synthesized [20], whose studies however showed that the activity was not enhanced as compared to (55).

Several other phosphonates with acyl, aroyl, chloroacyl (57) and their respective oxime derivatives (58) were prepared [Roy & Taneja, 21] and tested for fungicidal activity. The compounds exhibited moderate activity except trichloroacetyl diethylphosphonate which had the highest activity [Roy, unpublished].

$$R\text{-}\overset{\overset{O}{\|}}{C}\text{-}\overset{\overset{O}{\|}}{P}\text{-}(OR^1)_2 \quad ; \quad R\text{-}\overset{\overset{HON}{\|}}{C}\text{-}\overset{\overset{O}{\|}}{P}\text{-}(OR^1)_2 \qquad \begin{array}{l} R = \text{alkyl, aroyl and} \\ \text{chloroacyl} \\ R^1 = \text{methyl or ethyl} \end{array}$$

$$(57) \hspace{3cm} (58)$$

Phosphorothioates

A series of S-methyl diarylphosphorothiolates were reported by Roy and Bedi [22] having the general formula (59). Among these, 2,4,5-trichloro and pentachloro analogues (R = 2,4,5-Cl$_3$ & Cl$_5$) were found to be most effective against blast disease of rice.

By increasing the length of carbon chain attached to sulphur by one carbon (60), the fungicidal activity spectrum was not significantly changed.[Gupta & Roy, 23].

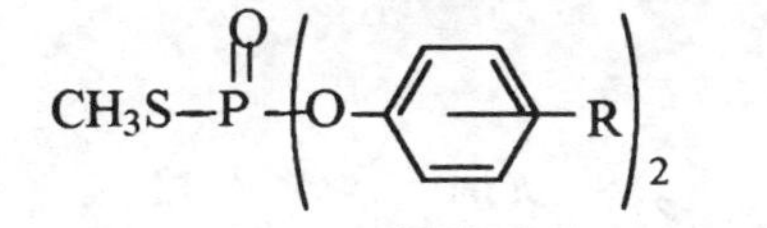

$$CH_3S-P \begin{pmatrix} O- \bigcirc -R \end{pmatrix}_2 \qquad ; \qquad C_2H_5S-P- \begin{pmatrix} O- \bigcirc -R \end{pmatrix}_2$$

(59) 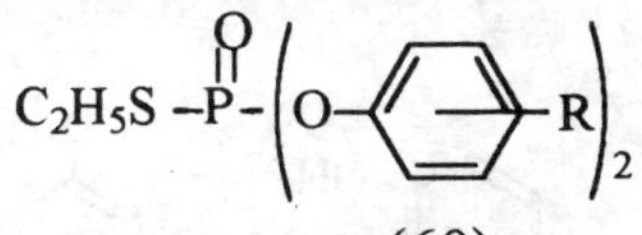   (60)

But on further lengthening the carbon chain i.e. in S-*iso*propyl diaryl phosphorothioates (61), the fungicidal activity was found to be drastically reduced [Gupta and Roy, 24].

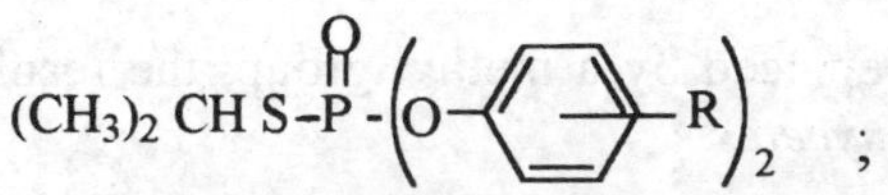

$$(CH_3)_2 CH S-P- \begin{pmatrix} O- \bigcirc -R \end{pmatrix}_2 \qquad ; \qquad R-X-P- \begin{pmatrix} S- \bigcirc -Y \end{pmatrix}_2$$

(61) 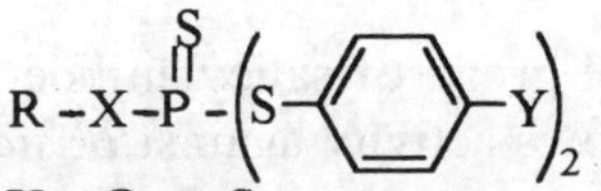   X = O or S
                                              Y = Cl, Br,   (62)
                                              R= alkyl groups

More developments led to the preparation of two more series of O-alkyl/S-alkyl, S,S-diaryl-phosphorotrithioates (62) [Gupta and Roy, 25, 26]. Amongst the trithioates, O-methyl S,S-di-(p-chlorophenyl)phosphorotrithioate(63) its p-bromo analogue and S-methyl S,S-diphenylphosphoro-trithioate were found to be most active against *R. bataticola* and *P. oryzae* respectively.

A series of phosphorylated monoterpenoids and related compounds [27] were synthesized (64, Fig. 3.173) out of which dimethyl phenethylphosphorothionate showed strongest fungicidal activity against *Pythium* spp.

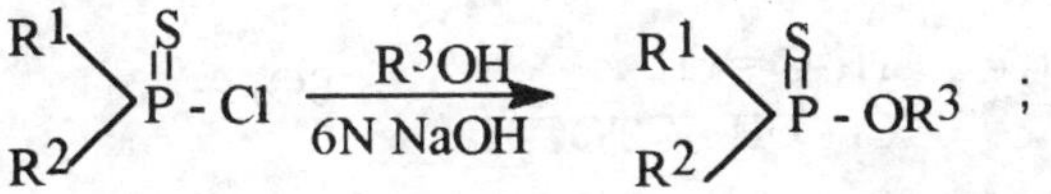

$$R^1 \underset{R^2}{\overset{S}{>}} P-Cl \xrightarrow[6N\ NaOH]{R^3OH} R^1 \underset{R^2}{\overset{S}{>}} P-OR^3 \ ; \quad \begin{array}{l} R^1 \& R^2 = MeO, EtO, Ph \\ R^3OH = Essential\ oil\ moiety \\ \qquad\qquad e.g.\ phenethyl \\ \qquad\qquad alcohol,\ and\ eugenol \end{array}$$

(64)

Fig. 3.173. Preparation of phosphorylated monoterpenoids.

Some developmental work on the rate of dissipation of dichloromethyl O,O-diphenyl phospho-nate and its p-nitro analogue was studied in rice plant by Bedi and Roy [28, 29]. The results indicated that these compounds persist moderately, with their half lives of 5.9 and 5.1 days respectively. However, the newly discovered fungicide dichloromethyldiphenylphosphonate did not leave any toxic residues on the grain and fodder. These observations show that organophos-phorus fungicides do not have any phytotoxicity as long as they are applied at the recommended dosages. Phytotoxicity studies on methyl O,O-bis (2,4,5-trichlorophenyl)phosphonate and other potential phosphonates showed no phytotoxicity even at 1000 ppm [Vasu and Roy, 30].

b. Systemic fungicides [31, 32, 33]

i) Carboxamide and carboximide compounds [34]

Hydroxybenzoic acid and particularly its ester and amide derivatives possess good fungicidal action. Salicylanilide, Shirlan (65, Fig. 3.174), a selective and protective fungicide is effective against certain *Ascomycetes* and *Phycomycetes* and powdery mildew and can be prepared as follows.

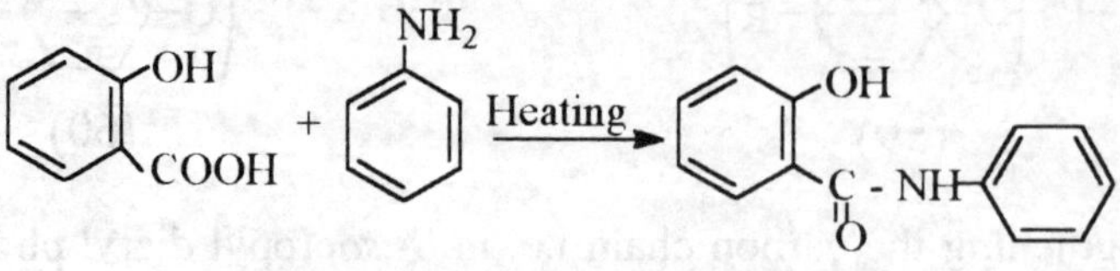

(65)

Fig. 3.174. Preparation of salicylanilide.

When the OH group of salicylanilide is replaced by a methyl group, the resultant compound **mebenil** (66) shows activity against *Basidiomycetes.*

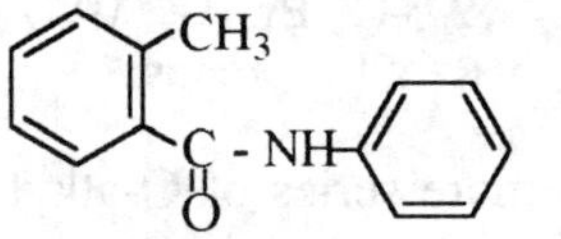

(66) 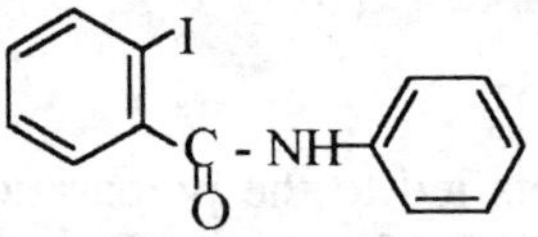   (67)

2-Iodo-N-phenylbenzamide, benodanil (67) more effectively controls rusts in cereals, than mebanil and, also shows appreciable activity against rust fungi of coffee, tobacco, vegetables and ornamental plants.

The most widely used fungicide of carboxamide group is **carboxin**, 5,6-dihydro-2-methyl 1, 4-oxathin-3-methyl carboxanilide (68) which can be prepared as shown in Fig. 3.175.

(i) $CH_3COCH_2COOC_2H_5 \xrightarrow{SO_2Cl_2}$ $\begin{array}{c} CH_3-C=O \\ | \\ Cl-CH-COOC_2H_5 \end{array}$ $+ SO_2$

(ii) $CH_3COCHClCOOC_2H_5 + HO-CH_2CH_2SH \longrightarrow$ $+ HCl$

(iii)

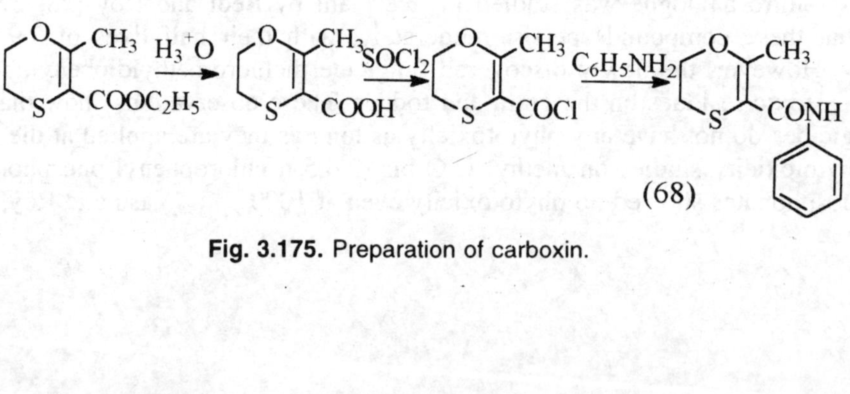

(68)

Fig. 3.175. Preparation of carboxin.

Carboxin is solid substance with LD_{50} 3820 mg/kg. It is a selective and systemic fungicide used against bunts of wheat, barley and loose smut disease. It is translocated through the xylem after being absorbed by the roots and accumulated mainly in the epicotyl. The sulfone of carboxin, **oxycarboxin**, 5,6-dihydro-2-methyl 1,4-oxathin-3-carboxanilide-4,4-dioxide (69a, 69b, Fig. 3.176), however possesses reduced activity and is prepared as follows.

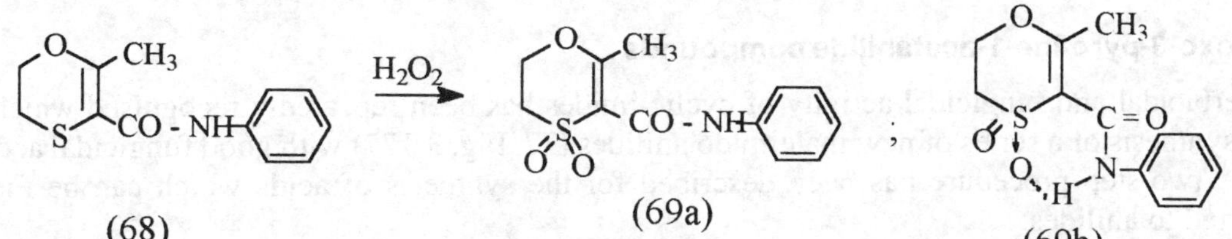

Fig. 3.176. Preparation of oxycarboxin.

It is more effective against rust fungi as it is more stable because of interhydrogen bonding

Pyracarbolid, 3,4-dihydro-6-methyl N-phenyl-2H-pirane-5 carboxamide (70), has action similar to that of the oxathins but is somewhat more effective as seed dressing agent against smut and rust fungi. It is primarily used for the protection of coffee and tea crops.

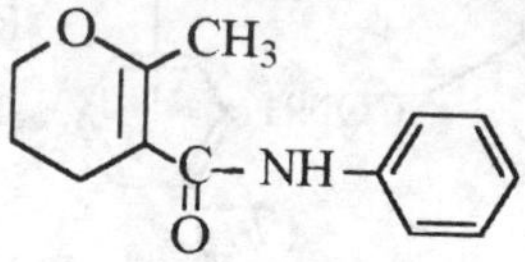

(70)

SAR studies showed that furan skeleton in position 2 is essential for fungicidal activity, against *Basidiomycetes* spp. The 2,5 Dimethyl substitution in the phenyl ring enhances the activity whereas 2,4,5-trimethyl substitution resulted in somewhat reduced efficacy. Methyl substitution in the aniline ring reduced the fungicidal action by 30-50% and substitution by Cl or Br atom by 70-100%.

A few more carboxamides with furan ring are; **Furcarbanil**, 2, 5-dimethyl N-phenyl furan-3-carboxamide (71), highly effective against loose and covered smuts of cereals, **Cyclafuramid**, N-cydohexyl-2, 5-dimethylfuran 3-carboxamide (72) with moderate activity, **Fenfuram**, 2-methyl-N-phenylfuran-3-carboxamide (73) an effective seed dressing agent against smut and bunt disease of temperate cereals and **Methfuroxam**, N-phenyl-2,4,5-trimethyl-3-furancarboxamide (74) active against *Basidiomycetes,* loose and covered smut of wheat, barley and also rust fungi. Carboxamides containing a thiazole ring, such as 2,4-dimethyl· N-phenyl-5-thiazole carboxamide (75), and 2-amino-4-methyl-N-phenyl-5-thiazole carboxamide (76) are also found effective against smut and rust fungi, similar to carboxin and oxycarboxin.

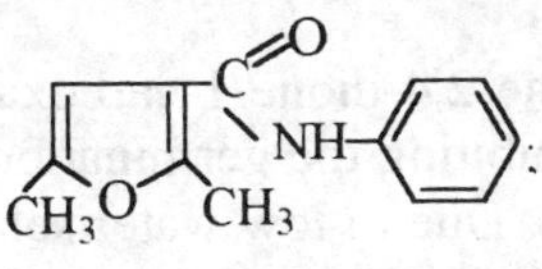

(71) 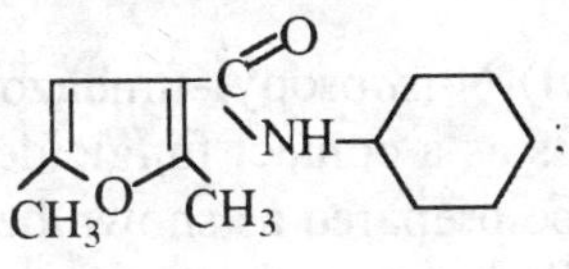   (72) 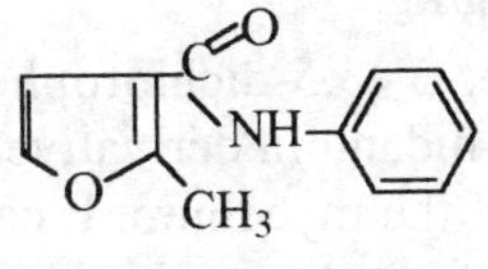   (73)

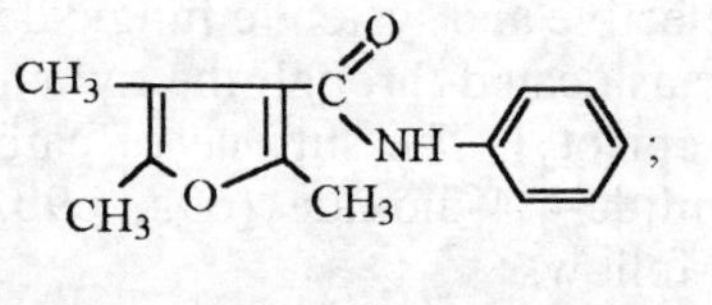

(74) 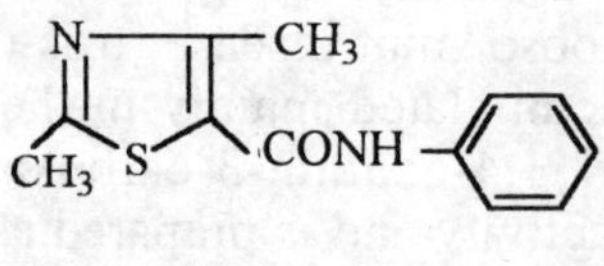   (75) 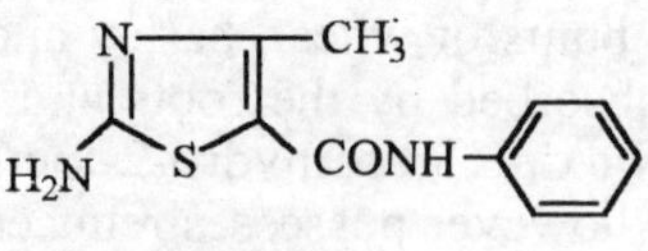   (76)

2,5-Dioxo-3-pyroline-1-acetanilide compounds

The herbicidal and fungicidal activity of cyclic imides has been repeatedly recognised which led to the synthesis of a series of new maleimido anilides (77, Fig. 3.177) with good fungicidal activity [35]. A two-step procedure has been described for the synthesis of acids which can be further converted to anilides.

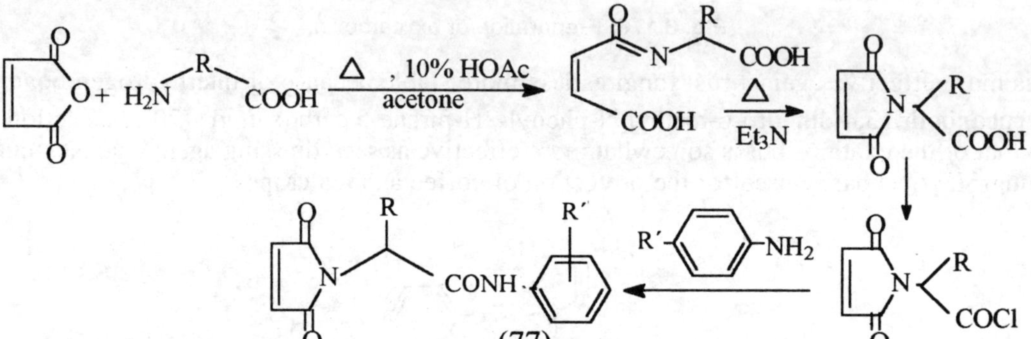

Fig. 3.177. Preparation of maleimido anilides.

In general, all carboxamides are effective against *Basidiomycetes,* of which smuts and rusts are prominent. They are also effective against *Rhizoctonia solani.* Carboxamides in general are not toxic to mammals, (acute oral $LD_{50} > 2000$ mg/kg). These compounds are ecofriendly because they rapidly decompose to form non-toxic metabolites.

Dicarboximide compounds

More recently a series of dicarboximide derivatives with excellent fungicidal action have been discovered in which the heterocyclic moiety is mostly pyrrolidine dione, oxazolidine dione or imidazolidine dione.

Procymidone, N-(3,5-dichlorophenyl)-1,2-dimethylcyclopropane 1,2-dicarboximide (78, Fig. 3.178) which has moderate systemic action, is prepared as follows.

It is photostable, white crystalline material which is not affected by heat and humidity with LD_{50} 6800 mg/kg.

Iprodione, 3-(3,5-dichlorophenyl)-N-*iso*propyl-imidazolidine-2,4-dione-1-carboxamide (79, Fig. 3.179) a hidantoin derivative acts as a contact fungicide inhibiting the germination of spores and growth of the mycelium. It can be prepared as shown below. Due to low water solubility and volatility and formation of biologically inactive degradatory products, it acts as an ecofriendly but active fungicide. A few more active molecules belonging to this class are **Vinclozoline**, 3-(3,5-

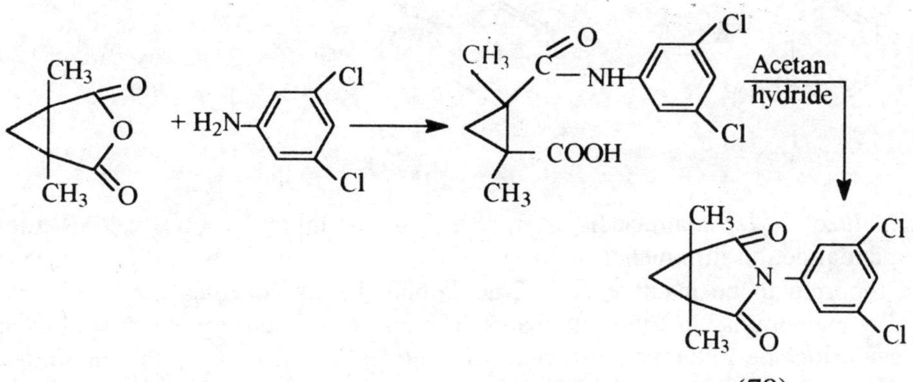

(78)

Fig. 3.178. Preparation of procymidone.

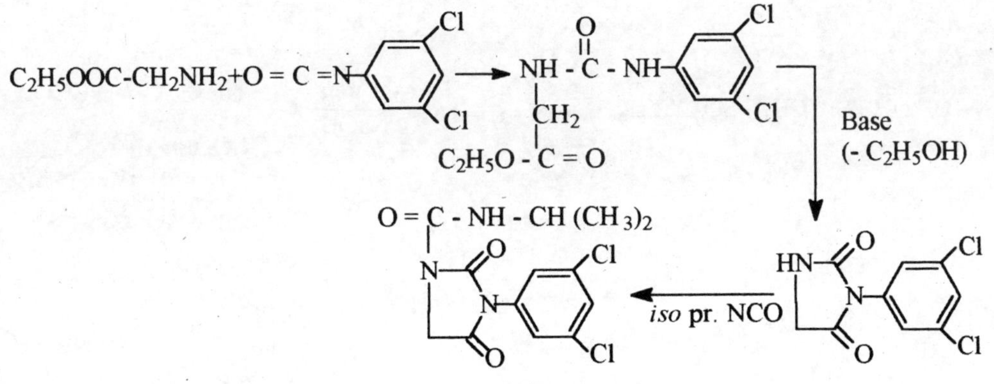

(79)

Fig. 3.179. Preparation of iprodione.

dichlorophenyl) 5-methyl-5-vinyl-1,3-oxazolidin-2,4-dione (80); **Dichlozoline**, 3-(3,5-dichlorophenyl) 5,5-dimethyl 1,3-oxazolidine-2,4-dione (81).

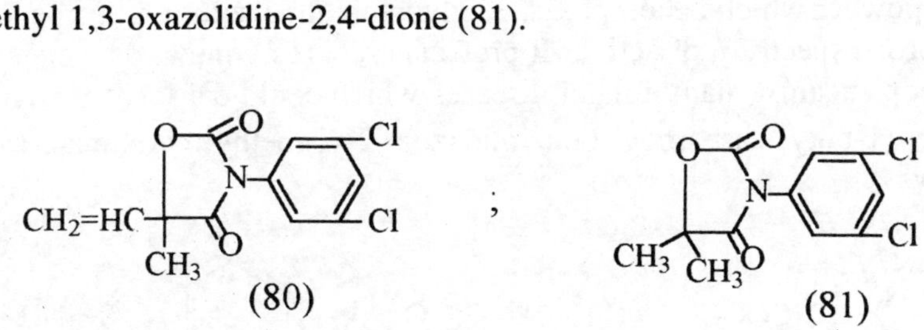

(80) ; (81)

Benzimidazole compounds and precursors [36]

The most important fungicides used in agriculture are 2-substituted benzimidazoles and amino benzimidazoles (82, 83)

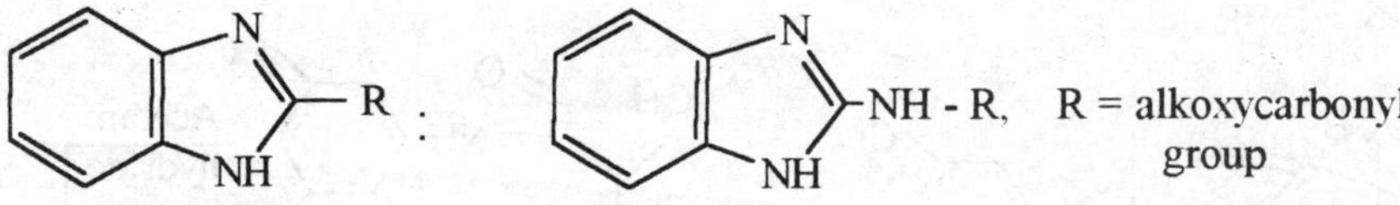

(82)   (83)   R = alkoxycarbonyl group

N-Benzimidazol-2-yl-carbamic acid esters exhibit good fungicidal activity. SAR studies showed that the fungicidal action diminished with increasing number of carbon atom as hexyl and octyl esters were reported to be inactive. The 2-acylamino benzimidazoles are less active than the analogous alkyl benzimidazol-2-yl-carbamates. The antifungal activity was lost after introduction of a methylene bridge between the benzimidazole ring and 2-methoxy carboxylamino group of the molecule. The methylation of either the carbamate nitrogen or the imidazole nitrogen also resulted in inactive molecules.

i) **Carbendazim, MBC**, methyl N-benzimidazol-2-yl-carbamate (84, Fig. 3.180) is prepared as follows.

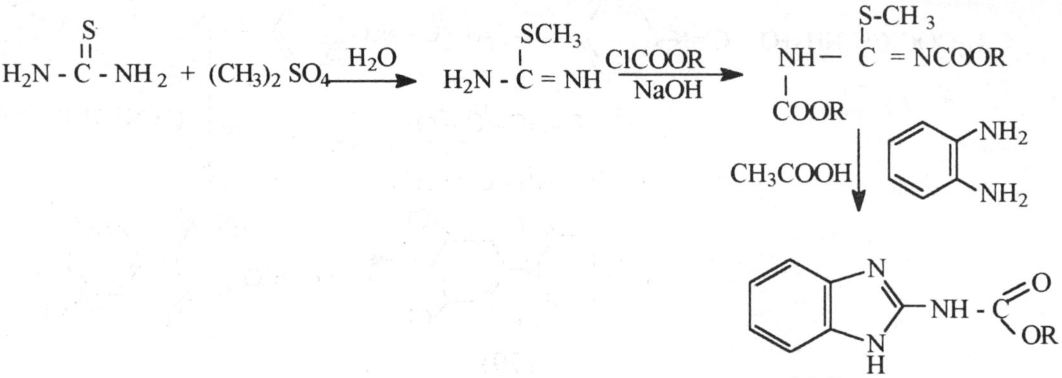

Fig. 3.180. Preparation of carbendazim.

It is light grey powder which decomposes slowly in alkaline solution. Carbendazim is a foliage fungicide with a broad spectrum of action. It protects orchards, vineyards, vegetables, ornamental plants and field crops against many fungal diseases which could cause very significant damages.

ii) **Benomyl**, N-(1-butyl carbamoyl benzimidazole-2-yl)methyl carbamate (85, **Fig.** 3.181) is prepared as follows.

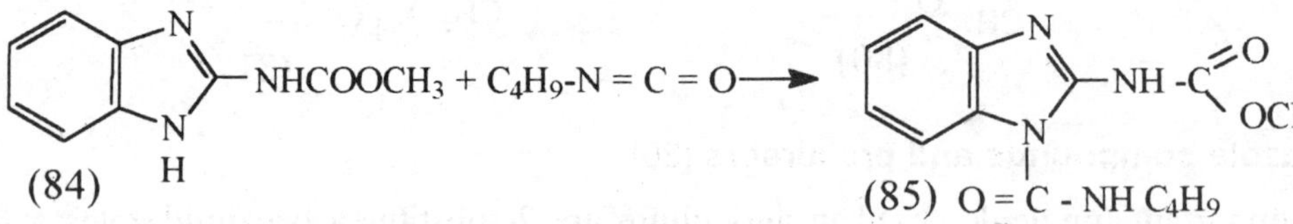

Fig. 3.181. Preparation of benomyl.

In aqueous media benomyl is hydrolysed rapidly to methyl N-benzimidazol-2yl-carbamate (MBC). It is soluble in organic solvents but insoluble in water. Benomyl is a foliage and soil fungicide also recommended for seed dressing. It is effective against grey mould, apple scab and powdery mildew.

iii) **Thiabendazole**, 2,-(4^1-thiazolyl) benzimidazole (86, Fig. 3.182) is prepared as shown below.

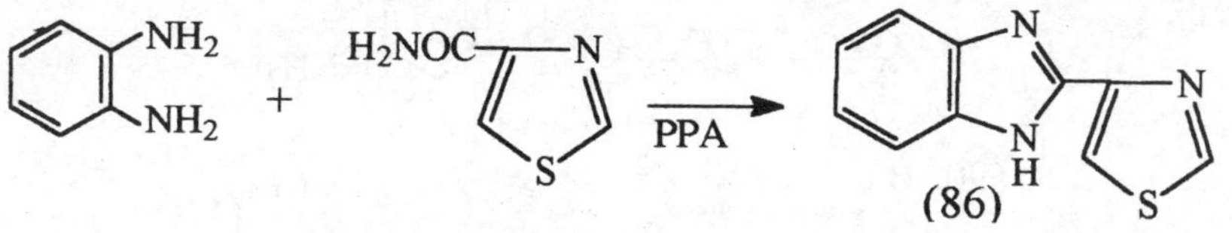

Fig. 3.182. Preparation of thiabendazole.

It is used for the protection of apples and pears against fungi causing rot disease during storage. The active substance has a systemic effect but it is translocated more slowly than benomyl or MBC.

iv) **Fuberidazole**, [2,-2^1-furyl) benzimidazole (87, Fig. 3.183) is prepared as follows.

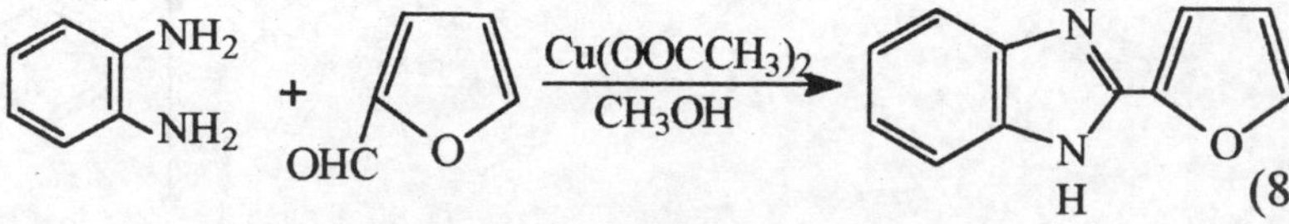

Fig. 3.183. Preparation of fuberidazole.

Its range of activity is similar to that of thiabendazole, effective for seed treatment against *Fusarium*, leaf rust and powdery mildew on barley.

v) **Cypendazole**, methyl-N[1-(5-cyanopenthylcarbamoyl)-2-benzimidazole]-carbamate (88, Fig. 3.184), can be prepared as follows. It is a fungicide with protective and eradicant action, similar to benomyl.

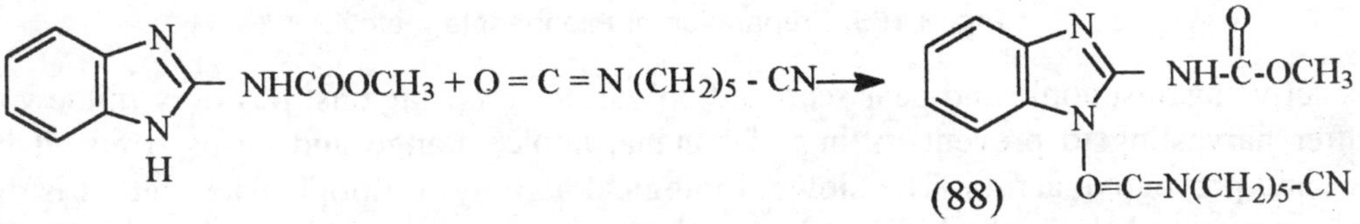

Fig. 3.184. Preparation of cypendazole.

vi) Alkoxy substituted benzothiazolones

A series of benzothiazolones (89) have been synthesised recently to screen against blast disease of rice. The compounds having substituents 7-methoxy, 7-methoxy ethoxy and 7-propynyloxy groups in the benzene ring showed good fungicidal activity [37].

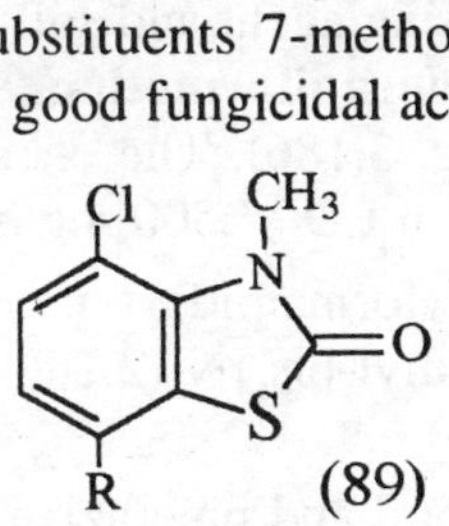

The most effective thioallophanic acid derivatives are thiophanate methyl (90), thiophanate ethyl (91) and NF-48 (92). These 1,2-disubstituted benzene derivatives are converted by chemical reaction in aqueous medium with NaOH or DMSO into N-benzimidazol-2-yl carbamates.

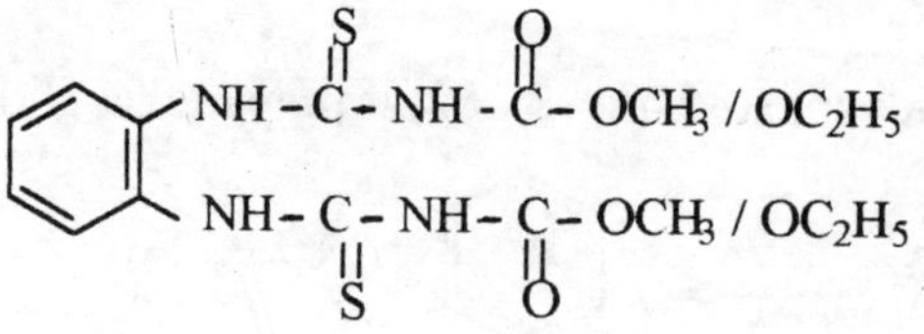

(90/91) 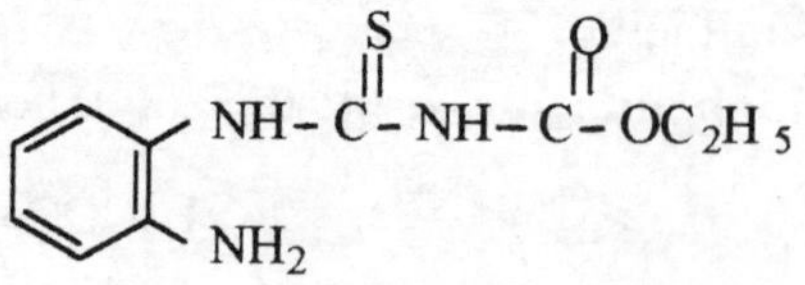 ; (92)

The preparation of **thiophanate methyl**, 1,2-bis(3-methoxycarbonyl-1-thioureido benzene (90, Fig. 3.185) is as follows.

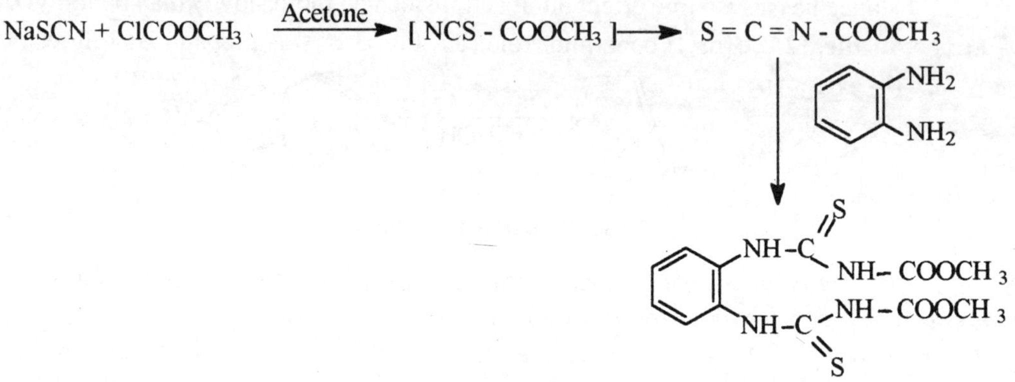

$$NaSCN + ClCOOCH_3 \xrightarrow{Acetone} [\,NCS\text{-}COOCH_3\,] \longrightarrow S=C=N\text{-}COOCH_3$$

(90)

Fig. 3.185. Preparation of thiophanate methyl.

It is active against apple and pear scab and on various crops against powdery mildew and also used after harvesting to prevent rotting of banana, apple, orange and citrus fruits. It has both curative and preventive action. The slower biological activity of thiophanate methyl is due to its slow conversion to MBC, which depends on pH and temperature and is catalyzed by sunlight.

1. Formamide type compounds [38]

Formalin (formaldehyde 40%) is known for its excellent disinfectancy. It exhibits fungicidal properties and is used mainly as seed dressing agent and soil fumigant. Based on this idea, a few chemicals have been developed like **chloraniformethan**, Imugan, N-[2,2,2-trichloro-1-(3,4-dichloroaniline) ethyl] formamide (93, Fig. 3.186). The chemical has a systemic action against powdery mildew, particularly in cereals with LD_{50} 2500 mg/kg.

Another powerful active substance of formamide type, with systemic activity **triforine**, a piperazine derivative, 1,1′-piperazine 1,4-diyl-bis [N-(2,2,2-trichloroethyl) formamide] (94, Fig. 3.187), is prepared as shown below.

Triforine is rapidly decomposed to chloral and piperazine in acid medium. However, in alkali

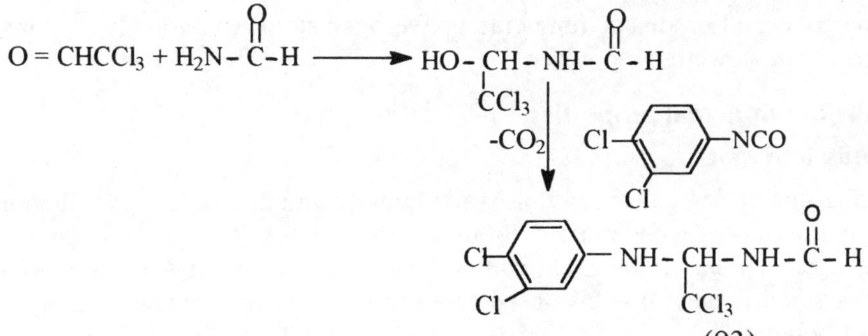

O = CHCCl₃ + H₂N– C–H ⟶ HO– CH– NH– C –H

Fig. 3.186. Preparation of chloraniformethan.

(93)

Fig. 3.186. Preparation of chloraniformethan.

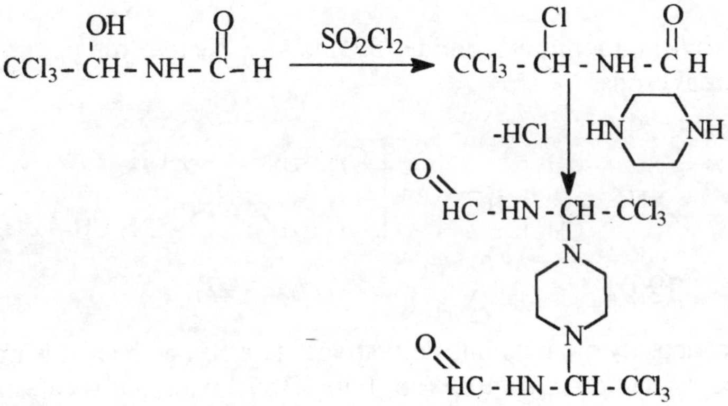

(94)

Fig. 3.187. Preparation of triforine.

and neutral aqueous solution, the decomposition is slow. It is non toxic to fish and bees and also does not contaminate the ground water as generally it is retained in the top layer of the soil from where it does not leach down.

The active substance absorbed through the roots is rapidly translocated to the aerial parts and accumulates in the leaves. The strong fungicidal action of triforine on powdery mildew of wheat can partly be attributed to the higher water solubility and chemical stability of these conjugates which are persistent and get rapidly absorbed and translocated.

Trimorphamide, N-(1-formamide-2,2,2-trichloroethyl) morpholine (95, Fig. 3.188) can be prepared as follows.

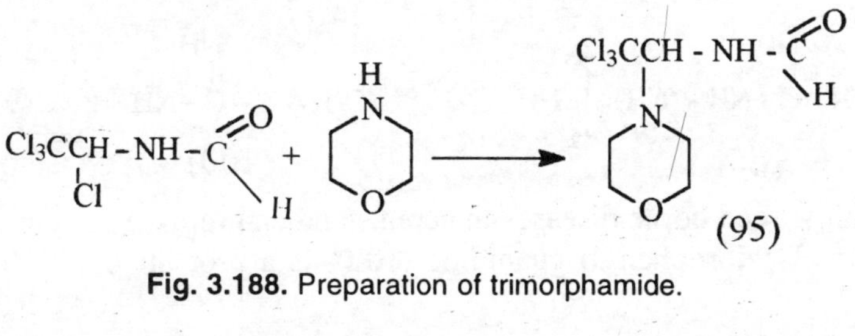

(95)

Fig. 3.188. Preparation of trimorphamide.

It is a protective and eradicant fungicide active against *Erysiphates*. It is more effective in emulsified form than as wettable powder.

Compounds with fungicidal properties

1. Alkyl amines and salts

i) To decipher the biological activity of N-alkylamines and their salts, several compounds with fungicidal properties have been identified. S-butylamine and its salts are actually used in agriculture for protection against *Penicillium digitatum* which causes citrus rot and *Penicillium expansum* parasitising pears and apples. It is also used as a fumigant for the control of gangrene, skin spot and diseases of potatoes.

ii) A few fungicides have also been developed in the group of quaternary ammonium compounds. **Deciquam**; didecyl-dimethyl ammonium bromide (96), has a protective and curative effect against apple scab.

Allyl didecyl methyl ammonium bromide (97), is an effective fungicide with wide range of action on various cereal crops.

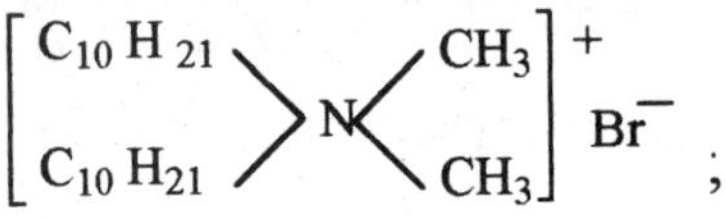

$$\left[\begin{array}{c} C_{10}H_{21} \\ C_{10}H_{21} \end{array} \!\!>\!\! N \!\!<\!\! \begin{array}{c} CH_3 \\ CH_3 \end{array} \right]^+ Br^- \quad ; \quad \left[\begin{array}{c} C_{10}H_{21} \\ C_{10}H_{21} \end{array} \!\!>\!\! N \!\!<\!\! \begin{array}{c} CH_3 \\ CH_2\,CH=CH_2 \end{array} \right]^+ Br^-$$

$$(96) \qquad\qquad\qquad (97)$$

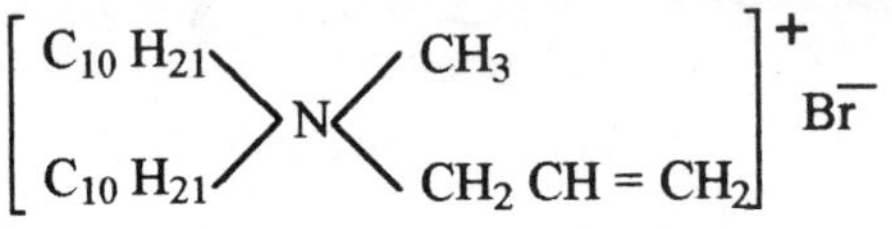

(iii) The biological activity of guanidine derivatives increases with increasing alkyl chain length. The acetate of dodecyl guanidine, **dodine** (98, Fig. 3.189) exhibited greatest fungitoxicity, with LD_{50} 1000 mg/kgs and is prepared as follows.

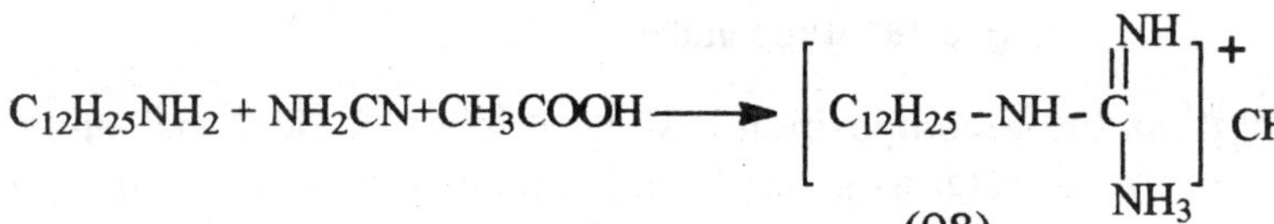

$$C_{12}H_{25}NH_2 + NH_2CN + CH_3COOH \longrightarrow \left[C_{12}H_{25}-NH-\overset{\overset{\displaystyle NH}{\|}}{\underset{\underset{\displaystyle NH_3}{|}}{C}} \right]^+ CH_3COO^-$$

$$(98)$$

Fig. 3.189. Preparation of dodine.

The water soluble acetate form is used against scab on apple and pear and cherry leaf spot. The active substance is metabolised in plant by the oxidation of alkyl chain and methylation into creatine.

iv) **Guazatine**, bis (8-guanidine octyl) amine (99), is used in the form of its water soluble triacetate with LD_{50} 260 mg/kg.

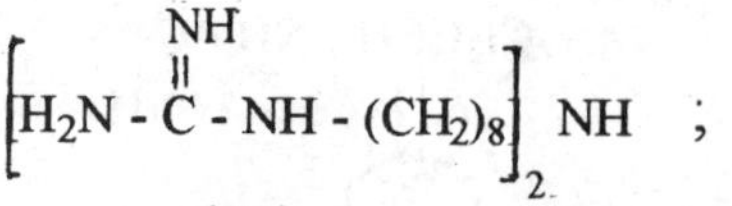

$$\left[H_2N-\overset{\overset{\displaystyle NH}{\|}}{C}-NH-(CH_2)_8 \right]_2 NH \quad ; \quad (CH_3)_2\,N-\overset{\overset{\displaystyle NH}{\|}}{C}-NH-\!\!\!\bigcirc\!\!\!-Cl$$

$$(99) \qquad\qquad\qquad (100)$$

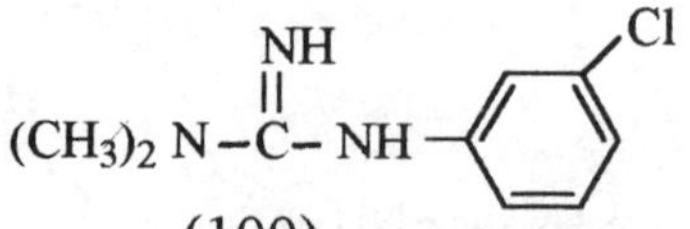

It is effective against seed borne diseases in cereals and can replace mercuric fungicides. **FDN** or N, N-dimethyl-N^I-(3-chlorophenyl) guanidine (100), is a protective and eradicant fungicide, having the same activity as dinocap against powdery mildew of cucumber. It is a soil fungicide, the active substance being absorbed by the roots of the plant with LD_{50} 420 mg/kg.

2. Alkane carboxylic acid compounds

N,N-(dimethyl thiocarbamoyl) thioacetic acid a crystalline compound protects against fungal infection. The activity of these compounds is generally explained by metabolic changes in conjunction with growth regulating activity. The two most effective compounds **propamocarb** or propyl N-(3-dimethylaminopropyl) carbamate (101) and **prothiocarb**; S-ethyl N-(3-dimethylaminopropyl)thiocarbamate (102, Fig. 3.190) specifically active against Phycomycetes fungi are prepared as follows.

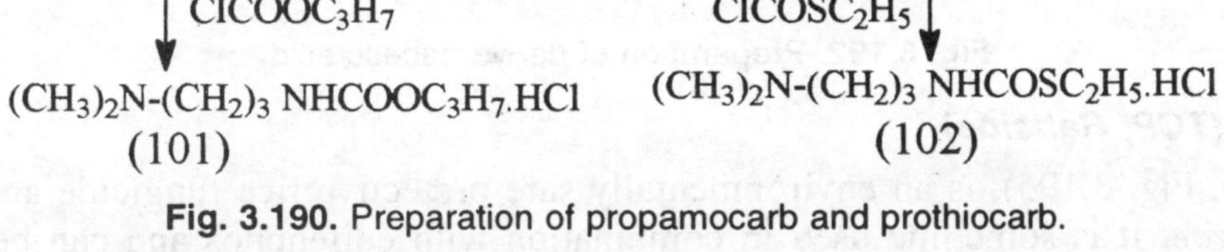

$$(CH_3)_2N\text{-}(CH_2)_3\ NH_2$$

$$ClCOOC_3H_7 \qquad\qquad\qquad ClCOSC_2H_5$$

$$(CH_3)_2N\text{-}(CH_2)_3\ NHCOOC_3H_7.HCl \qquad (CH_3)_2N\text{-}(CH_2)_3\ NHCOSC_2H_5.HCl$$
$$(101) \qquad\qquad\qquad\qquad (102)$$

Fig. 3.190. Preparation of propamocarb and prothiocarb.

Buthiobate, a dithiocarbonic acid derivative was found to be toxic to powdery mildew fungi and is named as butyl 4-t-butyl benzyl N-(3-pyridyl)-dithiocarbonimidate (103). It has both preventive and curative action against powdery mildews.

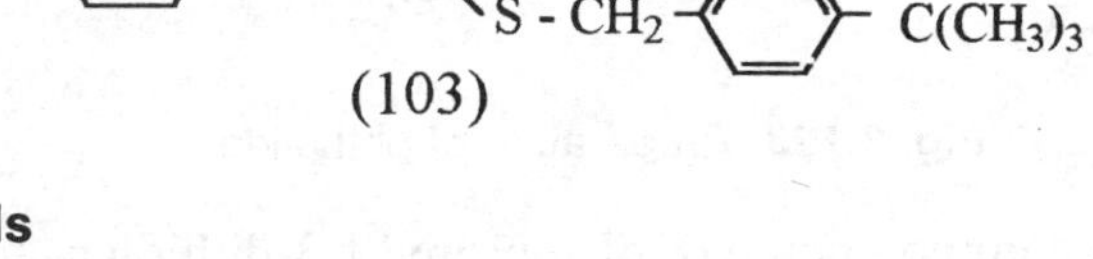

$$N = C \begin{array}{l} S - (CH_2)_3\ CH_3 \\ S - CH_2 - \text{(C}_6\text{H}_4) - C(CH_3)_3 \end{array}$$

(103)

3. Aromatic compounds

Some examples of active aromatic compounds are :

i) 4-Chloro-3,5-dimethylphenoxy ethanol is effective against *fusarium* wilt disease.

ii) Nitrothal-*iso*propyl and di*iso*propyl-5-nitro*iso*phthalate fungicides are not systemic but have specific activity against apple powdery mildew.

iii) Fenaminosulf, a diazo compound, the sodium salt of *p*-dimethylamino benzenediazosulfonic acid (104, Fig. 3.191), having LD_{50} 60 mg/kg, is toxic to mammals. It can be prepared as shown below.

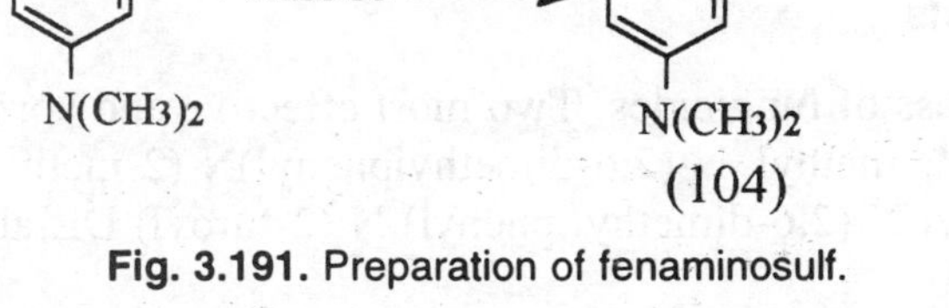

$$\text{N=N-Cl} \qquad + Na_2SO_3 \longrightarrow \text{N = N - SO}_3\text{Na}$$

$$\text{N(CH}_3)_2 \qquad\qquad\qquad \text{N(CH}_3)_2$$
$$(104)$$

Fig. 3.191. Preparation of fenaminosulf.

It is a promising seed and soil fungicide. It gives protection against seed and soil borne diseases particularly against *Phycomycetes*.

(a) Dehydroacetic acid

3-Acetyl-6 methyl 2,4-pyrandione (105, Fig. 3.192) inhibits mould growth on fresh and dried fruits and vegetables caused by mould fungi and bacteria. It is used for impregnation of wrapping papers for food stuffs and is prepared from diketene as shown below.

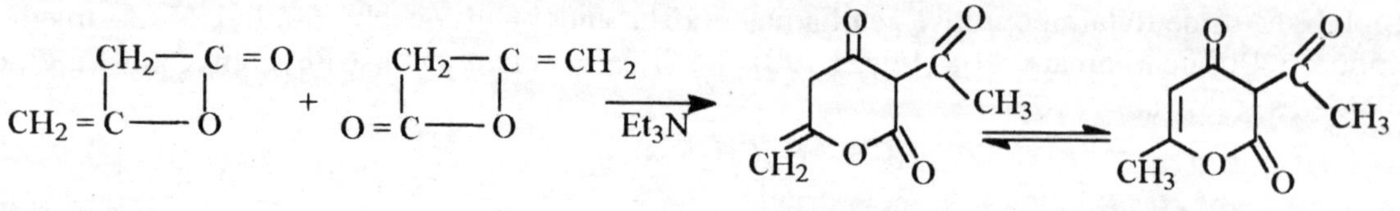

(105)

Fig. 3.192. Preparation of dehydroacetic acid.

(b) Phthalide (TCP, Rabcide)

Phthalide (106, Fig. 3.193), is an environmentally safe protective rice fungicide and is effective against *P. oryzae*. It is sometime used in combination with edifenphos and can be prepared as follows.

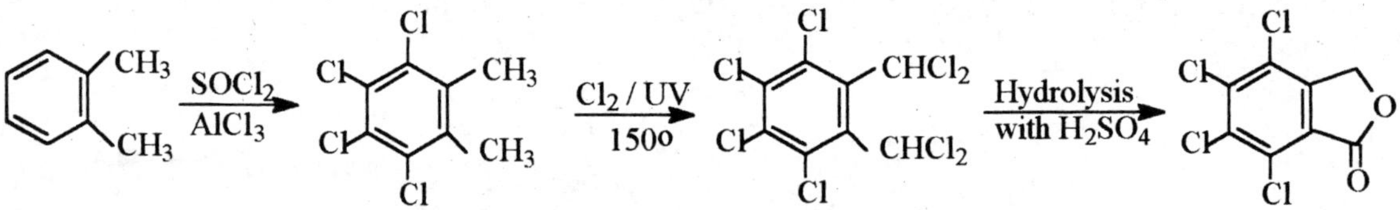

(106)

Fig. 3.193. Preparation of phthalide.

(d) **Isoprothiolane** is the common name of di*iso*propyl 1,3-dithiolan-2-ylidene malonate (107).

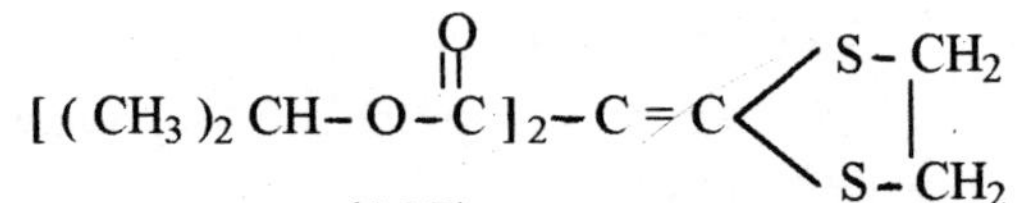

(107)

It is a selective fungicide with excellent systemic action against the fungus *Pyricularia oryzae*. The active ingredient from the granules applied to flooded field easily gets through the roots and is translocated into the leaves with LD_{50} 1190 mg/kg.

4. Acylalanine Compounds

Acylalanines are a new class of fungicides. Two most effective and commercialized members of this class are : **metalaxyl**, methyl N-(2,6-dimethylphenyl)N-(2-methoxy acetyl)-DL-alaninate (108) and **furalaxyl**, methyl N-(2,6-dimethyl phenyl) N-(2-furoyl) DL-alaninate (109).

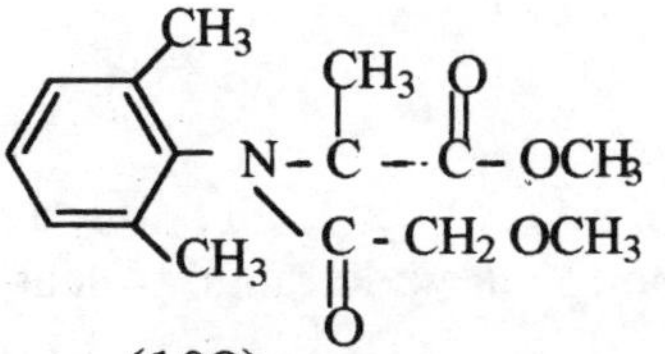

(108) 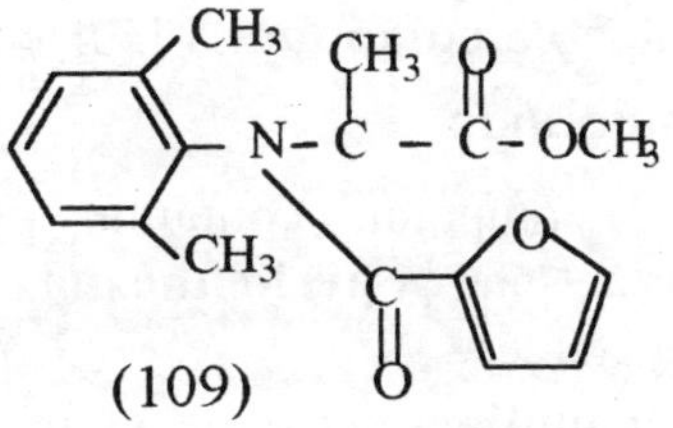   (109)

Both are highly effective at low concentrations for foliar and soil application against diseases, caused by both air and soil borne *Oomycetes* spp. in various agricultural and horticultural crops. These are protective fungicides with systemic properties and are slightly toxic to fish.

Their main field of application is against disease caused by *Phytophthora infestans* on potato and tomato and *Plasmopara viticola* on vine. Their range of action increases in conjunction with protective fungicides like dithiocarbamates.

ii) Acylamino oxazolidinones (oxadixyl, 110)

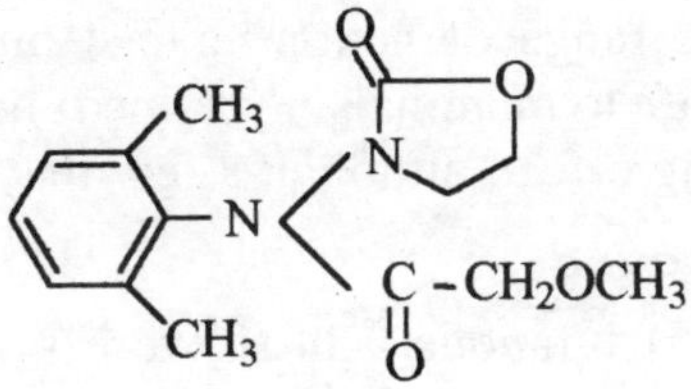

(110)

It is crystalline solid having mp 104-105°C, soluble in polar organic solvents but solubility in water is 3400 ppm. The product is non-volatile, and therefore, has no inhalation toxicity risk or rapid loss of efficacy. Also it does not accumulate in animal tissue since the octanol/water partition coefficient is very low.

CGA 80000, N-(chloro-2,6-dimethyl phenyl)-N-(methoxy methyl carbonyl)-3 amino-γ-butyrolactone, a fungicide was developed by CIBA-Geigy [39a, 39b] to control soil borne *Pythium* and *Phytophthora* spp. causing stem and root infections in tobacco, citrus fruit and ornamental plants. Because of the presence of chiral atom at the 3-position of the lactone ring and atropisomerism due to restricted rotation at phenyl-nitrogen bond, four stereo-isomers were obtained. The single isomer (111b) with absolute configuration (α S, 3R) was found to be 4 times more active than (111a).

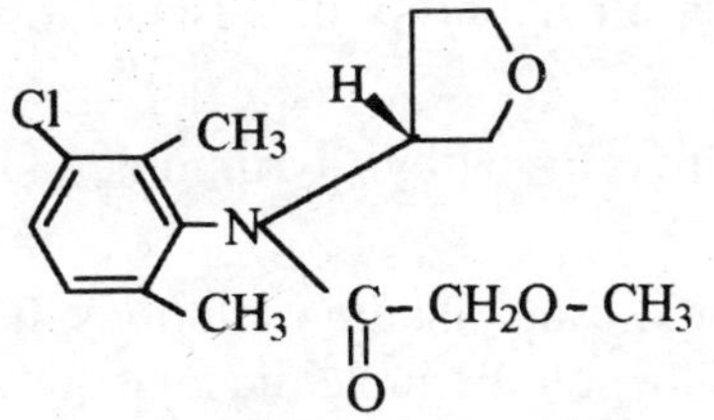

(111a) 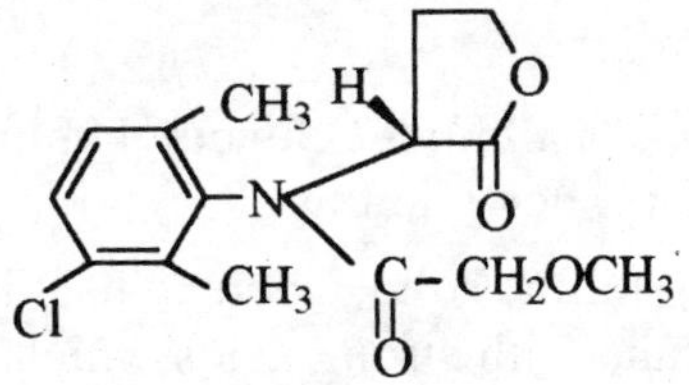   (111b)

Pyridine, pyrimidine, quinoline, quinoxaline, morpholine, piperidine, oxazole and thiazole compounds [40]

I) Pyridine compounds

A large number of compounds containing pyridine moiety have been found effective as fungicides. The most important one is **pyrifennox** (112) which controls a wide range of leaf spot pathogens of fruits and vegetables.

Pyridinyl pyrimidines (113), developed by Sumitomo Chemical Co Ltd., Japan are promising fungicides.

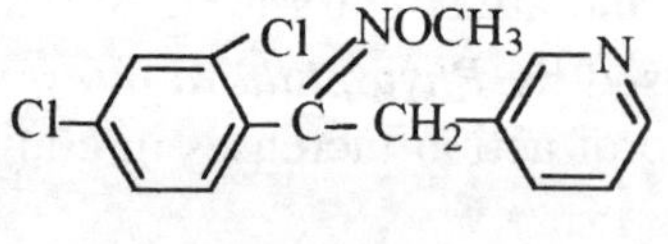

(112) 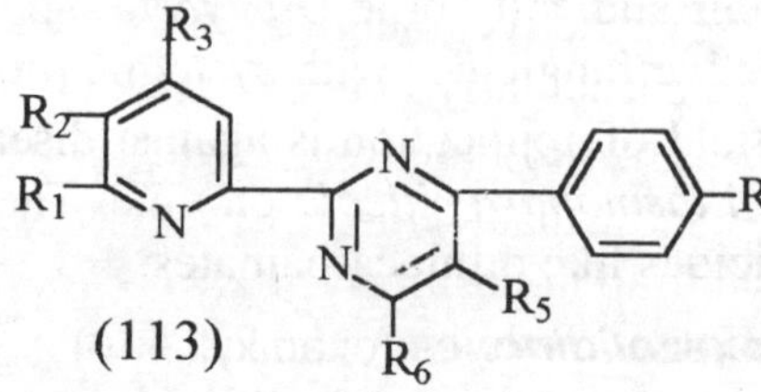   ;   (113)

Pyridinitrile (114), has a wide range of action against number of fungi. It is a protective fungicide with LD_{50} 5000 mg/kg, safe to mammals. Pyridine fungicides however cannot be applied on foliage due to their skin irritating effect, although a few may be applied as soil fungicides.

a) Pyridino 5(2H)-furanone compound

Recently a series of pyridino 5(2H) furanones (115), were synthesised having good fungicidal activity against *P. oryzae*.

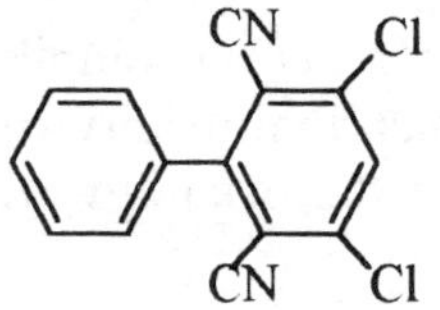

(114) 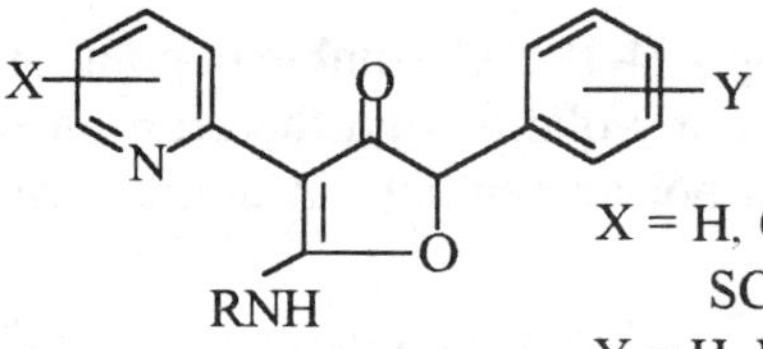   ;   (115)

X = H, Cl, Br, OCH_3, NO_2,
SCH_2CH_3, CN
Y = H, F, Cl, CH_3, CF_3
R = CH_3, $CH_2C≡CH$,
$CH_2CH=CHCl$

b) Pyrimidine compounds

Pyrimidine nucleus plays a significant role in biological systems and can be utilized to prepare a number of important drugs and pesticides. Two most active compounds of this group are **dimethirimol** and **ethirimol** which differ from each other only with respect to the alkylamino group in position 2.

Dimethirimol, 5-n-butyl-2-dimethylamino-4-hydroxy-6-methyl pyrimidine (116, Fig. 3.194), can be prepared as shown below.

It is white crystalline solid, stable to heat in both acidic and alkaline solutions. It is a weak base, forming stable salts with strong acids which are readily soluble in acidic aqueous solution.

Ethirimol, 5-n-butyl-2-ethylamino-4-hydroxy-6-methyl pyrimidine (117), is white crystalline solid stable to heat, both in acidic and alkaline solutions.

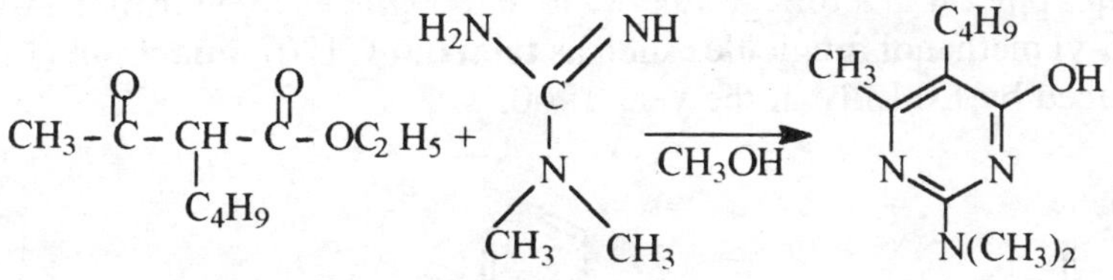

Fig. 3.194. Preparation of dimethirimol.

Both the above active mentioned compounds are specifically toxic to powdery mildew. Dimethirimol is particularly active against the powdery mildew of cucurbits while ethirimol is effective against powdery mildew of cereals, particularly barley.

SAR studies show that (i) the greatest efficiency is attained by the alkylamino group in position 2, (ii) the alkyl group in position 6 influences the biological spectrum, and this position is the most advantageous for a methyl or ethyl group, (iii) substituents in position 5 alter the lipophilic properties of the molecule and (iv) normal carbon chains containing 2-5 carbon atoms give the most efficient product. These compounds are relatively stable in soil.

Aliphatic aldehyde and ketone pyrimidinyl hydrazone (118), with wide range of organic radicals (R^1 to R^4) exhibit antifungal activity. The optimum activity against rice blast fungus (*P. oryzae*) was observed by the di*iso*butyl ketone-4, 6-dimethyl derivative $R=R^1=CH_3CH(CH_3)CH_2$-, $R^2=R^4=CH_3$; $R^3=H$

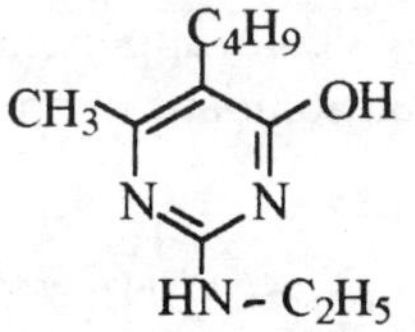

(117)

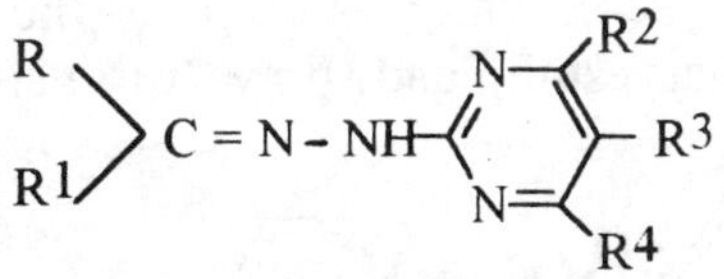

(118)

Another important fungicide of this group is **bupirimate**, 5-n-butyl 2-ethylamino-6-methylpyrimidine-4-yl-dimethylsulfamate (119, Fig. 3.195), which can be prepared as follows.

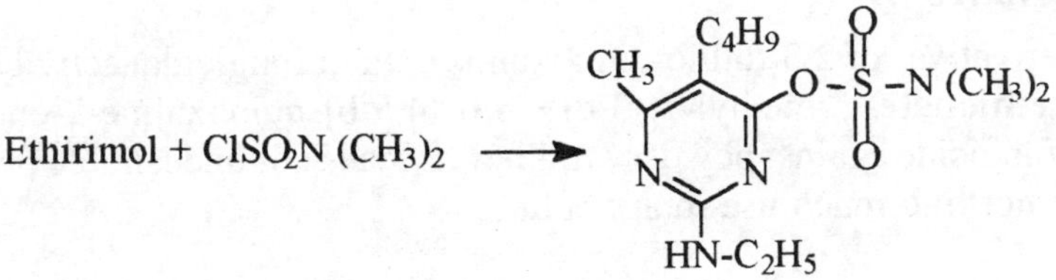

Fig. 3.195. Preparation of bupirimate.

Bupirimate is waxy solid, which decomposes on heating and is unstable on long term storage at temperature above 37°C with LD_{50} 4000 mg/kg. This substance is also effective only against powdery mildew of apple. It is a foliage fungicide and degrades to ethirimol. A series of 3-substituted pyrimidine-5-yl methanol fungicides such as **triarimol** (120), **nuarimol** (121) and **fenarimol** (122) were introduced by Eli Lilly in the year 1960.

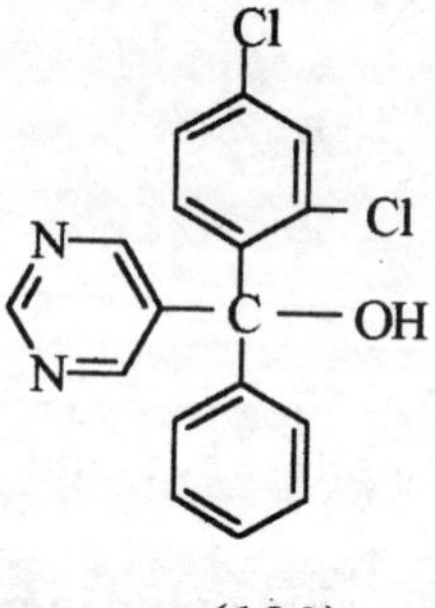

(120) 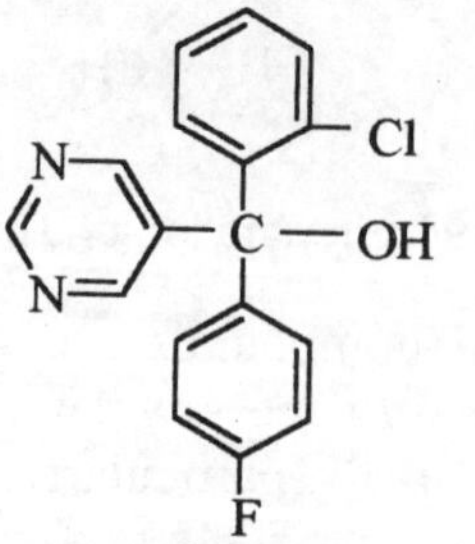   ;   (121) 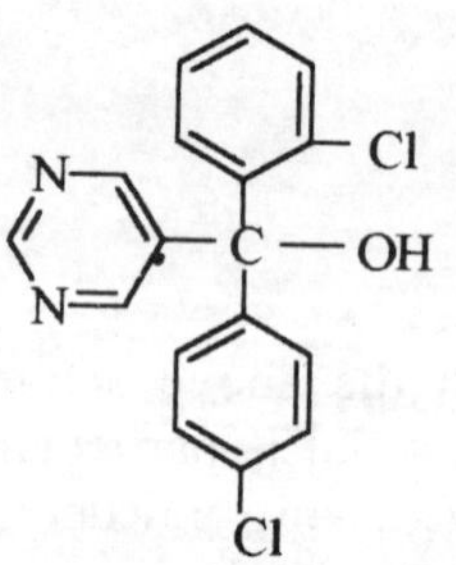   ;   (122)

All of them have curative, protective and eradicant action and are very effective against powdery mildew of barley and apples.

c) Quinoline derivatives

The first fungicide with quinoline moiety was 8-hydroxyquinoline having a systemic action. **Oxine copper**, bis(8-quinolinolate) copper (123), is used in agriculture as a seed dressing agent. It is an excellent seed dressing agent, if mixed with oxatin type of compound, for protection against loose smut of wheat and barley.

Another active commercial fungicide in this group is ethoxyquin, 1,2-dihydro-6-ethoxy-2,2,4-trimethyl quinoline (124), which is used for the control of scald in apples, and for preharvest treatment and post harvest dip and spray. It is also used as an antioxidant in spices and fish meal.

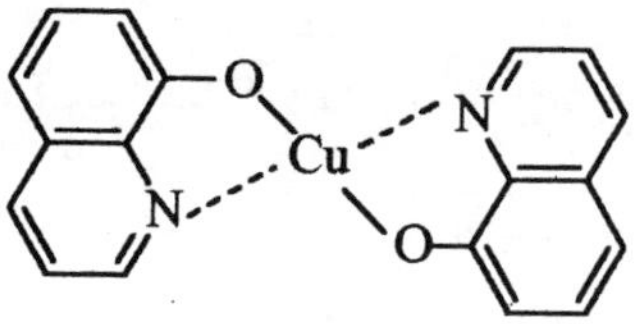

(123) 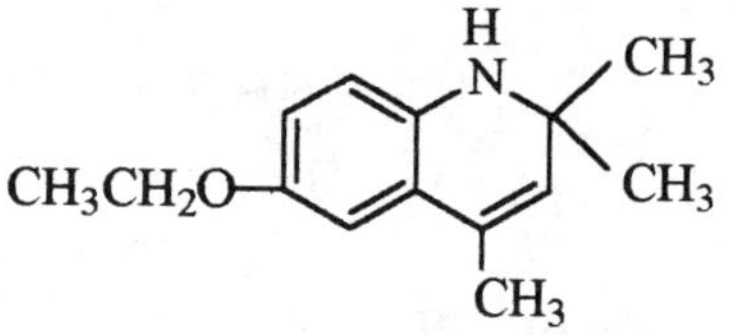   ;   (124)

d) Quinoxaline derivative

The carbonic acid derivatives of 2,3-dithio-quinoxaline exhibit fungicidal activity. The most active product is **Quinomethionate**, 6-methyl-1, 3-dithiolo [4,5-b] quinoxaline-2-one (125), which is effective as contact fungicide against powdery mildew. Because of undesirable properties however, this compound does not find much use in agriculture.

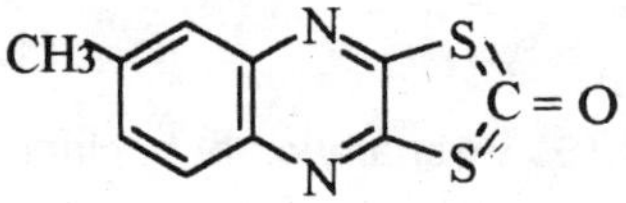

(125)

e) Morpholine compounds

Compounds containing morpholine moiety also possess fungicidal activity. Two most active compounds of this group are tridemorph and dodemorph.

Tridemorph, 2,6-dimethyl-4-tridecyl-morpholine (126), is obtained by the reaction of 2,6-dimethyl morpholine with tridecyl chloride. It is colourless oily liquid, miscible with water. It is an eradicant fungicide with systemic action, absorbed through foliage and roots, and used against powdery mildew disease.

Dodemorph, 4-cyclododecyl-2,6-dimethylmorpholine (127), is another important systemic fungicide of this class.

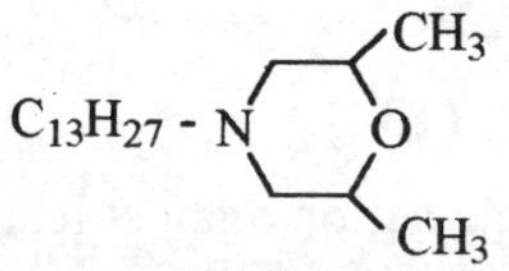

(126) 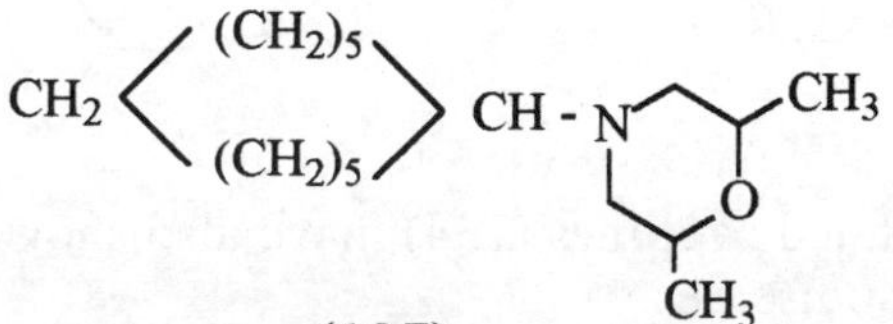   (127)

It is yellow liquid, miscible with water. The active substance is its acetate form which is stable as an eradicant fungicide and absorbed through foliage and roots. It is used mainly against powdery mildew of ornamental plants.

Fenpropiomorph, (128) developed by BASF (1980) has systemic action and is effective against powdery mildew and rust disease on cereals. Recently a few more fungicides (129 to 131) based on (128) were synthesized [41] and found effective against powdery mildew disease.

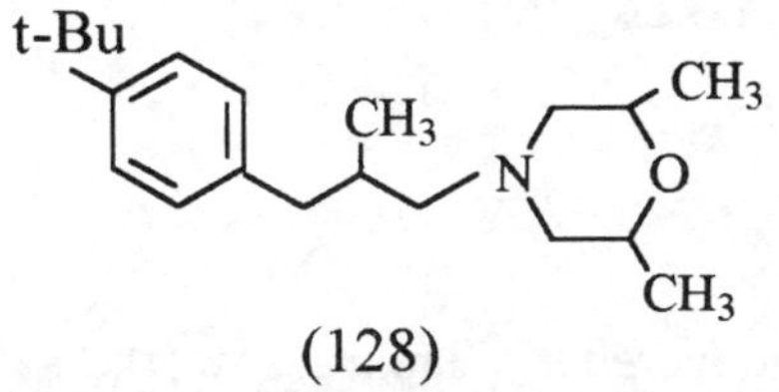

(128)   (129)   n = 1-3

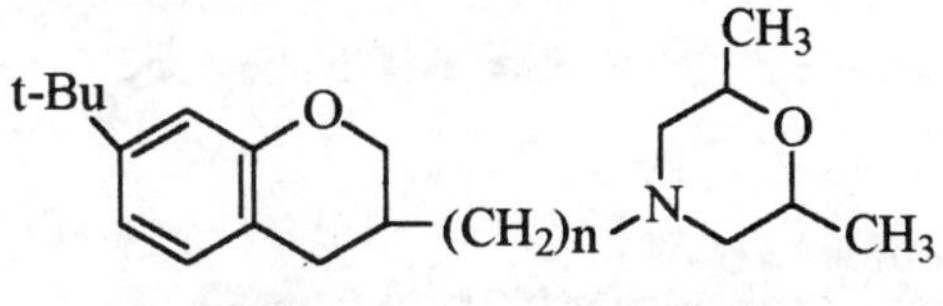

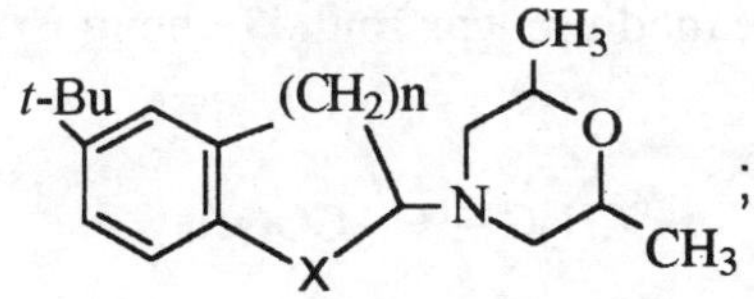

X = CH$_2$, O ;   n = 1,2

(130)   (131)

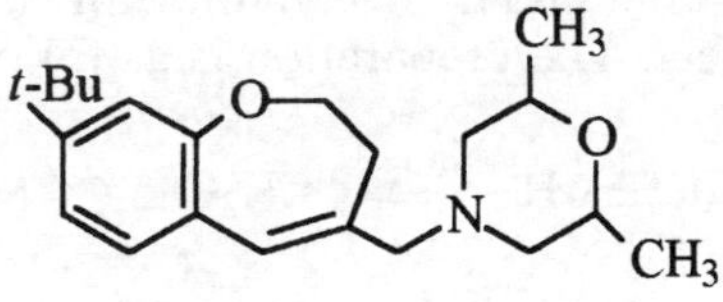

Two systemic fungicides in this group **aldimorph** (132) and **trimorphamide** (95) effective against powdery mildew disease are commercialized recently.

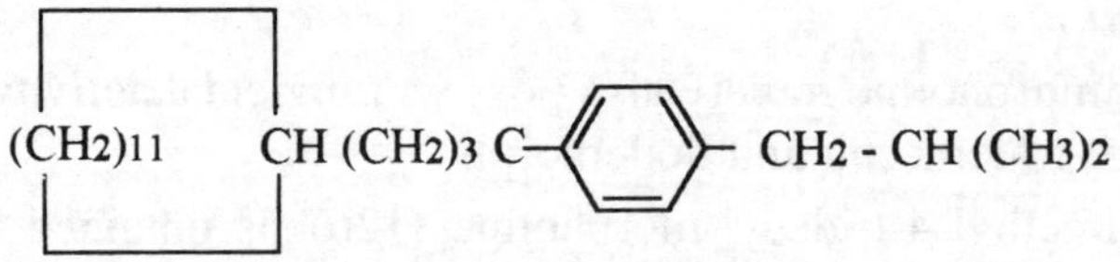

$$(CH_2)_{11} \quad CH\,(CH_2)_3\,C\!-\!\!\langle\!=\!\rangle\!-\!CH_2\text{-}CH\,(CH_3)_2$$

(132)

Piperidines

The only piperidine derivative is **fenpropidin** (133), developed and introduced in 1986 as new systemic fungicide effective against powdery mildew and rust on cereals.

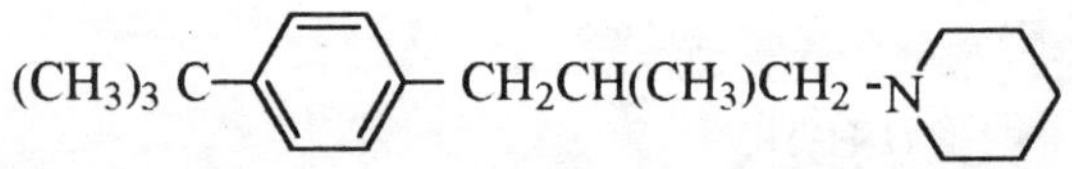

$$(CH_3)_3\,C\!-\!\!\langle\!=\!\rangle\!-\!CH_2CH(CH_3)CH_2\text{-}N\!\!\langle\ \rangle \qquad ; \qquad$$

NHR

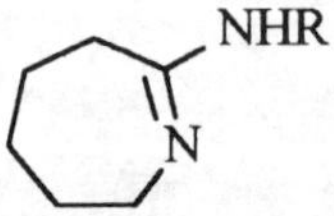

R = long chain alkyl group C_{10} - C_{12}

(133)

(134)

Some substituted **azepines** (134), have also shown systemic action against leaf spot, powdery mildew, and rust diseases.

i) Oxazole derivatives

1,2-Oxazole group of compounds exhibit high antifungal activity. The most effective compound, **drazoxolon**, 4-(2-chlorophenylhydrazono)-3-methyl5-*iso*xazolone (135), is effective mainly against powdery mildew. It is used as a seed dressing agent for the control of *Pythium* and *Fusarium* infestation of peas, maize, vegetables and cotton with LD_{50} 126 mg/kg. The analogue 4-(3-chlorophenyl hydrazino)-3-methyl 5-isoxazolone is less toxic to mammals.

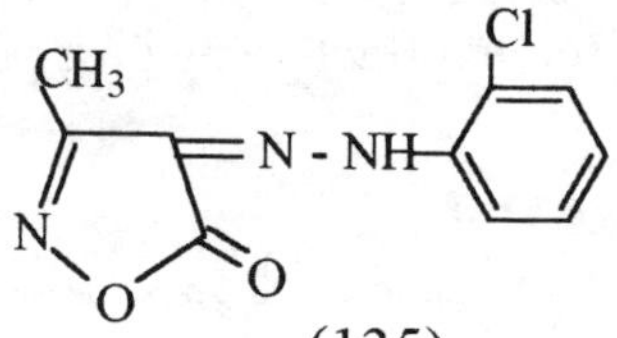

(135)

iii) Isoxazole compounds [42]

A series of 3-hydroxy*iso*xazole derivatives was discovered by Japanese workers as new soil fungicides. Amongst them, **hymexazol**, 3-hydroxy-5-methyl isoxazole (136, Fig. 3.196) prepared from tetrolic acid ester and hydroxylamine in alkaline medium was found to be most effective with no phytotoxic effect. The preparation is as follows.

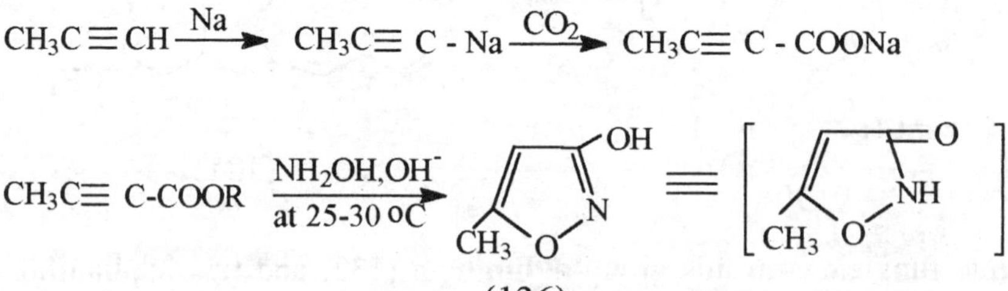

$$CH_3C\equiv CH \xrightarrow{Na} CH_3C\equiv C\text{-}Na \xrightarrow{CO_2} CH_3C\equiv C\text{-}COONa$$

$$CH_3C\equiv C\text{-}COOR \xrightarrow[\text{at 25-30 oC}]{NH_2OH,OH^-}$$

(136)

Fig. 3.196. Preparation of hymexazole.

Remarkable biological activities of hymexazol were observed besides controlling soil borne diseases caused by *Fusarium* spp., and *Rhizoctonia solani*. It was found to accelerate the root and shoot growth of many kinds of plants at seedling stages. The metabolic pathway of this compound was studied on soils and rats. The degraded products were found to be non-toxic in nature.

h) Thiazoles

Etridiazole, 5-ethoxy-3-trichloromethyl-1, 2,4-thiadiazole (137), is a soil fungicide with systemic action with LD_{50} 2000 mg/kg.

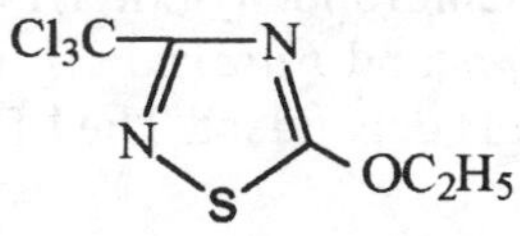

(137)

TCMTB, 2-(thiocyanomethylthio)benzothiazole (138), is an effective soil fungicide used as seed dresser with LD_{50} 1590 mg/kg. Another effective fungicide is 2-(carboxymethylthio) benzothiazole (139).

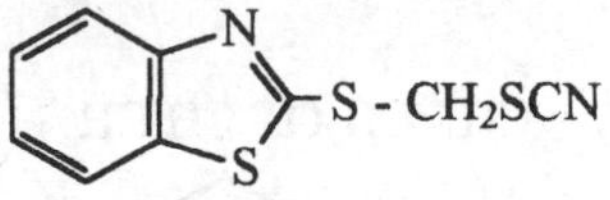

(138) 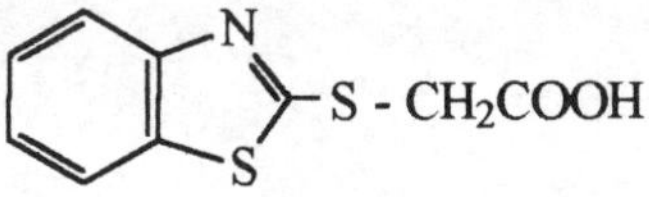   ;   (139)

Tricyclazol, 5-methyl, 1,2,4-triazolo [3,4-b] benzothiazole (140, Fig. 3.197), is an effective against rice blast disease (*P. oryzae*) and can be prepared as follows.

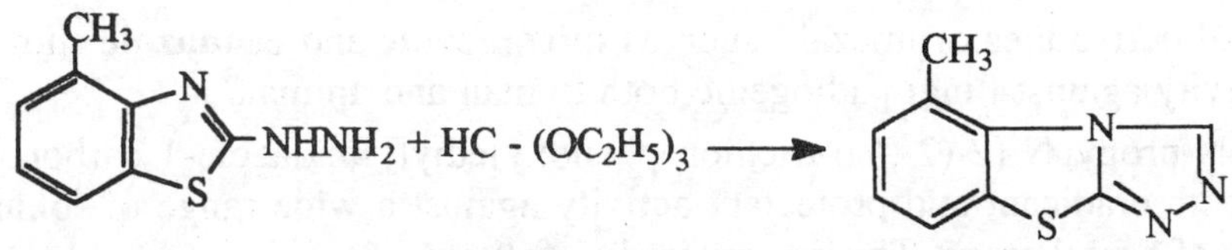

(140)

Fig. 3.197. Preparation of tricyclazol.

Azole compounds

a. Imidazole derivatives

The triarylmethyl derivatives of imidazole show fungistatic effects against powdery mildew and *Phytophthora infestans*. The most potent one is **clotrimazole**, 1-(2-chlorophenyl diphenyl-methyl)imidazole (141), which is used in general against human pathogenic fungi mycoses.

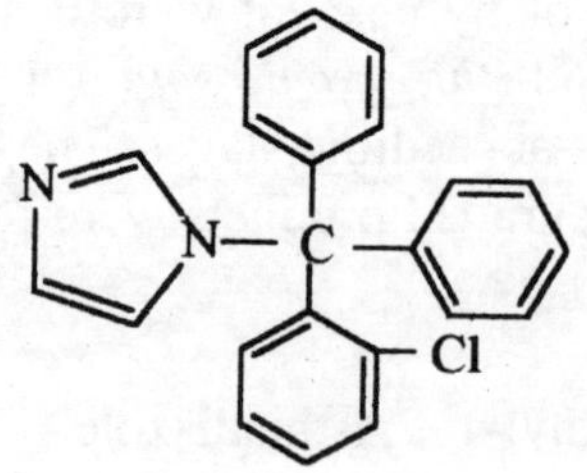

(141)

Imazalil, 1-(β-propenyloxy-2,4-dichlorophenylethyl)1-*H*-imidazole (142, Fig. 3.198) has systemic action, against *Helminthosporium* and *Fusarium*. It is used as seed dresser against diseases of cereals. The efficacy of imazalil is pH dependent. The LD_{50} (rat) is 320 mg/kg. It can be prepared as follows.

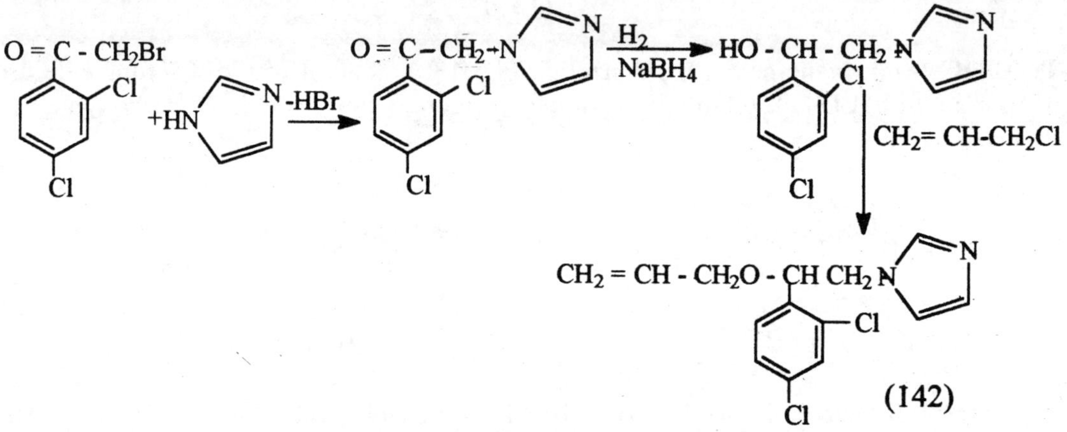

Fig. 3.198. Preparation of imazalil.

A few related derivatives of imazalil such as **micomazole** and **econazole** show *in vitro* broad antimicotic activity against fungi pathogenic both to man and animals.

Prochloraz N-propyl N-[2-(2-,4,6-trichlorophenoxy)ethyl] imidazole-1-carbonamide (143, Fig. 3.199). It has both eradicant and protectant activity against a wide range of foliar, stem and ear mycopathogens of cereal crops. This is prepared as follows.

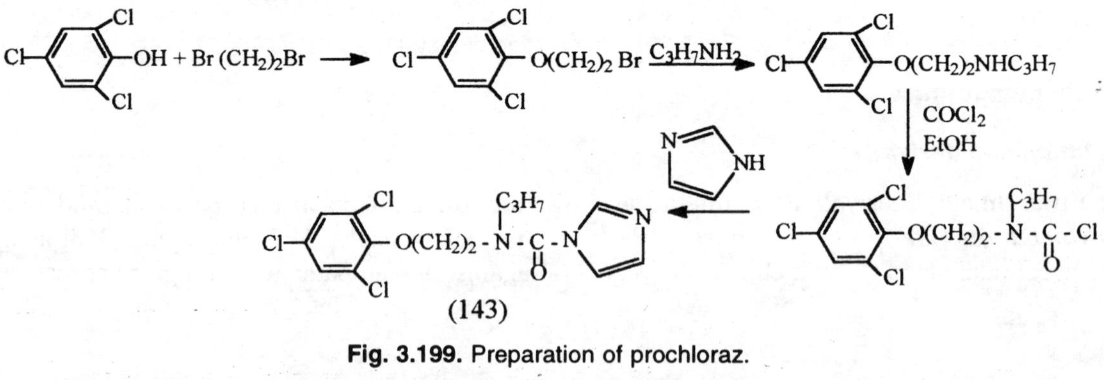

Fig. 3.199. Preparation of prochloraz.

Two more active molecules of this series are **baysan** (144) and **fenopanil** (145)

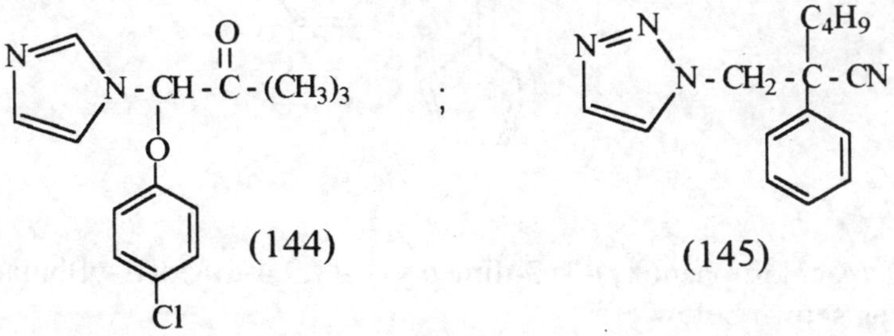

Compound (144) and (145)

Triazole derivatives

Buchel *et al.* [43a] synthesised several triazole compounds active against several fungi. A few important azole fungicides are discussed in the following text.

    i) **Fluotrimazole** (146, Fig. 3.200) can be prepared as shown below.

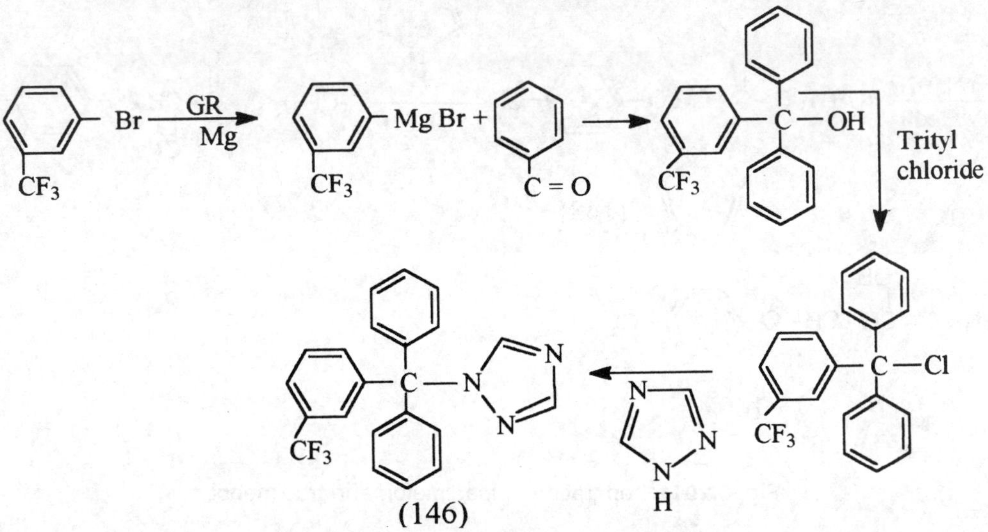

Fig. 3.200. Preparation of fluotrimazole.

Fluotrimazole gives effective protection against powdery mildew of cereals, apples, cucumber and roses and shows some activity against brown and yellow rusts with LD_{50} 5000 mg/kg. It has a selective action and favourable toxicological properties from the view point of environmental protection.

    ii) **Triazolyl O,N-acetals** (147) has good fungicidal properties similar to fluotrimazole. Unlike trityltriazole, these molecules have protective as well as systemic action. The systemic action and the range of action of these compounds depends on the number and nature of substituents in the phenyl nucleus. When 2-Cl atoms are incorporated in the phenyl ring the systemic action decreases but their spectrum of action is extended to other fungi also.

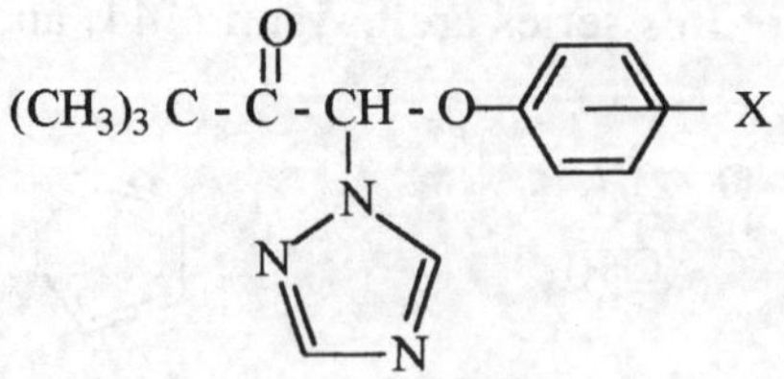

$$(CH_3)_3 \ C - \overset{\overset{\displaystyle O}{\|}}{C} - CH - O - \underset{}{\bigcirc} - X$$

(147)

Triadimefon, 1-(4-chlorophenoxy)-3, 3-dimethyl-1-(1,2,4-triazol-1-yl)butan-2-one (148, Fig. 3.201) is obtained as shown below.

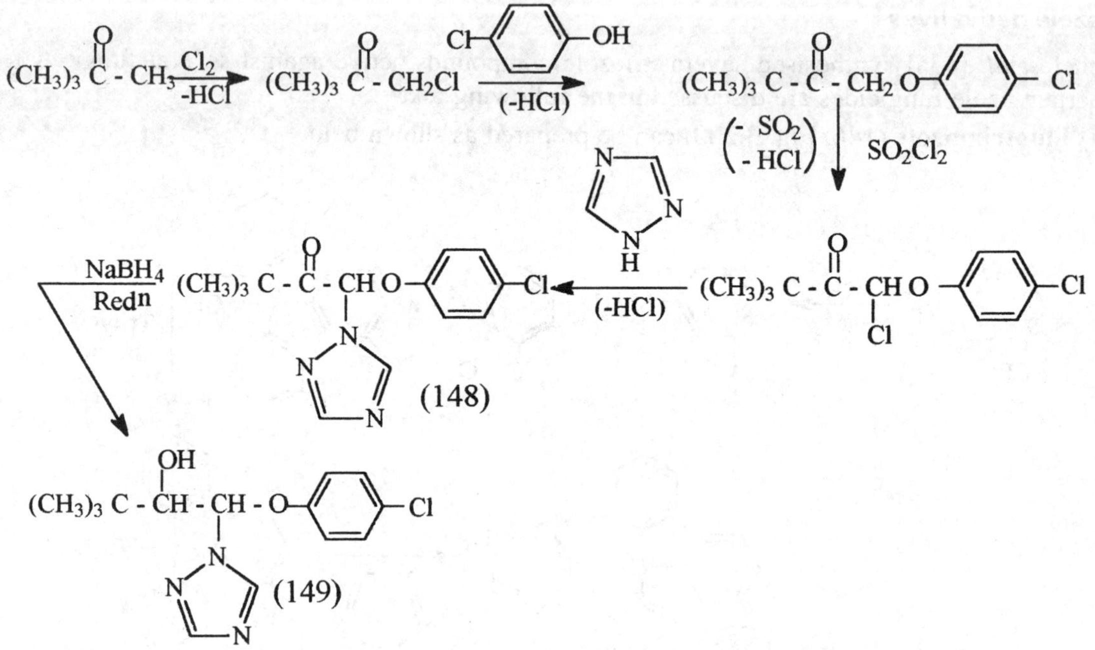

Fig. 3.201. Preparation of triadimefon and tradimenol.

This fungicide has excellent systemic action. It can be used as foliar spray, seed dresser and on soil. In addition to the control of powdery mildew, it protects cereals against all kinds of smuts. It has a powerful action against *Helminthosporium* spp. For the first time, the simultaneous protective, curative and eradicant control of rust fungi in cereals can be done by triadimefon. The disadvantage of triadimefon is that sometimes it may adversely affect germination and also growth of the host. Ecologically it is the safest fungicide with LD_{50} 568 mg/kg.

Triadimenol (149, Fig. 3.201) closely related to triadimefon has been identified within fungal and plant tissues as an effective metabolite. It is effective against a number of fungal disease with LD_{50} 1100 mg/kg. It is prepared by the reduction of triadimefon (148).

Dichlobutrazole, (2R, 3R) and (2S, 3S)-1-(2,4-dichlorophenyl)-4,4-dimethyl2-(1,2,4-triazol-1-yl) pentan-3-ol is strongly active against rust, powdery mildew and a variety of other fungal pathogens of plants. It has a systemic action with LD_{50} 400 mg/kg. The synthesis of dichlobutrazole (150) is similar to that of triadimefon.

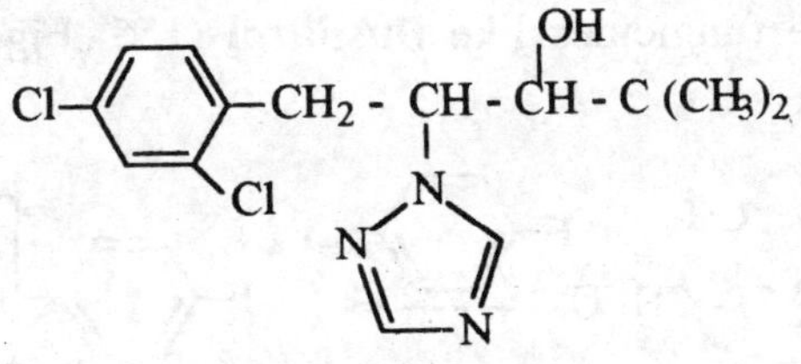

(150)

Propiconazole, 1-[2-(2,4-dichlorophenyl)-4-propyl-1,3-dioxolon-2-yl-methyl] 1,2,4-triazole (151, Fig. 3.202) and **Etaconazole,** 1-[2-(2,4-dichlorophenyl)-4-ethyl-1,3-dioxolon-2-yl-methyl)-1,2,4-triazole (152, Fig. 3.202) are found to be very effective fungicides for the control of powdery mildew and scab of fruits. They possess both curative and protective properties. These are prepared as follows.

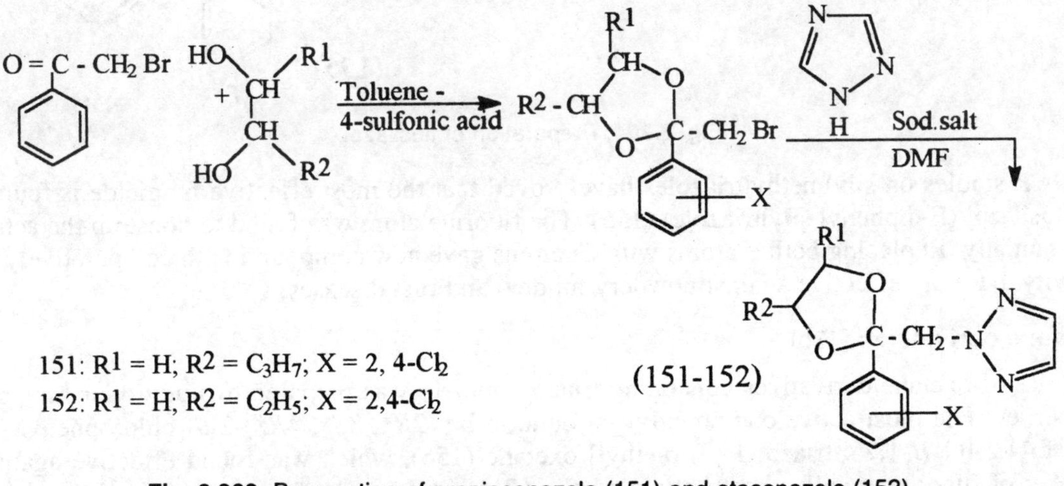

151: R^1 = H; R^2 = C_3H_7; X = 2, 4-Cl_2
152: R^1 = H; R^2 = C_2H_5; X = 2,4-Cl_2

(151-152)

Fig. 3.202. Preparation of propioconazole (151) and etaconazole (152).

Triazbutil, RH-124, 4-n-butyl-1,2,4-triazole (153, Fig. 3.203) is prepared as shown below. It has systemic action, useful for the control of wheat leaf rust and as seed dresser.

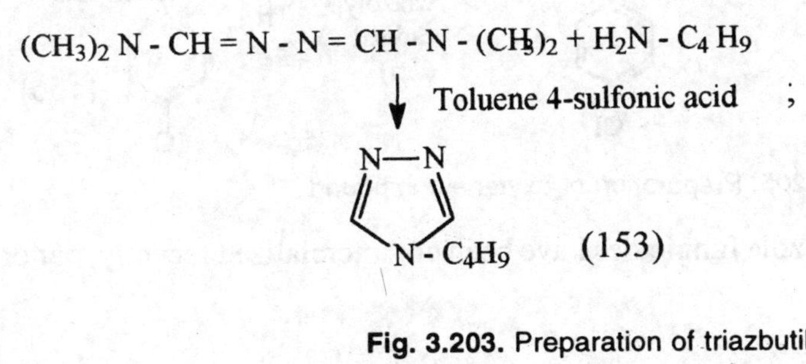

$$(CH_3)_2 N - CH = N - N = CH - N - (CH_3)_2 + H_2N - C_4H_9$$

↓ Toluene 4-sulfonic acid   ;

(153)

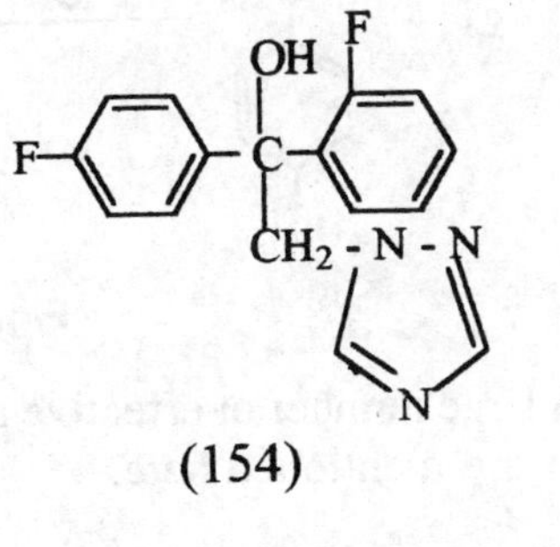

(154)

Fig. 3.203. Preparation of triazbutil.

Flutriafol (154), developed by ICI is a systemic foliar fungicide used to control cereal diseases. In combination with carbendazim, it exhibited broad spectrum of activity and controlled *Fusarium* spp. causing smut and bunt in cereals.

Silicon containing triazole fungicides like **flusilazol** (155, Fig. 3.204), can be prepared as follows.

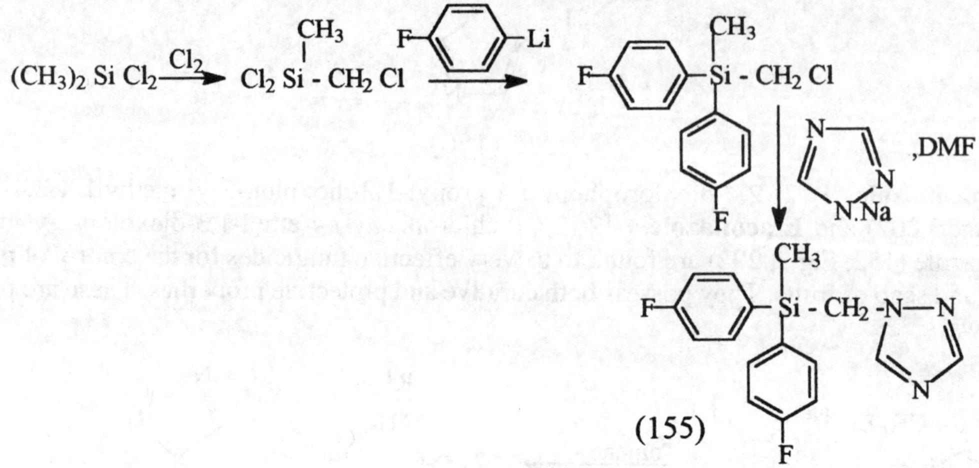

Fig. 3.204. Preparation of flusilazol.

SAR studies on silylmethyltriazoles have proved that the most effective fungicide is found to be flusilazol (F-diphenyl silyltriazole) (155). The fluorine atom was found to boost up the activity substantially. Replacing both F atoms with Cl atoms gave new compound with comparatively less activity, but very effective against powdery mildew and rust diseases.

Oxetane derivatives [43b]

A series of oxetane derivatives containing triazoyl moiety was prepared to study their fungicidal properties. The most active compound was found to be (2R*, 3S*, 4R*)-2-(4-chlorophenyl)-3,4-dimethyl-2-[(1-*H*-1,2,4-triazol-1-yl) methyl] oxetane (156), which was found effective against a number of fungi especially sheath blight of rice (*Rhizoctonia solani*). A stereoselective synthetic method using [2 + 2] photoreaction was followed to prepare this compound (Fig. 3.205).

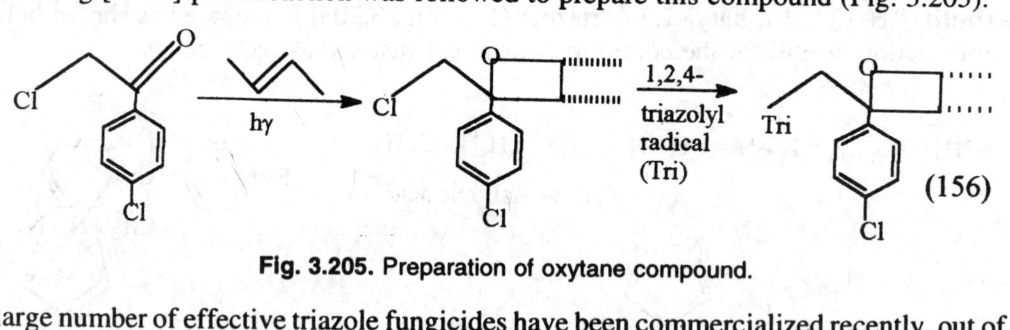

Fig. 3.205. Preparation of oxytane compound.

A large number of effective triazole fungicides have been commercialized recently, out of which a few are mentioned here.

Hexaconazole (157), a broad spectrum systemic fungicide developed by ICI, (1986) effective at low dosage (10-20 ppm) controls apple scab, coffee rust and vine powdery mildew. It has both curative and antisporulation activity.

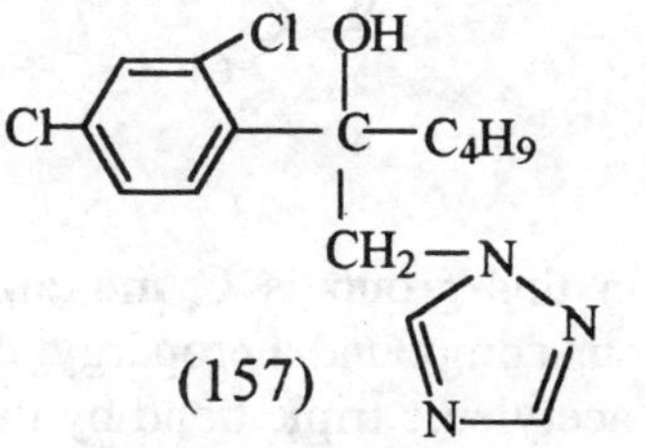

(157)

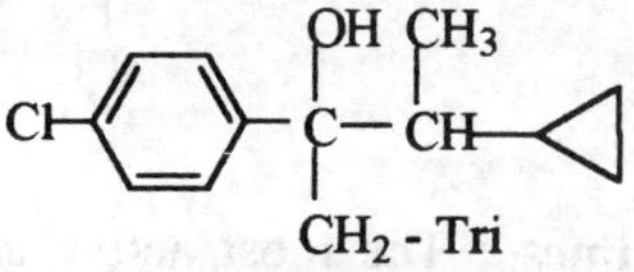

(158)

Tri = 1,2,4-triazolyl radical

A similar type of compound (158), was developed by Sandoz (1986) with excellent activity against powdery mildew, rust disease in cereals, vegetables and fruits.

Myclobutanil (159), developed by Rhom and Haas (1986) with broad spectrum of activity is used to control apple scab, black rot and powdery mildew. Another compound (160), developed by the same company in 1988, is an excellent seed dresser and controls specifically *Septoria* and *Puccinia* spp. Similarly a molecule (161), developed by Rhone-Poulene (1988) was found active against a number of fungi was specially *Ascomycetes* and *Basidiomycetes* and controls powdery mildew, rust and leaf spot diseases.

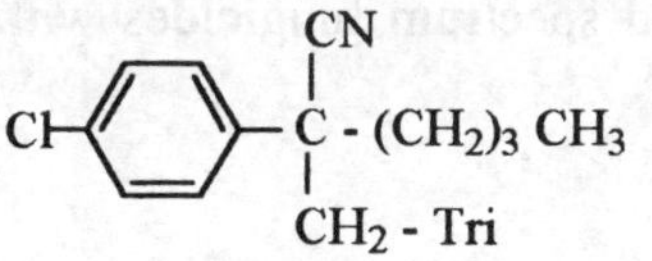

(159)

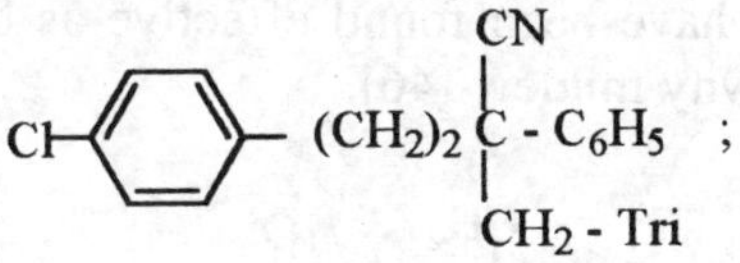

(160)

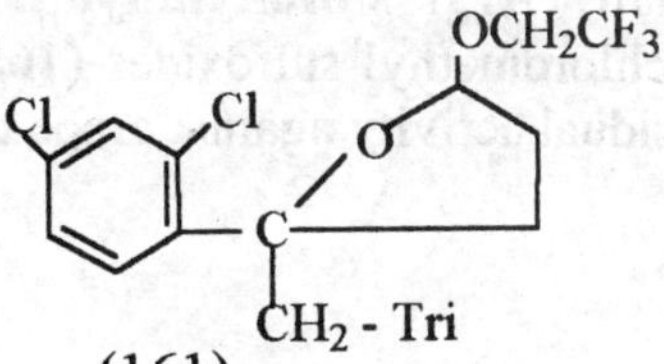

(161)

A Ciba-Giegy product (162), has shown broad spectrum of activity and is effective against all diseases of wheat, fruits and vegetables at low dosages.

The product (163), developed by Montedison (1988) exhibited activity to control powdery mildew of wheat and barley by a single application and with good persistence.

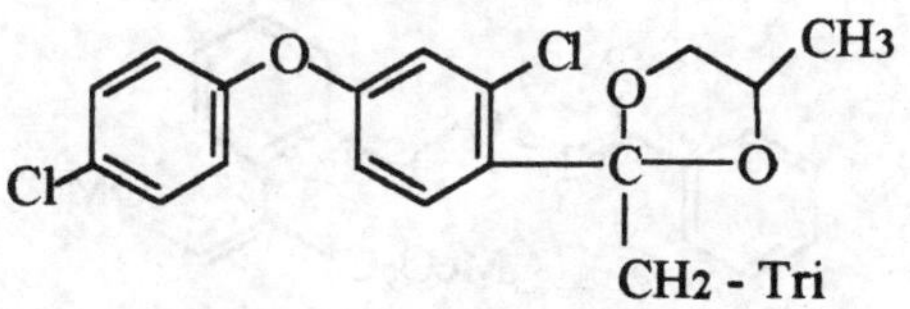

(162)

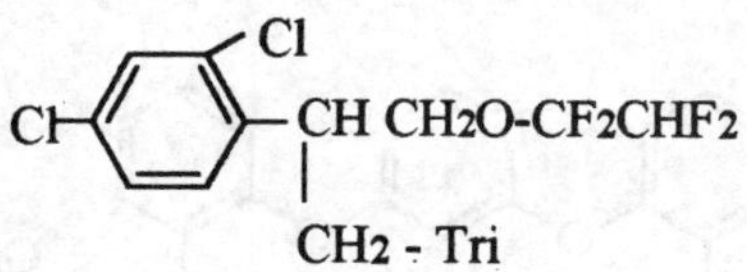

(163)

Recent potent fungicides for future use

i) **Chloroximes** (164), are very effective, broad spectrum fungicides. The substituents in the oxime moiety play a key role in changing the biological activity [44].

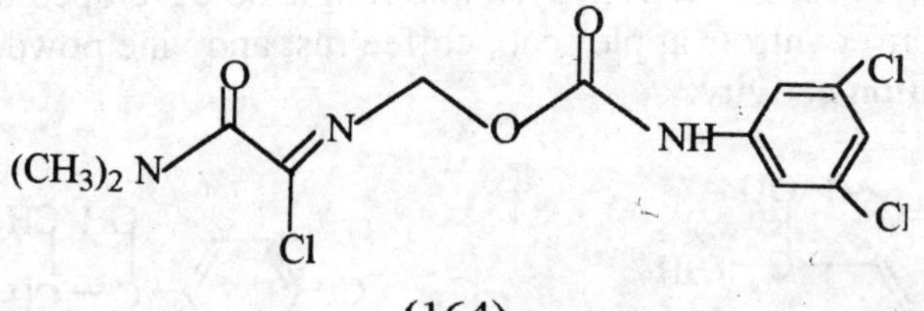

(164)

ii) **Cyano-oximes :** The most active compound in this group is **Cymoxanil** (165), which controls grape vine downy mildew disease. The analogous compound a propargyl derivative (166), is equally effective. After isosteric replacement of the acetylenic triple bond by the cyanide triple bond the resulting compound (167) showed enhanced activity against downy mildew [45].

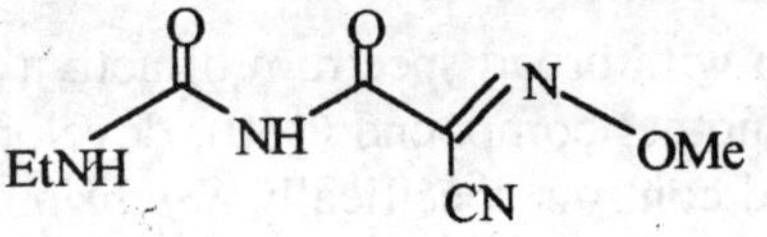

(165) 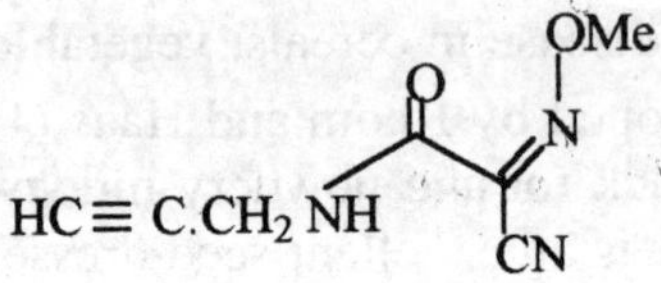   (166) 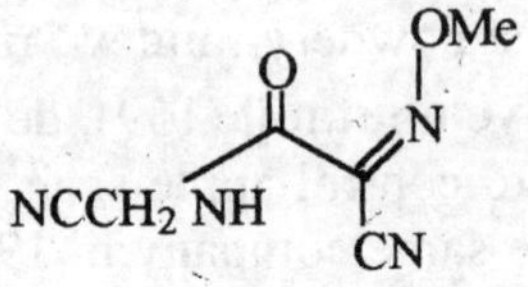   (167)

iii) **Aryl sulfonylallyl trichloromethyl sulfoxides :** A series of 2-aryl sulfonylallyl trichloromethyl sulfoxides (168), have been found effective as broad spectrum fungicides with residual activity against grape downy mildew [46].

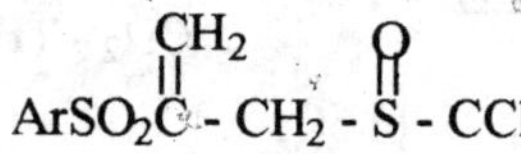

$$\text{ArSO}_2\overset{\overset{\displaystyle CH_2}{||}}{C} - CH_2 - \overset{\overset{\displaystyle O}{||}}{S} - CCl_3$$

(168)

iv) **β-methoxyacrylates with oxime ether side chain :** **ICIA5504**, Azoxystrobin (169), is a well known broad spectrum fungicide which facilitates the control of a wide range of major plant pathogens. If the central pyrimidine ring is replaced with an oxime ether moiety, it yields a highly effective fungicide (170).

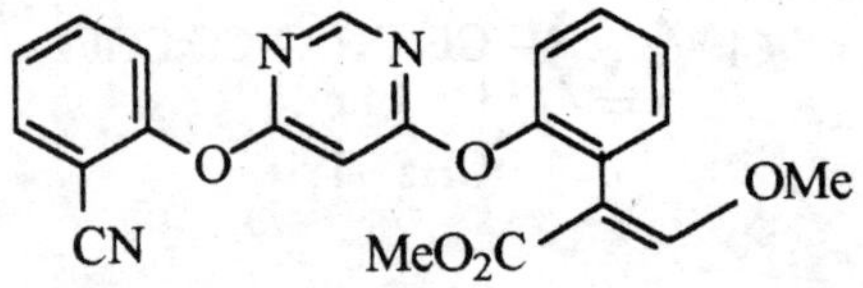

(169) 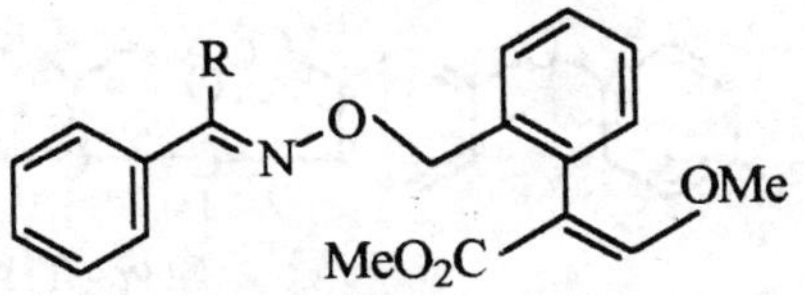   (170)

Analogous compounds containing a heterocyclic moiety instead of phenyl ring were also prepared to ascertain the fungicidal activity [47a, 47b].

v) **Pyrimidine derivatives :** A series of novel 2-anilinopyrimidine compounds, based on lead compound (171), were synthesized and introduced to the market in 1994 [47c]. The synthetic compound **mepanipyrim** (172), exhibited excellent activity against grey mould of vine and vegetables, scab of apple and pear and brown rot of peach.

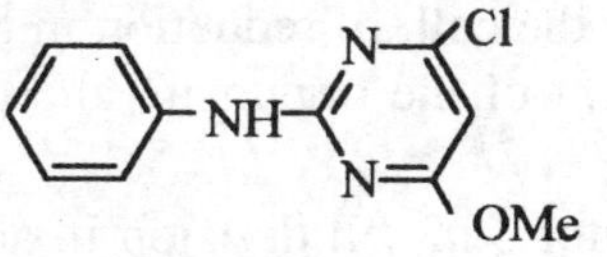

(171) 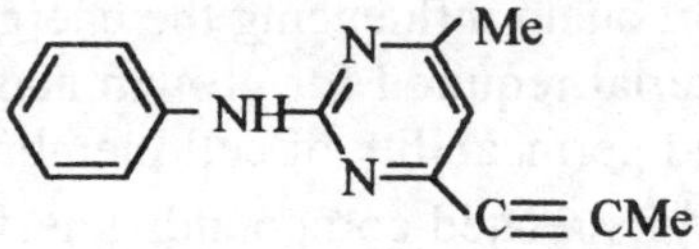   (172)

vi) **Pyridylthiophene carbamates :** Among the series of pyridylthiophene carbamates (173), [47d] a few compounds displayed high degree of activity against *Pyricularia oryzae* causing blast disease of rice.

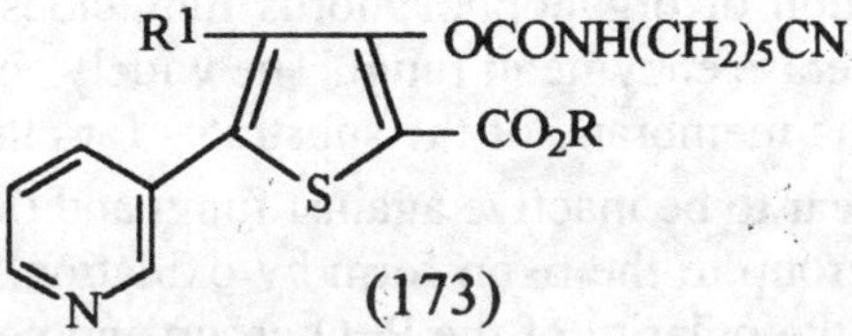

(173)

SAR study showed that the activity depends on the position of pyridine nitrogen and substituents at the 2 and 4 positions of the thiophene ring.

Biofungicides

These comprise of antibiotics and a few microbes such as *Pseudomonas cepacia, Peniophora gigantea* and *Trichoderma viride* which control a number of fungi associated with major crops and are discussed in Chapter 2 & 5.

Mode of action of fungicides

a) Non systemic [48]

The toxic action of sulphur in the cell is still not clear, however, several theories have been proposed from time to time. The theory accepted at present is that sulphur acts as hydrogen acceptor in metabolic systems to form H_2S, and in doing so disrupts the normal hydrogenation and dehydrogenation reactions in the cell. But in case of Cu-fungicides, the Cu ions precipitate or inactivate the proteins (enzymes of sulphydryl group) and thus kill the spores.

The mercury fungicides also act either as vapour or in ionic form and destroy sulphydryl group of (-SH) enzymes. Organomercurials are more toxic than the inorganic mercuric ones due to enhanced lipid solubility facilitating diffusion through the spore membrane to the site of action.

The mode of action of quinone derivatives may be due to binding of the quinone nucleus to -SH and -NH_2 groups in the cell leading to disturbance in the electronic transport systems. The activity of captan and related analogues may thus involve the role of Cl and S atoms of the molecule leading to inactivation of sulphydryl group of enzymes.

b) Systemic

The general mode of action of systemic fungicides is associated with a) interference with the electron transport chain influencing the energy budget of the cell, b) reduction in the biosynthesis of new cell material required for growth and development of the organism, and c) disruption of cell structure and permeability of cell membrane.

Benomyl and its related compounds interfere with mitosis in cell division in angiosperms and fungi [49]. Benzimidazoles, thiophanates, oxathins, phenylamides (metalaxyl derivatives) influence DNA synthesis and are also mitosis inhibitors [50]. The triazole group of fungicides interfere with the biosynthesis of fungal steroids and ergosterol which are important constitutents of the cell wall [51]. Pyrimidine derivatives inhibit purine biosynthesis and several pyridoxal dependent enzymes. The mode of action of morpholines is still not well understood but appears to be inhibition of sterol biosynthesis. The mode of action of organophosphorus fungicides is different from insecticides due to the absence of cholinesterase enzyme in fungi. The widely accepted theory is that it inhibits permeation through cytoplasmic membrane of the substrates for chitin synthesis.

The thiono compounds appear to be inactive against fungi and this may be due to fungus being unable to activate the thiono group to the oxon form by oxidation. The effect of penetration into the fungal hyphae depends on the polarity of the P=O group and needs to be balanced by a larger lipophilic group such as, the second thiophenyl group in case of **edifenphos**, cyclohexyl group in case of **cerezin**, benzyl mercaptan in case of **kitazin** or **kitazin**-P and the phenyl radical in the case of **inezin** [52]. However, **fosetyl-Al** degrades rapidly in plant and soil to give H_3PO_3 which is inhibitory to the mycelial growth of several fungi.

Metabolism

The stability of fungicides in soil depends on chemical structure, nature of soil and climatic conditions. In general, the fungicides are not as stable as organochlorine insecticides. The most versatile dithiocarbamate group of compounds, decompose in acidic soils to give non-toxic amines and carbon disulfide.

Metabolism of alkyltin compounds in liver microsomal monooxygenase system and in mammals leads to the following sequence of detannylation (carbon-tin cleavage) reaction (Fig. 3.206).

$$R_4Sn \rightarrow R_3SnX \rightarrow R_2SnX_2 \rightarrow RSnX_3 \rightarrow SnX_4 , \qquad X = anion$$

Fig. 3.206. Carbon-tin cleavage reaction.

The first step reaction product possessed increased toxicity and potency as inhibitors to mitochondrial respiration whereas in the subsequent steps the reaction product possesses less potency and has altered nature of biocidal activity.

The carboximides such as **captan, folpet, captafol** are hydrolysed under neutral and alkaline conditions. **Chloroneb** degrades to the phenolic derivative but reconversion to parent molecule is a microbial process and this might be the probable reason for long term effectiveness of chloroneb in soil.

Benzimidazole systemic fungicides like **benomyl, thiabendazole** and **thiophanate** methyl are first converted to carbendazim, an active ingredient at the site of action. These are finally degraded to non toxic compounds such as aniline, phenyl diamine and cyanoaniline. The breakdown of benomyl into MBC occurs by intramolecular process in slightly acidic or neutral media. A hydrogen bond is formed between the free electron pair of the N atom of the benzimidazole ring and hydrogen on the nitrogen of the butylcarbamoyl side chain, forming an unstable four membered ring which opens up to yield MBC and butylisocyanate. The cyanate rapidly forms butylcarbamic acid with water which in turn decomposes into CO_2 and butylamine.

The major metabolites of dimethirimol are ethirimol (117) and 2-amino derivative (117-a).·

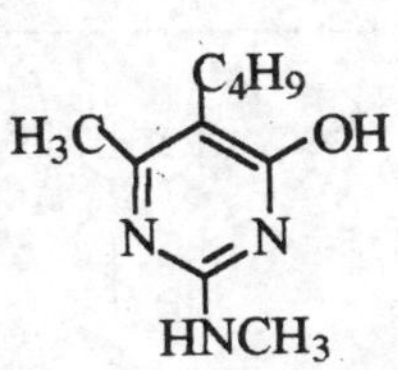

(117) 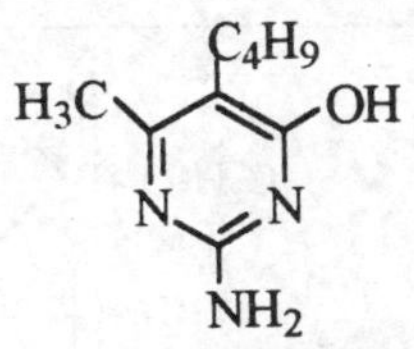   (117-a)

The metabolic fate of organophosphorus compounds is quite interesting and a number of metabolites from **kitazin-P** (32) were identified from rice plant (p). The major metabolite was O,O-di*iso*propyl hydrogen phosphorothioate along with several minor metabolites such as *m*-hydroxy derivative, di*iso*propyl hydrogen phosphate and *iso*propyl dihydrogen phosphate along with sulfides and disulfides (Fig. 3.207) [Tomizawa and Yasuhiko, 53].

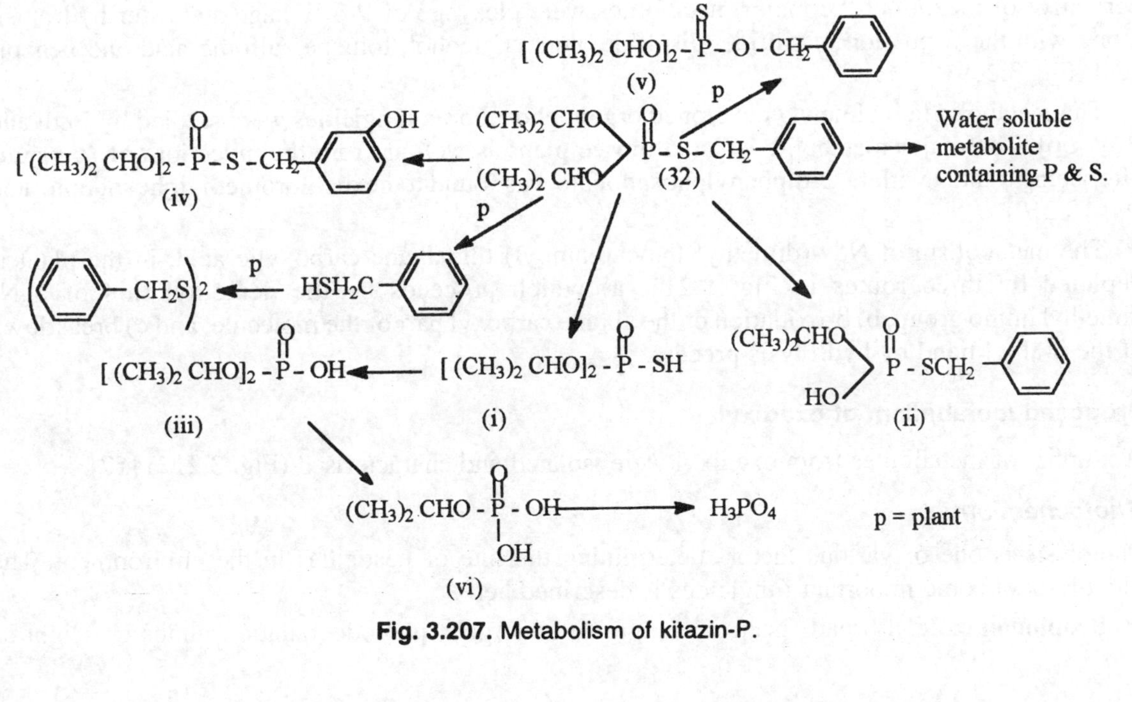

Fig. 3.207. Metabolism of kitazin-P.

The metabolic pathway of **edifenphos** (hinosan) (34) by *P. oryzae* was studied by Uesugi and Tomizawa [54]. A part of edifenphos was hydrolysed at *p*-position of phenyl radical but the major metabolites were the hydrolytic products involving P-S or ethyl ester linkage. But in rice plant (p) a few more metabolites such as diphenyl disulfide and triphenyl phosphorotrithiolate were detected (Fig. 3.208).

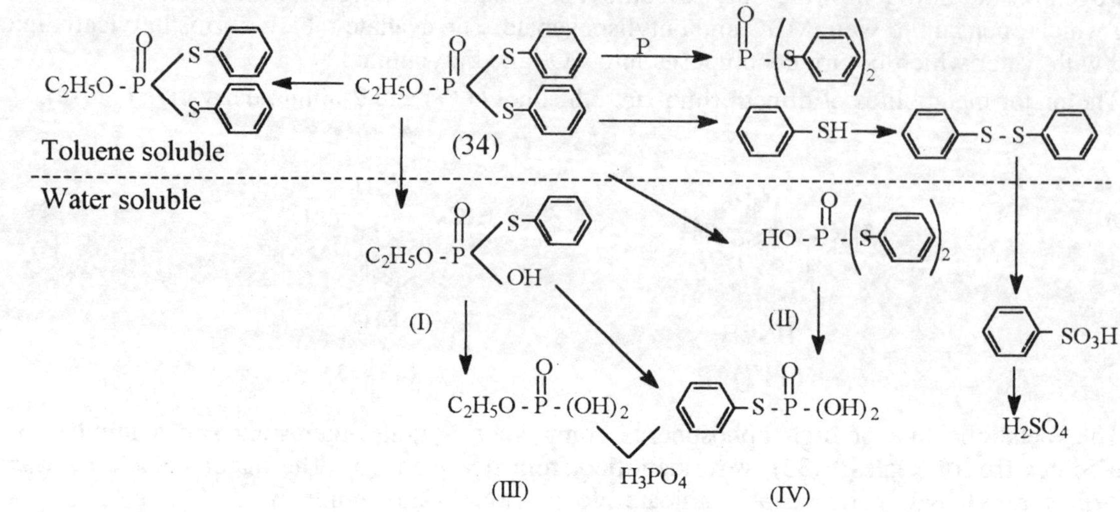

Fig. 3.208. Metabolism of edifenphos.

Studies on metabolism of **inezin**, S-benzyl, O-ethyl phenylphosphonothioate by mycelial cells of *P. oryzae* by Uesugi and Tomizawa [55] showed that a part of inezin is converted to hydroxy derivative of inezin but the major metabolites were cleavage of P-S linkage and ester hydrolysis along with the formation of sulfide, disulfide, benzyl alcohol, toluene sulfonic acid and benzoic acid (Fig. 3.209).

The metabolic fate of newly developed organophosphorus fungicides were studied by Bedi and Roy [56]. The major metabolite isolated in rice plant as well as from the mycelium of *P. oryzae* after being treated with O,O-diphenylphosphonate was found to be dichloromethylphosphonic acid (Fig. 3.210).

The metabolism of N,N-(dimethyl thiocarbamoyl) thioalkane carboxylic acid, in the plant is depicted by three routes in Fig. 3.211, a) which proceeds by the demethylation of N,N-dimethylamino group, b) β-oxidation of the alkane carboxyl part of the molecule, and c) breakdown of the S-alkyl bond by hydrolysis process.

Proposed metabolism of oxadixyl

A number of metabolites from oxadixyl were isolated and characterised (Fig. 3.212) [57].

Photochemistry

Photolysis is one of various factors determining the fate of pesticides in the environment. The photolysis of some important fungicides is described here.

Propioconazole, a broad spectrum fungicide undergoes photodegradation under UV light as

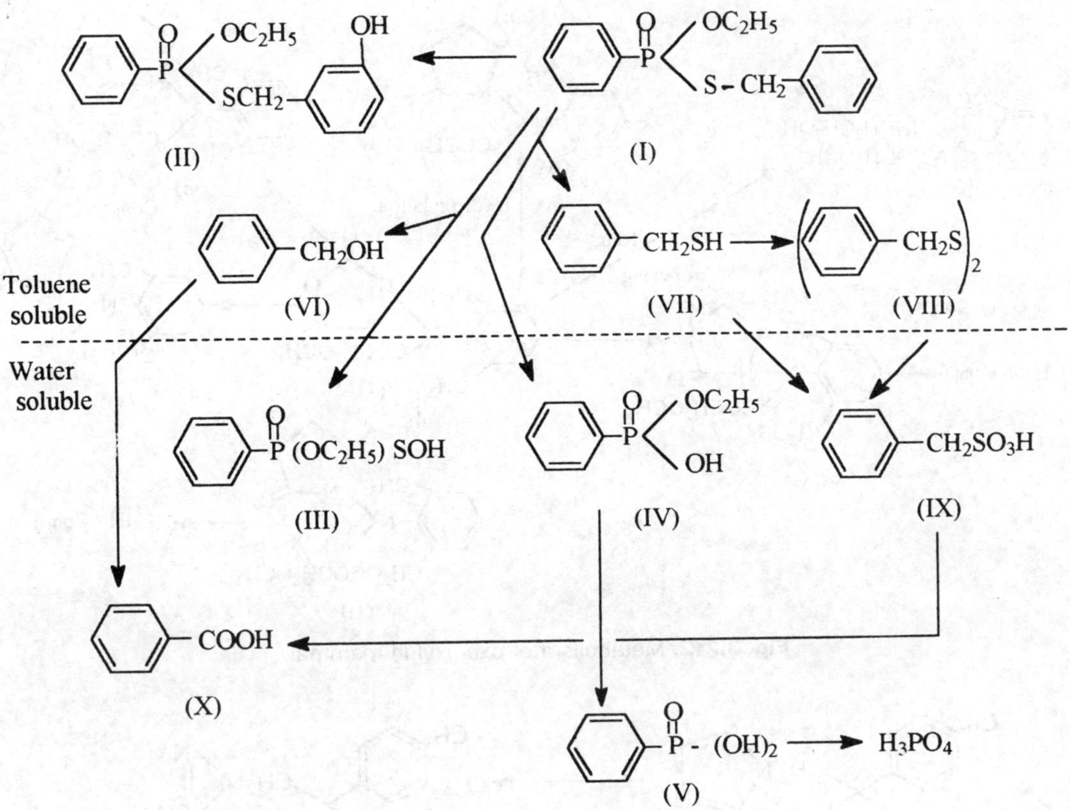

Fig. 3.209. Metabolism of inezin.

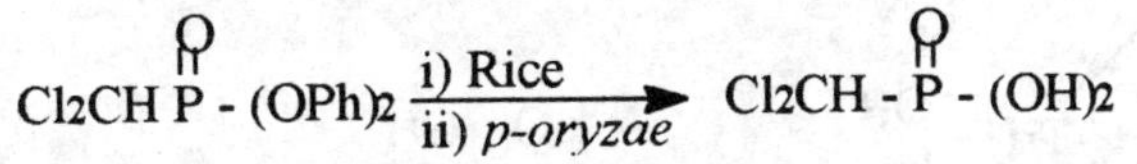

$$Cl_2CH \overset{O}{\overset{\|}{P}} - (OPh)_2 \xrightarrow[\text{ii) } p\text{-}oryzae]{\text{i) Rice}} Cl_2CH - \overset{O}{\overset{\|}{P}} - (OH)_2$$

Fig. 3.210. Metabolic fate of dichloromethyl O,O-diphenyl phosphonate.

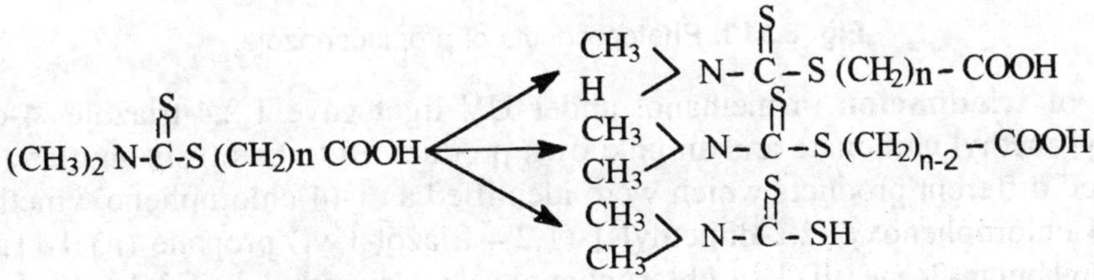

Fig. 3.211. Metabolism of dimethylthiocarbamoyl thioalkane carboxylic acid.

well as sunlight (Fig. 3.213). The major photoproducts include 1,2,4-triazole (i) and the alcohol (ii) after opening of dioxolone ring (iii). A minor metabolite aziridine, (iv) a carcinogenic product was also formed by loss of N-CH group from the triazole [Dureja *et al.* 58].

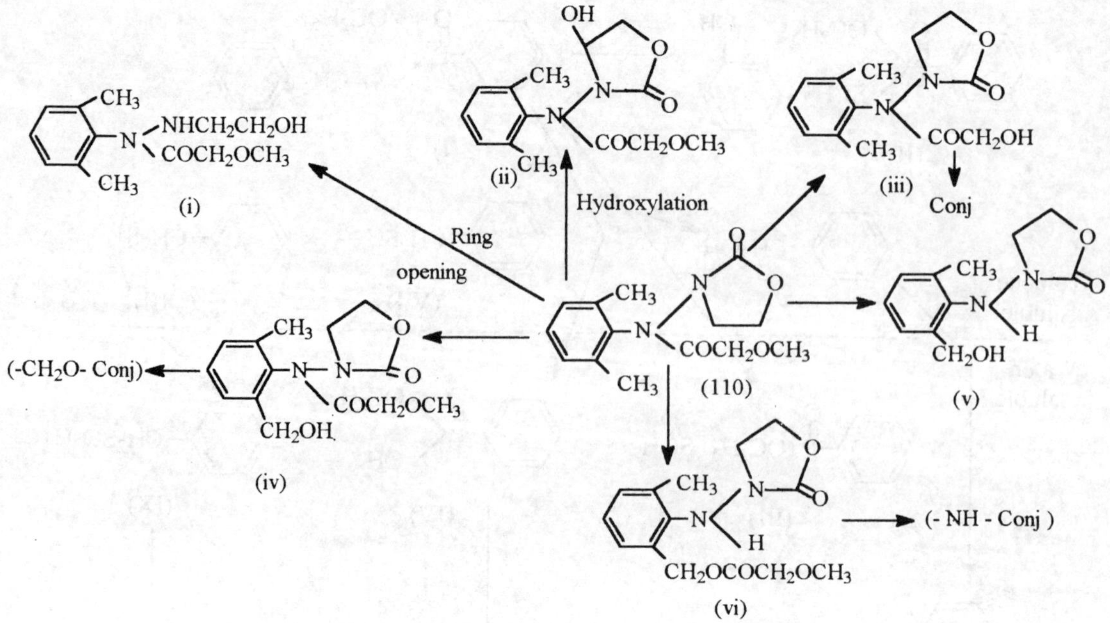

Fig. 3.212. Metabolism of oxadixyl in mammal.

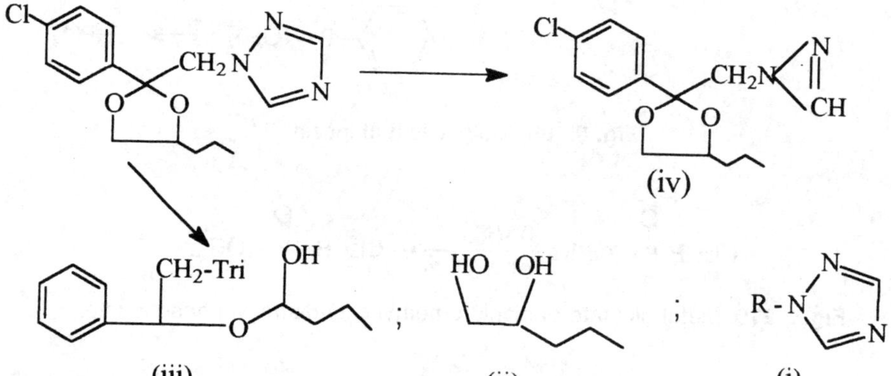

Fig. 3.213. Photoproducts of propioconazole.

Photolysis of **triadimefon** in methanol under UV light gave 1,2,4-triazole, 4-chlorophenol, 4-chlorophenyl methyl carbonate and an unknown product. But photolysis on glass(s) and soil(s) surface, yielded different products which were identified as 1-(4-chlorophenoxymethyl)-1,2,4-tri-azole (i), 1-(4-chlorophenoxy)-2,2-dimethyl-1-(1,2,4-triazol-1-yl) propane (ii) 1-(1,2,4-triazol-1-yl)-3,3-dimethyl butan-2-one (iii) 1-(4-chlorophenoxy)-3,3-dimethyl-1-(1,2,4-triazol-1-yl)butan-2-ol (iv),1-(4 chlorophenoxy)-(1-diazirin-1-yl-3,3-dimethyl-butan-2-one (v), 4-chlorophenoxy-(1,2,4-triazol-1-yl)methanol (vi) and 4-chlorophenol (vii). These products originated as a result of cleavage of the C-1 to C-2 bond, C-1 to triazole bond and by the process of hydrolysis, reduction and extrusion of C = O bond (Fig. 3.214) [Nag and Dureja, 59].

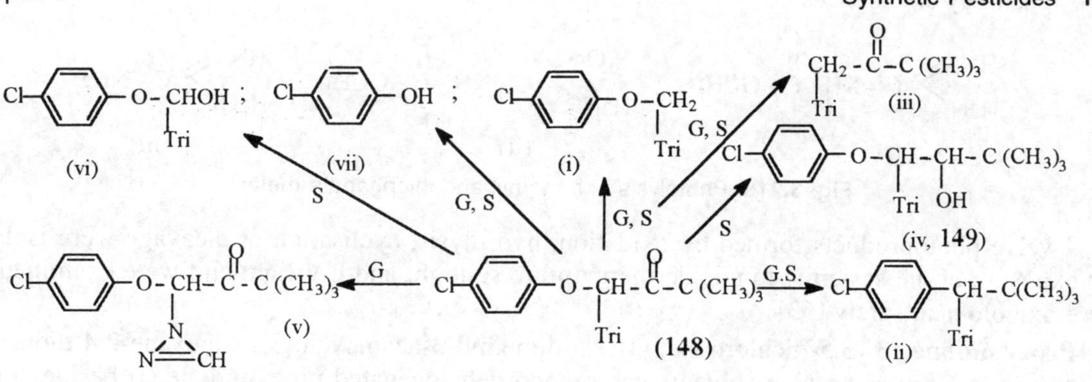

Fig. 3.214. Photolysis of triadimefon (S = Soil surface, G = Glass surface).

The major photoproducts isolated by photolysis of **fluotrimazole** in aqueous suspension/sprayed on barley leaves were identified as 3-(trifluoromethyl) triphenyl methanol (i), methyl 3-trifluoromethyltrityl ether (ii) and tri-fluoromethyl triphenyl methane (iii) [60].

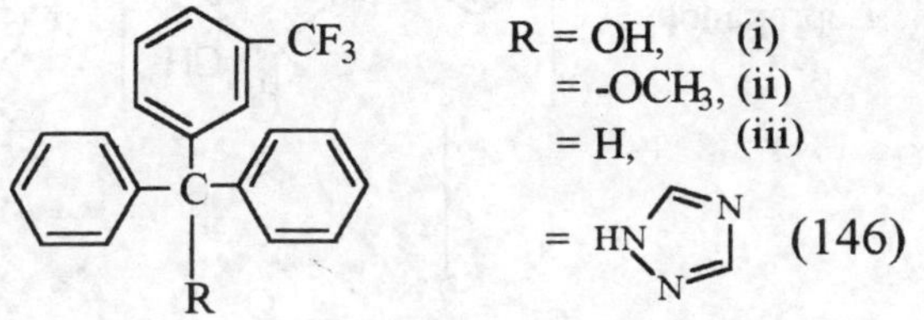

Photolysis of **vinclozolin** yielded five products [61] identified as 3, 5-dichloro-phenyl isocyanate (i), 3,5-dichloroanilino (ii), methyl 3,5-dichlorophenyl carbamate (iii), 3-(3-chlorophenyl)-5-methyl-5-vinyl-oxazolidine-2,4-dione (iv) and 3-(3-chlorobiphenyl)-5-methyl-5-vinyloxazolidine-2,4-dione (v) (Fig. 3.215).

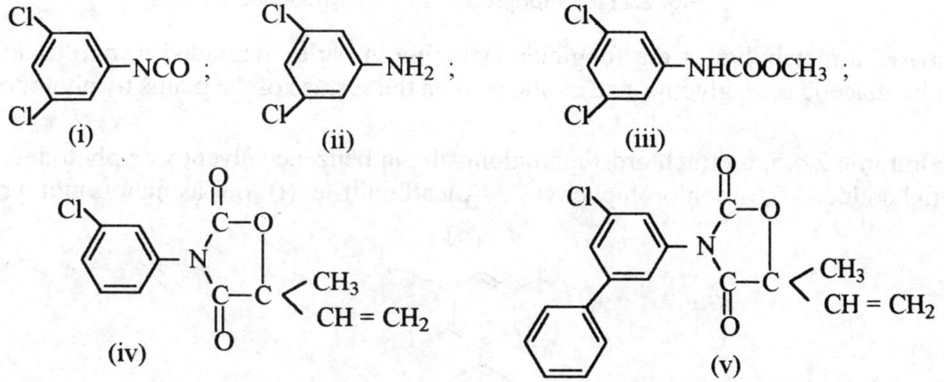

Fig. 3.215. Photolysis of vinclozolin.

Benomyl and **thiophanate-methyl**, in plants decomposes first to MBC which then gives photoproducts (Fig. 3.216) like carbomethoxyguanidine (i), carbomethoxyurea (ii) and guanidine (iii) [62].

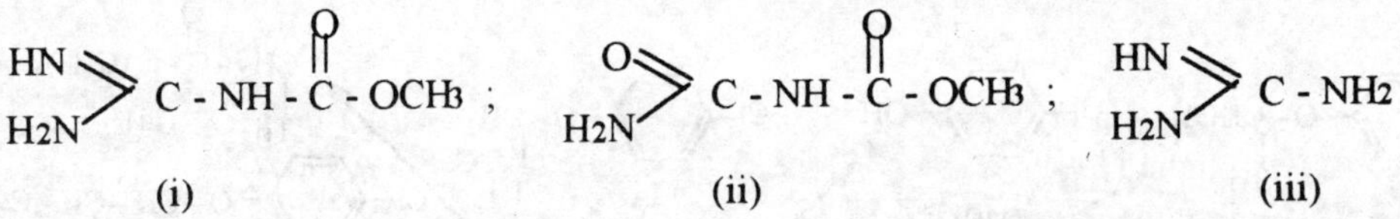

Fig. 3.216. Photolysis of benomyl and thiophanate-methyl.

Twelve photoproducts formed by oxidation, hydrolysis, cyclisation or cleavage were isolated on exposure of the systemic fungicide **mepronil** to sunlight and UV light and were examined for eco-toxicological safety [63].

Procymidone, 3-(3,5-dichlorophenyl)-1,5-dimethyl-3-azabicyclo[3,1,0]hexane-2,4-dione, on photolysis in various organic solvents gave mono dehalogenated procymidone (i) besides many other minor compounds [64] depending on the nature of the solvents used (Fig. 3.217).

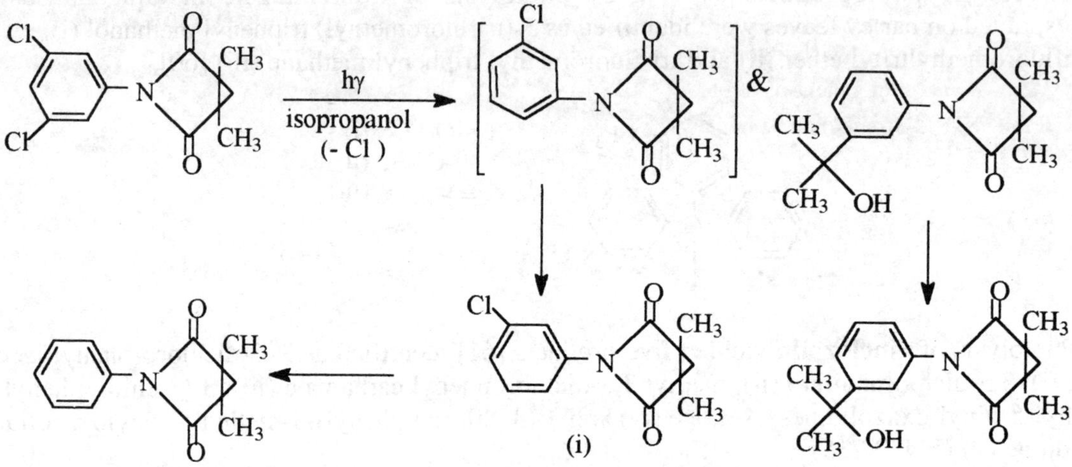

Fig. 3.217. Photoproducts of procymidone.

Piperazine, a metabolite of the fungicide **triforine** in barley degraded to non toxic products such as iminodiacetic acid, glycine, and oxalic acid on the surface of the plants by photodecomposition [65].

Chlorthalonil, 2,4,5,6-tetrachloro*iso*phthalonitrile, in benzene solvent was photodegraded to a monophenyl adduct 3,5,6-trichlorobiphenyl-2, 4-dicarbonitrile, (i) [66] as major photoproduct.

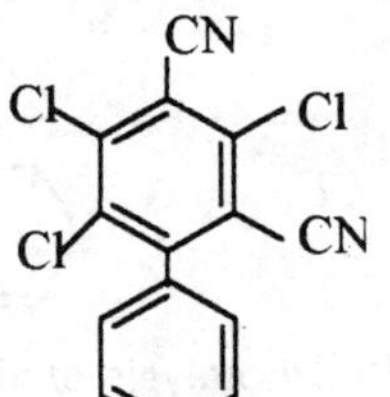

(I)

REFERENCES

1. Nene, Y.L. and Thapliyal, P.N. (eds., Ed. III). *Fungicides in plant disease control*, Oxfords IBH Publishing Co., New Delhi, India (1991).

2. Cremlyn, R.J. (ed.) Fungicides. In : *Agrochemicals*. John Wiley & Sons, New York, 157-216 (1990).

3. Matolcsy, G., Nadasy, M. and Andriska, V. Fungicides. In : *Pesticide Chemistry,* (G. Matolcsy, M. Nadasy and V. Andriska. eds.). Elsevier-Amsterdam, 277-486 (1988).

4. Horsfall, J.G. Fungicides and their action. *Chron. Bot. Comp.* Waltham, Mass, 239 (1945).

5. Coates, G.E. *Organo-metallic compounds*. 2nd Ed. Methuen, London (1960).

6. Tweedy, B.G. **In :** *Fungicides* (D.C. Torgeson ed.). Academic Press, New York, 119 (1969).

7. Matolcsy, G., Kovacs, J., Tuske, M., Kovacs, M., Nadasy, M., Enisz, J. and Nagy, B. Hungarian Pat. 178, 319 (1979).

8. Matolcsy, G., Bordas, B., Bokar, G.Y., Dombai, Z., Dudas, J., Fodar, I., Grega, J., Nagy, J. and Pinter, Z. Hungarian Pat. 171, 736 (1974).

9. Green, M.B. and Spilker, D.A. *Fungicidal chemistry : Advances and practical applications.* American Chemical Society, Washington, D.C. (1986).

10. Yesugi, Y. Development of organophosphorus fungicides. *Japan Pestic. Inf.* **2**, 11 (1971).

11. Roy, N.K. Chloroalkyl phosphonates and phosphorothioates—a new group of fungicides. *Proc. Indian natn. Sci. Acad. B-56(3)*, 305-310 (1990).

12. Bedi, S. and Roy, N.K. Preparation and evaluation of diaryldichloromethyl phosphonates against rice blast. *Indian J. Agric. Sci.* **48**, 248-251 (1978).

13. Roy, N.K. and Mukherjee, S.K. Synthesis of potential organophosphorus fungicides. Part 1 diaryl dichloromethylphosphonodithioates. *Pestic. Sci.* **6**, 497-500 (1975).

14. Dureja, P., Roy, N.K. and Mukherjee, S.K. Preparation and biocidal activity of alkyl phenyl (dichloromethyl) phosphonates. *Pestic. Sci.*, **11**, 685-688 (1980).

15. Roy, N.K., Lalljee, B. and Bedi, S. Synthesis and fungicidal activity of diaryl 1,1-dichloroethyl phosphonates. *Agr. Biol. Chem.* **44**, 2995-2997 (1980).

16. Vasu, K. and Roy, N.K. Synthesis and fungitoxicity diarylmethylphosphonates. *Agr. Biol. Chem.* **47**, 2657 (1983).

17. Nidiry, E.S.J. and Roy, N.K. Synthesis, fungitoxicity and quantitative structure activity relationship of O,O-bisaryl *iso*propylphosphonates. *Indian J. Chem.* **27B**, 1024 (1988).

18. Samanta, S. and Roy, N.K. Synthesis, fungitoxicity and QSAR study of O,O-bis (aryl) *iso*butyl phosphonates. *Indian J. Chem.* **37B** (6), 564-571 (1998).

19. Sanyal, D. and Roy, N.K. Synthesis and structure-activity relationships for the fungicidal activity of O,O-bisaryl *sec*-butylphosphonates. *Pestic. Sci.*, **50**, 85-90 (1997).

20. Panda, A.K. and Roy, N.K. Synthesis, fungitoxicity and quantitative structure activity relationship analysis of O-aryl-O-ethyl-2-chloroethylphosphonates. *Indian J. Chem.* **37B** (10), 1016-1020 (1998).

21. Roy, N.K. and Taneja, H.K. Synthesis of aroyl phosphonates and oximes and their antifungal activity. *Intern. J. Trop. Agri.*, **7**, 142-149 (1989).

22. Roy, N.K. and Bedi, S. A novel transformation to S-methyl phosphorodichloridothiolate and the fungitoxicity of diaryl S-methylphosphorothiolates. *Agr. Biol. Chem.* **46** (7), 1935-37 (1981).

23. Gupta, R.L. and Roy, N.K. Synthesis and fungitoxicity of O,O-diaryl S-ethyl phosphorothiolates. *Pestic. Sci.* **15**, 553-556 (1984).

24. *Ibid.* Synthesis and fungicidal activity of O,O-diaryl S-*iso*propyl phosphorothioates, *Proc. Indian natn. Sci. Acad.* **B54** (3), 287-290 (1988).

25. *Ibid.* Synthesis and quantitative structure activity relationships for the antifungal activity of O-alkyl S,S-diaryl phosphorotrithioates. *Indian J. Chem.* **29B** (9), 870-875 (1990).

26. *Ibid.* Synthesis and quantitative structure activity relationships of S-alkyl S,S-diaryl phosphorotrithioates. *Pestic. Res. J.* **5** (1), 16-22 (1993).

27. Tawata, S., Taira, S., Kobamoto, N., Ishihara, M. and Toyama, S. Synthesis and fungicidal activity of new thiophosphorylated monoterpenoids and related compounds. *J. Pestic. Sci.* **21**, 141-146 (1996).

28. Bedi, S. and Roy, N.K. Residues of diaryl dichloromethylphosphonates in/on rice plant and their chemical stability. *Indian J. Agric. Sci.* **48**, 701-704 (1978).

29. Bedi, S. and Roy, N.K. Metabolic fate of dichloromethyl O,O-diphenyl phosphonate(1) and dichloromethyl O,O-di(*p*-nitrophenyl) phosphonate (11) in/on rice plants. *Environ. Sci. Health* **B14** (4), 443-458 (1979).

30. Vasu, K. and Roy, N.K. Hydrolysis, phytotoxicity and photolysis studies on methyl O,O-bis (2,4,5-trichlorophenyl) phosphonate, a potential fungicide, *Agr. Biol. Chem.* **49**, 307-310 (1985).

31. Lye, H. (ed.) *Modern selective fungicides.* Longman, Harlow, London (1987).

32. Vyas, S.C. (ed.) *Handbook of systemic fungicides*, **I**, **II** & **III**, Tata McGraw-Hill Publishers Co. Ltd., New Delhi (1993).

33. Marsh, R.W. (ed.) *Systemic fungicides*, 2nd Ed. Longman, London (1977).

34. Woodcock, D. Chemotherapy of plant disease : progress and problems. *Chem. in Britain* **7** (10), 415-423 (1971).

34a. Pappas, A.C. and Fisher, D.J. A comparison of the mechanism of action of vinclozolin, procymidone, iprodione and prochloraz against *Botrytris cinerea. Pestic. Sci.* **10**, 239-246 (1979).

35. Patel, B. and Fleming, L. 2,5-Dioxo-3-pyrroline-1-acetanilides as fungicidal agents, *Eighth Internatn. Congr. Pestic. Chem.*, Washington D.C., abs. 803 (1994).

36. Wetty, R.E. and Rawlings, J.O. Effect of benomyl on Sclerotinia crown and stem rot of alfalfa. *Plant Dis.* **68**, 294-296 (1984).

37. Holah, D.S. and Mellor, M. Alkoxy-substituted benzothiazolones as rice blasticides. *Seventh Internatn. Congr. Pestic. Chem.*, Hamburg, abs. 1A-65 (1990).

38. Rouchaud, J., Decallonne, J.R. and Meyer, J.A. Metabolism of [2,5-^{14}C] piperazine in barley plants. *Pestic. Sci.* **9** (1), 139-145 (1978).

39a. Eckhardt, W. and Suess, H. Ciba-Geigy AG, US, Pat. 472, 1797 (1986).

39b. Eckhardt, W., Francotte, F. and Margot, P. Synthesis and fungicidal activities of CGA-80000 and of its four stereoisomers. *Seventh Internatn. Cong. Pestic. Chem.*, Hamburg, abs. 01B-11 (1990).

40. Mohr, G., Niethammer, K., Lut, S. and Schneider, G., E. Mark A.G. FRG Pat. Part I. 182, 896 (1964).

41. Selby, T.P. and Smith, B.K. Benzodioxan and benzoxepin amine fungicides, *Eighth Internatn. Congr. Pestic. Chem.*, Washington D.C., abs. 804 (1994).

42. Takahi, Y. and Tomita, K. Hymexazol a new plant protecting agent. *Ann. Sankyo Res. Lab.* **25**, 1-51 (1973).

43a. Buchel, K.H. and Singer, R.J. Farbenfabriken Bayer A.G., FRG Pat. 255, 3301, (1977).

43b. Takeshiba, H., Tobitsuka, J., Kajino, H., Itoh, H.; Oida, S. and Takahi, Y. Synthesis and fungicidal activity of oxetane derivatives. *Eighth Internatn. Congr. Pestic. Chem.*, Washington D.C., abs 806 (1994).

44. Brown, R.J., Adams, J.B. Jr., Campbell, C.L., Drumm, J.E., Erbes, D.L., Hartzell, S.L., Holliday, M.J., Kleier, D.A., Martin, M.J., Pember, S.O. and Ramsey, G.R. Oximes fungicides : Chemistry. *Eighth Internatn. Congr. Pestic. Chem.*, Washington D.C., abs. 787 (1994).

45. Cheetham, R.C., Crowley, P.J. and Evans, S.E. Synthesis and activity of some new cyano-oxime fungicides, *Seventh Internatn. Congr. Pestic. Chem.*, Hamburg, abs. 01A-43 (1990).

46. Canada, E.T., Miesel, J.L., Arnold, W. and Alexander, A. Synthesis and fungicidal activity of 2-aryl sulfonylallyl trichloromethyl sulfoxides and related compounds. *Eighth Internatn. Congr. Pestic. Chem.*, Washington D.C., abs. 788 (1994).

47a. DeFraine, P.J. β-methoxyacrylates; a new series of broad specturm fungicides with an oxime ether side chain. *Eighth Internatn. Congr. Pestic. Chem.*, Washington D.C., abs. 790 (1994).

47b. Joseph R. SI. The metabolism of the strobilurin fungicide azoxystrobin. *Ninth Internatn. Congr. Pestic. Chem.*, London, abs. **2** S.5.2 (1998).

47c. Masuda, K., Nagata, T., Itoh, S., Kojima, Y. and Maeno, S. Structure-activity relationship of 2-anilinopyrimidine derivatives. *Ninth Internatn. Congr. Pestic. Chem.*, London, abs. **I**, 1B-007 (1998).

47d. Mitchell, D., Riordan, P., Mellor, M. and Obourn, S. Synthesis and SAR of some pyridylthiophene carbamates active as rice blast. *Ninth Internatn. Congr. Pestic. Chem.*, London, abs. **I**, 1B-010 (1998).

48. Lukens, R.J.W. Chemistry of fungicidal action, Chapman and Hall, London (1971).

49. Sisler, H.D. Mode of action of benzimidazole fungicides. In : *Herbicides, fungicides, formulation chemistry* (A.S. Tahori ed.), Gordon and Breach, New York, 325-35 (1971).

50. Mathre, D.E. Mode of action of oxathin systemic fungicides. 111. Effect on mitochondrial activities. *Pestic. Biochem. Physiol.* **1**, 216-224 (1971).

51. Siegel, M.R. Sterol inhibiting fungicides, effect on sterol biosynthesis and sites of action. *Plant Dis. Reptr.* **65**, 986-989 (1981).

52. Maeda, T., Abe, H., Kakiki, K. and Misato, T. Studies on the mode of action of organophosphorus fungicides, kitazin. *Agr. Biol. Chem.* **34**, 700 (1970).

53. Tomizawa, C. and Uesugi, Y. Metabolism of S-benzyl O,O-diisopropyl phosphorothioate (kitazin-P.) by mycelial cells of *Pyricularia oryzae. Agr. Biol. Chem.*, **36** (2), 294-300 (1972).

54. Uesugi, Y. and Tomizawa, C. Metabolism of O-ethyl S,S-diphenyl phosphorodithioate (hinosan) by mycelial cells of *Pyricularia oryzae. Agr. Biol. Chem.* **35** (6), 941-949 (1971).

55. *Ibid.* Metabolism of S-benzyl O-ethyl phenyl-phosphonothioate (Inezin) by mycelial cells of *Pyricularia oryzae. Agr. Biol. Chem.* **36** (2), 313-317 (1972).

56. Bedi, S. and Roy, N.K. Metabolism of O,O-diphenyldichloromethyl phosphonate by mycelial cells of *Pyricularia oryzae.* CAV. *J. Environ. Sci. Health*, **B15** (3), 259-265 (1980).

57. Karapally, J. and Klotzsche, C. (eds.). *New oxazolidinone class of systemic fungicides.* In : *Bulletin* Published by Sandoz Agro Division, Basle, Switzerland.

58. Dureja, P., Walia, S. and Mukherjee, S.K. Photolysis of propiconazole. *Toxicol. Environ. Chem.* **16**, 61 (1987).

59. Nag, S.K. and Dureja, P. Phototransformation of triadimefon on glass and soil surfaces. *Pestic. Sci.* **48**, 247-252 (1996).

60. Clark, T., Watkins, D.A.M. and Weerasingha, D.K. Photolysis of fluotrimazole. *Pestic. Sci.* **14**, 449-452 (1983).

61. Clark, T. and Watkins, D.A.M. Photolysis of vinclozolin. *Chemosphere* **13** (12) 1391-1396 (1984).

62. Fleeker, J.R. and Morganhacy, H. Photolysis of methyl 2-benzimidazole-carbamate. *J. Agric. Food Chem.*, **25** (1), 51-55 (1977).

63. Yumita, T. and Yamamoto, I. Photodegradation of mepronil. *J. Pestic. Sci.* **7**, 123-131 (1982).

64. Scnwack, W., Bourgeois, B., and Walker, F. Fungicides and photochemistry : photodegradation of the dicarboxyimide fungicide procymidone. *Chemosphere* **31** (9), 4033-4040 (1995).

65. Rouchaud, J., Moons, C., Decallonne, J.R. and Meyer, J.A. Photodecomposition of piperazine in water by 'sunlight' ultraviolet radiation. *Pestic. Sci.* **9**, 305-309 (1978).

66. Khan, S.U. and Akhtar, M.H. Photodecomposition of chlorthalonil in benzene. *Pestic. Sci.* **14**, 354-358 (1983).

Section IX.   Chemistry of Herbicides

- Introduction
- Halogenated alkanoic acid compounds
- Aromatic acid compounds
- Phenoxy-alkanoic acid compounds
- Phenoxy phenoxy acid compounds
- Amide compounds
- Phenolic nitro compounds
- Dinitroaniline compounds
- Carbamate compounds
- Thiocarbamate compounds
- Urea compounds
- Arylalkoxy urea compounds
- s-Triazine compounds
- Pyridine compounds
- Pyridazine compounds
- Pyrazole compounds
- Quaternary ammonium salts
- Azole compounds
- Organophosphorus compounds
- Sulfonyl urea compounds
- Miscellaneous compounds
- Bioherbicides
- Herbicide safeners/protectants/antidotes
- Metabolism
- Mode of action
- Photochemistry
- *References*

In general, the unwanted plants growing along with the sown crop are called as weeds [1]. Mostly the weeds are removed manually or mechanically which turns out to be inefficient, labour intensive and largely impractical. Intensive agriculture involving increased plant nutrient input has necessitated the chemical control of these unwanted plants. In the early days, a number of inorganic chemicals such as crude rock salt, arsenic, borax, copper salts, cyanates, chlorides, azides and chlorates were used to control a number of weeds specially in the vicinity of the railway tracks, timber yards and golf courses. Some chemicals such as creosote, crude tar oils and petroleum oils were recommended as post-emergence herbicides. Since these inorganic chemicals and crude organic products are nonselective and phytotoxic, they are not found safe for post-emergence application for most of the eatable crops.

A new opening in the era of organic herbicides began after the discovery of DNOC, 2-methyl-4, 6-dinitrophenol, in France in the year 1933 which was found to be an effective and selective herbicide [2a, 2b, 3]. Various groups of selective herbicides developed since then, are described in brief in the following text.

Halogenated alkanoic acid compounds

Earlier halogenated alkanoic acids such as monochloroacetic acid, trichloroacetic acid were used as pre-emergence herbicides for the control of grass weeds in sugar beet, sugarcane and few vegetable crops. These were succeeded by **Dalapon**, 2,2-dichloropropionic acid, which was developed as an important contact selective herbicide for the control of grass weeds.

Aromatic acid compounds

Amongst aromatic acids, 2,3,6-trichlorobenzoic acid is widely used against broad-leaved annual and perennial weeds. The nitro/amino substituted chlorobenzoic acid also exhibits herbicidal activity. Both **dinoben**, 2,5-dichloro-3-nitro benzoic acid and **chloramben**, amiben, 2,5-dichloro-3-amino benzoic acid, are effectively used for selective control of annual grasses and broad leaved weeds in soybeans. The most important member of benzoic acid group of herbicides is **dicamba**, 3,6-dichloro-2-methoxy benzoic acid (1), which is prepared as follows (Fig. 3.218) .

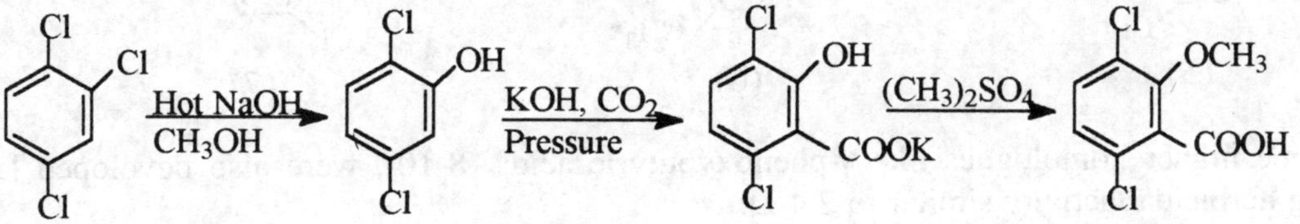

Fig. 3.218. Preparation of dicamba.

The most versatile group of herbicides, extensively used in agriculture is the phenoxyalkanoic acid series because of high specificity and cost effectivity. Three herbicides namely (i) **2,4-D**, 2,4-dichlorophenoxyacetic acid (2), (ii) **MCPA**, 4-chloro-2-methyl phenoxyacetic acid (3) and (iii) **2,4,5-T**, 2,4,5-trichlorophenoxyacetic acid (4) are commercially available. The salts and esters of 2,4-D are pre-emergence hormone type systemic foliage herbicides applied in cereals, maize and rice crops. It can be prepared as shown below (Fig. 3.219).

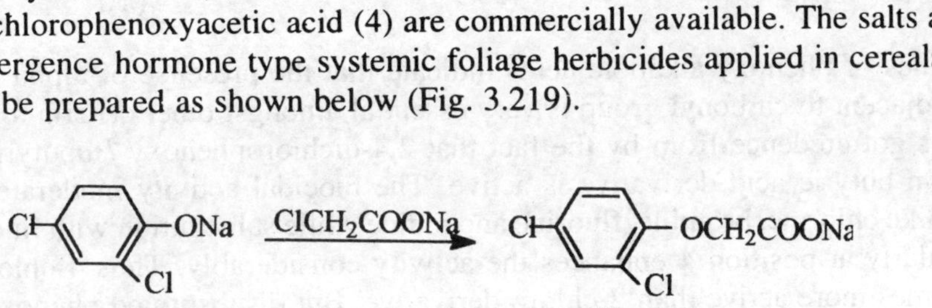

Fig. 3.219. Preparation of 2,4-D.

MCPA (3) is also prepared similarly (Fig. 3.220). The herbicidal activity of MCPA is similar to 2,4-D. 2,4,5-Trichlorophenol on condensation with sodium monochloroacetic acid gives 2,4,5-T (4) which is banned at present due to its carcinogenicity.

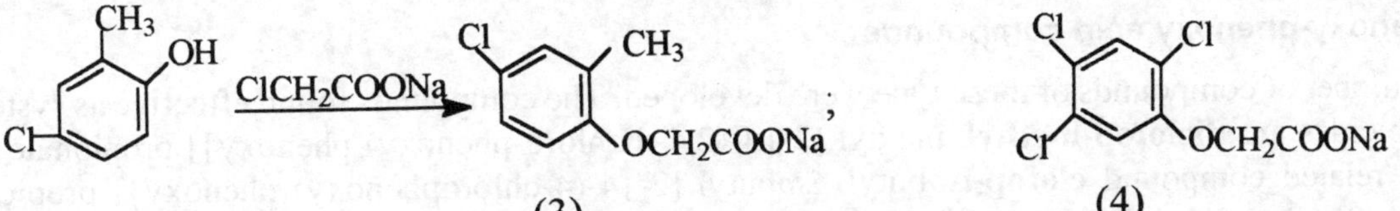

Fig. 3.220. Preparation of MCPA.

A number of α-methyl homologues of phenoxyacetic acid such as dichlorprop or **2,4-DP**, 2-(2,4-dichlorophenoxy) propionic acid (5), **MCPP** or CMPP, 2-(4-chloro-2-methylphenoxy) propionic acid (6) and **2,4,5-TP** or fenoprop, 2-(2,4,5-trichlorophenoxy) propionic acid (7), also possess auxin type herbicidal activity. These compounds have growth accelerating property at μM concentration but at higher concentrations they act as herbicides.

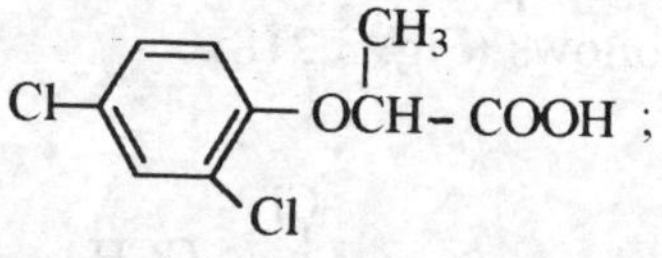

$$(5) \qquad\qquad (6) \qquad\qquad (7)$$

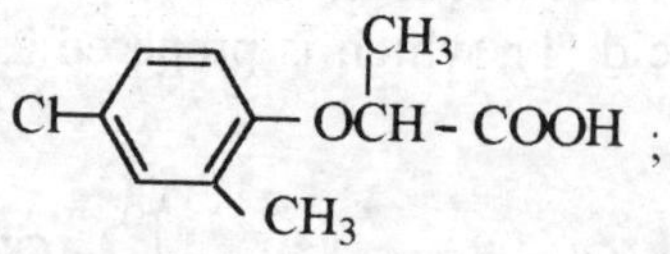

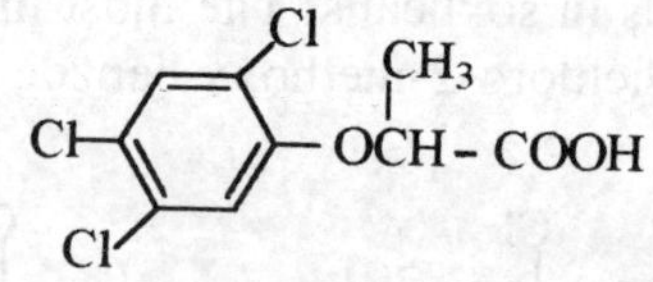

Some higher homologues like γ-phenoxybutyric acids (8-10), were also developed later on having herbicidal activity similar to 2,4-D.

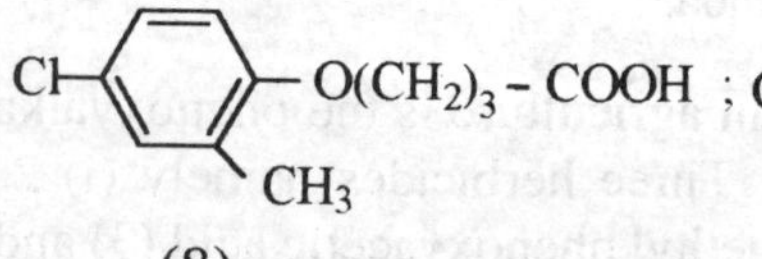

$$(8) \qquad\qquad (9) \qquad\qquad (10)$$

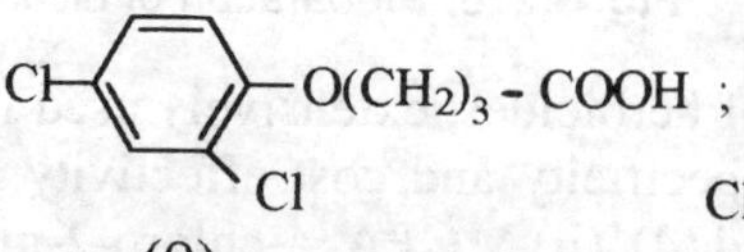

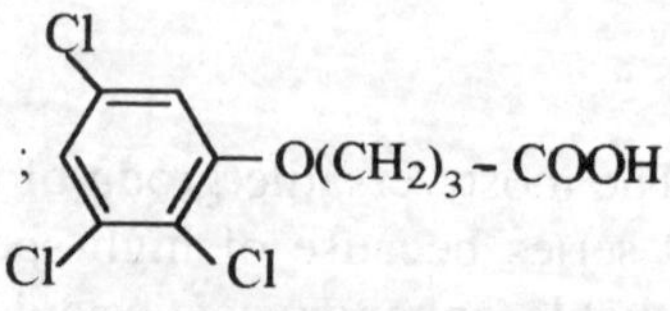

SAR studies of phenoxy alkanoic acids indicate that the presence of alpha hydrogen atom at the carbon adjacent to carbonyl group is very essential amongst other criteria to impart herbicidal activity. This got credence from by the fact that 2,4-dichlorophenoxy *iso*butyric acid is inactive whereas the n-butyric acid derivative is active. The biocidal activity moderately increases with substitution like chlorine, bromine, fluorine and iodine while substitution with fluorine and chlorine atom particularly at position 4 enhances the activity considerably. Thus 4-chlorophenoxy acetic acid is ten times more active than 2-chloro derivative. But disubstituted phenoxy acids are found to be more effective herbicides in the order 2,4 > 2,5 > 3,4 > 3,5 > 3,6 > 2,6. Among the tri-substituted chlorine derivatives, 2,4,5-T, is the most active, followed by 2,3,4 > 3,4,5 > 2,3,5 > 2,4,6 > 2,3,6.

Phenoxy ethanols and their derivatives such as **2,4-DES**, sodium 2-(2-4-dichlorophenoxy) ethyl sulphate and **erbon**, 2-(2,4,5-trichlorophenoxy) ethyl 2,2-dichloro-propionate were also developed both as pre- and post-emergence herbicides.

Phenoxy-phenoxy acid compounds

A number of compounds of these type were developed. The compounds found effective as systemic herbicides are **diclofop-methyl**, methyl [2-{4-(2,4-dichloro-phenoxy) phenoxy}] propionate (11) and related compound **clofop-***iso*butyl, *iso*butyl [2-{4-(4-chlorophenoxy) phenoxy}] propionate (12). Similarly developed active phenoxy herbicides of this class are **quinofop-methyl**, ethyl [2-{(4-chloro-2-quinoxalinyloxy) phenoxy}] propionate (13), **fluazifop-butyl,** butyl [2-{4(trifluoromethyl-2-pyridyloxy) phenoxy}] propionate (14a) and **Cyhalofop-butyl**, butyl (*R*)-2-[4-(4-cyano-2-fluorophenoxy) phenoxy] propionate (14b).

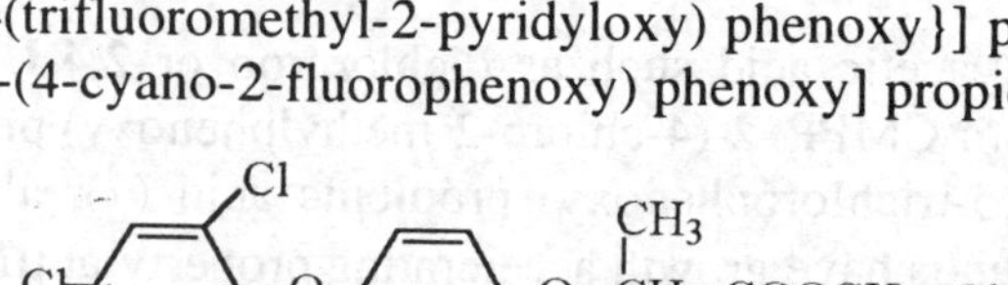

$$(11) \qquad\qquad\qquad\qquad (12)$$

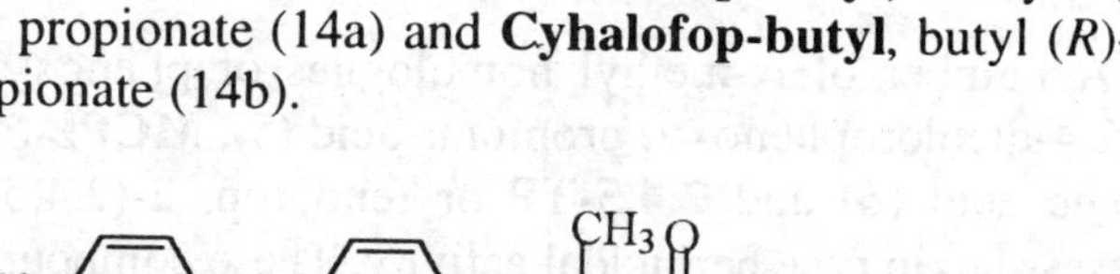

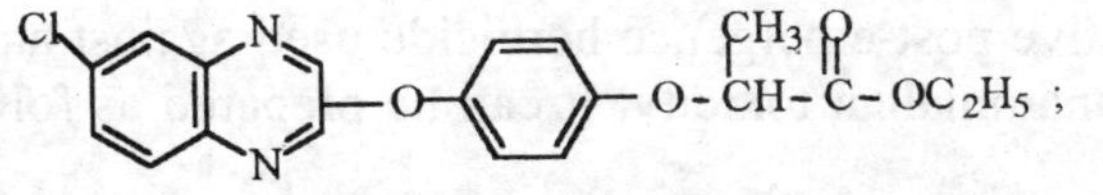

(13)

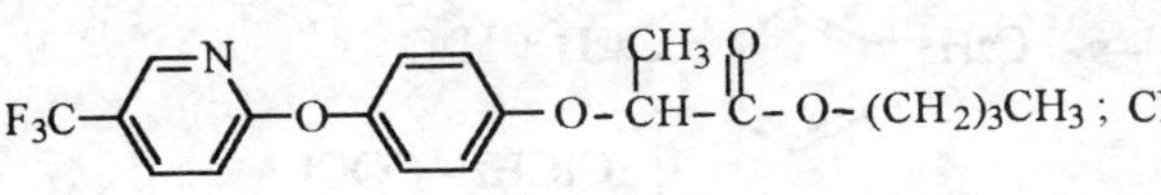

(14a) 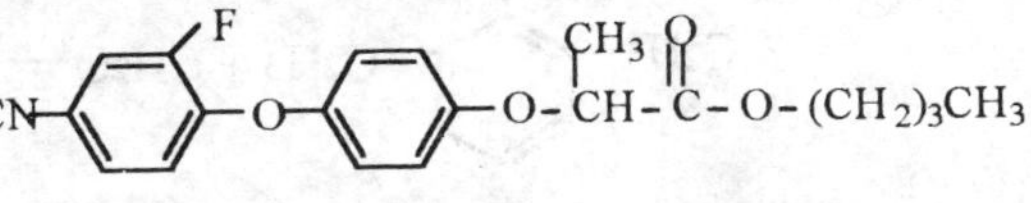   (14b)

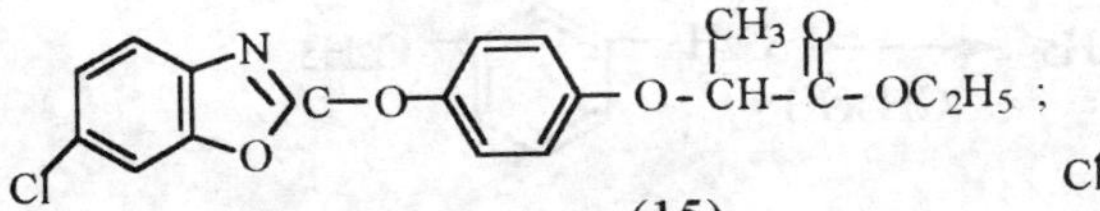

(15) 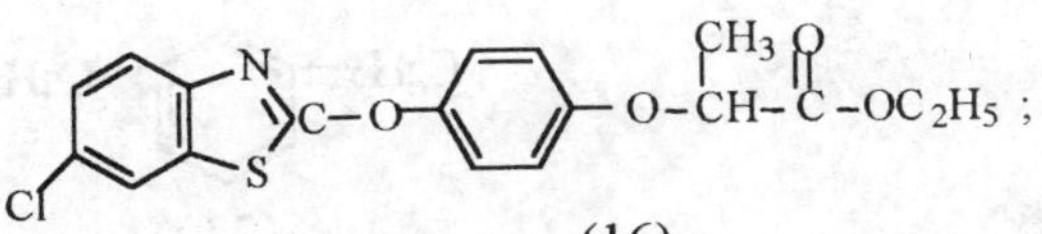   (16)

All these compounds exhibit systemic action and are very active against annual and perennial grasses and selective to broad-leaved crops, with low mammalian toxicity. Two more active molecules developed recently after structural modifications are **fenoxaprop-ethyl** (15) and **fenthiaprop-ethyl** (16) which possess systemic pre-emergence activity [Randte, 4].

Amide compounds

Several amides such as N-substituted α-chloroacetamides, particularly substituted anilides are found useful as pre- and post-emergence selective herbicides. Thus **CDEA**, N,N-diethyl chloroacetamide (17) and **CDAA**, allidochlor, N,N-diallylchloroacetamide (18) exhibit good activity against annual grass weeds. Similarly, an amide with carbonyl group is **propachlor**, 2-chloro-N-*iso*propyl acetanilide (19) whose preparation is as follows (Fig. 3.221a).

$$Cl\ CH_2CO-N(C_2H_5)_2 ;\qquad ClCH_2-CON-(CH_2-CH=CH_2)_2 ;$$

(17)   (18)

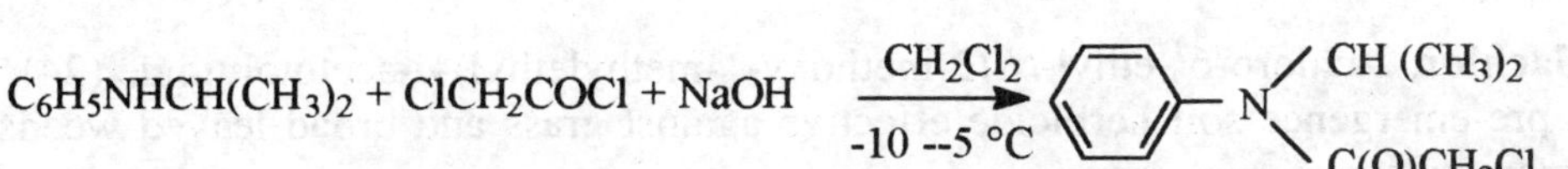

$$C_6H_5NHCH(CH_3)_2 + ClCH_2COCl + NaOH \xrightarrow[-10--5\ °C]{CH_2Cl_2}$$

(19)

Fig. 3.221a. Preparation of propachlor.

It is a pre-emergence selective herbicide, effective against annual grass and broad-leaved weeds in maize and cotton. A close analogue of propachlor is **prynachlor**, 2-chloro-N-(1-methyl-propynyl) acetanilide (20), having similar activity to that of propachlor.

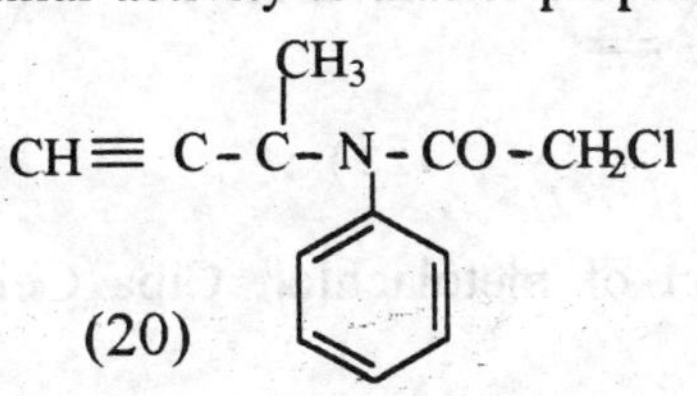

$$CH\equiv C-\overset{\overset{\displaystyle CH_3}{|}}{C}-N-CO-CH_2Cl$$

(20)

The most effective molecule belonging to this group is **alachlor**, 2-chloro-2, 6-diethyl-*N*-methoxy methyl acetanilide (21), which is a selective post-emergence herbicide used against annual weed grass and broad-leaved weeds with low mammalian toxicity. It can be prepared as follows (Fig. 3.221b).

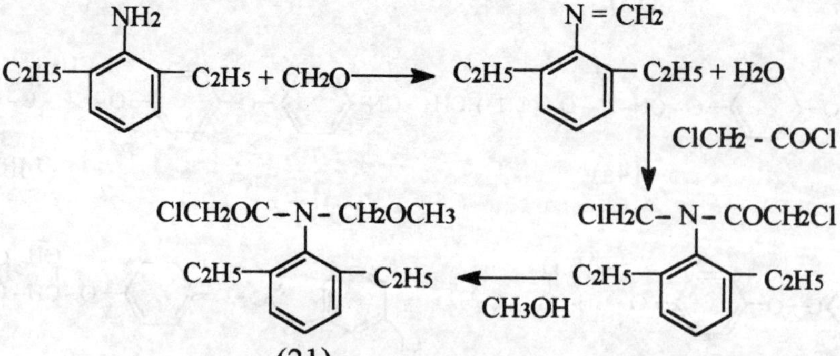

Fig. 3.221b. Preparation of alachlor.

The butyl homologue, **Butachlor** (22), is a pre- and early post-emergence selective herbicide for the control of grass weeds and a few broad-leaved weeds. It is not persistent, degrades in ten weeks in the soil and is moderately toxic.

Recently Monsanto Co. developed a safener, **Mon-4606**, commonly called flurazole; benzyl 2-chloro-4-trifluoromethyl-5-thiazole carboxylate (23), which is used along with the herbicide to save crop from herbicide injury.

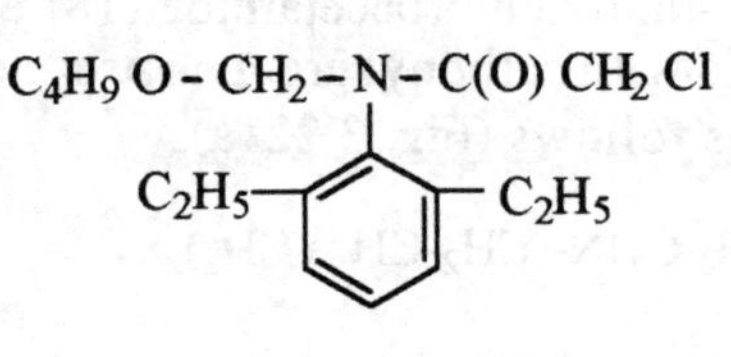

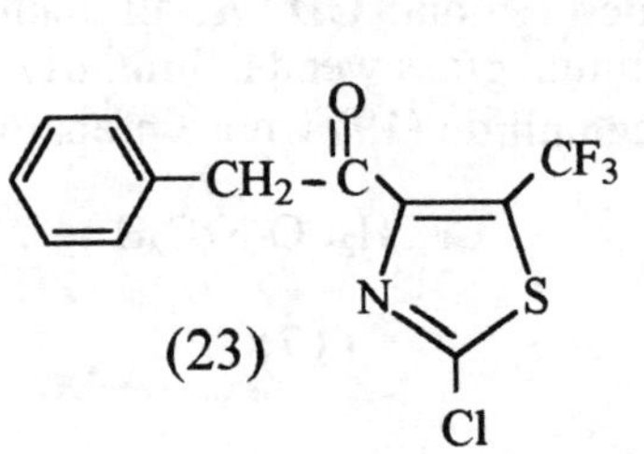

Metolachlor, 2-chloro-6′-ethyl-*N*-(2-methoxy-1-methylethyl)-*o*-acetotoluidide (24), is another selective pre-emergence soil herbicide effective against grass and broad-leaved weeds and non-toxic to mammals.

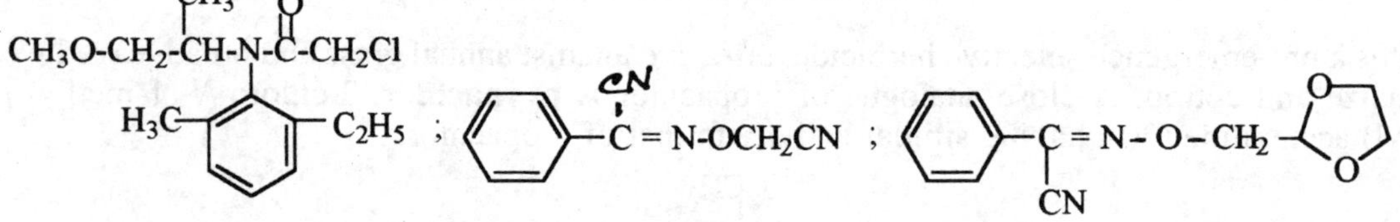

To counteract the toxic effect of metolachlor, Ciba Geigy AG developed two safeners,

cyometrinil, cyanomethoxyimino (phenyl) acetonitrile (25) and **oxabetrinil**, N-(1,3-dioxolan-2-yl)methoxyimino (phenyl) acetonitrile (26), the latter having better efficiency.

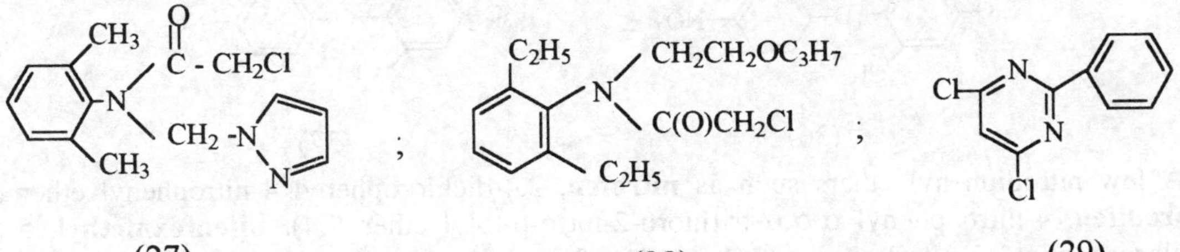

(27)   (28)   (29)

Metazachlor, 2-chloro-2,6'-dimethyl-N-(1-H-pyrazol-1-yl-methyl) acetanilide (27), is a pre-emergence herbicide recommended for **brassica** spp. crops including rapeseed. The latest added to the lists is **pretilachlor**, 2-chloro-2',6'-diethyl-N-(2-propoxyethyl) acetanilide (28), having broad spectrum activity [5]. The safening agent applied to protect rice seedlings from herbicide damage is **CGA 123407**, a pyrimidine derivative (29).

Phenolic nitro compounds [6]

The first compound of this series is **DNOC** (30), which can be prepared as follows (Fig. 3.222) but its use is restricted because of high mammalian toxicity.

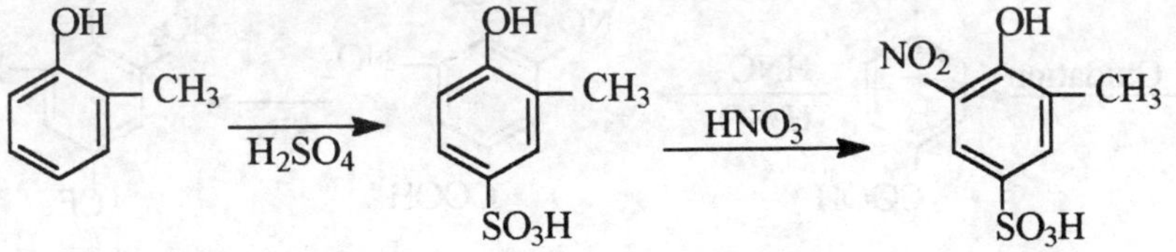

Fig. 3.222. Preparation of DNOC.

The analogous compound **dinoseb**, 2-sec-butyl-4,6-dinitrophenol (31) and **dinoterb**, 2-t-butyl-4,6-dinitrophenol (32) also showed considerable toxicity to mammals. However, the acetate derivative of dinoseb, **dinoseb acetate**, was found to be less toxic and is being used as selective pre-emergence herbicide.

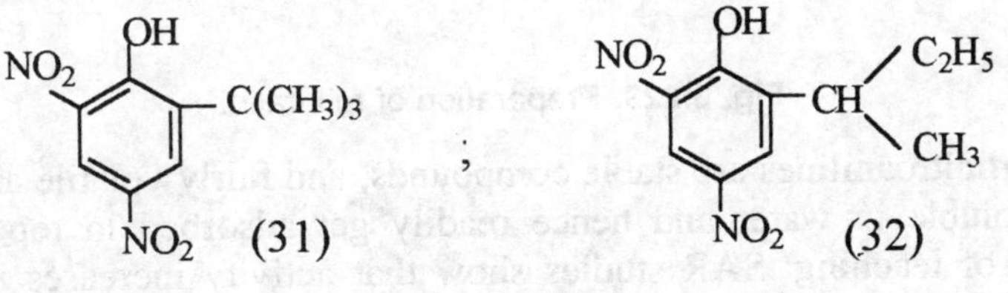

(31)   (32)

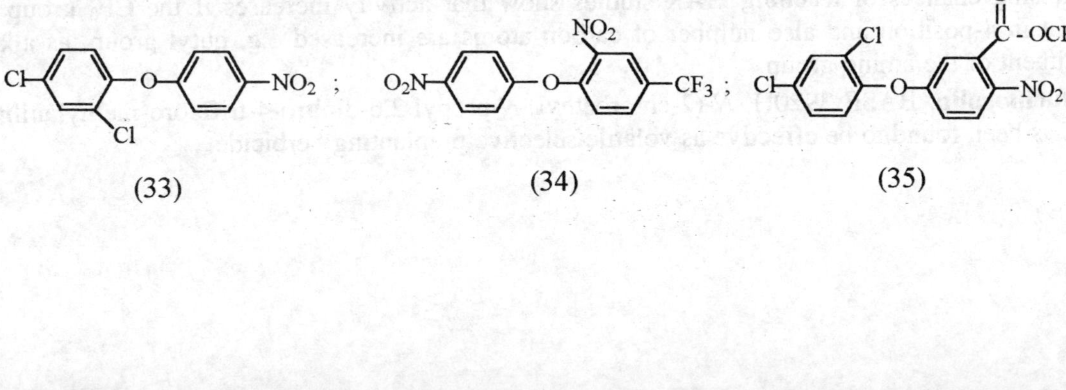

(33)   (34)   (35)

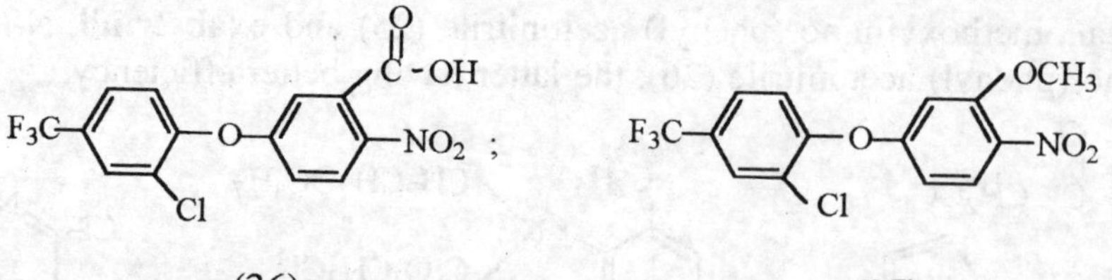

(36)   (37)

A few nitrodiphenyl ethers such as **nitrofen**, 2,4-dichlorophenyl 4-nitrophenyl ether (33), **fluorodifen**, 4-nitro phenyl α,α,α-trifluoro-2-nitro-p-tolyl ether (34), **bifenfoxmethyl**, 5-(2,4-dichlorophenoxy)-2-nitrobenzoate(35), **acifluorfen**, 5-(2-chloro-4-trifluoromethyl) phenoxy-2-nitrobenzoic acid (36) and **oxyfluorpen**, 2-chloro-4-(trifluoromethyl) phenyl 3-ethoxy-4-nitro-phenyl ether (37), were developed. All of them possess low mammalian toxicity and are applied as pre-emergence herbicides on soybean, cotton, tomato and green pepper.

Dinitroaniline compounds

A large number of substituted 2,6-dinitroaniline derivatives were synthesised as selective weed control chemicals. **Trifluralin**, α,α,α-trifluoro-2,6-dinitro-N,N-dipropyl-p-toluidide (38), has been found most effective and is commonly used as a selective weed control agent. The preparation is as follows (Fig. 3.223).

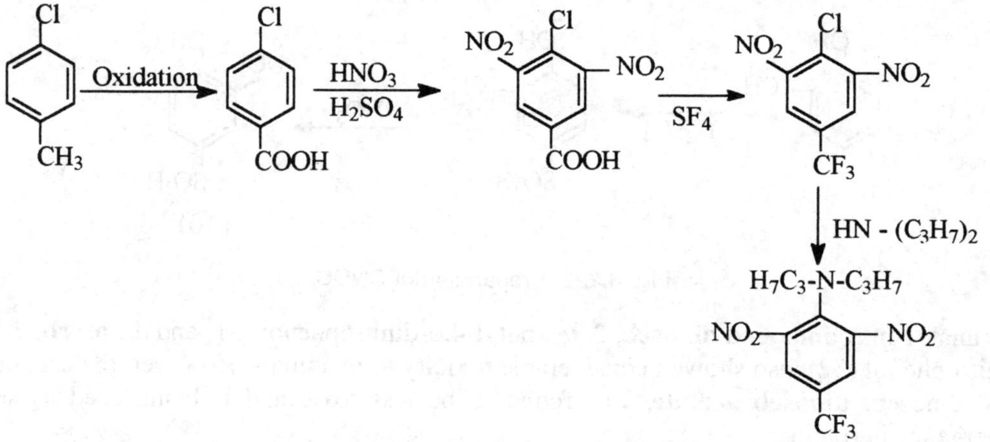

(38)

Fig. 3.223. Preparation of trifluralin.

In general, these dinitroanilines are stable compounds, and fairly volatile at room temperature. They are slightly soluble in water and hence readily get adsorbed in top layers of the soil, minimising chances of leaching. SAR studies show that activity increases if the CF_3 group is present in 4-position and also number of carbon atoms are increased e.g. butyl group, as alkyl substituent of the amino group.

Fluchloralin, BASF 3920H, N-(2-chloroethyl)-N-propyl 2,6-dinitro-4-trifluoro-methylaniline (39) has been found to be effective as volatile selective preplanting herbicide.

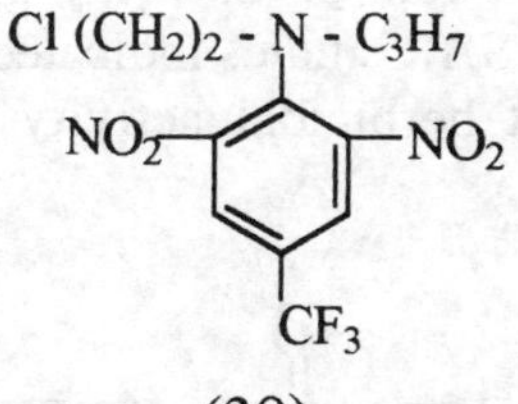

Cl (CH$_2$)$_2$ - N - C$_3$H$_7$

NO$_2$ — NO$_2$

CF$_3$

(39)

CH$_3$

CH$_3$

NO$_2$ — NO$_2$

NHCH(C$_2$H$_5$)$_2$

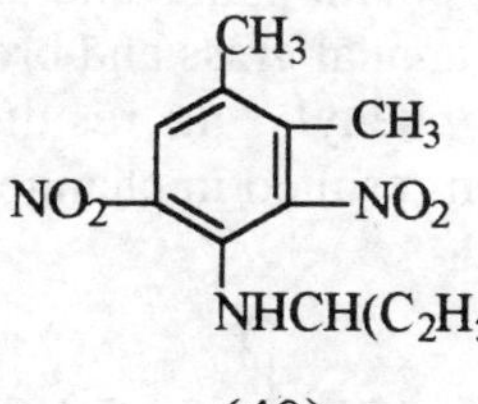

(40)

Pendimethalin, N-(1-ethylpropyl)-3,4-dimethyl-2,6-dinitrobenzen-amine (40), has low mammalian toxicity and controls most annual grasses and many annual broad-leaved weeds in cereals, vegetables, cotton, oil seeds and tobacco as both pre- and post-emergence herbicide.

Nitrile compounds

The important compound in this group is **dichlobenil**, 2,4-dichloro benzonitrile (41).

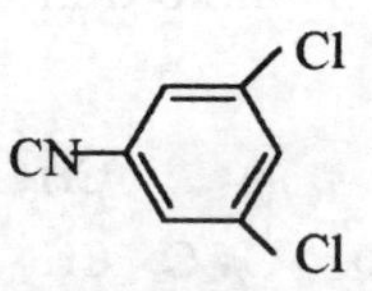

Cl

CN

Cl

(41)

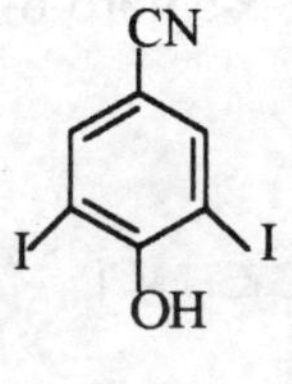

CN

I — I

OH

(42)

CN

Br — Br

OH

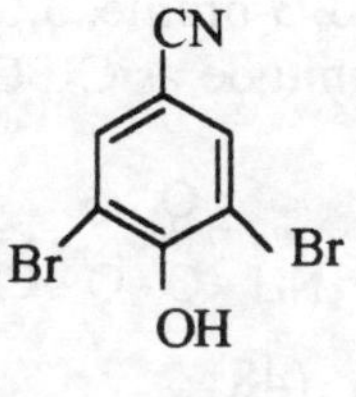

(43)

It is a selective pre- and post-emergence herbicide effective against several annual and perennial weeds and has mild mammalian toxicity. Two active compounds, **ioxynil**, 4-hydroxy-3,5-diiodo-benzonitrile (42) and **bromoxynil**, 4-hydroxy-3,5-dibromo benzonitrile, (43) have been developed as post-emergence contact herbicides to control broad–leaved weeds in cereals and grass crops. These are however moderately toxic to mammals.

Carbamate compounds [7a]

Carbamate group of compounds such as N-alkyl carbamates and N-aryl carbamates possess herbicidal activity. **Terbutol**, terbucarb, 2,6-di-t-butyl-4-methylphenyl 4-methyl carbamate (44), is a soil post-emergence herbicide effective against grass weeds, crab grass in turf and *Digitalia* spp.

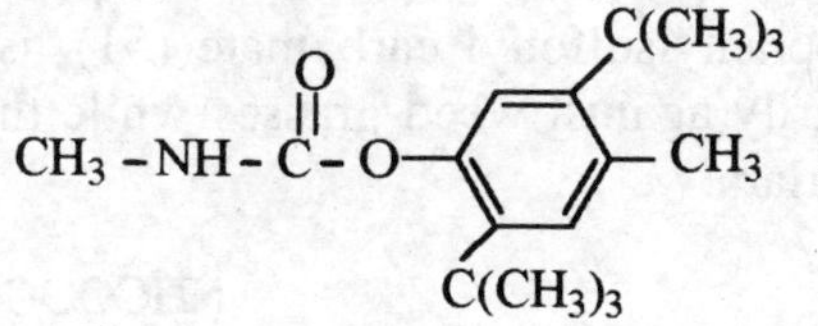

C(CH$_3$)$_3$

O
‖
CH$_3$ – NH – C – O — CH$_3$

C(CH$_3$)$_3$

(44)

O
‖
CH$_3$ – NH – C – O – CH$_2$ — Cl

Cl

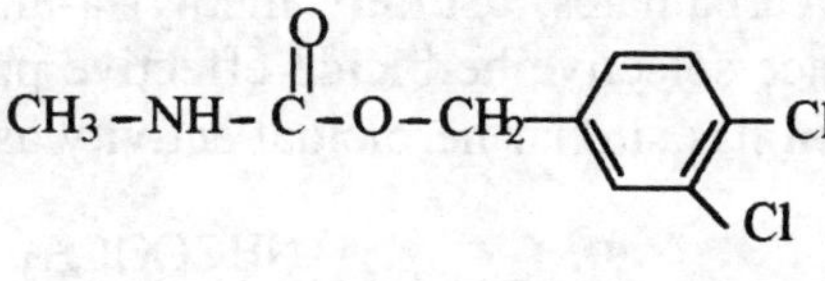

(45)

Dichlormate, 3,4-dichlorobenzyl N-methyl carbamate (45), was developed for use as pre- and post-emergence herbicide. As pre-emergence application, it controls monocot and dicot weeds in cotton, soybean and potato plants and as post-emergence it controls barnyard grass (*Echinochloa crusgalli*) and other annual grass and broad leaved weeds. SAR studies indicated that 3,4-dichloro substitution in the phenyl ring resulted in the highest herbicidal activity and variation of substituent's position resulted in change in activity.

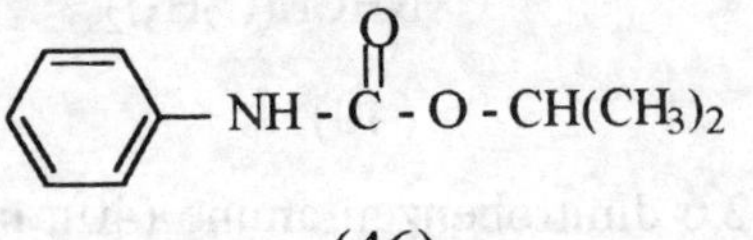

(46) 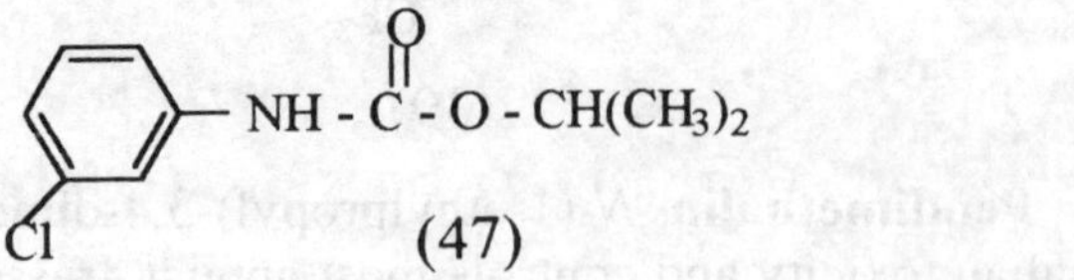 (47)

Amongst carbamate esters, **IPC**, propham, *iso*propyl N-phenyl carbamate (46), is very effective as pre-emergence soil herbicide for the control of grass weeds in beet, cabbage, peas, lettuce etc. The herbicidal activity can be enhanced by adding halogen substitution as seen in **CIPC**, chlorpropham, 3-chlorophenyl analogue (47). This is effective against several grass weeds (*Digitalia* spp.) whereas IPC is ineffective. The activity of other phenyl substitutions such as 3-methyl, 3-fluoro, 3-bromo, 3,6-dichloro, 3-chloro-6-methyl and 3-chloro-6-methoxy compounds was of same magnitude as CIPC.

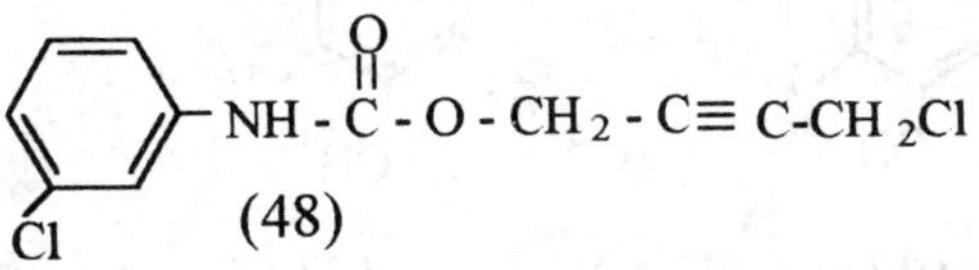

(48) 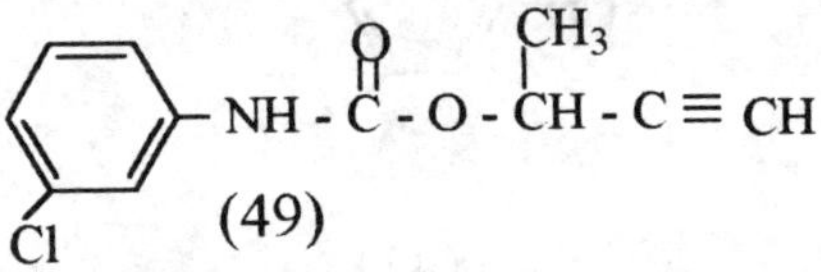 (49)

Carbamate herbicides containing alkynyl esters like **barban**, 4-chlorobut-2-yn-yl-*N*-(3-chlorophenyl) carbamate (48) and **chlorbufam**, BiPC, 1-methylprop-2-yn-yl-N-(3-chlorophenyl) carbamate (49), were developed to enhance the selectivity of herbicides. These compounds were effective as pre-emergence selective herbicides for the control of wild oats (*Avenafatuo*) and other monocot weeds.

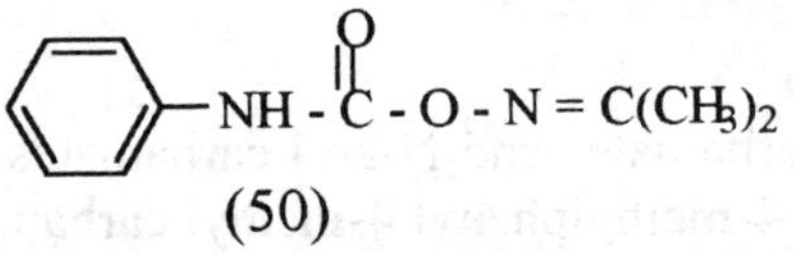

(50) 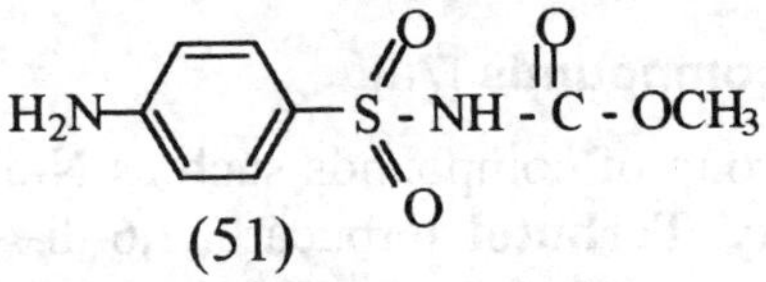 (51)

A large number of N-aryl carbamates were subsequently developed and patented but all of them did not find application in agriculture. **Proximpham**, phenyl carbamoyl propanone (50), is another selective soil herbicide meant for use in beet, onion, vegetables and ornamental plants. Amongst the sulfonyl carbamates, **asulam**, methyl 4-aminophenylsulfonyl carbamate (51), is a pre- and post-emergence selective herbicide effective primarily against weed grasses while the nitro analogue **nisulam** has similar herbicidal activity as asulam.

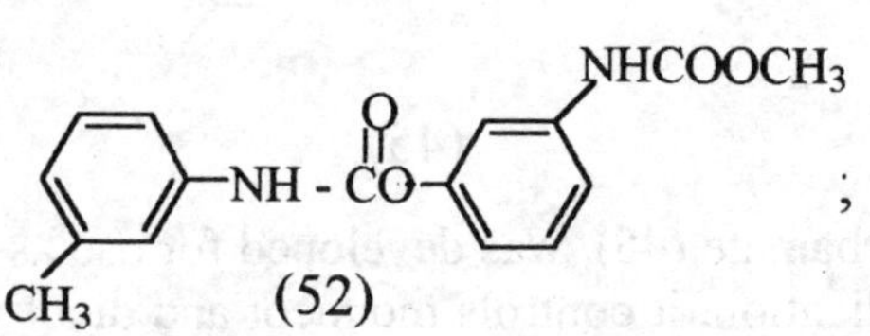

(52) 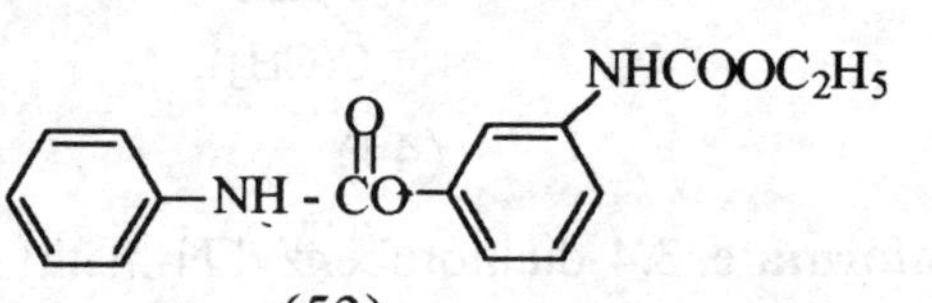 ; (53)

Two biscarbamate compounds namely, **Phenmedipham**, methyl (3-m-tolylcarbamoyloxy-phenyl)carbamate (52), and **desmedipham**, ethyl (9-phenyl carbamoyloxy-phenyl)carbamate (53) were developed. These are selective, post-emergence herbicides effective in sugar beet and many dicot weeds.

Thiocarbamate compounds

A large number of thiocarbamates were synthesised to obtain active herbicides. Amongst the dialkyl thiocarbamate group of herbicides, **EPTC**, S-ethyl N,N-dipropylthiocarbamate (54a) was developed by Stauffer Chemical Co. [6].

$$C_2H_5-S-CO-N-(C_3H_7)_2 \; ; \qquad Cl_2CH-CON \qquad O=C-O-C=O$$

$$
\begin{array}{ccc}
(54a) & (54b) & (55)
\end{array}
$$

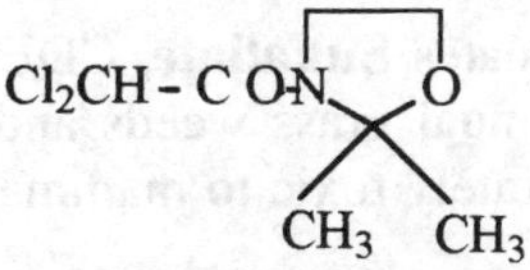

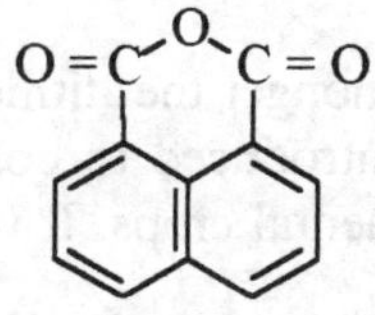

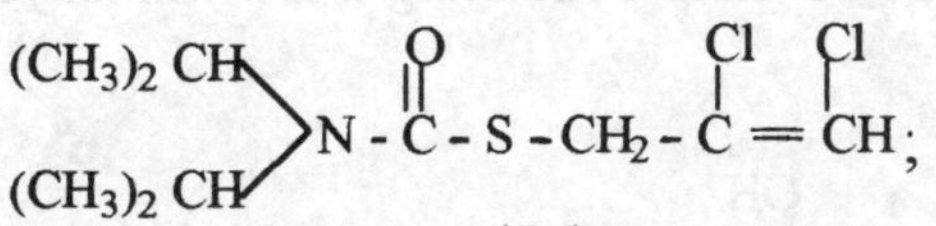

$$(56) \qquad\qquad (57)$$

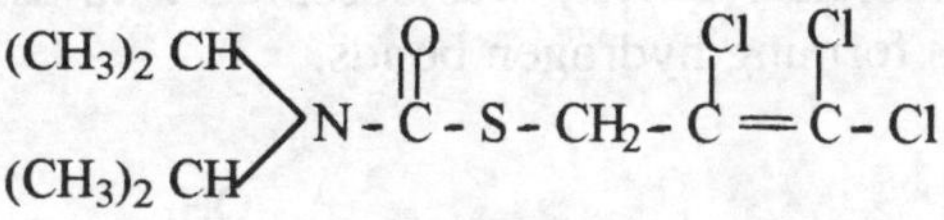

It is a selective pre-emergence herbicide effective against several monocot and dicot weeds in many crops, orchards and vineyards specially against grass weeds. Stauffer Chemical Co.developed a few antidotes viz. 3-dichloro-acetyl-2, 2-dimethyl oxazolidine (R-29148, 54b) and 1,8-naphthalic anhydride (55), to counteract the damages caused by EPTC and its analogous products. Further work on development of safer and effective molecules led to the preparation of halogen alkenyl thiocarbamate herbicide **diallate**, S-2, 3-dichloroallyl di*iso*propyl thiocarbamate (56), which is a mixture of both Z-(cis) and E-(trans) isomer where the isomers have different level of activity. The synthesis of **triallate**, S-2, 3, 3-trichloroallyl di-*iso*propyl thiocarbamate (57), is another important development in this direction. This compound is used for the control of wild oat and black grass and in order to get higher activity, these compounds have to be placed deep at 5 cm in the soil. Triallate is found selective in cereals, wheat and barley while root crops like sugar beet and maize and leguminous plants, are comfortable with diallate.

Benthiocarb, S-(4-chlorobenzyl) diethyl thiocarbamate (58) a selective rice herbicide is also effective for weed control in cotton, soybean, tomatoes, potatoes, peanut, beet and beans. It is moderately toxic and remains active for more than a month in the soil.

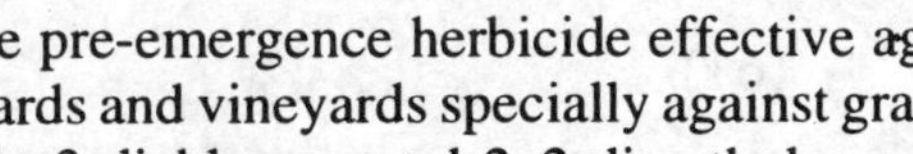

$$Cl-\langle\!\!\!\!\!\!\!\!\!\!\!\!\bigcirc\!\!\!\!\!\!\!\!\!\!\!\rangle-CH_2S-CO-N-(C_2H_5)_2 \; ; \qquad (CH_2)_6N-C-S-C_2H_5$$

$$(58) \qquad\qquad\qquad (59)$$

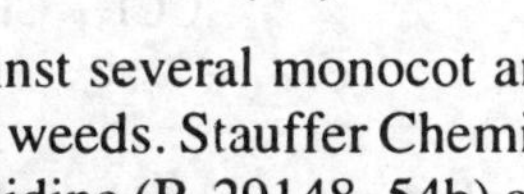

A heterocyclic amine thiocarbamate, **molinate**, S-ethyl, N,N-hexamethylene thiocarbamate, (in

C.A. usage, S-ethyl hexahydro-1 H-azepine-1-carbothioate) (59) was developed and is used as pre- and post-emergence soil herbicide mainly for the control of monocot weeds.

Recently a new concept has been emerged in the weed control management. It is reported that the thiophosphorus ester (R-33865) when added to the formulation of EPTC and molinate, enhances the persistence leading to prolonged effect.

$$(C_2H_5)_2-N-\overset{\overset{S}{\|}}{C}-S-CH_2-\overset{\overset{Cl}{|}}{C}=CH_2$$

(60)

Amongst the dithiocarbamates **Sulfallate**, CDEC, 2-chloroallyl diethyl dithiocarbamate (60), was introduced to control annual grass weeds and a few broad-leaved weeds of vegetable and ornamental crops. It is moderately toxic to mammals but non toxic to wild fowl.

Urea compounds

A large number of urea herbicides have been commercialized and used extensively in agriculture. Good herbicidal activity was observed with trisubstituted ureas containing a few iminohydrogen moieties forming hydrogen bonds.

$$CCl_3-\overset{\overset{OH}{|}}{CH}-NH-CO-NH-\overset{\overset{OH}{|}}{CH}-CCl_3$$

(61)

Amongst the aliphatic urea derivatives, **dichloral urea**, DCU, 1,3-bis (2,2,2-trichloro-1-hydroxyethyl) urea (61), exhibited selective action to control weed grasses and is non persistent in nature.

Cycluron, 3-cyclooctyl-1,1-dimethyl urea, an experimental herbicide introduced by BASF controls monocot germinating annual grass. Subsequently, bicyclo-aliphatic urea, *isonoruron* an isomer of noruron was introduced by BASF for the control of grass weeds.

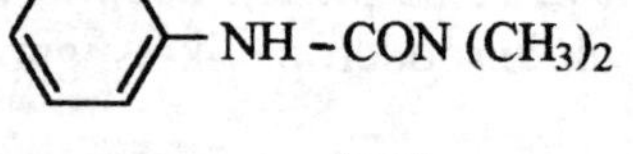

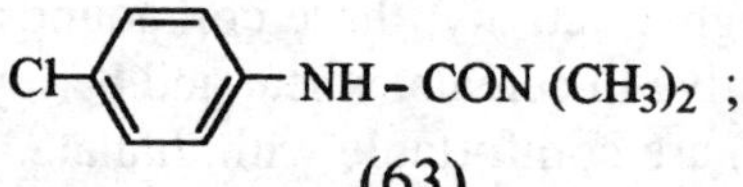

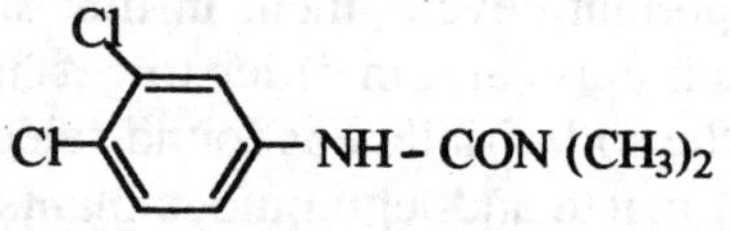

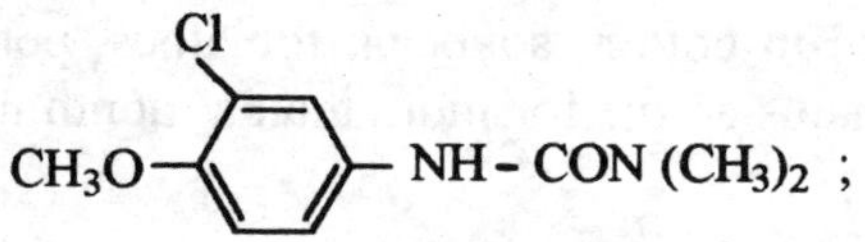

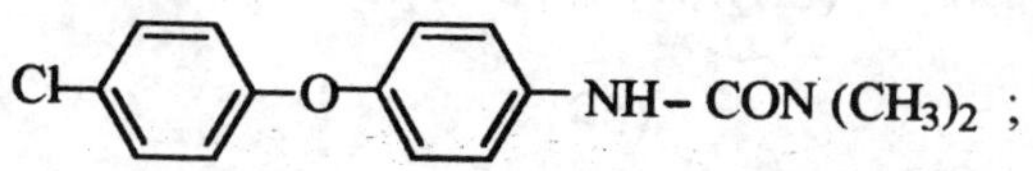

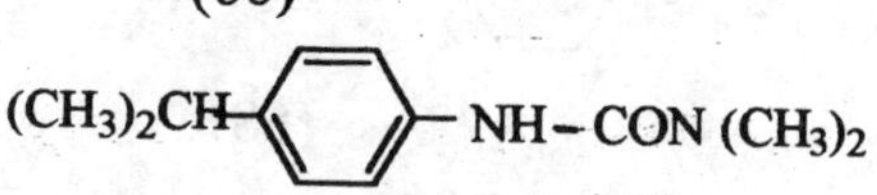

Amongst aryl urea derivatives, the most effective molecules are **fenuron**, 1,1-dimethyl-3-phenylurea (62), **monuron**, 3-(4-chlorophenyl)1-1-dimethylurea (63), **diuron**, 3-(3,4-dichlorophenyl)-1,1-urea (64), **fluometuron**, 1-1-dimethyl 3-(-trifluoro-m-tolyl) urea (65), **metoxuron**, 3-(3-chloro-4-methoxyphenyl) 1,1-dimethyl urea (66), **chloroxuron**, 3-(4-chlorophenoxyphenyl) 1,1-dimethylurea (67) and *iso*proturon, 3-(4-*iso*propylphenyl)-1,1-dimethylurea (68). The general methods of preparation of urea compounds are shown in Fig. 3.224.

(i) Aryl isocyanates are prepared by acylation of aniline derivatives with phosgene followed by reaction with dimethyl amine to yield the urea compounds.

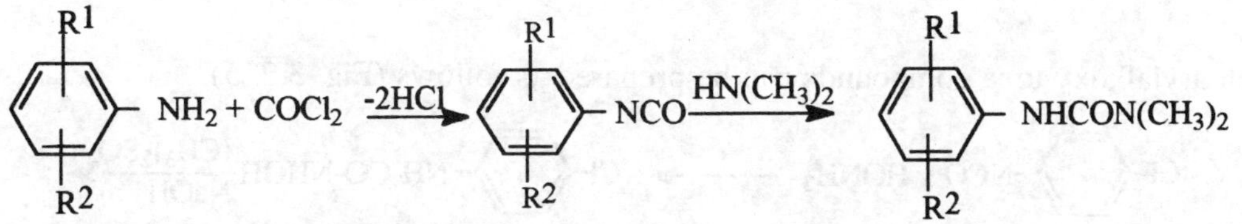

(ii) A CF$_3$ group is introduced in the phenyl moiety which can be then converted to yield the trifluromethyl urea compound.

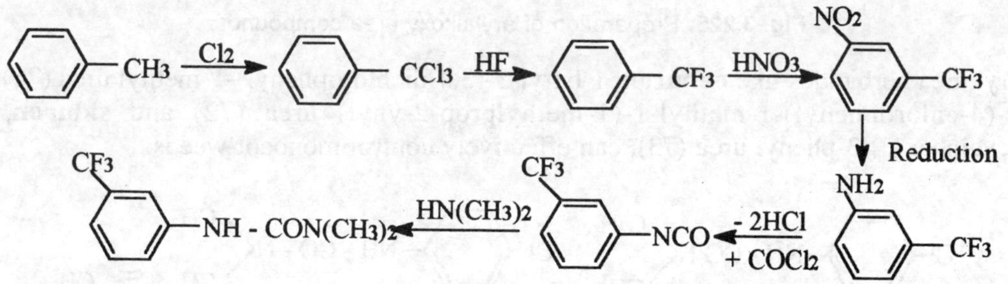

(iii) Alternatively the above compound can also be prepared as follows

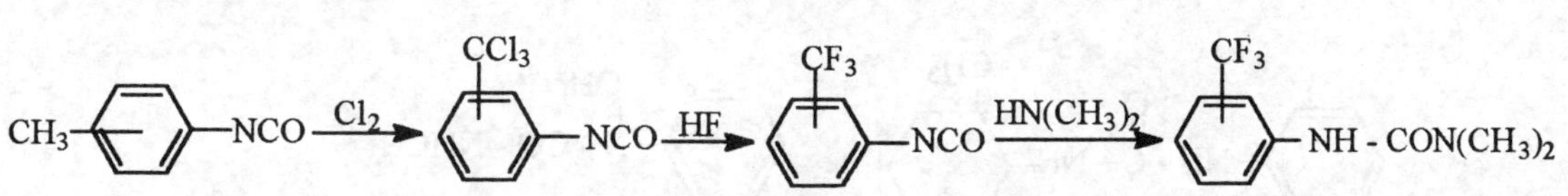

Fig. 3.224. Different routes (i, ii, iii) for the preparation of urea compounds.

Of all the urea derivatives, the herbicide fenuron has the highest solubility, hence it is least adsorbed and readily leaches through the soil. **Diuron**, is the most persistent one and mildy toxic whereas **monouron**, is nontoxic to bees. **Fluometuron**, has medium persistence and is applied as pre- and post-emergence for the control of annual grass and broad-leaved weeds. **Metoxuron**, used against grass weeds and has low persistence. **Chloroxuron**, is used for the pre- and post-emergence control of annual grass and broad leaved weeds. **Isoproturon**, is used both as pre- and post-emergence in winter cereals.

Arylalkoxy urea compounds

Amongst arylalkoxy urea compounds, **linuron**, 3-(3,4-dichloro)-1-methoxy-1-methylurea (69) and

chlorbromuron, 3-(3-chloro-4-bromophenyl)-1-methoxy-1-methyl urea (70), are the important ones which are used as pre- and post-emergence herbicide against annual weeds in vegetables, orchards and cereal crops These are non-persistent and less toxic to mammals.

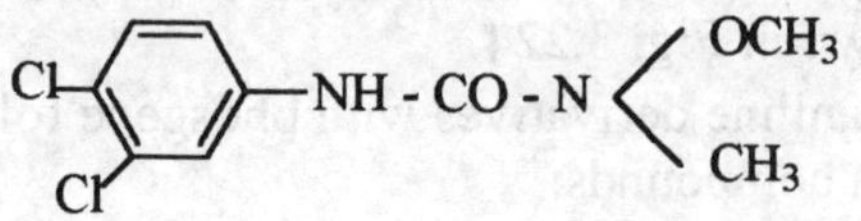

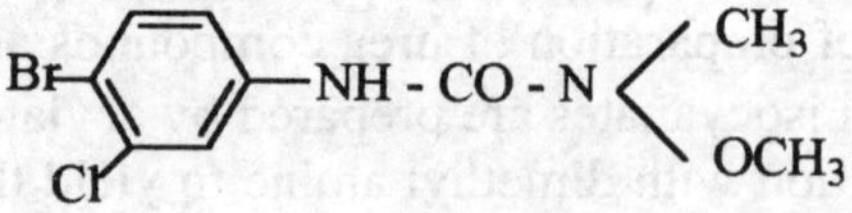

Such arylalkoxy urea compounds can be prepared as follows (Fig. 3.225).

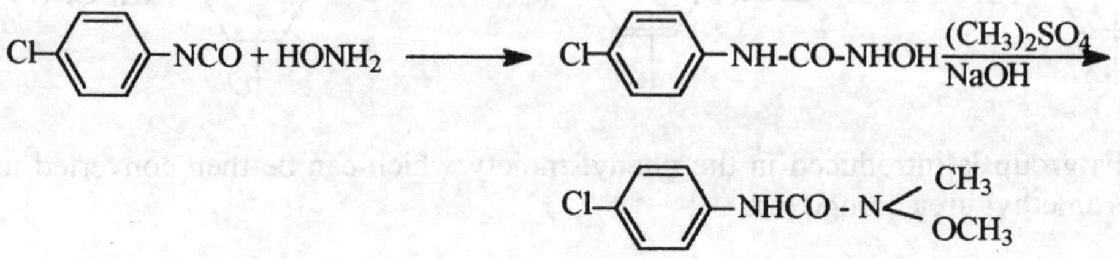

Fig. 3.225. Preparation of arylalkoxy urea compounds.

3-Aryl urea herbicides like **neburon**, 1-butyl-3-(3,4-dichlorophenyl)-1-methyl urea (71), **buturon**, 3-(4-chlorophenyl)-1-methyl-1-(1-methylprop-2-ynyl) urea (72) and **siduron**, 1-(2-methylcyclohexyl)-3-phenyl urea (73), can effectively control monocot weeds.

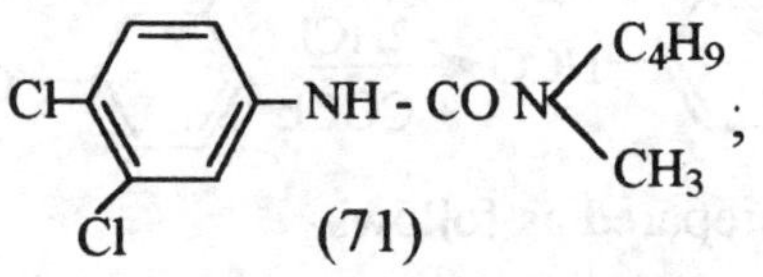

The fact is that the presence of hydroxyl group attached to the nitrogen atom reduces the activity but at the same time it enhances its selectivity and was used in the formulation of a selective herbicide **meturin**, 1-phenyl-1-hydroxy-3-methylurea (74), a target specific weedicide in cotton and potatoes.

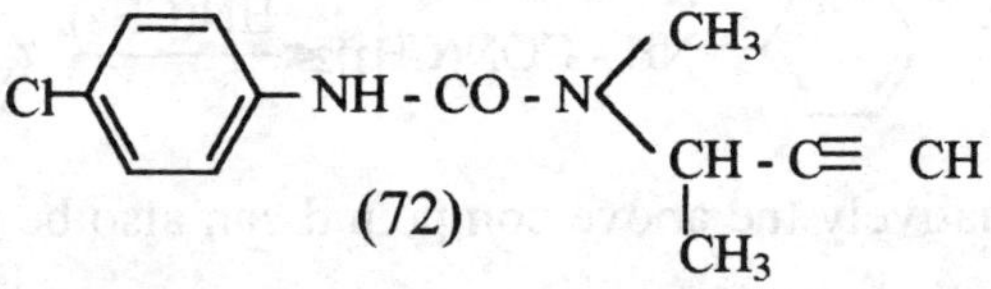

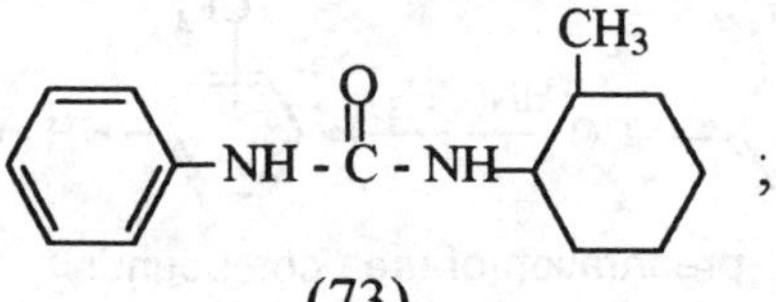

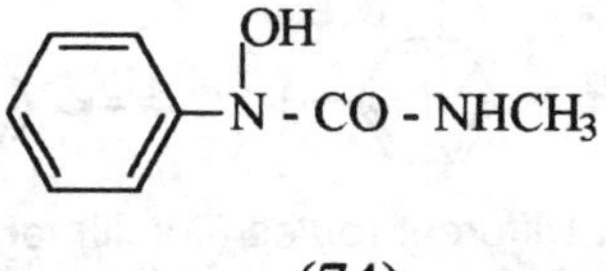

A number of urea derivatives with heterocyclic moiety have been prepared and studied for herbicidal action. Amongst them 1,2 and 1, 3 thiazolyl, thiadiazolyl, isothiazolyl, pyridyl, benzthiazolyl, benzimidazolyl and 1,3,4-thiadizolyl urea, have found useful as herbicides. Preparations like **tebuthiuron**, 1-(5-t-butyl-1,3,4-thiadiazol-2-yl)-1,3-dimethylurea (75), **thiazafluron** 1,3-dimethyl-1-(5-trifluoro-methyl-1,3,4-thiadiazol-2-yl) urea (76) and **sulfadiazole**, 1,3-dimethyl-1-(5-ethyl sulfonyl-1,3,4-thiadiazol-2-yl) urea (77) were found to be most effective.

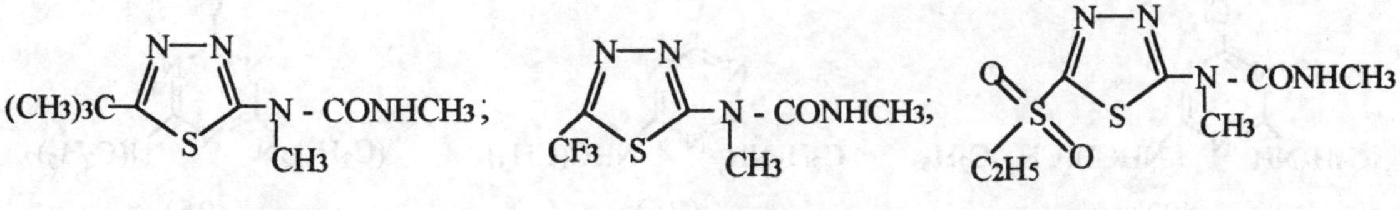

(75)   (76)   (77)

Tebuthiuron is a persistent herbicide which may be retained in the soil for 3-4 months while **Thiazafluron** is a pre-emergence herbicide for total vegetation control and possesses medium toxicity. **Sulfadiazole** is an experimental herbicide for the pre- and post-emergence application for total weed control.

A number of thiourea derivatives were also synthesised and patented, however, they exhibit poor activity. **Methiuron** 1, 1-dimethyl-3-(3-methyl phenyl) thiourea (78) and **ortho 11413**, 1-(4-chlorophenyl)-3-methyl-3-phenyl) thiourea (79), show promising activity.

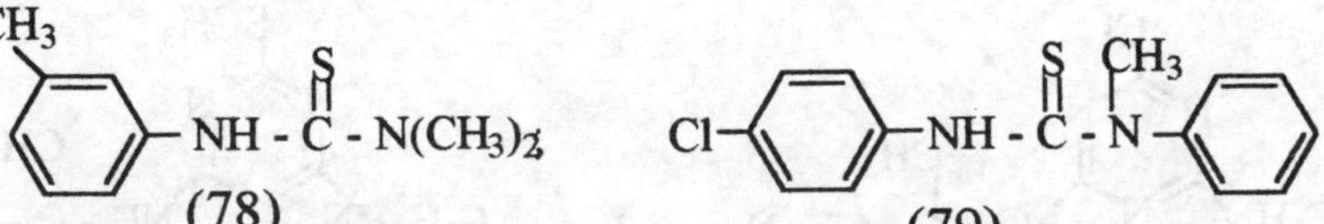

(78)   (79)

S-Triazine compounds

This group of herbicides are of great practical importance. Several triazine compounds have been synthesised by reacting cyanuric chloride with alkylamine (Fig. 3.226).

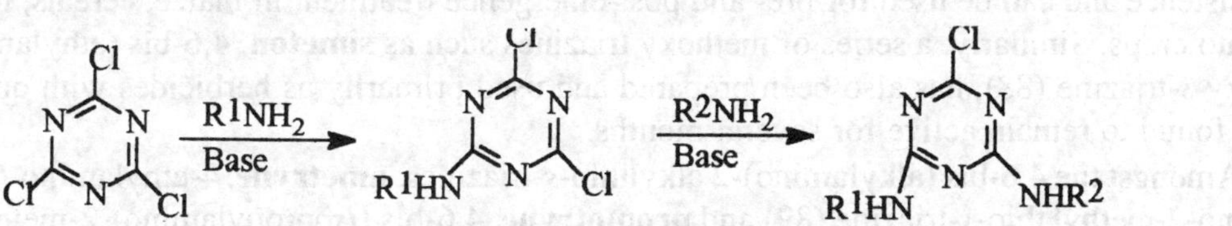

Fig. 3.226. Synthesis of triazine compounds.

Both symmetrical and asymmetrical substituted derivatives can be prepared by using same or different alkylamines. The most important member of this group discovered is **simazine**, 2-chloro-4,6-*bis* (ethylamino)-s-triazine (80), **propazine**, 2-chloro-4,6-*bis* (*iso*propylamino)-s-triazine (81), **atrazine**, 2-chloro-4-ethylamino-6-isopropylamino-s-triazine (82), **sebuthylazine,** 2-chloro-4-ethylamino-6-*sec* butylamino s-triazine (83), **terbuthylazine**, 2-chloro-4-ethylamino 6-*t*-butyl-amino-s-triazine (84) and **chlorazine**, 2-chloro-4, 6-bis (diethylamino)-s-triazines (85).

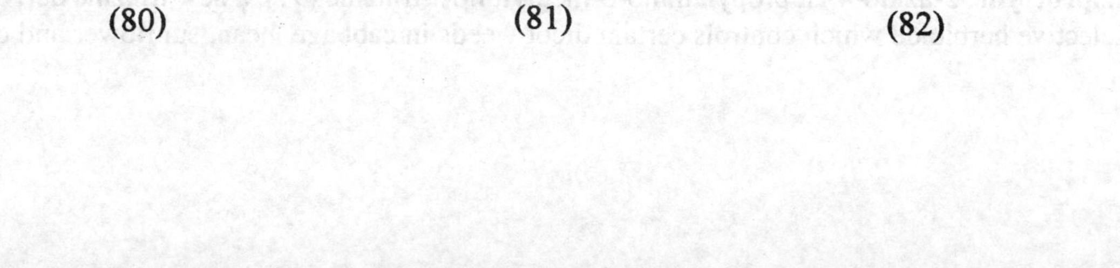

(80)   (81)   (82)

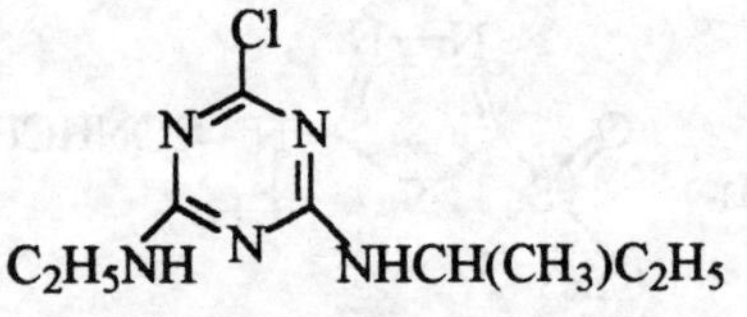

(83) 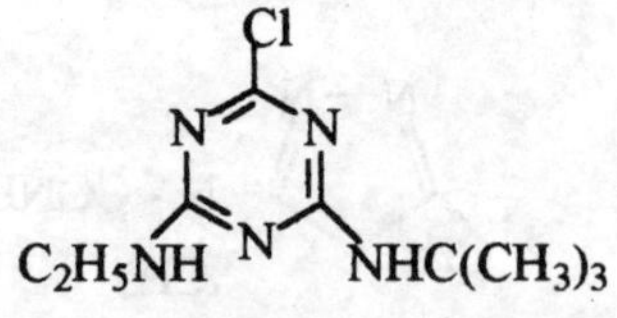   (84) 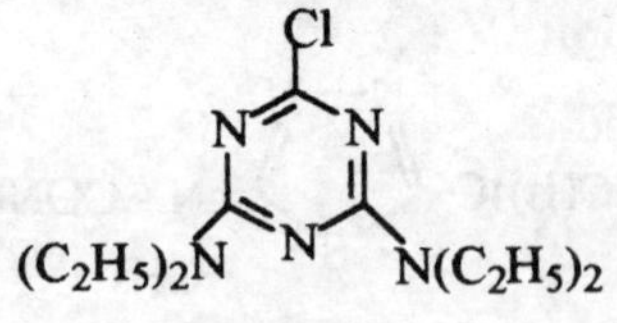   (85)

Simazine has the lowest solubility in water while atrazine has the highest. Both simazine and propazine have long persistence in soil. From the practical point of view atrazine is widely used in maize cultivation since the spectrum of activity is similar to that of symmetric 2-chloro-4, 6-*bis* (alkyl amino)-*s*-triazines and it shows strong action both on monocot and dicot weeds.

Change of substitution in alkyl side chain with cyano group made them more water soluble as observed in **cyanazine**, 2-chloro-4-(1-cyano-1-methylethylamino)-6-ethylamino-*s*-triazine (86) and **cyanatryn**, 4-(1-cyano-1-methylethylamino)-6-ethylamino-2-methylthio-*s*-triazine (87).

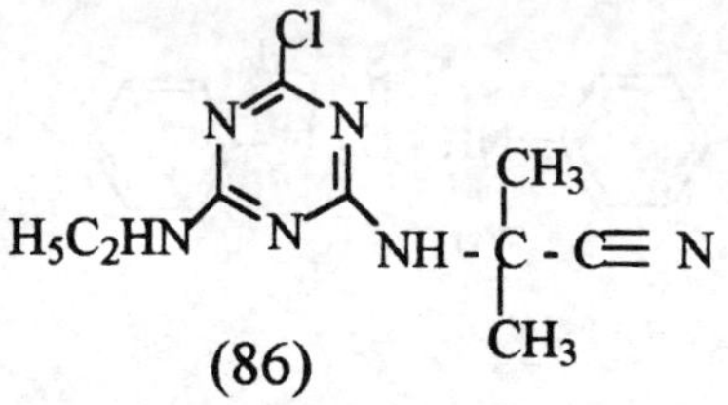

(86) 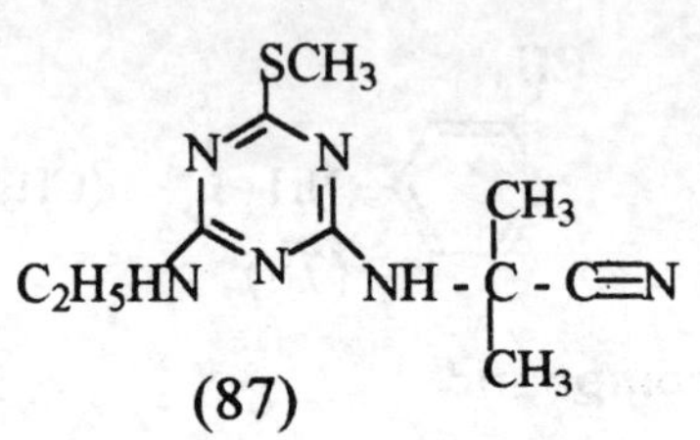   (87)

Cyanatryn is effective against aquatic weeds at a concentration of 0.05 to 0.5 ppm. It has low persistence and can be used for pre- and post-emergence treatment in maize, cereals, legumes and potato crops. Similarly, a series of methoxy triazines such as **simeton**, 4,6-bis (ethylamino)-2-methoxy-*s*-triazine (88), has also been prepared and used primarily as herbicides with quick activity and found to remain active for several months.

Amongst the 4,6-bis (alkylamino)-2 alkylthio-*s*-triazines, **ametryne**, 4-ethylamino-6-*iso*propyl-amino-2-methyl thio-*s*-triazine (89) and **promotryne**, 4,6-bis (*iso*propylamino)-2-methylthio-*s*-triazine (90) were found effective.

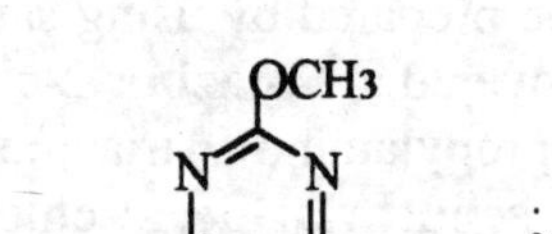

(88) 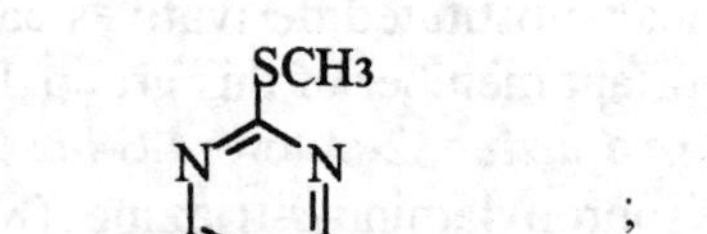   (89) 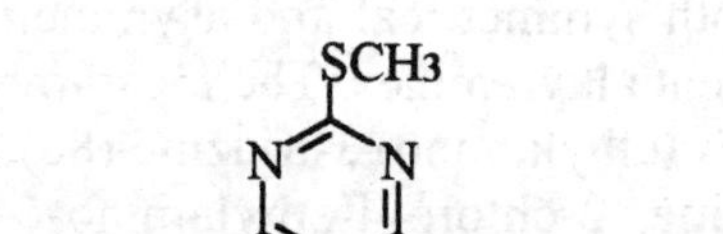   (90)

Both were used in orchards, vine and citrus plantation forest, sugarcane and pineapple culture as well as in cotton, potato, peas and carrots. These herbicides have a broad spectrum activity and are selective as compared to the corresponding methoxy derivatives with short duration of action.

Aziprotryne, 2-azido-4-*iso*propylamino-6-methylthio-*s*-triazine (91), a new triazine derivative, is a selective herbicide which controls certain dicot weeds in cabbage, bean, sunflower and onion.

Pyridine compounds

Pyridine derivatives with several halogen atom substituents show significant herbicidal activity. Two such active molecules are **pyrichlor**, 2,3,4-trichloro-4-pyridinol (92) and **picloram**, 4-amino-3,5,6-trichloropicalinic acid (93).

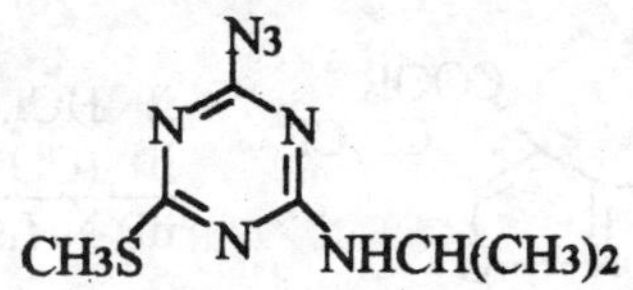

(91) 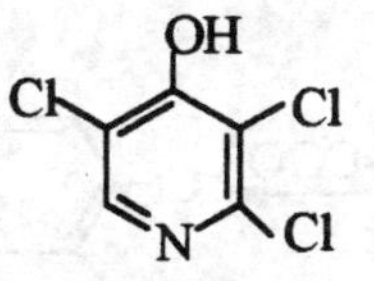   (92) 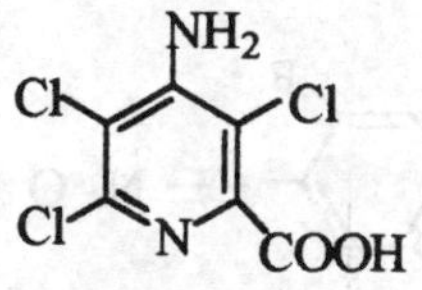   (93)

These pre-emergence herbicides are persistent in soil having low mammalian toxicity and are effective against annual grass weeds and broad-leaved weeds.

Pyridazine compounds

The first herbicidal compound in this group is **pyrazon**, 5-amino-4-chloro-2-phenyl-3-pyridazone (94) marketed under the trade name **pyramin** is a pre-emergence herbicide effective against broad-leaved weeds and moderately toxic to mammals.

A few related compounds with good activity include **metflurazon**, 4-chloro-5-dimethylamino-2(α,α,α-trifluoro-*m*-tolyl) pyridazin-3-one (95) and **ceredazine** (trade name, Kusikara), 3-O-tolyloxypyridazine (96). These are used as selective pre-emergence soil herbicides to control grasses and broad-leaved weeds in cotton and beet. In Japan ceredazine is also used in rice and potato crops.

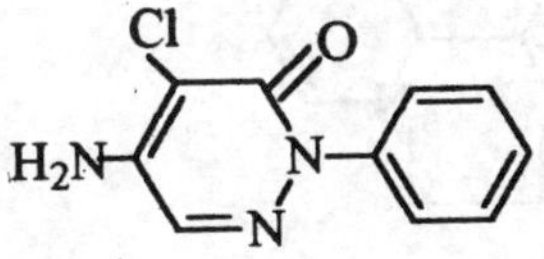

(94) 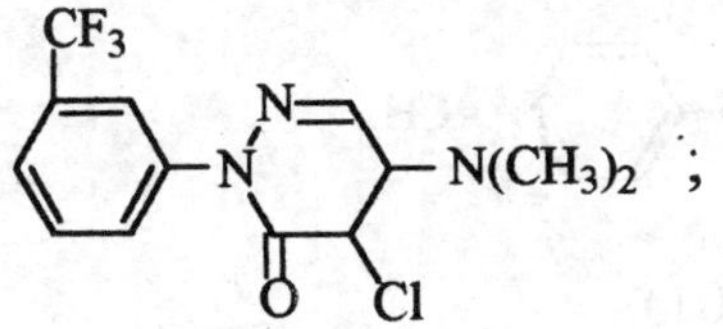   (95) 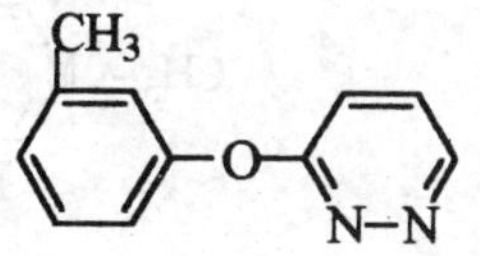   (96)

Pyrazole compounds [7b]

3-Aryl-4-halo-5-haloalkylpyrazoles and 3-aryl-4-halo-5-alkylsulfoxyl pyrazoles were developed as pre-emergence and post-emergence ecofriendly herbicides for the control of broad-leaved and narrow leaf weeds, since these herbicides are required in very low quantities (g/ha). Other active pyrazole derivatives having herbicidal activity are sulfonyl pyrazole (97), pyrazole phenyl ethers (98) and phenyl pyrazoles (99).

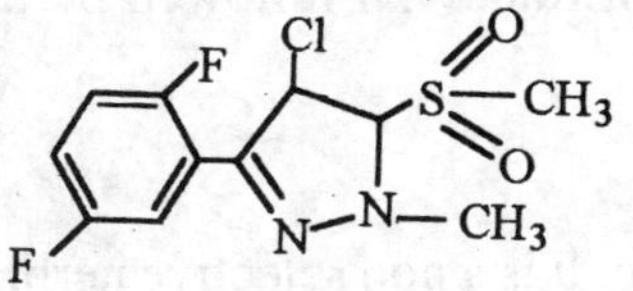

(97) 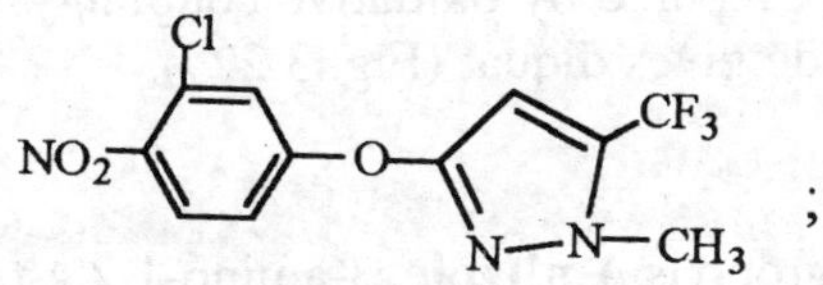   (98) 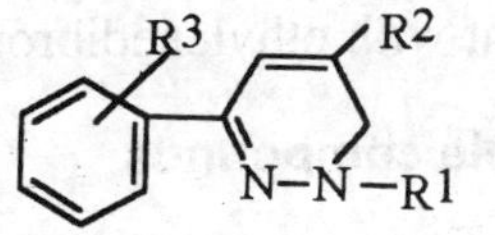   (99)

A broad spectrum arylisoxazoles having carbon linkage between two rings were also prepared by cyclo addition of nitrile oxides to acetylenic esters. The resultant isoxazole carboxylate esters were then converted to carboxamides (100) by a series of reactions (Fig. 3.227). SAR studies have shown that 2,4 or 2,4,5 substituted phenyl ring in the 3 position, a primary or secondary amide in the 4-position and a CF_2Cl group in the 5-position of the isoxazole ring enhance the herbicidal activity.

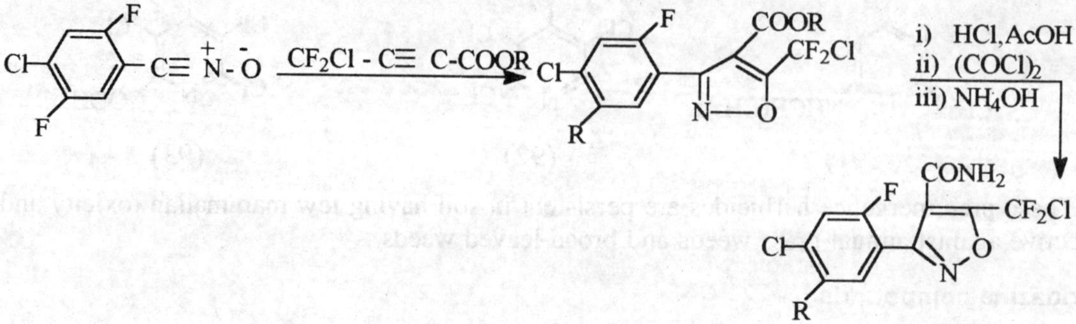

Fig. 3.227. Synthesis of aryl isoxazole carboxylate and carboxamide.

Quaternary ammonium salts

Paraquat, 4,4'-bipyridyl (101) and **diquat**, 2,2'-bipyridyl (102), are the most important quaternary salts. Diquat is used to control aquatic weeds, defoliation in potato and desiccation in rice, maize and sunflower.

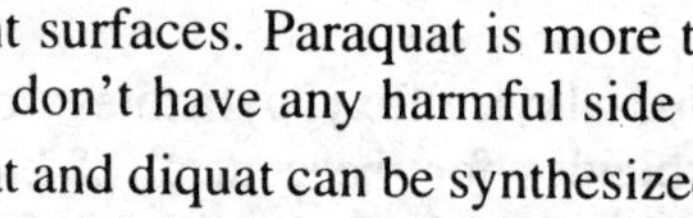

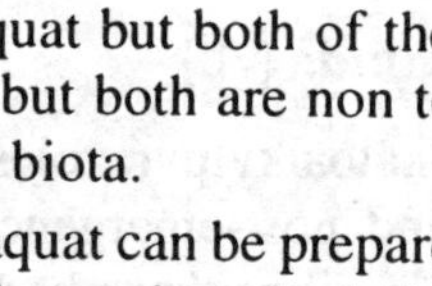

(101)   (102)

Paraquat is also found useful for pre-sowing and pre-emergence weed control in orchards and vineyards. Paraquat is more strongly adsorbed by soil than diquat but both of them are adsorbed strongly on plant surfaces. Paraquat is more toxic than diquat but both are non toxic to bees and earthworms and don't have any harmful side effect on the soil biota.

Both paraquat and diquat can be synthesized as follows. Paraquat can be prepared from pyridine by the Dimroth reaction. This involves reduction of pyridine in acetic anhydride and Zn-dust to give the amide which on oxidation followed by hydrolysis gives 4,4'-bipyridyl. This on further treatment with methyl chloride yields paraquat (Fig. 3.228).

Similarly 2,2'-bipyridyl prepared by oxidative coupling over hot Raney Ni followed by treatment with ethylenedibromide gives diquat (Fig. 3.229).

Azole compounds

The first compound of this group is **Amitzole**, 3-amino-1,2,4-triazole. It is a non selective herbicide hence does not find much use in agriculture.

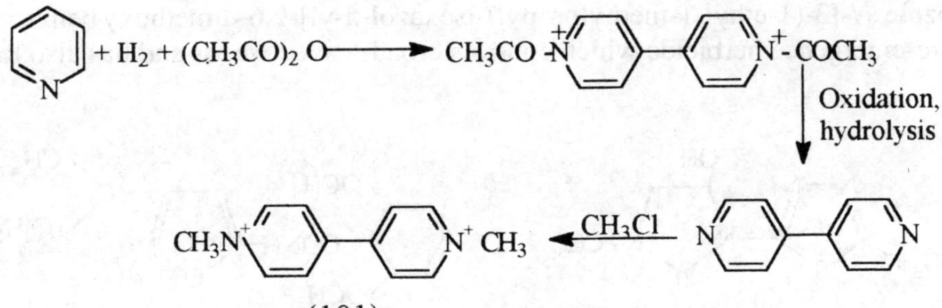

(101)

Fig. 3.228. Preparation of paraquat.

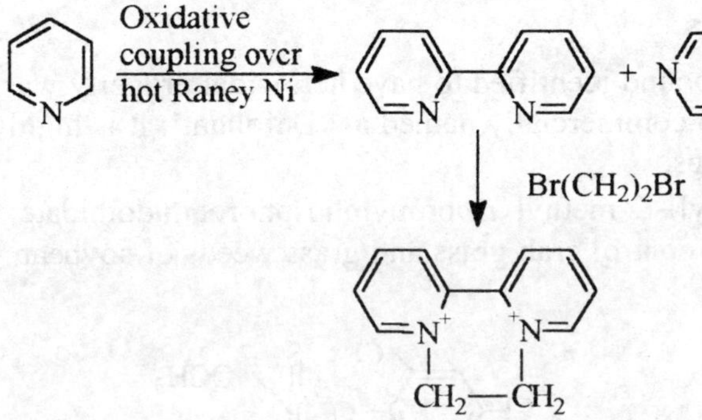

(102)

Fig. 3.229. Preparation of diquat.

The herbicide **methazole** (103), containing oxadiazolidine ring can also be put under azole group. It has selectivity and contact foliage action and is effective against certain grasses, many broad-leaved weeds. It is used as a pre-emergence weedicide in cotton, potato and garlic.

Oxadiazon, 5-*t*-butyl-3-(2,4-dichloro-5-*iso*propoxy phenyl)-1,3,4-oxadiazol-2-one (104), is another selective pre- and post-emergence herbicide of medium persistence active against several grasses and broad-leaved weeds.

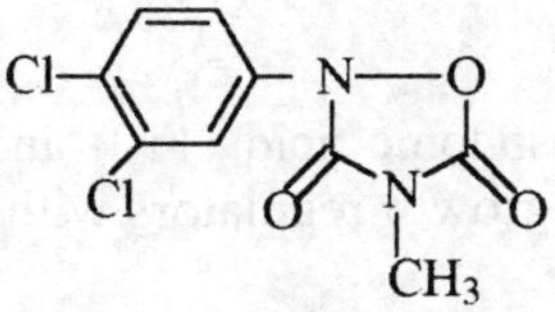

(103) 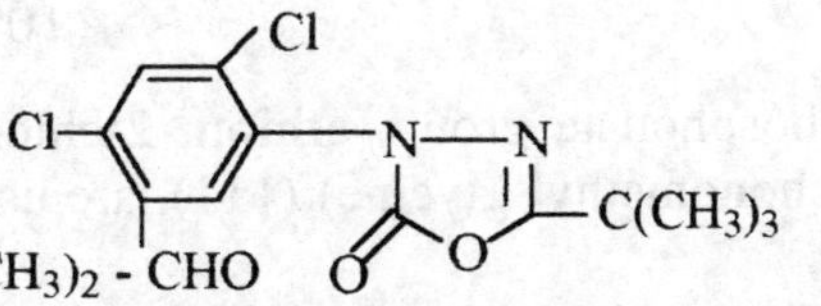   (104)

Buthidazole, 3-(5-*t*-butyl-1,3,4-thiadiazol-2-yl)4-hydroxy-1-methyl-2-immidazolidone (105), is a systemic, moderately toxic and fairly persistent herbicide for pre- and post-emergence use to control most annual grasses and broad leaved weeds. The compound is absorbed both by the roots and leaves and translocated through the phloem and the xylem.

Benzamizole, N-[3-(1-ethyl-1-methylpropyl)-isoxazol-5-yl]-2,6-dimethoxy benzamide (106), is a selective pre-emergence herbicide which controls broad-leaved weeds and has also low mammalian toxicity.

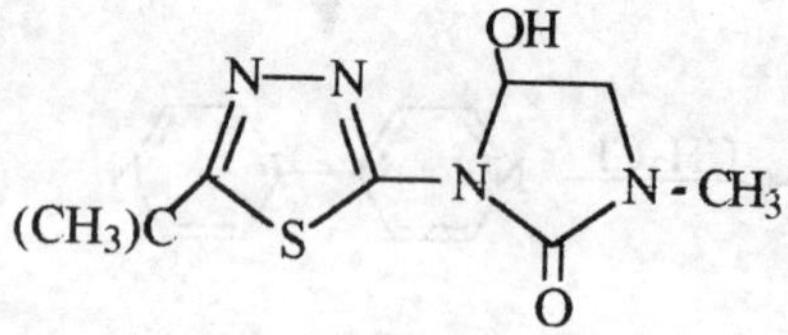

$$(105) \qquad\qquad (106)$$

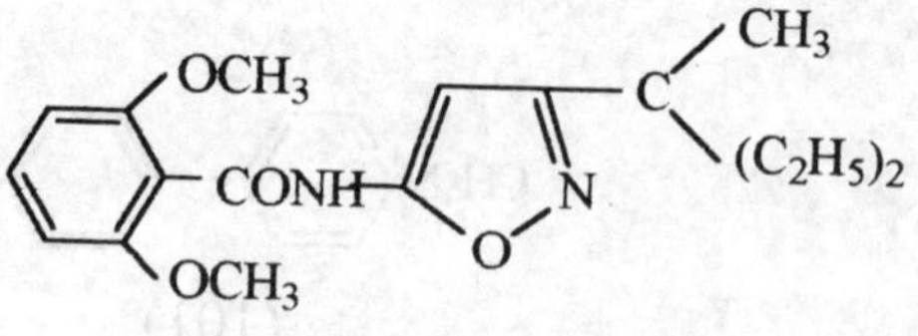

Organophosphorus compounds

The first organophosphorus compound identified to have herbicidal property was **tribufos**, S,S,S-tributyl phosphorotrithioate (107) commercially named as 'Defoliant'. It is highly phytotoxic and used for defoliation in cotton crops.

DMPA, O-(2,4-dichlorophenyl)-O-methyl *iso*propylphosphoroamidothioate (108), is a pre-emergence soil herbicide used to control crab grass and grass weeds of soybean, peas, beans and onion.

$$(C_4H_9S)_3P{=}O \qquad ;$$

$$(107) \qquad\qquad (108)$$

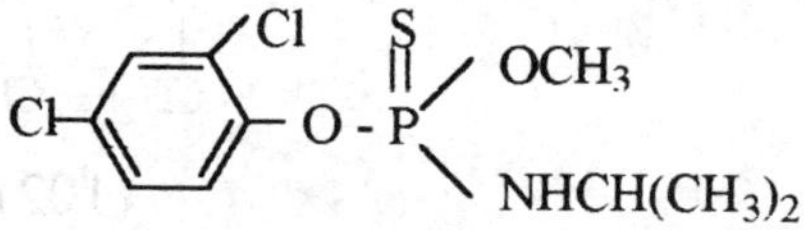

Bensulide, O,O-diisopropyl-S-(2-phenylsulfonyl amino) ethyl phosphoro-dithioate (109), is moderately persistent and toxic and is used as a pre-planting and pre-emergence herbicide in cotton, cucurbits, lettuce, and brassica spp.

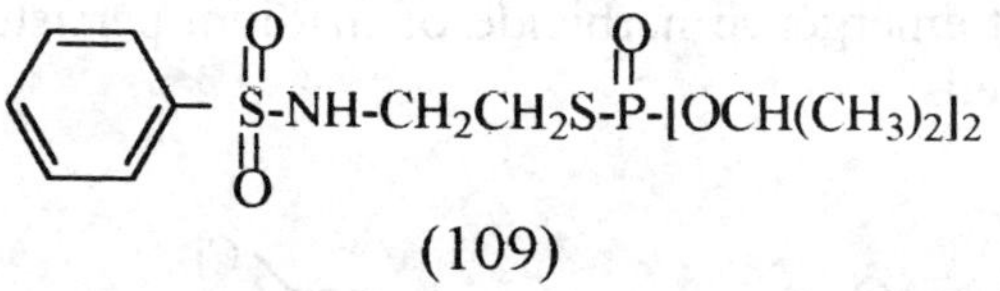

$$(109)$$

Amongst phosphonate group, **ethion**, 2-chloroethyl phosphonic acid (110) and **glyphosine**, N,N-bis (phosphonomethyl glycine) (111), are used as plant growth regulators with no herbicidal action.

$$Cl(CH_2)_2 - P(O) - (OH)_2 \quad ; \qquad [(HO)_2 - \overset{\overset{\displaystyle O}{\|}}{P} - CH_2{-}]_2 - NCH_2COOH$$

$$(110) \qquad\qquad (111)$$

Glyphosate, N-phosphonomethyl glycine (112), can be prepared as follows (Fig. 3.230). It is a non selective, post-emergence foliar spray herbicide, effective against monocot and dicot annual and perennial weeds. This is rapidly translocated from leaves to roots and is moderately toxic to mammals.

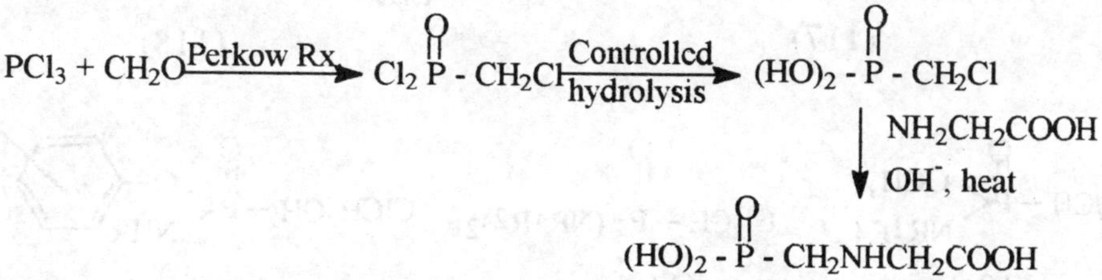

$$PCl_3 + CH_2O \xrightarrow{Perkow\ Rx} Cl_2\overset{\overset{O}{\|}}{P} - CH_2Cl \xrightarrow[hydrolysis]{Controlled} (HO)_2 - \overset{\overset{O}{\|}}{P} - CH_2Cl$$

$$\downarrow \begin{array}{c} NH_2CH_2COOH \\ OH^-,\ heat \end{array}$$

$$(HO)_2 - \overset{\overset{O}{\|}}{P} - CH_2NHCH_2COOH$$

$$(112)$$

Fig. 3.230. Preparation of glyphosate.

Fosamin-ammonium, ammonium ethylcarbamoylphosphonate (113), is a contact herbicide used in autumn before shedding of leaves, in non crop area. It is also used as a shrub killer and is non-toxic in nature.

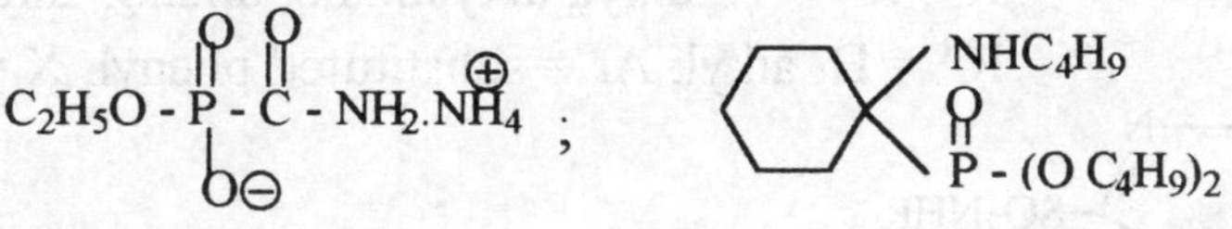

$$C_2H_5O - \overset{\overset{O}{\|}}{\underset{\underset{O^{\ominus}}{|}}{P}} - \overset{\overset{O}{\|}}{C} - NH_2.N\overset{\oplus}{H}_4$$

$$(113)$$

$$\begin{array}{c} NHC_4H_9 \\ \overset{\overset{O}{\|}}{P} - (O\ C_4H_9)_2 \end{array}$$

$$(114)$$

Aminophon, O,O-dibutyl-1-butylaminocyclohexyl phosphonate (114), is also used as defoliant and desiccant.

Isophos-3, chloromethyl O-(2-chloro-4-tolyl)N-sec-butyl phosphonamidothioate (115), is reported to be effective against grassy weeds.

Piperophos, S-(2-methyl piperidinocarbonyl) methyl O,O-dipropyl, phosphorodithioate (116), is a selective herbicide for pre- and post-emergence application in rice crop for the control of monocot annuals.

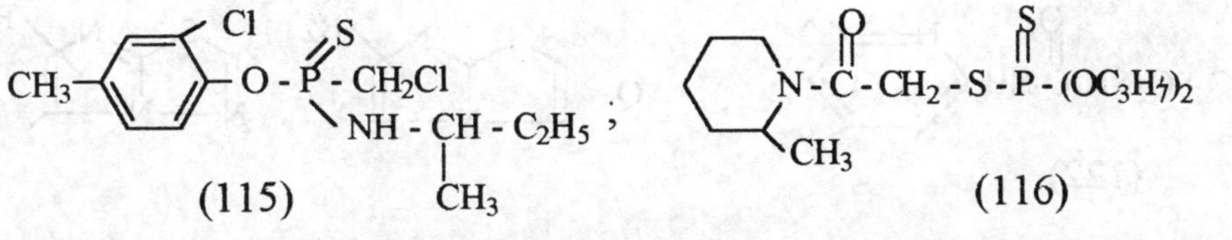

$$CH_3 - \overset{Cl}{\underset{}{\bigcirc}} - O - \overset{\overset{S}{\|}}{\underset{\underset{NH - CH - C_2H_5}{}}{P}} - CH_2Cl$$

$$(115)\qquad CH_3$$

$$\bigcirc N - \overset{\overset{O}{\|}}{C} - CH_2 - S - \overset{\overset{S}{\|}}{P} - (OC_3H_7)_2$$

$$CH_3$$

$$(116)$$

Amiprofos-methyl, O-methyl O-(2-nitro-4-tolyl) isopropyl phosphoramidothioate (117) and **butamifos,** O-ethyl-O-(6-nitro-3-tolyl) sec-butyl phosphoramidothioate (118), are being developed as selective emergence herbicide [8a].

A series of novel phosphonamidates and diamides (119a, 119b, 120) have been synthesized by Khazanchi and Roy [8b] and Das and Roy [8c] respectively which are known to possess moderate herbicidal properties.

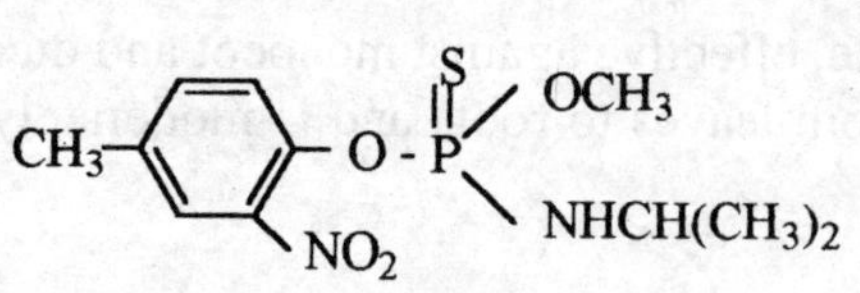

(117)

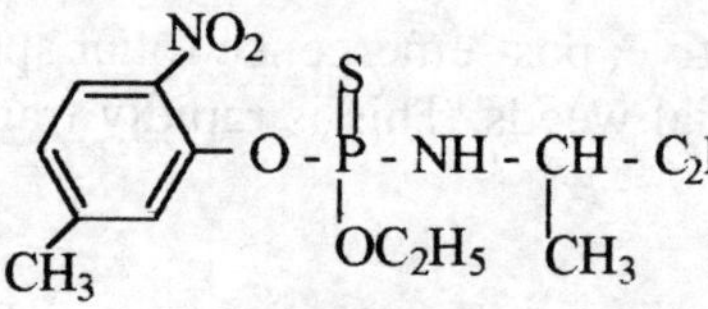

(118)

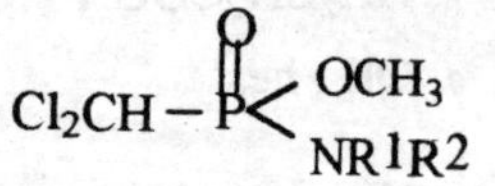

(119a)

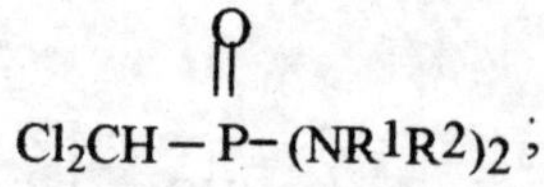

(119b)

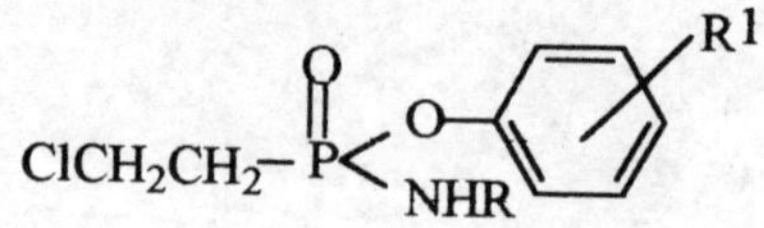

(120)

Heteroaryl-1,2,4-triazole-3 sulphonanilides compounds (121) under glass house conditions, control broad-leaved weeds such as *Viola* spp., *Galium* spp., *Polygonum* spp. and *Solanum* spp. as pre- and post-emergence with application rates between 10 and 300 g a.i./ha.

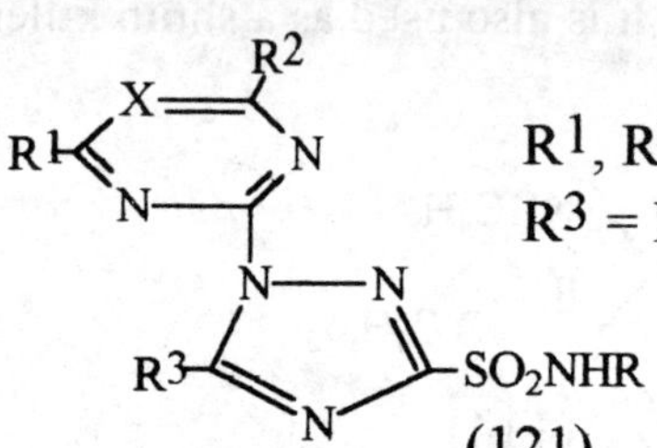

R^1, R^2 = H, alkyl, alkylamino, amino, alkoxy, halogen

R^3 = H, alkyl; Ar = substituted phenyl, X = CH or N

(121)

Sulfonylurea compounds [9, 10, 11a, 11b]

Recently, a new group of selective herbicides, the sulfonylureas have been developed with excellent herbicidal properties both as pre- and post-emergence at very low rate of application. A few most effective ones are discussed and some recent developments are mentioned in Table 3.1.

Novel sulfonylureas containing heterocyclic moieties (122), were synthesized. The compounds containing 2-halogenoimidazo-[1,2-a]pyridin-3-yl groups were very effective for controlling various kind of paddy crop weeds at low application rate without any injury to rice.

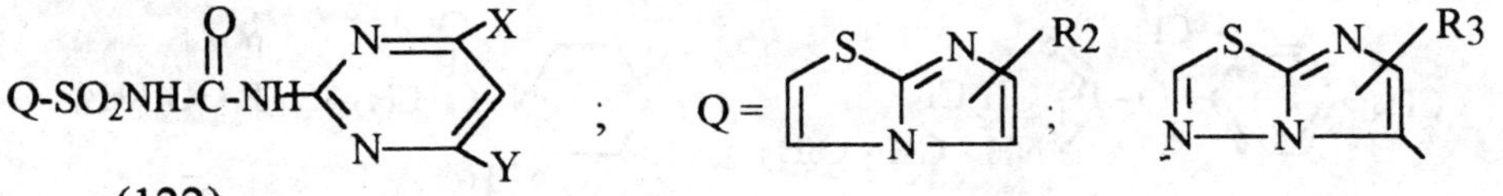

(122)

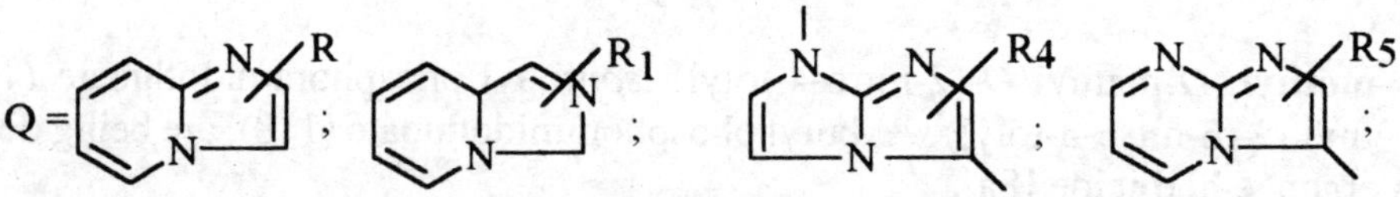

DPX 4189, chlorsulfuron, 1-(2-chlorophenylsulfonyl)-3-(4-methoxy-6-methyl-1, 3,5-triazin-2-yl) urea (123), controls broad leaved weeds in small doses at 10-40 g/ha. It is non volatile and taken up through both foliage and root system. The preparation is as follows (Fig. 3.231).

Table 3.1. Recently developed sulfonylurea herbicides

Sl. No.	Chemical Structure	Common name (Manufacturer)	Trade name	Primary use	Dosage range (g a.i./ha)
1.	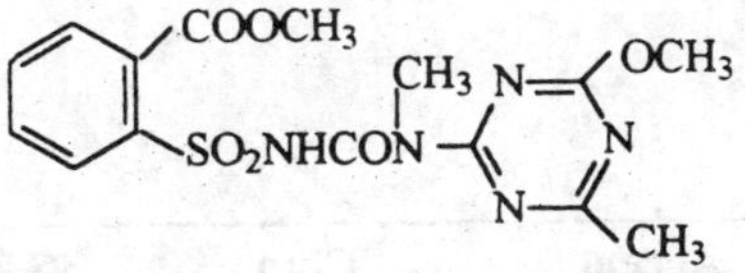	Tribenuron-methyl (Du Pont)	Express, DPX-L5300	Cereals	5-30
2.	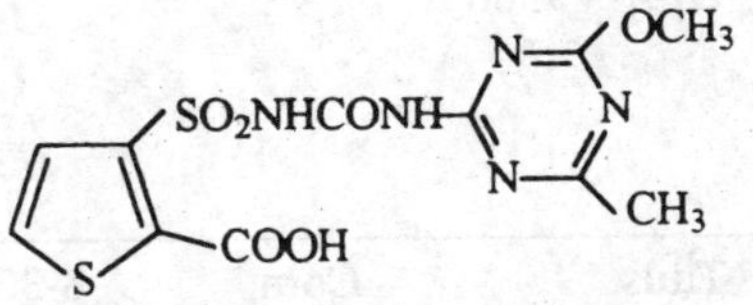	Thifensulfuron, Thiameturon (Du Pont)	Harmony	Cereals	7-12
3.	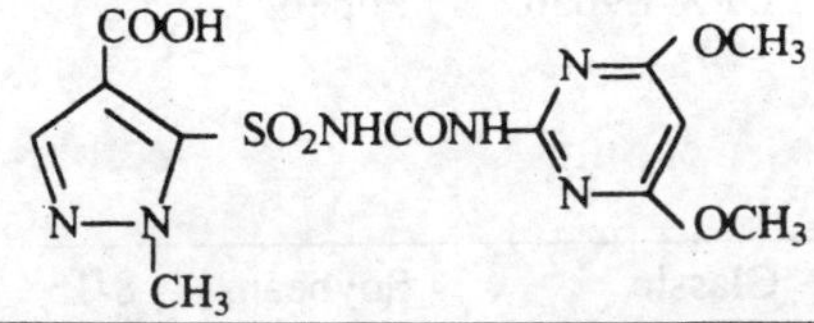	Pyrazosulfuron (Nissan)	Agreen	Cereals	8-15
4	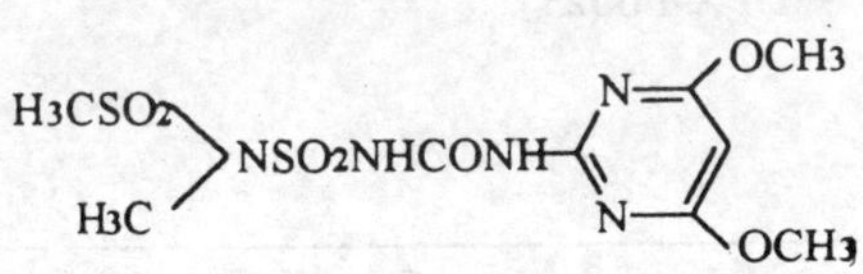	Amidosulfuron (Hoechst)	HOE-075032, Gratil	Cereals	45
5.	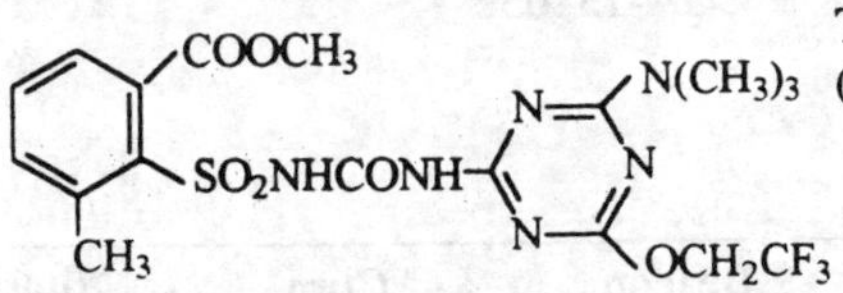	Triflusulfuron-methyl (Du Pont)	DPX-66037	Sugarbeets	10-20
6.	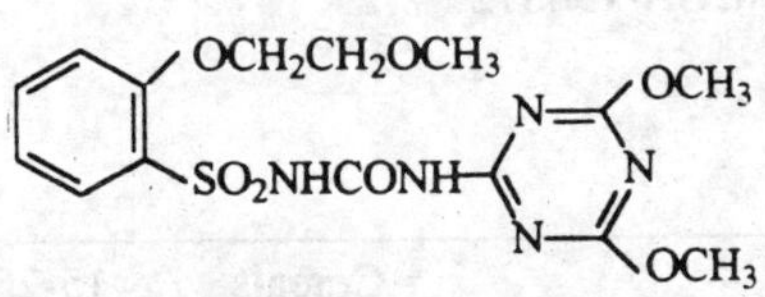	Cinosulfuron (Ciba-Geigy)	Setoff	Cereals	10-15
7.	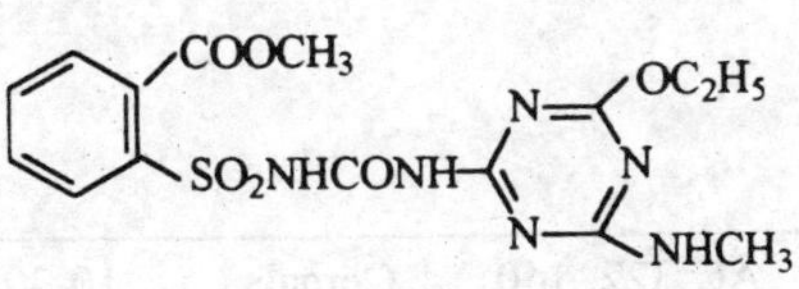	Ethametsulfuron-methyl (Du Pont)	Muster, DPX-A7881	Canoia (oilseed rape)	15-20

(Contd.)

Sl. No.	Chemical Structure	Common name (Manufacturer)	Trade name	Primary use	Dosage range (g a.i./ha)
8.	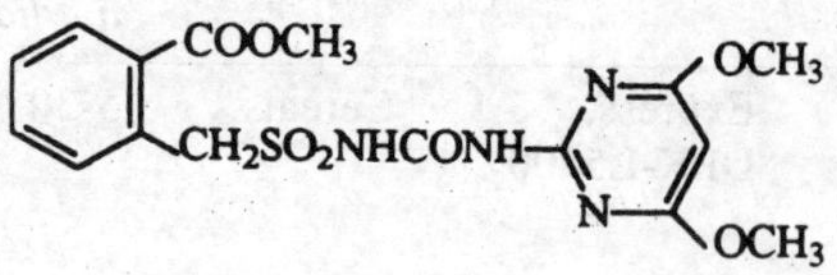 COOCH₃, CH₂SO₂NHCONH— (pyrimidine) OCH₃, OCH₃	Bensulfuron-methyl (Du Pont)	Londax, DPX-F5384	Rice	20-75
9.	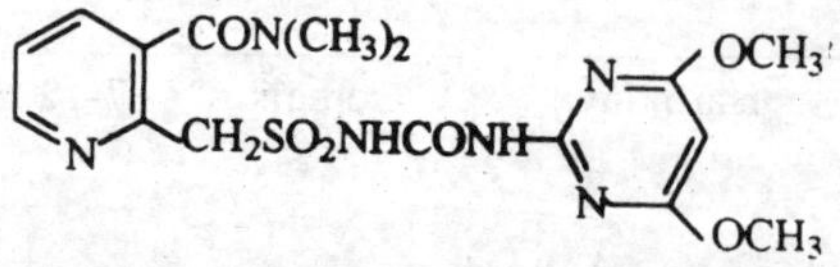 CON(CH₃)₂, CH₂SO₂NHCONH (pyrimidine) OCH₃, OCH₃	Nicosulfuron (Du Pont)	Accent, DPX-V9360	Corn	35-70
10.	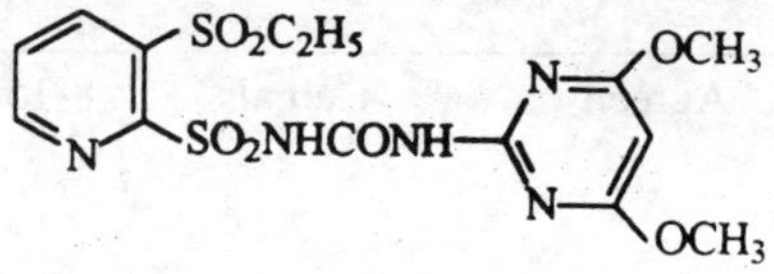 SO₂C₂H₅, SO₂NHCONH— (pyrimidine) OCH₃, OCH₃	Rimsulfuron (Du Pont)	Titus, DPX-E9636	Corn, Potato	8-35
11.	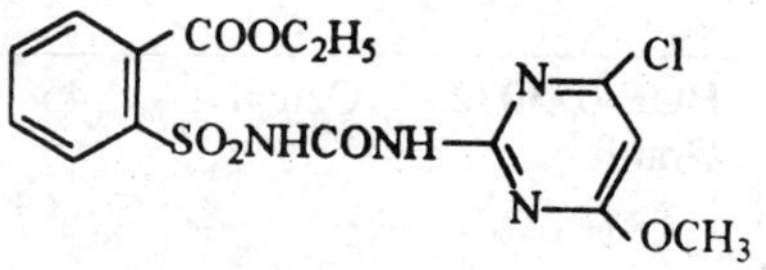 COOC₂H₅, SO₂NHCONH— (pyrimidine) Cl, OCH₃	Chlorimuron-ethyl (Du Pont)	Classic, DPX-F6025	Soybeans	8-13
12.	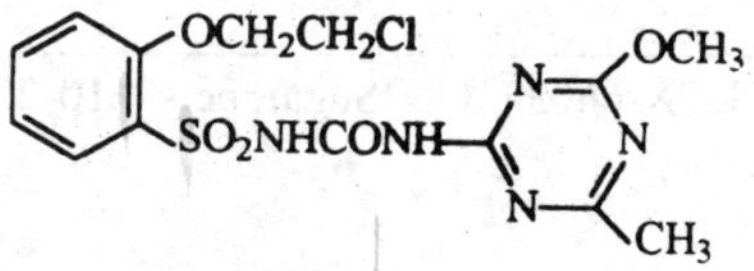 OCH₂CH₂Cl, SO₂NHCONH— (triazine) OCH₃, CH₃	Triasulfuron (Ciba-Geigy)	Amber, CGA-131036	Cereals	10-40
13.	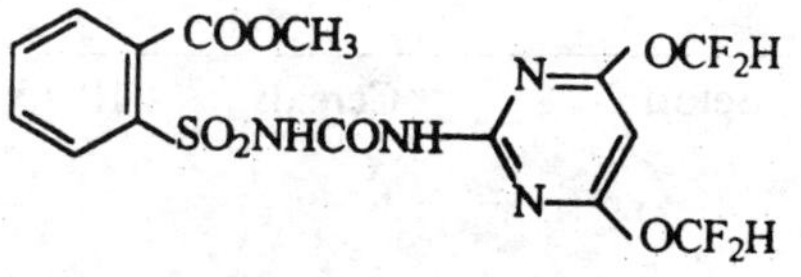 COOCH₃, SO₂NHCONH— (pyrimidine) OCF₂H, OCF₂H	Primisulfuron methyl (Ciba-Geigy)	Beacon, CGA-136872	Corn	20-40
14.	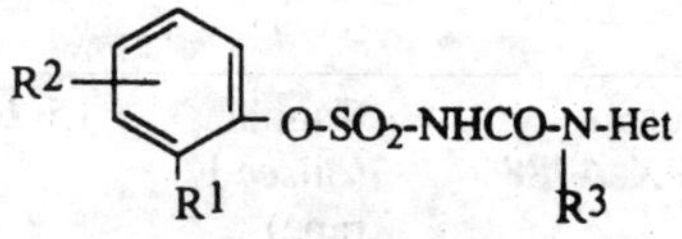 R²—(benzene)—O-SO₂-NHCO-N-Het, R¹, R³. R¹= O-alkyl; R² & R³= H, alkyl, CN. F	Phenoxy sulfonylurea (Hoechst AG)	—	Cereals	15-25
15.	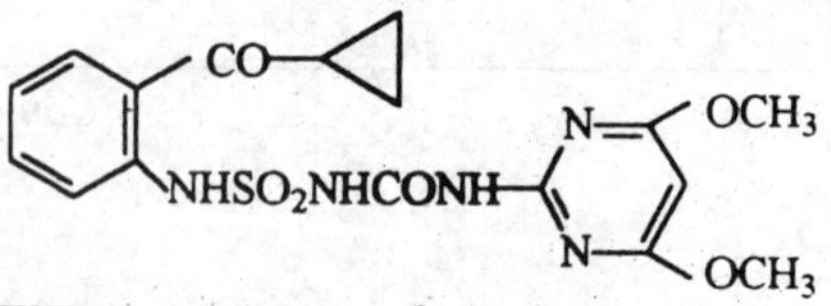 CO (cyclopropyl), NHSO₂NHCONH— (pyrimidine) OCH₃, OCH₃	Cyclosulfamuron	AC-322, 140	Cereals	10-20

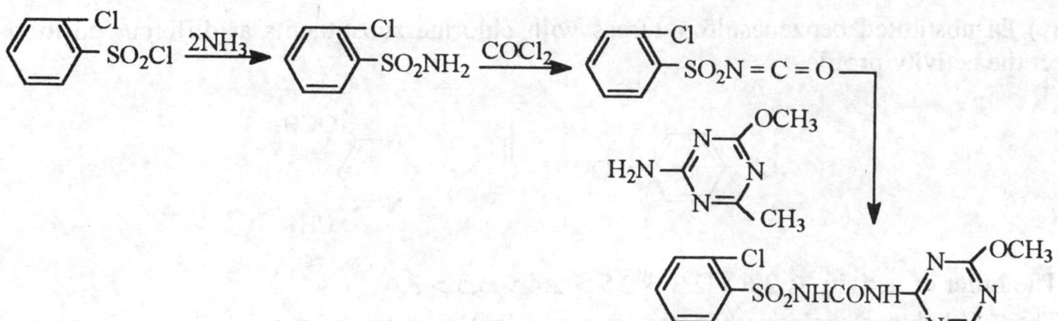

Fig. 3.231. Preparation of chlorsulfuron.

DPX-T 6376, metsulfuron-methyl, 2-(4-methoxy-6-methyl-1,3,5-triazin-2-yl-carbamoyl-sulfamoyl) methyl benzoate (124) and **DPX-T 5648, sulfometuron-methyl,** 2-(4,6-dimethyl-pyrimidin-2-yl-carbamoylsulfamoyl) methyl benzoate (125) were discovered for use in spring cereals to control a number of weeds with the advantage of low rate of application.

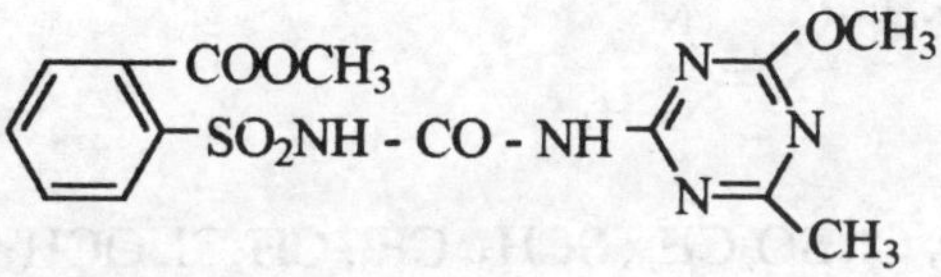

SAR studies on sulfonylurea herbicides were carried out to investigate the activity of these type of molecules.

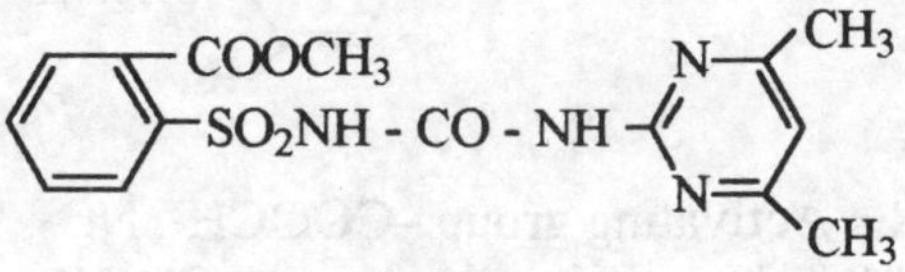

i) Effect on activity with no substitution in the phenyl ring but variation in X and Y of heterocyclic moiety.

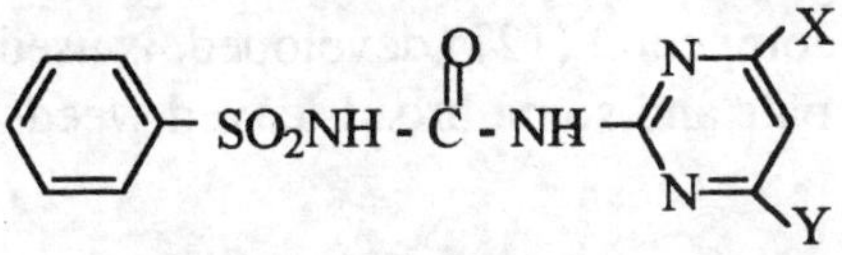

X	Y	Activity profile
CH$_3$	Cl	Growth retardant
Cl	Cl	Inactive
CH$_3$	CH$_3$	Improved growth retardant

ii) Change in monosubstitution in the phenyl ring resulted in change in the activity profile.

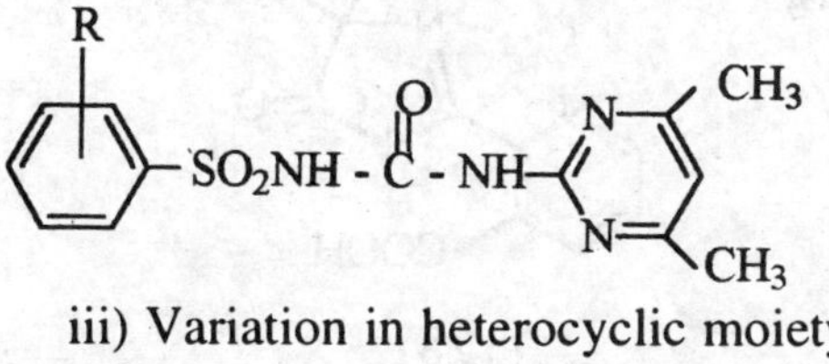

R	Activity profile
4-CH$_3$	Inactive
3-CH$_3$	Active herbicide
2-CH$_3$	Highly active herbicide

iii) Variation in heterocyclic moiety affects the activity profile

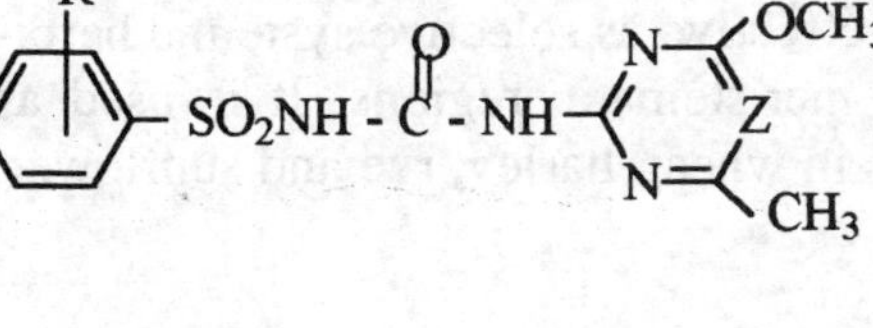

R	Z	Activity profile
H	N	Broad leaf herbicide—wheat safety
Cl	CH	Highly active herbicide, phytotoxic to wheat
Cl	N	Chlorsulfuron—highly active and safe in wheat, barley

iv) Disubstituted benzenesulfonylureas with chlorine substituents at different positions also affect the activity profile,

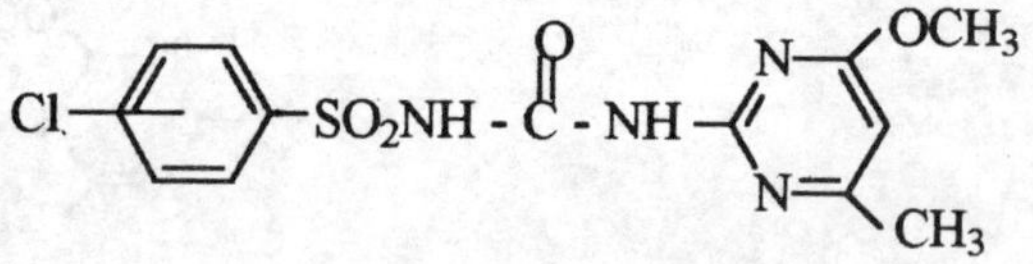

The order of activity is 2,6 > 2,3 > 2,5 > 2,4 > 3,5 > 3,4

The 2,6-dichloro analogue had almost equivalent activity compared to the 2-chloro analogue (monosubstituted product). A decrease in activity is observed if the second Cl-atom from 6 position is shifted to 3,5 and 4, position. When both the Cl-atoms are shifted to 3,5 and 3,4 positions, more reduction in activity is observed.

Many functional groups in the molecule can potentiate activity if present in 2 position. A variety of electron withdrawing and electron donating groups have exhibited potentiating effect.

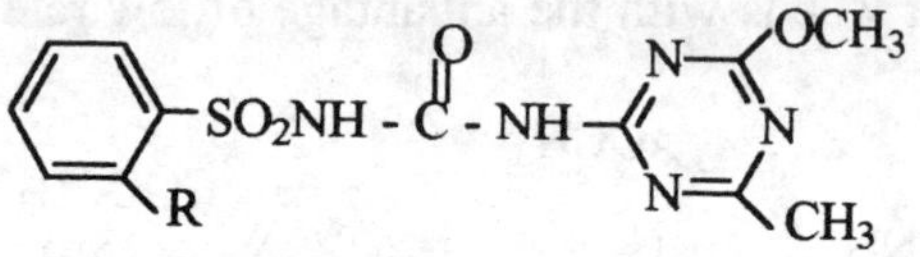

where, R = Activating group $-COOCH_3$, NO_2, F, Br, Cl, SO_2CH_3, SCH_3, CF_3, CH_2Cl, OCH_3, OCF_3, CH_3 and R = Non-activating group $-COOH$

Miscellaneous compounds [12a]

Imidazolinones

The important compounds of this group are **imazaquin** (126), **imazethapyr** (127) and **imazapyr** (128) which have been found effective as pre- and post-emergence herbicides to control grasses and dicot weeds in soybeans and other leguminous crops. The compound (127) developed, is used as a selective post-emergence herbicide effective against wild oats and some broad-leaved weeds in barley, wheat and sunflower crops.

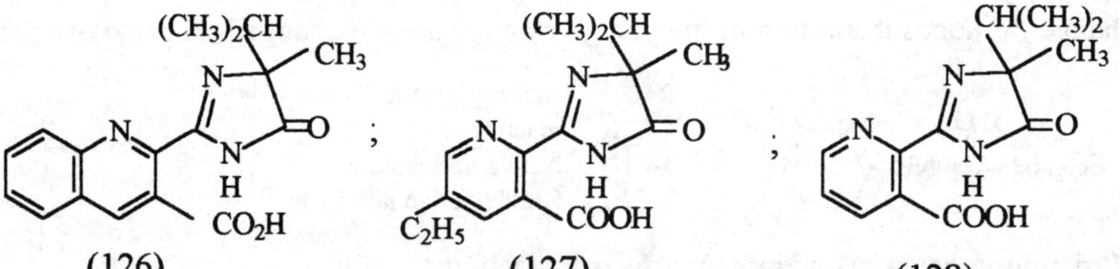

(126) ; (127) ; (128)

Imazamethabenz, AC 263810 was developed and found effective as selective systemic herbicide which is absorbed through roots and translocated to meristematic regions. It is used as post-emergence to control *Avena* spp., and some dicot weeds in wheat, barley, rye and sunflower. It is a mixture of two isomeric derivatives (129, 130).

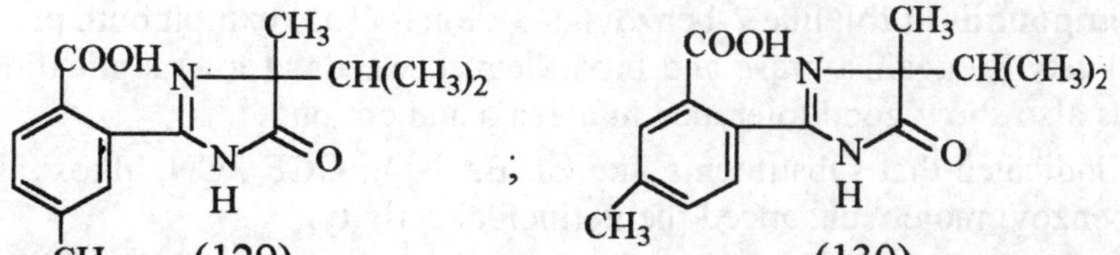

(129)   ;   (130)

R-4024-4, fluorochloridone, 1-(3-trifluoromethyl)phenyl-3-chloro-4-chloro-methyl-2-pyrrolidinone (131), is a novel pre-emergence herbicide used against broad-leaved weeds in number of crops. It is a powerful inhibitor of carotenoid synthesis having low toxicity to mammals and exhibits good selectivity in carrots, potatoes, sunflowers and winter wheat.

Bentazone, Basagran, 3-*iso*propyl-1*H*-2,1,3-benothiadiazin-4(3*H*)-one 2,2-dioxide (132), is a contact selective herbicide for post-emergence use mainly against dicot weeds. Many graminaceous crops such as rice, maize, barley and soybean are resistant to bentazone. However, broad-leaved weeds and sedges are sensitive and moderately toxic. But the activity of bentazone can be enhanced by mixing with hormone like herbicides MCPB.

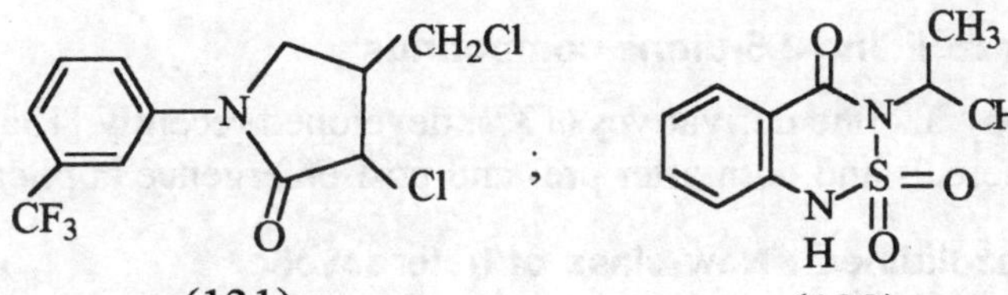

(131)   (132)

Cyclohexane derivatives

It has been reported that the substituents in cyclohexane ring and of the side chain strongly influence herbicidal activity and selectivity. The most effective compound, is **allyloxydim sodium**, methyl 3-(1-allyloxyimino)butyl-4-hydroxy-6,6-dimethyl-2-oxocyclohex-3-ene-carboxylate (133), is a pre-emergence herbicide used for the selective control of grass weeds in dicot crops such as sugar beet, soybean, peas, cotton and peanut. It is moderately persistent and slightly toxic to mammals.

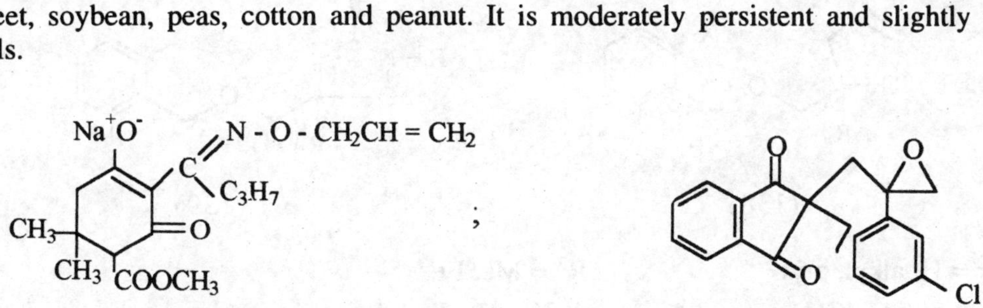

(133)   ;   (134)

A new herbicide **MK-243** (134), has been reported recently for control of annual grass weeds in transplanted rice in Japan. It has good selectivity to control weeds such as *Echinochloa crus-galli* with pre- and post-emergence application.

Triketone compounds [12b], like 2-benzoyl-1-3-diones (135) exhibit both pre- and post-emergence herbicidal activity against grass and broad-leaved weeds at low application rate. Some of these compounds also show good tolerance in cereals and cotton.

SAR studies indicated that substituents like Cl, Br, NO_2, OCF_3, CN, alkoxyalkyl groups at 2 position of the benzoyl moiety enhanced the herbicidal activity.

N-phenyltetrahydrophthalimides [12c], exhibit both pre- and post-emergence activity at low rates. The most active molecule **flumipropyn, S-23121**, 2-[4-chloro-2-fluoro-5-[(1-methyl-2-propynyl)oxy]phenyl]4,5,6,7-tetrahydro-1-*H*-*iso*indole-1,3(2*H*)-dione (136) controls most broad-leaved weeds in wheat at 10-20 g a.i./ha by early post-emergence.

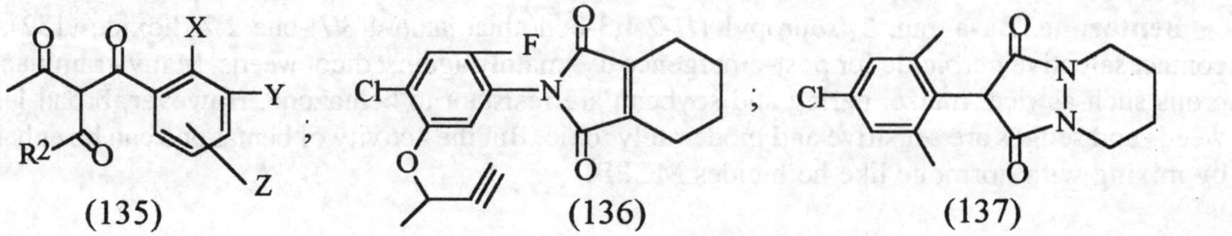

(135) ; (136) ; (137)

CGA 2712112, Aryl pyrazolidine-3,5-dione compounds

A series of 4-aryl-5-oxopyrazoline derivatives (137), developed recently [13a] have proved active against grass weeds in cereals and corn after pre- and post-emergence application.

Isoxazolines and isoxazolidines : New class of heterocyclic benzyl ether compounds [13b]

Isoxazolines (138) and isoxazolidines (139) derivatives developed were found to exhibit good herbicidal activity (136).

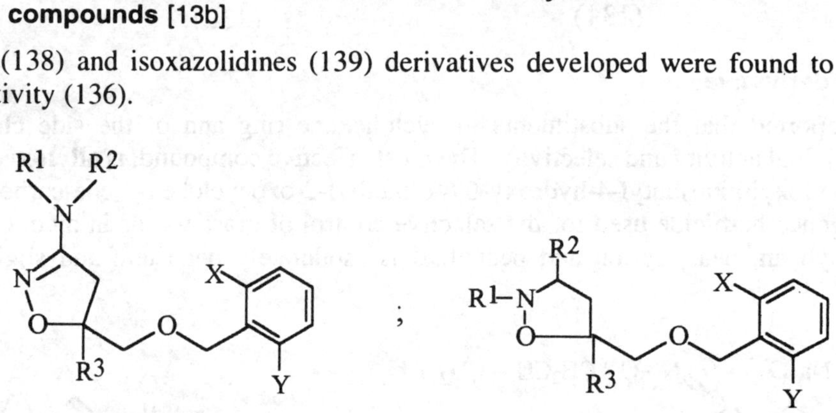

(138) ; (139)

R^1, R^2 = H, alkyl, aryl R^3 = Me, Et.
R^1, R^2 = alkyl, -$(CH_2)n$, (n=3,4)   X,Y = H, alkyl, halogen

Pyrimidine compounds [13c]

A series of pyrimidine compounds (140, 141) synthesized showed good pre- and post-emergence herbicidal activity towards soybean and barnyard grass in paddy fields.

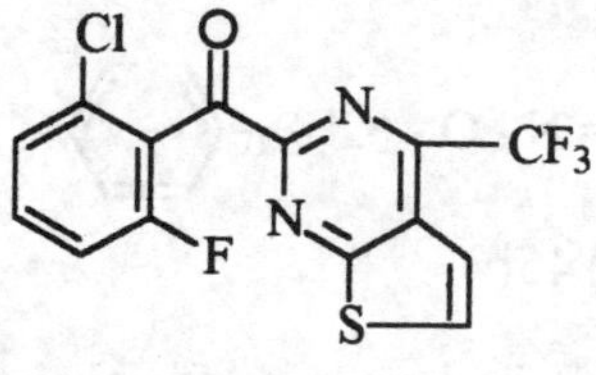

(140) 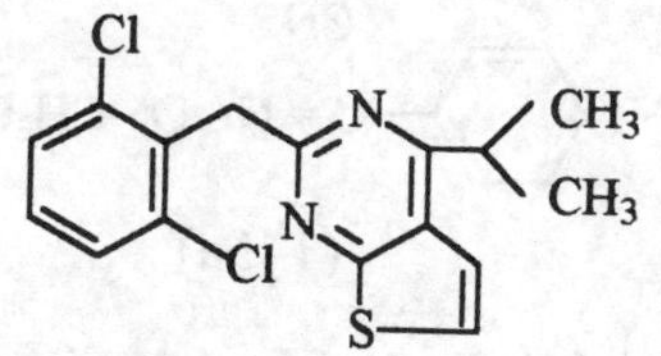   (141)

Bioherbicides

Herbicides of microbial origin have been discussed in Chapter 2.

Herbicide safeners/protectants/antidotes [14a, 14b]

The term antidotes or safeners is used to enhance the tolerance of the crop towards herbicides and protect the plants against damages by the action of herbicide that could otherwise be toxic. These are thus used to treat crops before or along with the herbicide application. The antidotes or safeners provide an opportunity of increasing the crop selectivity of many existing non selective herbicides. A large number of such compounds have been synthesised which depending upon the nature of the compound can be divided into following groups.

i) **Naphthopyranone compounds : Naphthalic anhydride** (NA) (142) and **phthalic anhydride** (PHA, 143) provide protection to maize crop plants against increased doses of EPTC or other thiocarbamate herbicides. These also effectively protect other crops like sorghum against alachlor, rice against molinate and oats against barban.

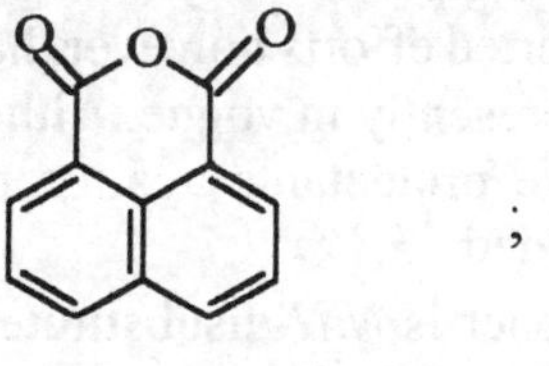

(142) 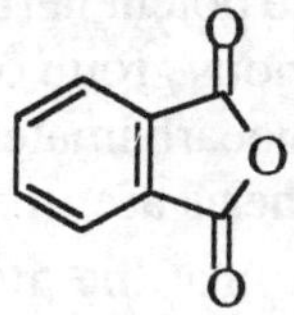   (143)

ii) **Chloroacetamide** compounds such as **allidochlor** or CDAA, (144a) and **dichlormid DCCA** (144b) possess, the dichloroacetamide moiety which affords maximum safening activity in corn, sorghum and rice against injury from chlorosulfuron.

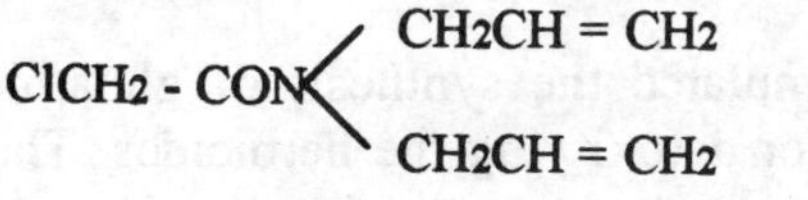

(144a) 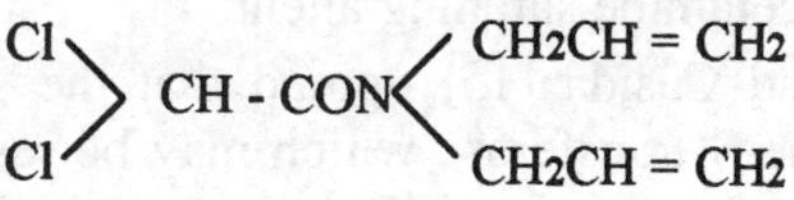   (144b)

iii) **Oxime ethers**, like CGA-43089, **cyometrinil** (145a), and the pyridine aldoxime ether (n = 1 or 2) (145b) similarly, protect sorghum crop against metolachlor.

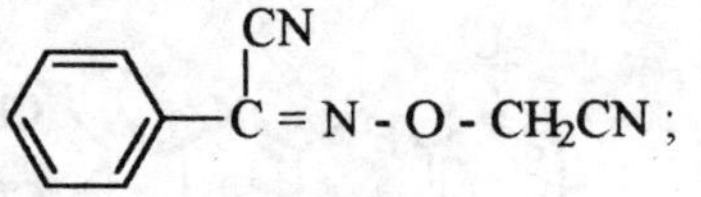

(145a) 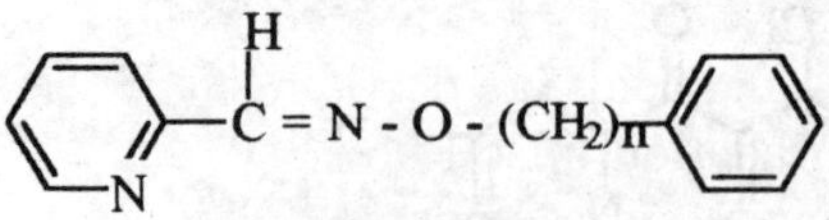   (145b)

iv) **Flurazol**, 2,4-disubstituted 5-thiazolecarboxylate (146) is the most effective antidote compound is used to save crops damaged by herbicides. A large number of fungicides have also been tested but only a few are found effective [14c].

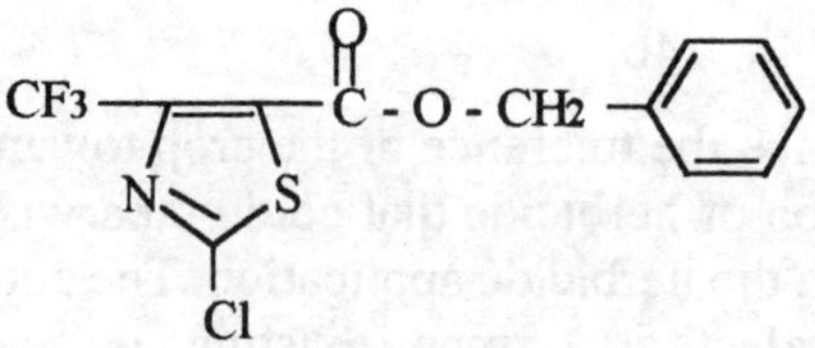

(146) 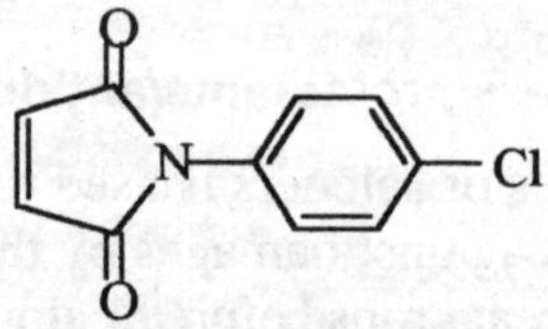   (147)

Prosafeners are also used because after biotransformation these chemicals exert safening action. An example of such compounds are N-arylmaleimide compound CPMI (147) and *iso*maleimide *in vivo* are hydrolysed to maleamic acid derivative which acts as safener in sorghum against alachlor.

The concept of crop safeners for herbicides has opened up a new era in chemical weed management. They allow existing chemicals to be applied safely at higher doses thus permitting wider application of more potent herbicides. Concerted efforts however, have to be made to develop new safeners against various group of herbicides presently in vogue. Although quite a few safeners are available against thiocarbamates but effective protectants against paraquat and glyphosate, which impair photosynthesis are immediately needed.

SAR studies indicate that the most active safener is N,N-disubstituted-2, 2-dichloroacetamide and converse is true for monochloroacetamides. The substituents on the nitrogen atom including alkyl, haloalkyl, alkenyl and heterocyclic groups impart various degrees of protective activity. Usually, compounds having disubstitution on the nitrogen atom are more active than monosubstitution.

The precise information about the mode of action of safeners is still not clear. However, it is believed that the structural and chemical similarity might exist between thiocarbamates and the dichloroacetamide safening agent.

Lay and Casida [15] opined that the safener stimulated the synthesis of glutathionate and glutathione S-transferase which may be responsible for detoxifying the herbicides. The mode of action of safeners with sulfonylurea herbicides in wheat and maize may be associated with accelerated metabolism induced by the herbicidally inactive degradation products. Experiments showed that safeners enhanced the rate of metabolism of chlorsulfuron in wheat and corn while maintaining moderate efficiency against the sensitive broad-leaved weeds. A cytochrome P-450 enzyme system is probably the reason for metabolic inactivation of sulfonylurea group of herbicides.

Metabolism

Phenoxy ethanols and acids are decomposed by enzymatic β-oxidation process to inactive phenol (Fig. 3.232). Degradation in soil however is microbe mediated which in turn is regulated by the organic matter content, pH, temperature and water content of the soil (Fig. 3.233).

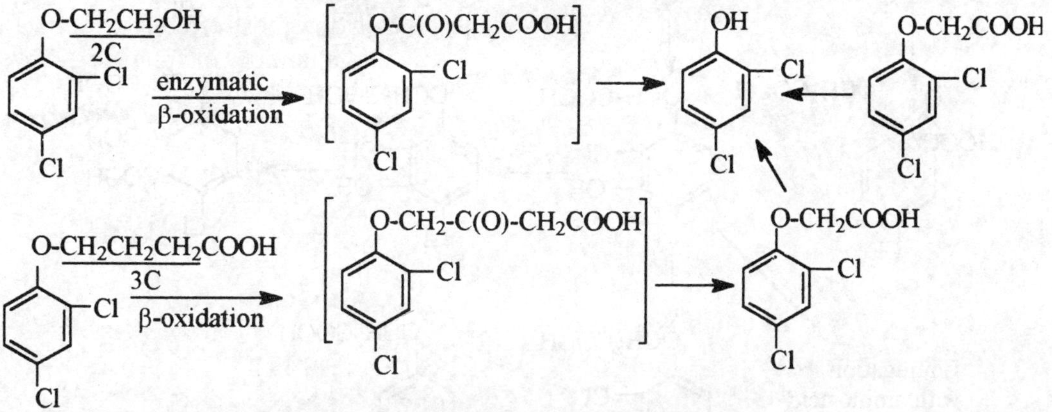

Fig. 3.232. Metobolites from phenoxyethanols/alkanoic acids.

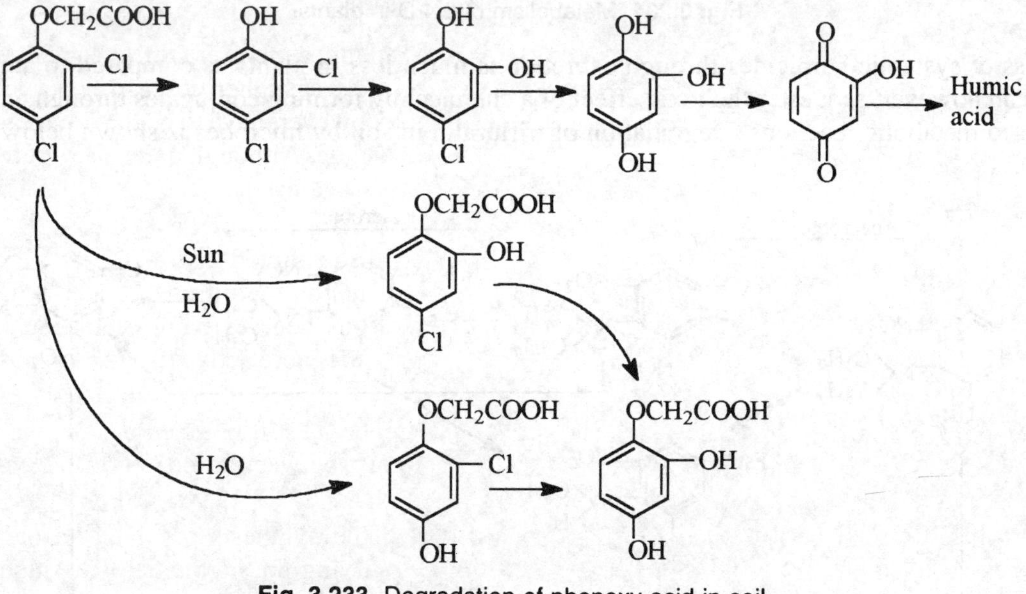

Fig. 3.233. Degradation of phenoxy acid in soil.

Certain soil microorganisms work however selectively. Some may decompose 2,4-D but not MCPA and 2,4,5-T. Of the phenoxy acetic acids, 2,4,5-T- is degraded slowly but the higher alkanoic acids are more stable than 2,4-D. Metabolism of 2,4-D in plants proceeds via hydroxylation. The ether bond in 2,4-D remains intact even after cleavage of the ring (Fig. 3.234).

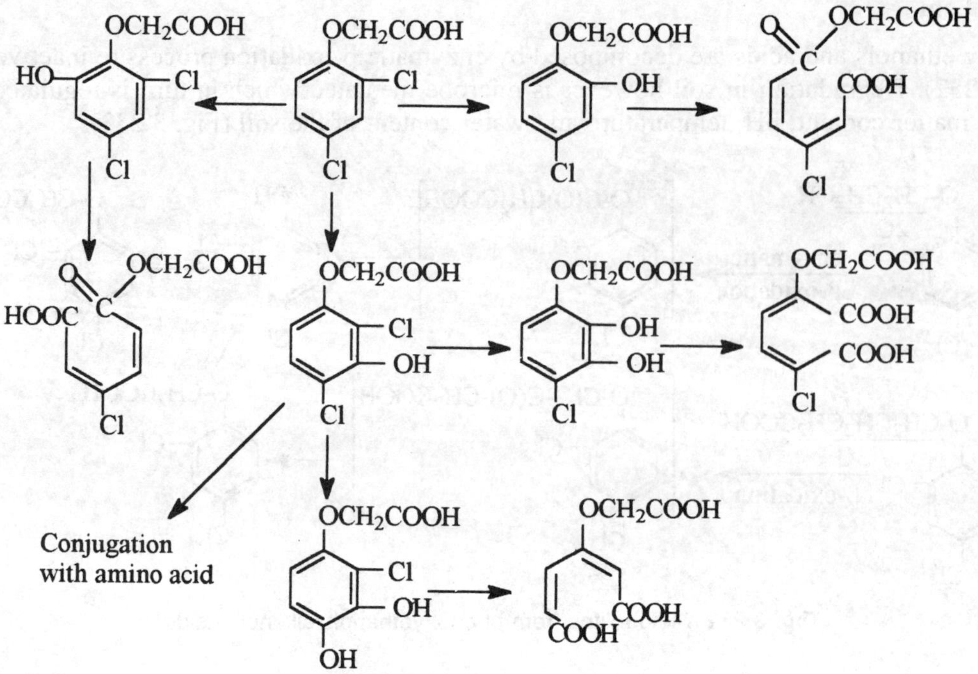

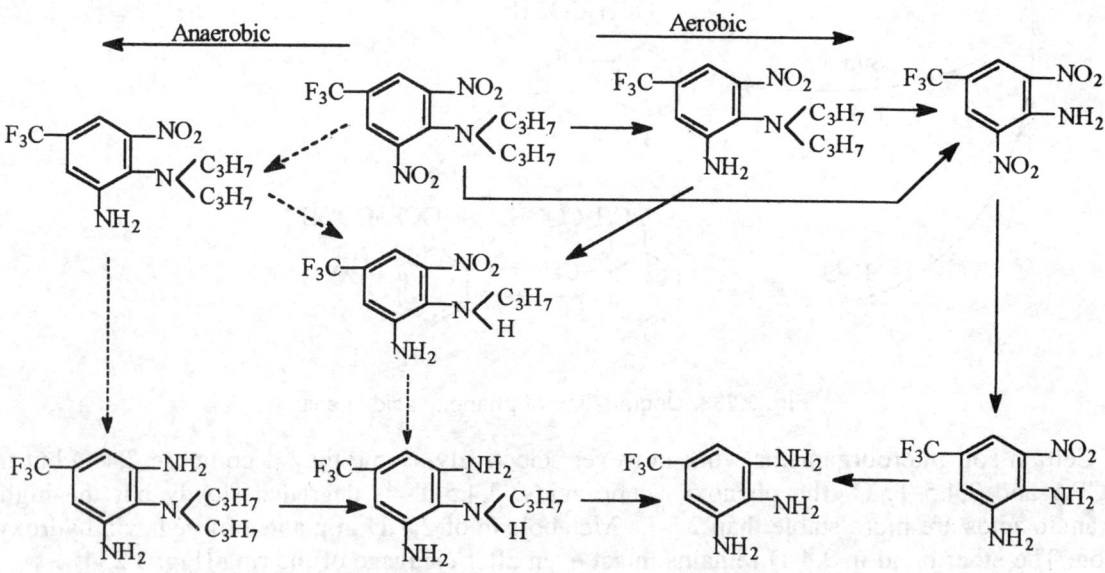

Fig. 3.234. Metabolism of 2,4-D in plants.

Loss of systemic herbicides through excretion is much less in plants as compared to animals. They can however, sequester the toxic effect of a chemical by forming conjugates through enzyme catalysed metabolic reactions. Degradation of trifluralin in soil by microbes is shown below (Fig. 3.235).

Fig. 3.235. Degradation of trifluralin in soil.

The metabolism of **pendimethalin** in plants and soil involves the oxidation of the 4-methyl group in benzene ring is oxidised to carboxylic acid via alcohol, reduction of nitro groups, cyclisation and conjugation [Zulian, 16]

The microbial degradation of **chloropropham** is accomplished as shown in (Fig. 3.236).

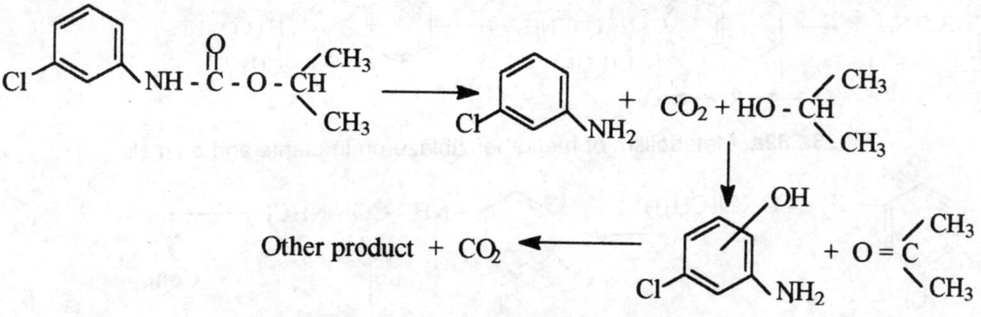

Fig. 3.236. Microbial degradation of chloropropham.

The metabolism of **benthiocarb**, yields a number of degradation products (Fig. 3.237).

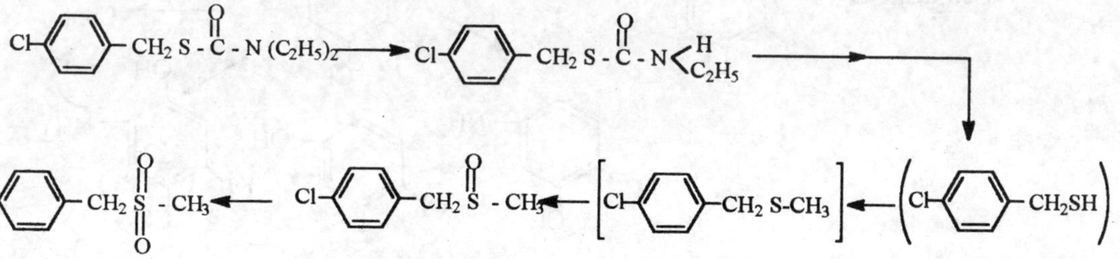

Fig. 3.237. Metabolites of benthiocarb.

The **urea compounds** used as soil herbicides are readily degraded and metabolised by the soil microflora. The microbial degradation of 3-aryl-1,1-dimethyl urea proceeds by stepwise demethylation into 3-aryl ureas which is finally degraded into aryl amines, CO_2 and NH_3 (Fig. 3.238).

$$R - NHCON(CH_3)_2 \longrightarrow R - NH\ CON \Big\langle {{CH_3} \atop H} \longrightarrow R - NHCONH_2 \longrightarrow R - NH_2 + CO_2 + NH_3$$

Fig. 3.238. Degradation of dimethyl urea compound.

Methabenzthiazuron is metabolised in plants of higher order in a different way than N-phenyl urea derivatives. A stable hydroxymethyl MTB is formed which is then gradually demethylated at the urea moiety. Moreover, conjugated products are formed with glucose from which the methabenzthiazuron metabolite can be liberated by metabolic process (Fig. 3.239a).

Diuron in animal system gives a number of metabolites (Fig. 3.239b).

The metabolism of **atrazine** in maize, is due to enzymatic inactivation which involves conjugation of atrazine with glutathione to form glutamyl-S-cysteine derivative (Fig. 3.240).

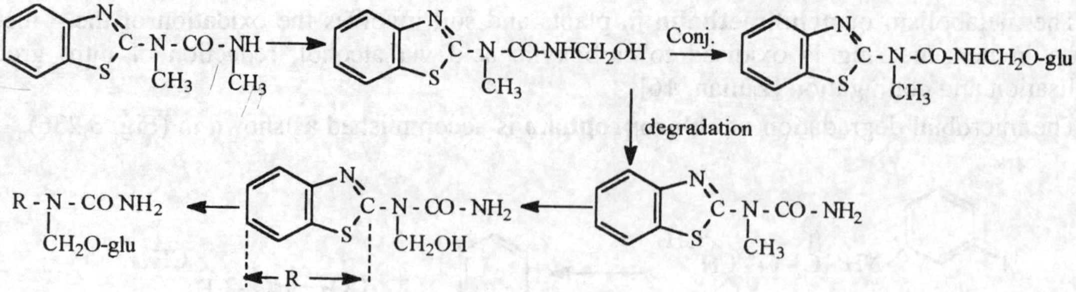

Fig. 3.239a. Metabolism of methabenzthiazuron in plants and animals.

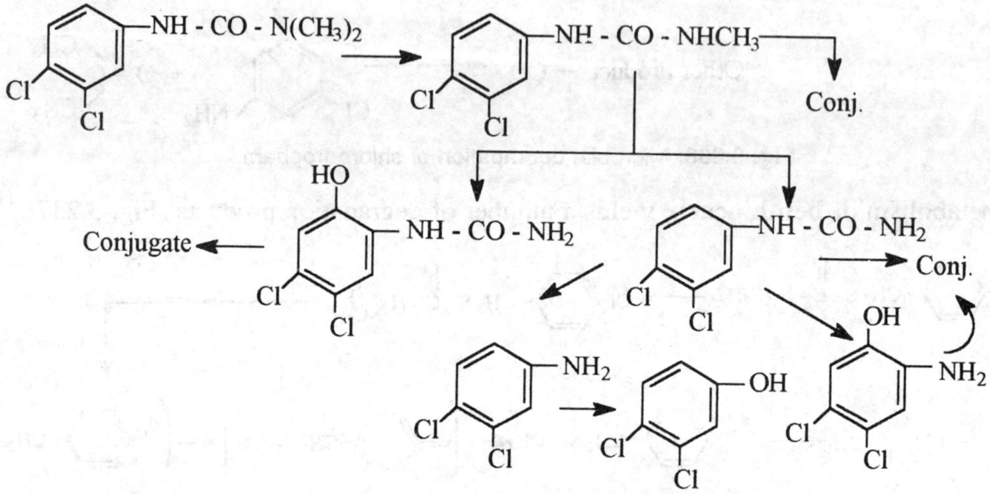

Fig. 3.239b. Metabolism of diuron in animals.

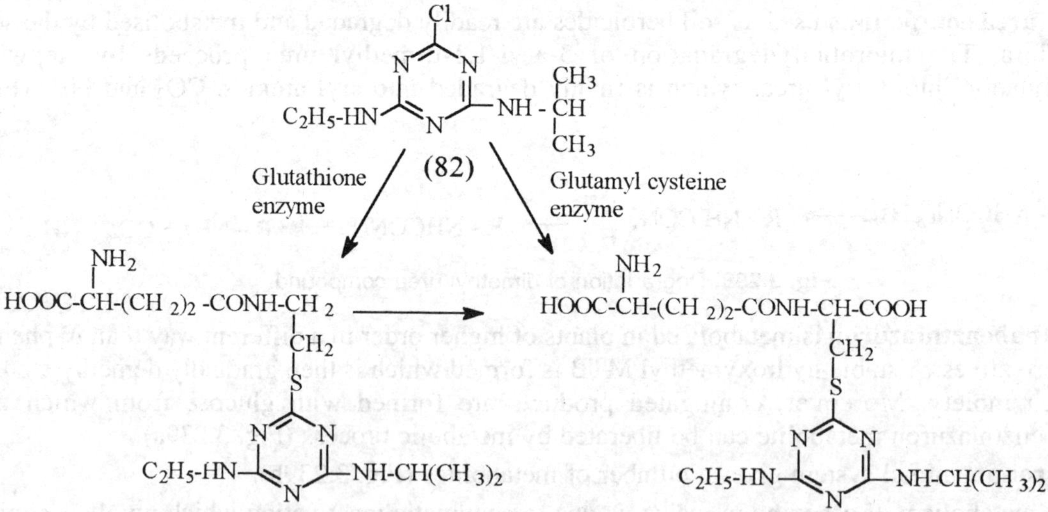

Fig. 3.240. Metabolism of atrazine.

In soil atrazine gives three phytotoxic metabolites (i, ii, iii) and is finally micro biologically degraded to cyanazine amide (iv).

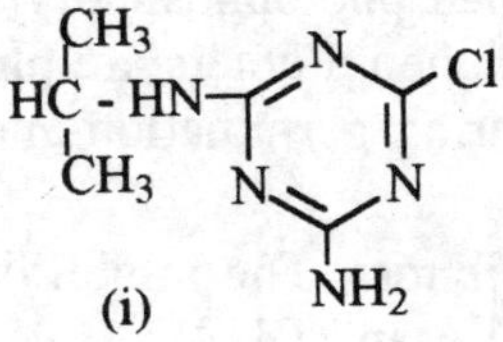

(i) 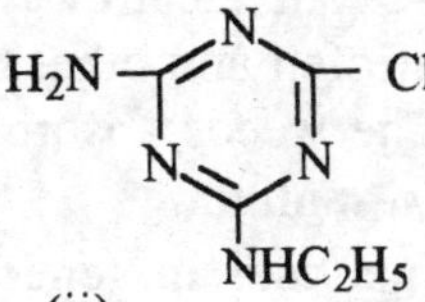 ; (ii)

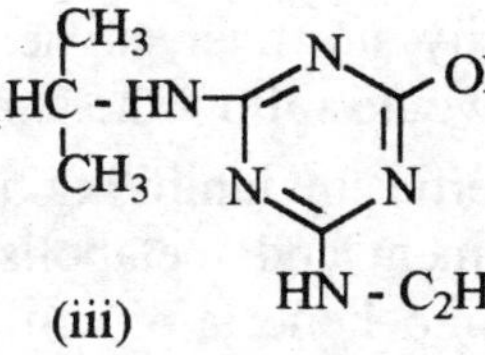

(iii) 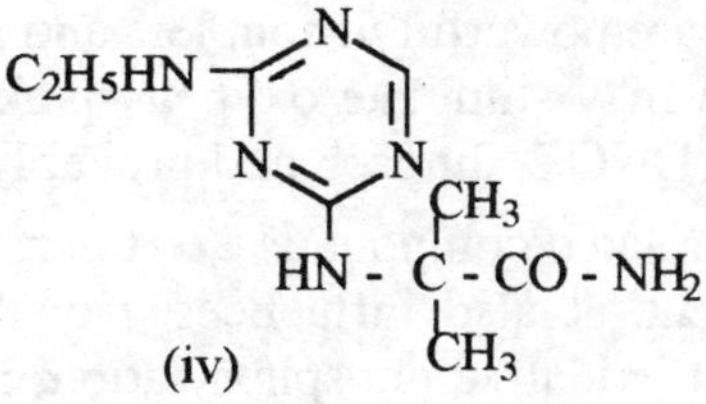 ; (iv)

Mode of action

Several theories have been propounded on the mode of action of phenoxy-alkanoic acids [17]. Phenoxy acids enter the plant through foliage, stem and roots and translocated through phloem and xylem. They affect almost all the biochemical processes of plants. The first manifestation is abnormal growth of the axis, followed by reduction in root and shoot growth, wilting and finally collapse of the plant.

The PGR auxin action of phenoxy acids is due to changes in water and mineral balance, vitamin, oil content, respiration photosynthesis, nitrogen and phosphorus metabolism and in enzyme system. The modern biochemical theory on the phenoxy acids action relates to its effect on nucleic acid synthesis and enzymes. Protein synthesis is regulated by enzymes, the necessary formation of which is regulated by repressors according to the biorhythm. The synthesis and functioning of enzyme starts when plant hormones stop the repressor effect. This increases the RNA level of the cells and DNA fragments. The herbicides with auxin action interfere at the gene level in the specific protein synthesis of plants. In the case of phenoxyethanol, the action starts after its oxidation by the enzyme to the active acetic acid (Fig. 3.241).

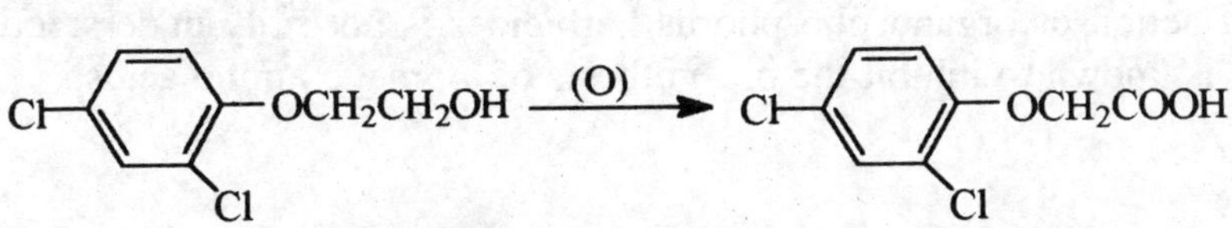

Fig. 3.241. Oxidation of phenoxy ethanol to acid.

Studies on the mode of action of fluazifop-butyl have shown that it moves in the xylem and the phloem showing that its effect does not depend on the growth stage of grasses. After treatment with the herbicide, there is a cessation of growth within 48 h while the nodes and buds become necrotic. Young leaves also show chlorosis, and necrosis, followed by discoloration of older leaves. Death of weeds occurs usually after three weeks. Allidochlor inhibits the germination in grass weeds, root and shoot growth. It also inhibits the functioning of certain enzymes, protein synthesis and respiration. The biological mode of action of alachlor is probably the inhibition of protein synthesis. Propyzamide inhibits mitotic cell division and also causes increase in DNA, RNA protein and cellulose activity. The nitrodiphenylethers interfere with the photosynthetic system of plants. They impair the non-cyclic electron transport and coupled photophosphorylation in chloroplasts and mitochondria. Nitrofen and other orthosubstituted diphenyl ethers are phytotoxic when these are applied to soil. The weeds absorb the chemicals during germination and elongation but die soon after exposure to sunlight.

Dinitroaniline compounds in general are growth inhibitors. The inhibition of mitosis in both mono and dicot plants affecting the synthesis of both RNA and DNA.

Dinitrophenols are powerful uncouplers and interfere with oxidative phosphorylation in plants. Low concentration may stimulate oxygen uptake temporarily, however, higher concentration results in inhibition. DNOC, dinoseb and probably dinosam would follow the similar mechanism.

Trifluralin and related compounds exert herbicidal properties by inhibiting both root and shoot growth development. It also influences growths development and metabolism. It inhibits the oxygen uptake and oxidative phosphorylation in isolated mitochondria of corn and, soybean and interferes with RNA and DNA synthesis which upsets the nucleic acid balance.

The triazine herbicides, like the substituted phenyl urea, acylanilines and carbamates, are classified as powerful inhibitors of photosynthesis by interrupting the light driven flow of electrons from water to nicotinamide adenine dinucleotide phosphate (NADP). Some of the triazine derivatives drastically inhibit the Hill reaction involving the evolution of oxygen from water. Chlorophyll is thought to be the principal pigment involved in this triazine phytotoxicity in plants which increases with the light intensity. The carbamate herbicides inhibit mitosis, oxidative phosphorylation and synthesis of ATP, RNA and protein. As a result abnormal cells are formed in sensitive plants which ultimately prove fatal. Diquat and paraquat after absorption through the plant roots cause severe stunting and chlorosis of the aerial parts. Diquat penetrates the leaf and enters the tissue during darkness causing only minimal herbicidal effects however, continued illumination exposure leads to localised concentration of the herbicide causing rapid tissue damage. The phytotoxic effects of pyridine derivatives include epinasty, plugging and browing of xylem vessels, wilting, necrosis and finally death. Picloram also interfers with the basipetal movement leading to the death of plants due to lack of food. The high effectivness of this herbicide may be due to its high mobility and resistance to break down within the plant.

The mode of action of organosphosphorus herbicides is not fully understood [Jawarski, 18], only glyphosate is known to inhibit the biosynthesis of aromatic amino acids.

Photochemistry

Photodecomposition of phenoxy-alkanoic herbicides, extensively used in agriculture is very important from safety view point. A number of hydroxylated photoproducts (i, ii, iii & iv) were isolated either by cleavage or replacement of ring chlorines when sodium salt of 2,4-D was irradiated with UV light at 254 nm (Fig. 3.242).

Photolysis of anilides

A number of products were identified after irradiation of propanil. These include hydroxylated propanil, dichloroaniline and tetrachloroazobenzene (Fig. 3.243).

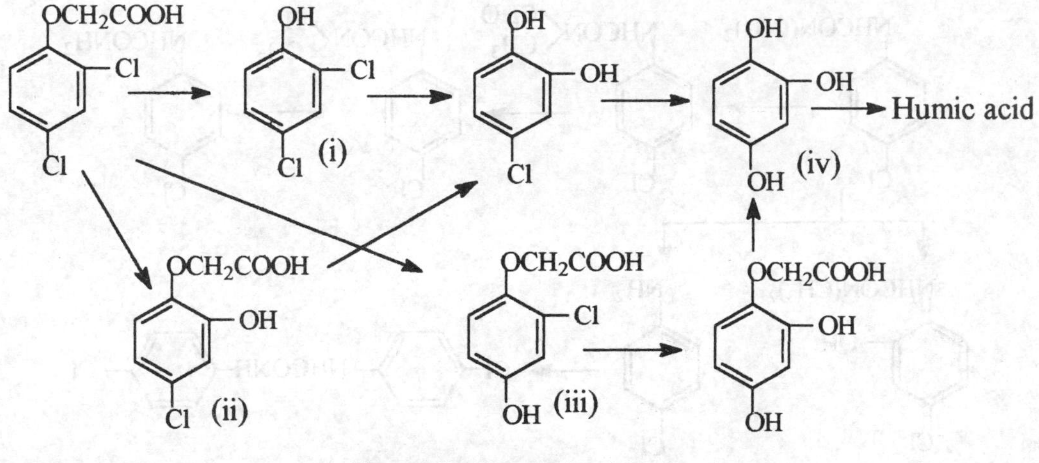

Fig. 3.242. Photolysis of 2,4-D.

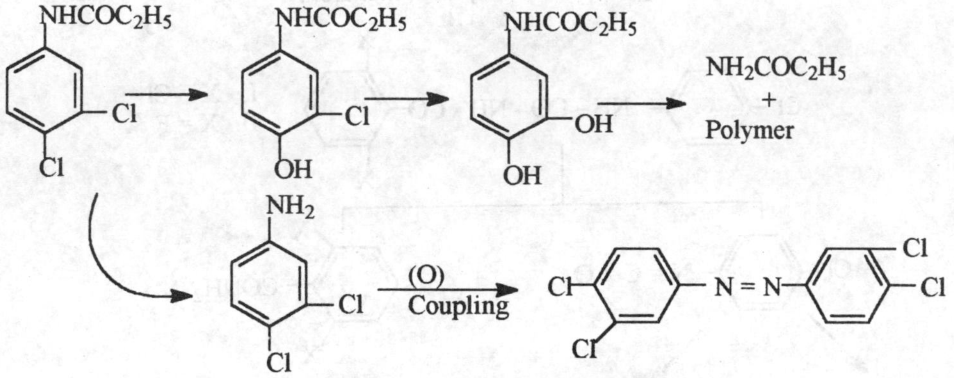

Fig. 3.243. Photolysis of propanil.

Photolysis of ureas

A number of photodecomposition products were characterised from monuron after irradiation in aqueous solution (Fig. 3.244).

Photolysis of 1-(4-chlorophenyl)-3-(2,6-dichlorobenzoyl) urea (I) and 1-(4-chlorophenyl)-3-(2,6-difluorobenzoyl) urea (II), a system containing more than one amide bond, was carried out in methanol. The major photoproducts identified were *p*-chlorophenylisocyanate, *N*-4-chlorophenyl methyl carbamate, 2-chlorobenzamide, 2,6-dichlorobenzamide and 2,6-diflurobenzamide (Fig. 3.245).

Photolysis of isoproturon [19]

Isoproturon, 3-(4-isoproylphenyl)-1,1-dimethylurea, a selective herbicide on photolysis in aqueous medium gave a number of photoproducts by the process of *N*-demethylation, *N*-oxidation, ring hydroxylation and dimerization. Nine products identified were shown in Fig. 3.246.

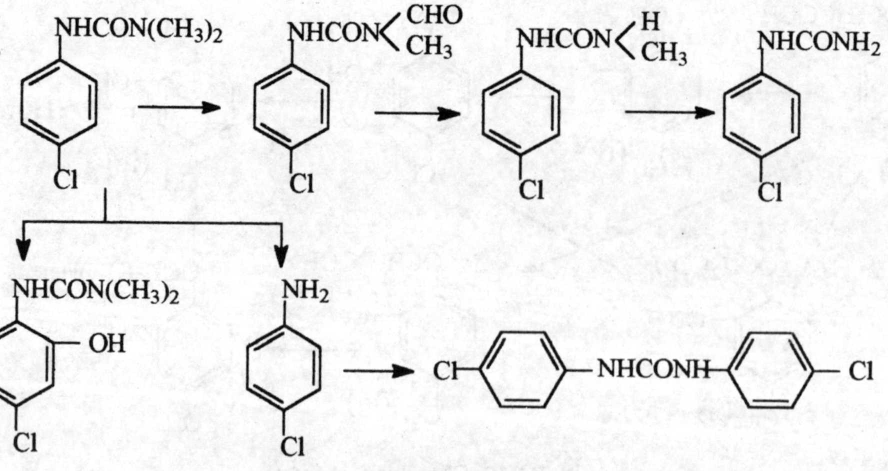

Fig. 3.244. Photolysis of monuron.

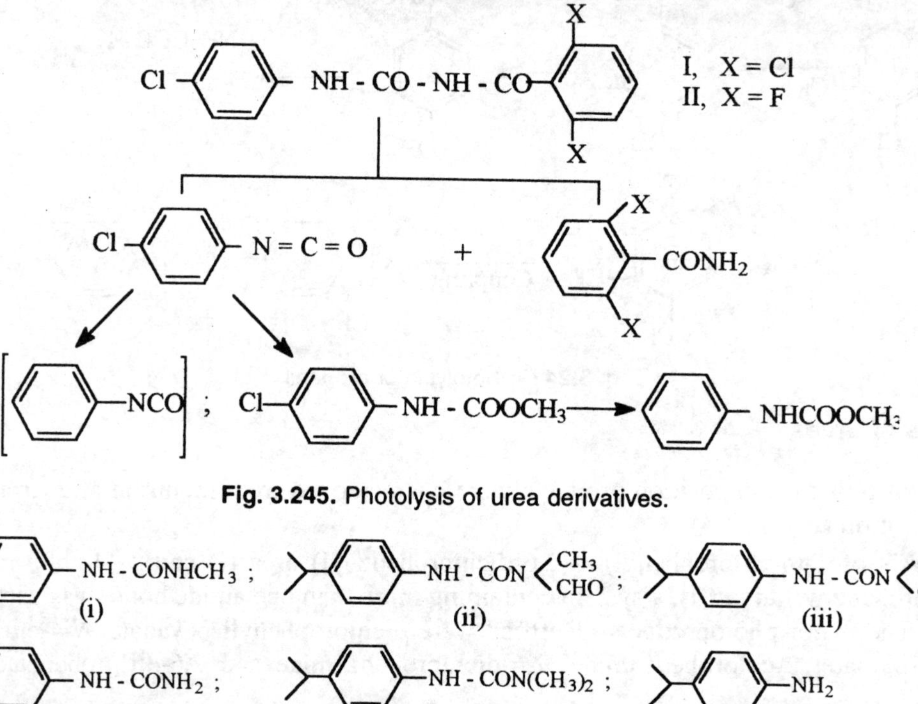

I, X = Cl
II, X = F

Fig. 3.245. Photolysis of urea derivatives.

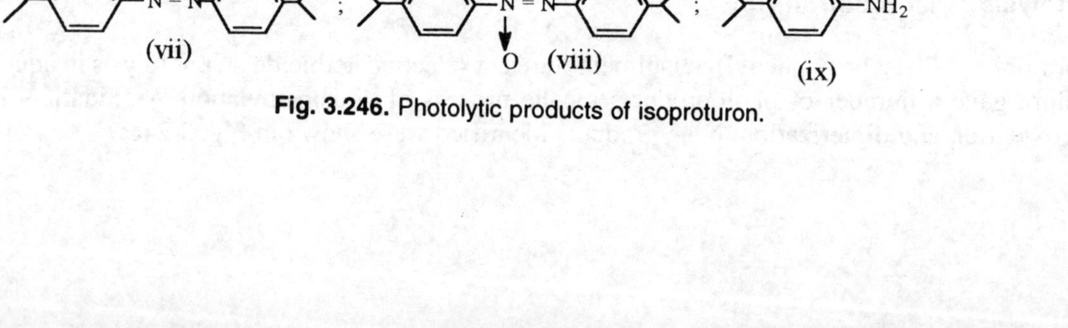

Fig. 3.246. Photolytic products of isoproturon.

Phenylalkylethers such as bifenfox (35), nitrofen (33) and fluorodifen (34) on photolysis gave 2,4-dichlorophenol, p-nitrophenol and the amino derivative of nitrofen as major products.

Photolysis of dinitroamine derivatives

The photodecomposition of N^3,N^3-diethyl-2,4-dinitro-6-trifluoromethyl-m-phenylenediamine in methanol and in water was investigated. The major photoproducts identified were 6-amino-1-ethyl-2-methyl-7-nitro-5-trifluoromethyl benzimidazole and related products (Fig. 3.247).

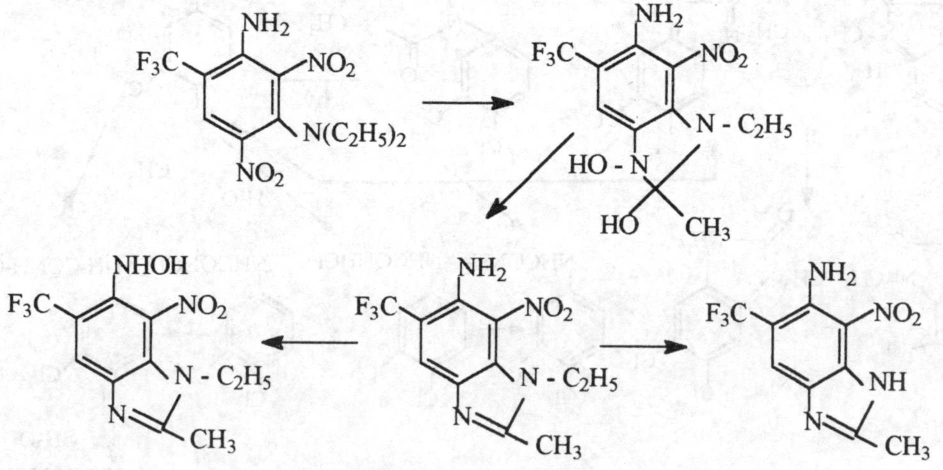

Fig. 3.247. Photolysis of dinitroamine derivative.

Photolysis of pendimethalin [20]

Pendimethalin, N-(1-ethyl propyl)-2,6-dinitro-3,4-xylidine, on irradiation in methanol and on soil sufaces yielded many photoproducts like 2-amino-6-nitro-N-(1-ethylpropyl)3,4-xylidine (i), 2-amino-2-nitroso-6-nitro-3,4-xylidine (ii), hydroxylated pendimethalin (iii) and cyclised products (iv, v) besides minor dealkylated products (Fig. 3.248).

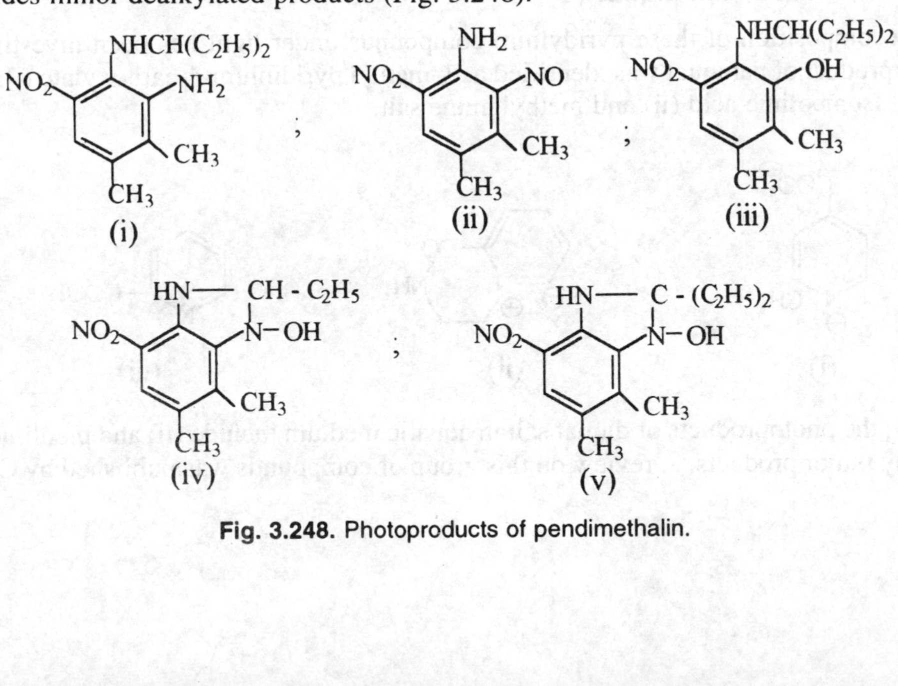

Fig. 3.248. Photoproducts of pendimethalin.

Photolysis of methazole

Photodecomposition of methazole, 2-(3,4-dichlorophenyl)-4-methyl-1,2,4-oxadiazolidine-3, 5 dione, with UV in methanol or sunlight in water resulted in loss of carbon dioxide from the oxadiazolidine ring and produced several minor products. The main photoproducts identified include 3,4-dichloro nitrobenzene, 1-(3,4-dichlorophenyl)-3-methyl urea, 1-(3,4-dichlorophenyl) urea and two isomeric dichloro-1-methyl-2-benzimidazolinones (Fig. 3.249).

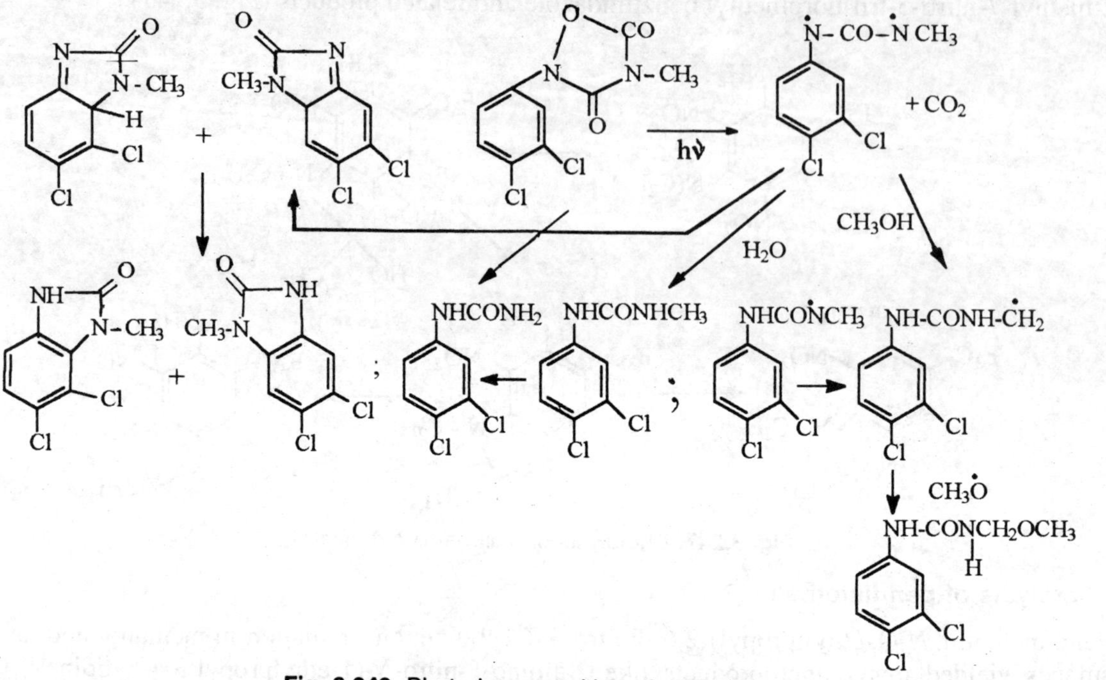

Fig. 3.249. Photodecomposition of methazole.

Photolysis of paraquat and diquat

The photodecomposition of these pyridylium compounds under light has been investigated. The major photoproduct of paraquat was identified as 1-methyl pyridinium-4-carboxylate (i), the betain of N-methyl *iso*nicotinic acid (ii) and methylamine salt.

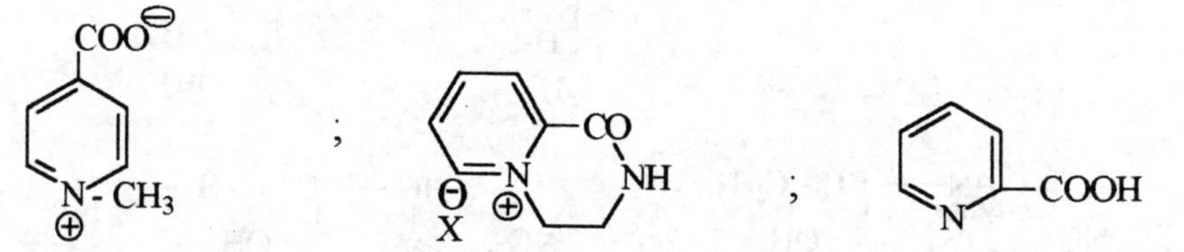

Similarly, the photoproducts of diquat salt in aquatic medium include (ii) and picolinic acid (iii) besides many minor products. A review on this group of compounds was published by Calderbank [21].

Photolysis of bentazone, basagran

Photolysis of basagran, 3-isopropyl-1H-2, 1,3 benzothiadiazin-(4)-3H-one 2,2-dioxide under sunlight irradiated conditions has been investigated in aqueous solution, thin film and on soil (Fig. 3.250). The major routes of photodecomposition include oxidative dimerization and nonconcerted loss of sulphur dioxide [22].

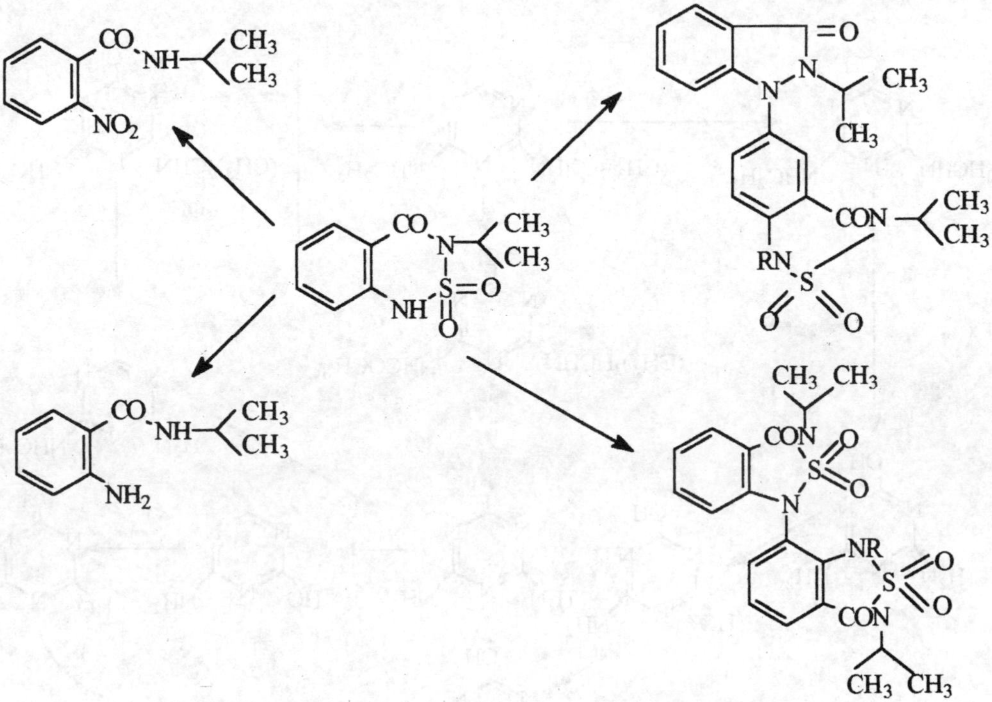

Fig. 3.250. Phototransformation of bantazone/basagran.

Phototransformation of fluridone

Two benzaldehydes, two benzoic acids and N-methylformamide were identified when fluridone, 1-methyl-3-phenyl-5-[3-(trifluoromethyl)-phenyl]-4(1H)-pyridinone (Fig. 3.251) was photolysed in aquatic medium [23].

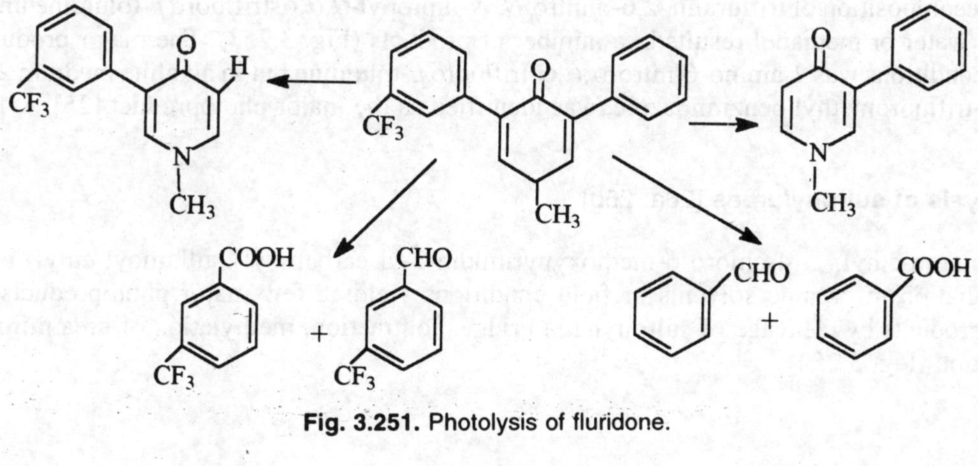

Fig. 3.251. Photolysis of fluridone.

Photodegradation of atrazine

The photodegradation of s-triazines such as atrazine, simazine, trietazine, prometon and prometryn was investigated in the presence of catalyst TiO_2 under simulated solar light. A number of photoproducts were obtained (Fig. 3.252) and the final degradation product was identified as cyanuric acid [24].

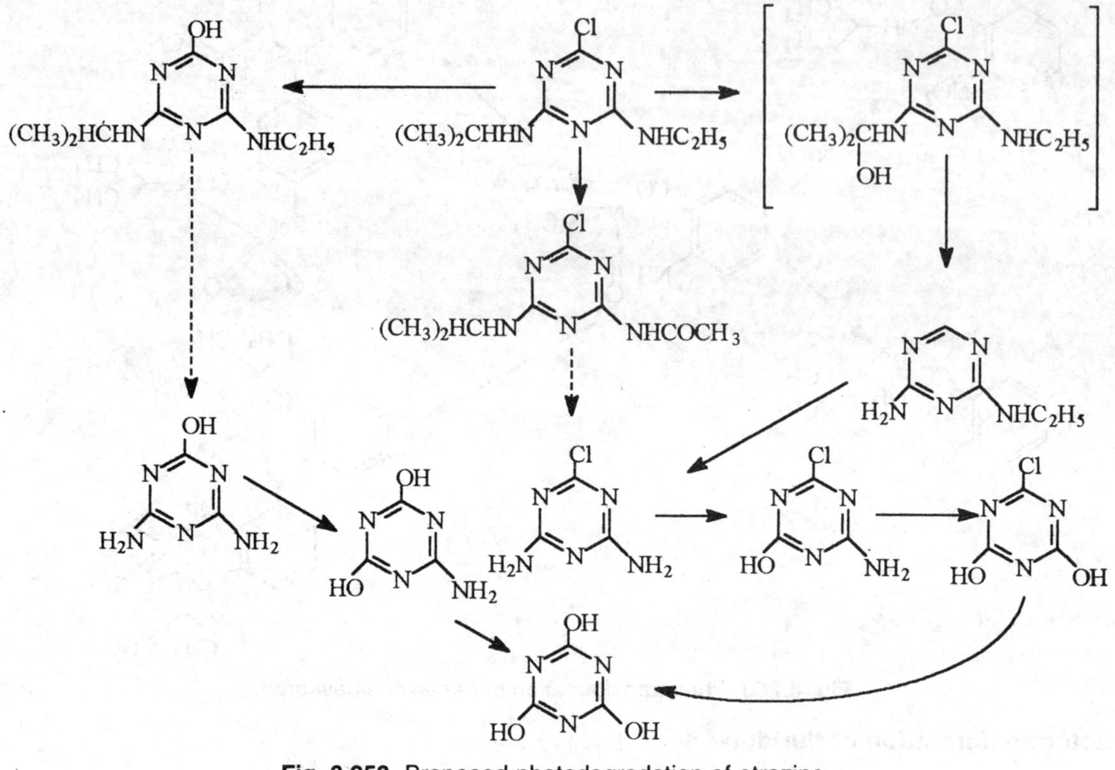

Fig. 3.252. Proposed photodegradation of atrazine.

Photodecomposition of trifluralin

Photodecomposition of trifluralin, 2,6-dinitro-N, N-dipropyl-α,α,α-trifluoro p-toluidine under sunlight in water or methanol results in a number of products (Fig. 3.253). The major product under acidic conditions was 2-amino-6-nitro-α,α,α-trifluoro p-toluidine but in alkaline medium 2-ethyl-7 nitro-5-trifluoromethyl benzimidazole, was identified as the major photoproduct [25].

Photolysis of sulfonylureas [26a, 26b]

Chlorimuron-ethyl, 2-(4-chloro-6-methoxypyrimidin-2-yl carbamoyl) sulfamoyl ethyl) benzoate, on irradiation in organic solvents or field conditions yields a few major photoproducts besides minor products by cleavage of sulfonylurea bridge, contraction, methylation of urea nitrogen and cyclisation (Fig. 3.254).

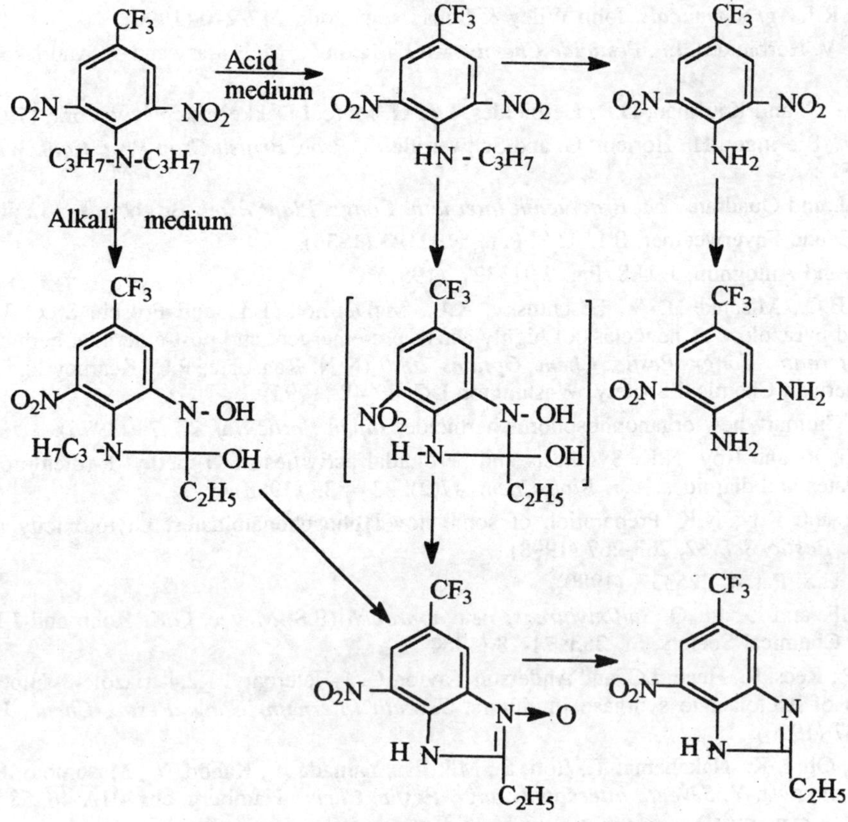

Fig. 3.253. Photodecomposition of trifluralin.

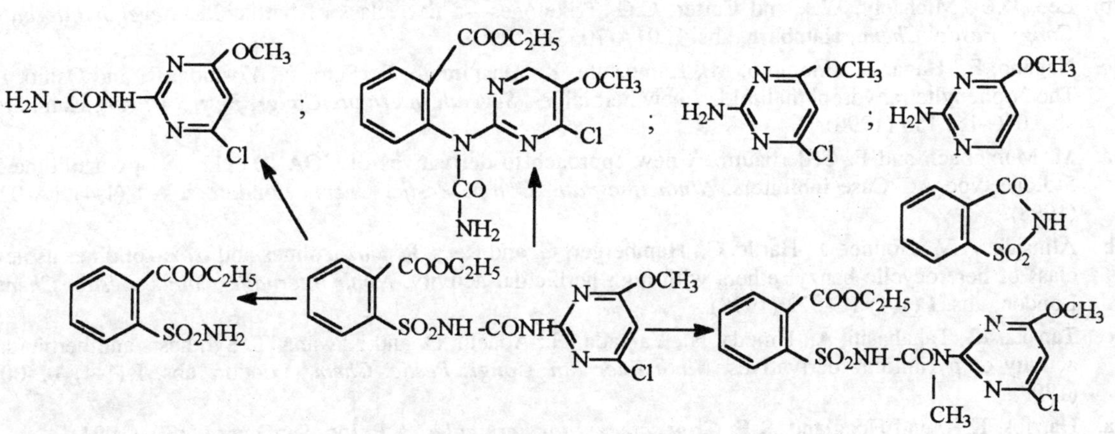

Fig. 3.254. Photolysis of chlorimuron-ethyl.

REFERENCES

1. Wain, R.L. and Weightman, F. Studies on plant growth regulating substances XI. Auxin antagonism in relation to a theory on mode of action of aryl and aryloxy-alkane carboxylic acids. *Ann. Appl. Biol.* **45**, 140 (1957).

2a. Cremlyn, R.J. *Agrochemicals.* John Wiley & Sons, New York, 217-270 (1990).

2b. Andriska, V. Herbicides. In : *Pesticide Chemistry* (G. Matolcsy, M. Nadasy and V. Andriska, eds.) 487-785 (1988).

3. Kearney, P.C. and Kaufman, D.D. Herbicides, **I & II**, Marcel Dekker, Inc. New York, (1970).

4. Randte, R., Bieringer, H., Horlein, G. and Schwerdtle, F. *Proc. British Crop Prot. Conf. Weeds*, Brighton, **I**, 19 (1982).

5. Rufener, J. and Quadranti, M. *Proc. tenth Internatn. Congr. Plant Prot.*, Brighton **I**, 332 (1983).

6. Mills, L.E. and Fayerweather, B.L. U.S. Pat. 2,192197 (1936).

7a. Tilles, H. and Antognini, J. U.S. Pat. 2,913327, (1956).

7b. Hamper, B.C., Mischke, D.A., Leschinsky, K.L., McDermott, L.L. and Prosch, S.D., 3-Aryl-4-halo-5 substituted pyrazoles : A new class of highly active pre-emergent and post-emergent herbicides. *Proc. Sr. Eighth Internatn. Congr. Pestic. Chem. Options 2000* (N.N. Ragsdale, P.C. Kearney and J.R. Plimmer, eds.). American Chemical Society, Washington DC. 42-48 (1995).

8a. Ueda, M. Cremart, new organophosphorus herbicide, *Japan Pestic. Inf.* **23**, 23 (1975).

8b. Khazanchi, R. and Roy, N.K. Synthesis and herbicidal activities of O-methyl P-(dichloromethyl) phosphonamidates and diamides. *Agr. Biol. Chem.* 47(2), 331-335 (1983).

8c. Das, S.K. and Roy, N.K. Preparation of some novel phosphonamidates, phytotoxicity and herbicidal properties. *Pestic. Sci.* **52**, 263-267 (1998).

9. Levitt, G. U.S. Pat. 4, 225337, (1980).

10. Sauers, R.F. and Levitt, G. *Sulfonylureas; new approach* (P.S. Magee, G.K. Kohn and J.J. Menn eds.), American Chemical Society, Sr. **253**; 21-28 (1984).

11a. Wegner, P., Rees, R., Head, J.C. and Anderson-Taylor, G. 1-Heteroaryl 1,2,4-triazole-3-sulphonanilides—a new class of acetolactate synthase inhibitors. *Seventh Internatn. Cong. Pestic. Chem.*, Hamburg, abs. 01A-45, 37 (1990).

11b. Ishida, Y., Ohta, K., Nakahama, T., Itoli, S., Mik, H., Yamada, J., Kando, Y., Masumoto, K., Kamikado, T. and Yoshikawa, Y. *Seventh Internatn. Congr. Pestic. Chem.*, Hamburg, abs. 01A-46, 58 (1990).

12a. Withenbank, C.S. *Fifth Internatn. Congr. Pestic. Chem.*, Tokyo, abs. **II**, C-7 (1982).

12b. Lee, D.L., Michaely, W.J. and Carter, C.G. Triketones—a new class of herbicides. *Seventh Internatn. Congr. Pestic. Chem.*, Hamburg, abs. **I**, 01A-70, 82 (1990).

12c. Nagano, E., Haga, T., Enomoto, M., Kamoshita, K., Hashimoto, S., Seto, P., Yoshida, R. and Oshio, H. The N-phenyltetrahydrophthalimides, new herbicides. *Seventh Internatn. Congr. Pestic. Chem.*, Hamburg, abs. 01C-15, 155 (1990).

13a. M. Muhlebach and F. Cederbaum. A new approach to derivatives of CGA 271312 aryl pyrazolidine-3, 5-dione type AC Case inhibitors. *Ninth Internatn. Congr. Pestic. Chem.*, London, abs. **I** (1-4) 1A-021 (1998).

13b. Almsick, A.V., Bohner, J., Harde, C., Hamberger, G. and Rees, R. *Iso*xazolines and *iso*xazolidines as new class of heterocyclic benzyl ethers with high herbicidal activity. *Ninth Internatn. Congr. Pestic. Chem.*, London, abs. **I** (1-4) 1C-022 (1998).

13c. Tanaka, K., Tahahashi, A., Umeda, N., Yamada, H., Adachi, H. and Kawana, T. Synthesis and herbicidal activity of pyrimidine derivatives. *Ninth Internatn. Congr. Pestic. Chem.*, London, abs. **I** (1-4) 1C-007 (1998).

14a. Hatzios, K.K. and Hoagland, R.E. *Crop safeners for herbicides.* A.P. Inc. SanDiego, USA (1991).

14b. Parker, C. Herbicide antidotes—a review. *Pestic. Sci.* **14**, 40-48 (1983).

14c. Phillips, B.A., Bhagsari, A.S. In : *Chemistry and Action of Herbicide antidotes* (F.M. Pallos; J.E. Casida, eds.), A.P. New York, 21-34 (1978).

15. Lay, M.M. and Casida, J.E. Influence of glutathione and glutathione S-transferase in the action of dichloro acetamide antidotes for thiocarbamate herbicides. In : *Chemistry and Action of Herbicide antidotes* (F.M. Pallos and J.E. Casida, eds.) 151-160, New York, USA (1978).

16. Zulian, J. Study of the absorption, excretion, metabolism and residues in tissues in rats treated with carboxy 14 labelled pendimethalin, PROWL. *J. Agric. Food Chem.* **38**, 1743 (1990).

17. Cast, A. The phenoxy herbicides. *Weed Sci.* **23**, 253-263 (1975).

18. Jaworski, E.J. Mode of action of N-phosphonomethylglycine : Inhibition of L-aromatic amino acids biosynthesis. *J. Agric. Food Chem.*, **20**, 1195 (1972).

19. Dureja, P., Walia, S. and Sharma, K.K. Photolysis of isoproturon in aqueous solution. *Toxic. Env. Chem.*, **34**, 65-71 (1991).

20. Dureja, P. and Walia, S. Photodecompsion of pendimethalin. *Pestic. Sci.* **25**, 105-114 (1989).

21. Calderbank, A. The bipyridylium herbicides. *Adv. Pest Control Res.* **8**, 127-229 (1968).

22. Nilles, G.P. and Zabik, M.J. Photochemistry of bioactive compounds. Multiphase photodegradation and mass spectral analysis of basagran. *J. Agric. Food Chem.* **23** (3) 410-415 (1975).

23. Saunders, D.G. and Moseir, J.W. Photolysis of aquatic herbicide fluridone in aqueous solution. *J. Agric. Food Chem.*, **31**, 237-241 (1983).

24. Pellzetti, E., Maurino, V., Minero, C., Carlin, V., Pramauro, E., Zerbinati, O. and Tosato, M.L. Photocatalytic degradation of atrazine and other *s*-triazine herbicides. *Environ. Sci. Technol.*, **24**, 1559-1565 (1990).

25. Leitis, E. and Crosby, D.G. Photodecomposition of trifluralin. *J. Agric. Food Chem.* **22**(5), 842-847 (1974).

26a. Choudhury, P. and Dureja, P. Photolysis of chlorimuron-ethyl. *Toxic. Env. Chem.* **63**, 71-81 (1997).

26b. *Ibid.* Photodegradation of chlorimuron-ethyl in soil. *Pestic. Sci.* **51**, 201-205 (1997).

CHAPTER 4

New Improved Pesticide Formulations

- Introduction
- Liquid formulations
- Dry formulations
- New generation formulations
- Biopesticide formulations
- Neem formulations—commercialized
- Quality control of formulated products

- Methods of analysis
- Shelf life
- Safety aspects
- Use of modern tools in pesticide formulations
- Environmental aspects
- Conclusion
- *References*

A pesticide formulation [1, 2, 3] can be defined in a broad sense as a physical mixture of one or more biologically active chemicals with inert ingredients in a definite proportion so as to make it more effective, safe, economical and easy to use. It is a process through which a small quantity of an active ingredient (a.i.) is formulated into a final product to be used by the farmers/consumers to control insect pests and diseases. The prime purpose of formulation is the dilution of high concentration pesticide down to an applicable level at which it is toxic to target pests but non toxic to non target species and environment.

Pesticide formulations are generally classified into liquid or dry form [4]. The conventional liquid formulations can be emulsifiable concentrate, oil concentrate, oil solution and aqueous concentrate while the dry powder type formulations include dust, water dispersible powder, granules and water soluble powder (SP). While these formulations effectively and efficiently control plants pathogens, insect pests and weeds, they pose a potential hazard during their storage, transportation and application. Many liquid formulations are highly inflammable and being petroleum based products amount to resource depletion. On the other hand, the dust formulations pollute the air and ground water and endanger the existence of non-target organisms including man. A considerable amount of these preparations miss the target and amount to substantial waste. It has been estimated that even with careful application only 10-20% of the dust and 25-50% of the sprays are deposited on the plant surfaces and rest pollute the environment. This calls for further improvement in the application technology and developing more ecofriendly formulations. A few new safe or eco-friendly formulations have been developed recently [5] and are discussed in brief here. Among new solid formulations, one can mention driftless-dust, dry flowables or dispersible granules, floating granules, fine granules and encapsulated granules with controlled release. The new liquid formulations developed include wet flowable (suspension concentrate), aqueous emulsions, microemulsions and suspoemulsions.

Liquid formulations

a) **Oil concentrates (OC)** contain high concentration of active ingredients. They can be used undiluted for ultra low volume (ULV) applications but can also be diluted conveniently with hydrocarbon solvents as and when needed.

b) **Emulsifiable concentrates (EC)** are similar to oil concentrates but in addition contain a surfactant or emulsifier for quick and easy dilution with water for spray application. The solvent system must be immiscible with water to produce an uniform emulsion lasting throughout the spraying period. Commonly used solvents are xylene and solvents of aromatic naphtha type (aromax) or aliphatic kerosene type. Solvents may be chosen based on flash points so as to reduce possible risks of fire during transportation and use. This is a most convenient and popular form of formulation usually available for most of the recommended pesticides.

c) **Aqueous concentrates (AC)**. Some pesticides readily dissolve in water. Salts of certain herbicides are soluble in water. They are formulated as AC concentration and are generally expressed in terms of amount of acid equivalent per unit volume.

d) **Oil solutions (OS)** contain pesticides in low concentration usually below 5% by weight. These formulations are generally used for household or institutional insect/pest control measures. These are odourless, colourless and contain non-staining high flash point solvents to minimise the fire hazard.

e) **Invert emulsifiable concentrates (IEC).** These formulations differ from normal EC by the fact that their dilution with water provides an emulsion in which the external or continuous phase of the emulsion is the oil phase whereas the internal or discontinuous phase of the emulsion is water. The oil soluble herbicidal esters are in general formulated as IEC. These formulations have several advantages over normal EC as they require less quantity of water at the time of dilution and also oil which has a low vapour pressure due to which the evaporation of continuous phase is minimum.

Dry formulations

Conventional dry formulations include dust, granules and wettable powders which are mixed with water during application.

a) **Dust bases or concentrates** are dry, free flowing powder containing a high concentration of active ingredients varying from 25 to 75 per cent. These are mixed with a suitable inert material before field application.

b) **Dust.** These are finely powdered pesticides which are formulated to field strength varying from 1-10 per cent depending upon the potency of the pesticide and the rate of application.

The particle size is usually less than 30 μm diameter. The use of dust has been limited by their tendency to drift downward and they are mostly used for seed dressing.

c) **Wettable powder (WP).** These are similar to dust bases except that they are formulated for dilution with water into final spray. The quality is judged by the rapidity of wetting and stay in suspension when mixed with water for field application. The proper choice of wetting agents is to enhance the wetting power and good suspensibility which can be maintained by reducing the particle size. Surfactants of the dispersant class are added to prevent the agglomeration of the particles resulting in sedimentation. WP is frequently used for the slurry treatment of seeds. In general wettable powder formulations are not compatible with other types of formulations specially with emulsions causing sedimentation.

d) **Granules (GR).** The granular pesticides are different from powdered pesticides according to the mesh size. The mesh size starts from 4 mesh (U.S. standard) to 80 mesh. The granular pesticides are generally free flowing and do not cake during storage and there is no problem of drift during application, hence easy to handle. Since the activity depends on the release of the a.i., the granules should have fast or slow disintegration after entering the system. Granules are more effective as prophylactic application when weather conditions are unfavourable for spray.

New generation formulations [6a, 6b, 7]

There are many problems associated with conventional formulations of pesticides. WP formulations are dusty and not easy to measure. Further, the dust clouds from WP are not only very fine but also have high concentration of pesticides endangering human health and contaminating the environment. The organic solvents used in EC formulations enhance the percutaneous toxicity of the pesticide by altering the dermal penetration. Moreover these are inflammable and expensive. Due to these disadvantages, the ecofriendly new generation formulations have been developed during the last decade. The most important of these are discussed in the following text.

i) **Water-dispersible granules (WG).** It is an improvement over wettable powder preparations (WP). The product looks like small spheres, flows like a liquid and can be measured by volume. The product may be sometimes called as dry flowable (DF). Most of the sulfonylurea herbicides are now available in this formulation. For safer use, these products are now available in water soluble polyvinyl alcohol bags in pre-measured quantity to avoid direct contact with the workers.

ii) **Water emulsifiable gels (GL).** It is an improved version of emulsifiable concentrate (EC). These gel formulations can be packaged in water soluble polyvinyl alcohol bags in pre-measured quantities to avoid exposure to pesticides.

iii) **Floating granules.** These formulations are characterised by the release of the active ingredient from the granules floating on the water surface and function as an efficient aquatic pest control measure.

iv) **Fine granules (ordinary & FG).** Both these formulations are prepared with active material like systemic pesticides which may be absorbed and translocated through the plant tissue. The ordinary granules have particle size range of 105-297 μm (150-48 mesh) whereas 90% of the fine granules lie in 62-210 μm (250-65 mesh) range. These formulations are suitable for use in situation where drift of the dust needs to be checked. These formulations are very efficient and environmentally safe, convenient to handle and less hazardous to workers.

v) **Dust driftless (DL-dust).** This formulation overcomes the drift problems associated with the conventional dusts. It has mean particle diameter of 20-30 µm which differentiates it from the conventional dusts having mean diameter of only 10-12 µm. It has good floatability which enables the coverage of lower side of the leaf also. It is environmentally safe, floats less during application and has good coverage of the target. Many organophosphate, carbamate and cartap insecticides are recently being formulated as DL-dust for use in Japan.

vi) **Controlled release formulations (CR)** [8, 9]. This formulation is a relatively new development aimed at providing controlled release of a.i. for specific types of biological actions. The controlled release may be defined as a method by which an active agent is released on to an intended target at a slow rate so that the contamination of environment is minimised. The primary aim of all the CR systems is the employment of a suitable natural and biodegradable polymer to act as a rate controlling device, container or membrane, for the a.i. to be released at the desired rate helped by the moisture or soil microorganisms. The rapid progress in CR technology has been possible because of the fast advancement in polymer science which provided materials for preparation of CR. Today a variety of natural and biodegradable synthetic polymers and elastomers, often suitably modified by co-polymerization, cross-linking, degradation and chemical reactions, are being employed in this technology. In general, there are two types of CR : (1) Chemical type—a chemical linkage exists between a.i. and polymer

$$M + A \longrightarrow MA$$

$$MA \xrightarrow{\text{Polymerise}} \begin{matrix} M - M \\ | \quad\quad | \\ A \quad\quad A \end{matrix}$$

A - a.i.;   M - monomer

and (2) Physical type—where no chemical linkage exists. These can be further classified as : a) Micro and Macro capsules, b) Hollow fibres, c) Monolithic matrices, d) Laminated structures

a) **Microcapsules [Encapsulated Suspension (CS)]** are small particles of solid or liquid droplets of the a.i. enclosed in a thin polymeric wall material spherical in shape (5-50 microns) and normally dispersed in water for use. The capsules bigger in size are known as macrocapsules.

b) **Hollow fibres.** The utility of pheromones has been realised through the use of well defined controlled release delivery system. The hollow fibres comprise of fine capillary tubes sealed at one end which hold the liquid a.i. inside it by capilliary action. The device is found suitable as insect attractant baits as well as insect mating disruption aids.

c) **Monolithic matrices** are made by incorporating a.i. into a polymeric or elastomeric matrix by extrusion, injection molding or casting. These are used for the preparation of CR larvicides, herbicides and molluscicides.

d) **Laminated structures** consist mainly of three layers of laminated plastic material. The control layer is a reservoir for the a.i. and is sealed between two outer plastic layers to control its release. The pheromone is thus released by diffusion from the reservoir through the membrane. The rate is controlled by the membrane composition and thickness.

The advantages of these formulations include increased persistence of biological activity, reduced phytototoxicity, low mammalian toxicity with reduction in environment contamination. The highly toxic pesticides formulated as CS are mainly parathion, phorate and aldicarb.

vii) **Flowable (Suspension concentrates OF/SC).** This involves dispersion of micronised technical grade solid pesticide in water. Hydrolytically stable, high melting point, friable crystal compounds having low solubility in water are formulated in this form. It has several advantages over the EC formulations and WDP such as non-dustiness, reduced operational hazards, minimum nozzle blockages, easier ULV application, non-inflammability, easy handling and transportation, low phytotoxicity and better biocidal activity. It is thus suitable for both soil and foliage application.

viii) **Concentrated emulsions (emulsion concentrate, EW).** This formulation is produced by dispersing or emulsifying the technical grade liquid pesticide in water or dissolving in minimum quantity of solvent. The advantages of this formulation are minimum use of solvents, reduced operational hazards and toxicity to mammals, less phytotoxicity and enhanced bioactivity due to fine droplet size.

ix) **Microemulsions.** This formulation is transparent liquid forming micelles of liquid pesticide (technical). It requires selection of suitable solvent and surfactant package. Thermodynamically stable microemulsions provide an alternative approach to conventional kinetically stable, coarse emulsions. These preparations have low phytotoxiciy, are non-corrosive, do not vaporise at ambient conditions and have almost no toxicity to mammals, fish and birds.

x) **Suspoemulsions (SG).** This comprises of suspension of a pesticide emulsified in water as the dispersing medium. The preparation and stabilisation of this multiple phase suspension concentrate is rather a difficult task. The oil phase containing the highest possible content of a.i. (ideally a liquid) having no water solubility product, exists side by side in a dispersed solid phase whose a.i. is insoluble in water as well as in the dispersed oil phase.

xi) **Ultra-low volume (ULV).** An ULV undiluted formulation is more potent per unit volume for a large scale aerial spray on forest or river terrain. This formulation must contain sufficient anti-evaporant to prevent the evaporation of spray droplets during their free fall to reduce drift and the formulation must be non-corrosive to aircraft and spray equipment.

xii) **Briquitte BR** is a solid block formulation and is made by mixing the a.i. with low density, inert granules and binding agents. It is convenient for the manual application of pesticides in aquatic environment where spray application is ineffective.

xiii) **Smoke.** The pesticide is mixed with an oxidant and combustible material which generates large amount of hot gas. A special form of smoke generator is the mosquito coil which is used as a mosquito repellent.

Biopesticide formulations [10]

The types of formulations applied to microbial agents are quite similar to that of chemical pesticides. The aim is to produce a stable product that exhibits optimum effectiveness and economical use for a particular environment and circumstances, with special reference to shelf stability. The main difficulty is that the hydrophobicity and particulate nature of microbes effects the wettability, dispersibility and suspensility of the preparation. In general the microbials are much more sensitive to temperature and their half life is short. The preparation based on *B. thuringiensis* and nuclear polyhederosis virus remain stable only under freezing temperatures. The stability of the protozoan *N. locustoe* is extremely poor. Amongst all microbials Bt is most widely used because of its stability.

$$H_2O + O_2 \longrightarrow O\overset{\bullet}{H} + HO_2$$

$$PH + \overset{\bullet}{O}H \longrightarrow \overset{\bullet}{P} + H_2O$$

$$\overset{\bullet}{P} + O_2 \longrightarrow PO_2$$

$$2PO_2 \longrightarrow 2P\overset{\bullet}{O} + O_2$$

$$P\overset{\bullet}{O} \longrightarrow F_1 + F_2$$

The stability of the formulation also depends on field application. Under aerobic conditions, the biologically active protein may be inactivated by the cleavage of the peptide chain yielding protein fragments. This happens due to the formation of OH radicals which react with the biomolecules to yield peroxy radicals and oxyradicals which ultimately decompose to fragments.

Attempts to block the free radical formation by various means, like encapsulation have successfully enhanced the microbial persistence. The research on improving the field persistency of microbial agents is still under investigation.

Formulation of microbials (Table 4.1) available for general use are, water dispersible granules (WG), wettable powder (WP), granules and dust. Liquid formulations are suspension concentrates (SC), ultra low volume (ULV), oil miscible flowable concentrate (OF) and briquette (BR).

Table 4.1. Commercialized Bt strains

Subspecies	Crystal toxin types	Target pests
Kurstaki	1Aa, 1Ab, 1Ac, 11A, 11B	Lepidoptera
Aizwai	1Aa, 1Ab, 1B, 1C, 1D, 1F	Lepidoptera
Israelensis	IV	Diptera mosquitoes, blackflies
Tenebrionis	III	Coleoptera, Colorado potato beetles

Neem formulations

The development of a formulated neem seed extract for insecticidal use has been described by Larson [11]. A formulation containing 3000 ppm azadirachtin with reasonable shelf life has been prepared and marketed earlier as Margosan-O for use in greenhouse, nurseries, forests and crops. At present a number of neem formulations are available in India and abroad (Table 4.2).

Table 4.2. Commercially produced neem formulations available in India for use on crops

Product	Active ingredient(s)	Activity claimed	Manufacturer
Azadit	—	—	M/s Pesticides (India) Ltd., Udaipur-313001
Biosol	Oil	—	—
Godrej	Azadirachtin	Antifeedant, repellent/deterrent, male sterility, molt/chitin inhibitor, change in sexual behaviour, growth regulator/disruptant, ovipositional deterrent, juvenile homone, interference in biosynthesis, toxicant, ovicidal, oviposition deterrent, altered egg hatchability	M/s Bahar Agrochem & Feeds Pvt. Ltd., E-24, Lote Parshuram, Mumbai-415722; M/s Godrej Agrovet Ltd., Mumbai 400 079
Field marshal	Azadirachtin	Antifeedant/repellent	M/s Khetiwadi Corner, Shiyabave, Vadodara-390001, Gujarat
Jawan crop protector	Extracts containing azadirachtin	Antifeedant/repellent, disturbs growth and reproduction	M/s McDA Agro Pvt. Ltd., Mumbai-400001
Margocide-CK*	Azadirachtin and its 15 derivatives	Antifeedant, growth inhibitory, ovicidal, oviposition deterrent, nematicidal, chemosterilant	M/s Monofix Agroproducts Ltd., 52-53, First Floor, Swimming Pool Complex, Hubli-580029
Margocide-OK*	Azadirachtin (major ingredient)	Insecticidal	-do-
Moskit	Oil	Mosquito repellent	M/s Investment and Commerical Enterprises, Mumbai-400022
Neem based emulsifiable concentrate, dust, water	Kernel, oil	Pesticidal	Laboratory technology of IARI, New Delhi-110012
Neemhit	—	Pesticidal	M/s Skylark Agro-Chem, W 92, M.I.D.C. Phase II, Dombivali (East), Thane-421 204, Maharashtra
Neem oil emulsion	Oil	Pesticidal	M/s Sio Agro Research Laboratories, Dhiraj Apartments, Mulund (West), Mumbai-400 080
Neem plus	—	—	M/s B.D. Khaitan & Co., Mumbai, Calcutta
Neem Top	—	—	M/s Sri Krishna Company, Coimbatore, Tamil Nadu
Neemark	Azadirachtin	Antifeedant/repellent, nematicidal, synergistic	M/s West Coast Herbochem Pvt. Ltd., V.S. Marg, Mumbai-400 025

(Contd.)

Product	Active ingredient(s)	Activity claimed	Manufacturer
Neemasol	—	—	M/s E.I.D. Parry (India) Ltd., Dare House, P.B. No. 12, Chennai-600 001
Neemgold	Azadirachtin, kernel extract	Antifeedant	M/s Southern Petro-Chemical Industries Corporation Ltd., Chennai-600 032
Neemguard	Extract concentrate	Repellent, metamorphosis disruption (safe to natural enemies of insects)	M/s Akshay Chemicals, Mumbai-400052
Neemrich	Extracts	Neemrich I-Warehouse pests Neemrich II-Antifeedant	Technology of NPL, Pune (synonymous with products Margocide CK and Margocide OK)
Neemol	—	—	M/s Agro Links, Trichy, Tamil Nadu
Neemosan	—	—	M/s Agronule Industries, Tamil Nadu
Neempourn	—	—	M/s Prabhakar Oil Mills, Maharashtra
Neemta 2100	Extract concentrate	Repellent, metamorphosis disruptor (safe to natural enemies of insects)	M/s A.J. Chemicals (Mkted by M/s Kisan Brothers Pvt. Ltd., Ahmedabad-380002)
Nimba	Kernel based powder	Pesticidal	Laboratory technology of IARI, New Delhi-110 012
Nimbecidine*	Azadirachtin	Antifeedant, repellent metamorphosis disruptor, synergist	M/s T. Stanes and Co. Ltd., Coimbatore-641 018, Tamil Nadu
Nimbosol	—	—	M/s Victoria Laboratories, Salem, Tamil Nadu
Nimlin	Extract concentrate	Repellent, metamorphosis disruptor (Safe to natural enemies of insects)	M/s Sunline Agro Chemicals, Sakri Road, Dhulia, Maharashtra; M/s Phyto Products, Thiruthuraipoondi, Tamil Nadu
Phytowin (Vembu)	—	—	M/s Phyto Products, Thiruthuraipoondi, Tamil Nadu
Replin 555			M/s Micro Chemicals, Madurai, Tamil Nadu
RD-9 Repelin*	Neem, *Pongamia, Annona* and castor products (Azadirachtin 300 ppm)	Antifeedant, repellent (safe to pollinators)	M/s ITC Limited, Rajahmundry, Andhra Pradesh
Sukrina	Azadirachtin, meliantriol and other actives of *A. indica* and *Galedupa indica*	Antifeedant, repellent reduces depression	M/s Conster Chemicals Pvt. Ltd., Chennai-600 116

(Contd.)

Product	Active ingredient(s)	Activity claimed	Manufacturer
Swaticure	—	Pest repellent	M/s Swati Industries Pvt. Ltd., Bandra, Mumbai-400 050
Vapacide	—	Antifeedant	Technology of NICT, Hyderabad-500 007
Wellgro	Neem kernel powder	Repellent, fungus inhibitory, antiviral, plant nutrition, N-loss prevention	M/s ITC Limited, Rajahmundry, Andhra Pradesh

Source : Neem Research Development, (N.S. Randhawa and B.S. Parmar eds.) SPS, India, pp. 272-274 (1993).

* Registered with C.I.B., Faridabad, Haryana.

Quality control of formulated products [12]

The quality control of formulations can be defined as an effective mechanism of co-ordinating the maintenance and improvement of quality at different stages of formulation and production, so that the finished product with assured quality is made available to the consumer at competitive price.

The control has to be effected at every stage of the manufacturing process starting from the raw material, processing to packaging. This can be done by defining the quality of the raw material and devicing quick analysis methods so that the product can be tested at crucial stages along the downstream process. In most cases the suppliers of raw materials follow their own specifications. If the supplier is reliable, the raw materials can usually be accepted on a warranty basis, from the supplier. But still it is advisable to carry out, on the spot checks to determine the quality. It is observed that the presence of unknown contaminants in the raw materials can cause degradation and sometimes enhance toxicity in the pesticide formulation. Excess acidity, alkalinity or traces of moisture might aggravate the problem. Frequent checks should be made on raw materials after procurement to keep the contaminants within the permissible limits. In case dustiness arises in the formulation, the screen size of clay granules must be checked. Dust diluents and granules must also be examined for the presence of foreign matter such as nails, twine from bagging operations, sticks, stones or any material likely to interfere with the product application.

Some insecticides which are stable in technical form and in liquid formulations show marked decomposition/degradation if the commercial mineral carriers are used in dust and water dispersible powder formulations. The carriers like clay diluents and emulsifiers must be pre-tested before use in formulations. The shelf life expectancy and satisfactory physical characteristics of a particular formulation depend on the quality of carriers, diluents and emulsifiers. It has been observed that traces of metallic ions and metallic oxides, if present on the surface of carriers may enhance degradation of pesticides after prolonged storage conditions. Deactivators may be added in these cases in order to counter effect the situation and with the hope to improve the quality of the product. Specifications of solvents with reference to water, acidity or alkalinity, colour, specific gravity also must be examined before use.

The assay results of the technical grade pesticides supplied by the manufacturer must be studied thoroughly before preparing the formulations. Information on formulation compatabilities and uses should also be procured from the suppliers.

The specifications of packaging material are also important and must be clearly defined. Since moisture content is important for stability of dusts and water dispersible powders, there should be a layer of material in the bag wall as moisture trap. Care must be taken to use polyethylene bottles for packaging to keep away the air moisture. Sometimes polyethylene bottles and caps may be affected by the solvents or components of the formulations. These are also to be inspected for chips, cracks or brakes. Many liquid pesticides of technical grade and formulations are found unstable when kept in metal container because of interactions. In order to prevent this, a non-reactive resin material is coated inside the metal container before packaging.

Pesticides requiring refrigeration should be stored at about 0-5°C. Organophosphorus pesticides during storage under cold conditions show signs of condensation or precipitation. Such preparations which require refrigeration, must be allowed to reach room temperature before removing the cap for use. If instrumental method is used, the analyst must obtain a standard of performance for reference material for purpose of comparison specially retention time, number of peaks and instrument response if the same quantity of standard is used.

The writing and printing of labels on packages should conform to all government and international rules and regulations. Instructions for use, precautions for storage and use of specific antidotes in case of mishandling must be clearly printed on the packages before it goes to the market.

In recent years the problem of cross contamination or accidental mixing of two pesticides in the formulation has been reported. The presence of a herbicide in an insecticide formulation for use on an agricultural crop might cause the loss of some or the entire treated crop in the field. In order to prevent cross contamination it is necessary to keep the equipment in the plant and operating lines thoroughly clean. Flushing with steam or hot water before switching over to another formulation may help to a great extent.

Some factors which influence the quality of the formulations are listed below :

a) **Surfactants.** Surfactants reduce the interfacial tension between immiscible liquids or between liquids and solid surfaces. In pesticide formulation, the characteristics of the surfactant are easy wetting dispersion and emulsifiability for water dispersible powders and emulsifiable concentrates. The chemical structure of the surfactant comprises of two parts, one is oriented towards one phase while the other faces the second phase. If the system is oil and water, one portion of the molecule should be soluble in oil and the other in water. The surfactant molecules are generally non ionic and anionic types. In formulating pesticides both anionic and non-ionic surfactants are important. Some examples of surfactants which are commonly used are (Fig. 4.1) :

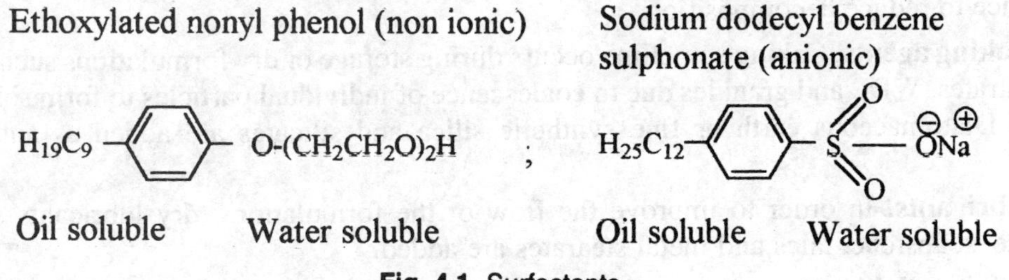

Ethoxylated nonyl phenol (non ionic)   Sodium dodecyl benzene sulphonate (anionic)

$H_{19}C_9$ —⟨ ⟩— $O\text{-}(CH_2CH_2O)_2H$; $H_{25}C_{12}$—⟨ ⟩— $\overset{\displaystyle O}{\underset{\displaystyle O}{S}}$—$\overset{\oplus}{\underset{\ominus}{O}}Na$

  Oil soluble   Water soluble   Oil soluble   Water soluble

Fig. 4.1. Surfactants.

The wetting agents used in water dispersible powders (WDP) are usually of the anionic type. Among them, the largest number are probably sodium salts of alkyl benzene sulfonates.

Dispersing agents used in addition to the wetting agents in WDP to impart some electrical charge to all the particles in suspension. The effect is that the individual particles repel each other and consequently, resist flocculation and agglomeration. Dispersants used in the WDP may be of lignosulfonates type with cations such as sodium or calcium sulphonates of polymeric phenols. Mostly these are dry, powdered solids which facilitate their incorporation in WDP.

b) **Emulsifiers.** The selection of proper emulsifiers to maintain the quality of formulations is very important. In order to select the proper emulsifier, the manufacturer prefers a paired emulsifier system. The paired emulsifier system (Fig. 4.2) represents a blend of anionic and non ionic emulsifiers with different hydrophilic and lipophilic characteristics. One part of the pair with lipophilic nature emulsifies easily with the pesticide solvent while the other part favours the emulsification of the hydrophilic pesticide solvent system.

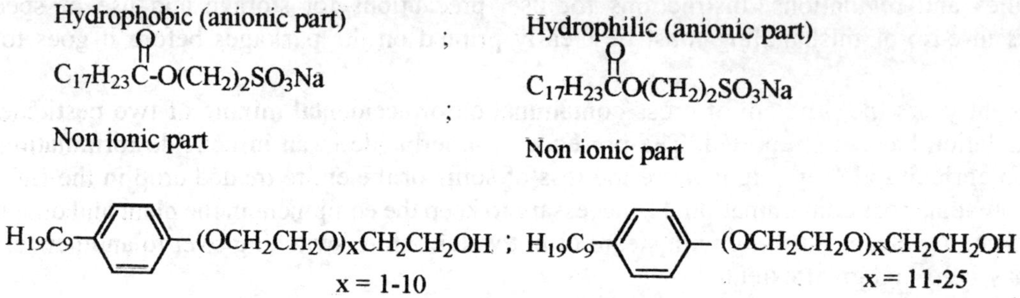

Hydrophobic (anionic part)

$$C_{17}H_{23}\overset{O}{\overset{\|}{C}}\text{-}O(CH_2)_2SO_3Na$$

Hydrophilic (anionic part)

$$C_{17}H_{23}\overset{O}{\overset{\|}{C}}O(CH_2)_2SO_3Na$$

Non ionic part

Non ionic part

$$H_{19}C_9\text{-}\langle\bigcirc\rangle\text{-}(OCH_2CH_2O)_xCH_2CH_2OH$$
$$x = 1\text{-}10$$

$$H_{19}C_9\text{-}\langle\bigcirc\rangle\text{-}(OCH_2CH_2O)_xCH_2CH_2OH$$
$$x = 11\text{-}25$$

Fig. 4.2. Paired emulsifier system.

c) **Adjuvants.** These are added to pesticide formulations to improve quality, effectiveness and user safety. A variety of adjuvants are added to perform different functions. The penetrant additives help the formulation to penetrate through the protective lipid membrane of the insect pests and enhance the rate of reaction of the pesticide at vital reactive sites.

d) **Deactivators.** The carriers and diluents of dry formulation such as clays and minerals have the property of surface acidity which catalyze the deactivation of pesticidal molecules. Thus deactivators are generally employed organic molecules such as ethers, ketones, esters, amine and glycol, which contribute a pair of electrons to the acid sites of the catalytically active substance to reduce decomposition.

e) **Anti-caking agents.** Caking sometime occurs during storage of dry formulations such as dust concentrates, WDP and granules due to coalescence of individual particles to form solid hard lumps. Diatomaceous earth or fine synthetic silica and silicates are added as anticaking agents.

f) **Dry lubricants.** In order to improve the flow of the formulations, dry lubricants such as graphite, soapstone, talcs and metal stearates are added.

g) **Protective colloids.** Protective colloids derived from polymeric materials like polyvinyl pyrrolidine, sodium carboxymethyl cellulose and collagen are added to improve the quality and inhibit the agglomeration and sedimentation of liquid formulations or aqueous dilution of WDP formulations.

h) **Stickers.** These are adjuvants sometimes added to spray tank prior to the pesticide application in order to prevent the run off of spray solution when applied to crops.

i) **Anti-dusting agents.** WDP and granular formulations contain a.i. in finely powdered form hazardous to operators to minimize undesirable effects, antidusting agents such as glycerine are added to prevent the dustiness.

j) **Anti-foaming agents.** To prevent excessive foaming during dilution of the formulation, antifoaming agents such as liquid silicone or higher homologues of alcohol are added to reduce froth formation and maintain proper spraying conditions by the operators.

Methods of analysis [13-16]

The analytical methods to determine the active ingredient (a.i.) content are readily available with Collaborative International Pesticide Analytical Council (CIPAC) and World Health Organisation (WHO). The conventional methods of analysis of formulations are volumetric, potentiometric, gravimetric, colorimetric and spectnophotometric. These methods are still being used commonly in spite of some limitations. Alternatively, one can use for thin layer chromatography (TLC) which can be employed even for chemical identification and cross contamination tests. The recent analytical tools are gas liquid chromatography (GLC) and high performance liquid chromatography (HPLC) which are rapid and give reproducible and accurate data. A number of senstive detectors of GLC and HPLC such as FID, NPD, ECD and UV can analyse minute quantities of formulated products with great accuracy [6]. The chiral column (CC) analysis both by GC and HPLC are now becoming popular. The supercritical fluid extraction (SFE) analysis of environmental samples for various compounds can also be recommended for extraction of pesticides from solid formulations. Another improved extraction procedure known as accelerated solvent extractor (ASE) can also be employed for analysis of environmental samples which takes very less time (10-15 min) and energy. This extractor can be employed for extraction of solid formulations. The most recent one is capillary electrophoresis (CE). which is extremely suitable for pesticide residue analysis. However, its use on formulation analysis might create a problem because capillary column can accomodate only microgram quantity of analyte which means large dilution of the samples is required resulting in an increase in error. Other analytical methods use infrared spectrophotometer (IR), nuclear magnetic resonance spectrometer (NMR) [2] and atomic absorption spectrophotometer (AAS) which also give dependable accurate results. The realistic methods of analysis using other properties such as specific gravity, specific rotation, refractive index, emulsification, apparent density (for dusts and granules) colour, pH, particle size, dispersibility and acidity or alkalinity are also recommended by recognised institutions and research and development organisations.

The analytical control laboratory should be headed by a technical officer who maintains a detailed inventory for each formulated product comprising specifications, methods of analysis and analytical results.

Shelf life

Crops may be adversely affected if pesticides are applied after the expiry date. Partial or total damage of the crops may occur because of the presence of toxic degradation products formed during storage. The stability of formulations depends on weather and storage conditions, quality of packaging and kind of transport. The inerts added in the pesticide formulations should not have any adverse effect on the active ingredient. The acid and basic sites, invariably carried by the clay dilutents cause decomposition of pesticides. Hence, taking proper precautions at the time of formulation and production become imperative for the stability of the pesticide. The minimum shelf life of the formulation should be 1.5-2.0 years. This would effectively contain transportation and storage loss and facilitate creation of a reasonable stock ensuring off the shelf availability of the product to the consumer. Hence, before recommending a formulation for mass production, samples should be tested for a.i. content after storing at 25°C for 18 months, 37°C for 12 months and 45°C for 3 months. The auxilliary attributes such as emulsification, appearance, suspensibility wetting time and caking of the formulations must be taken as adjuncts.

Safety aspects

Persons engaged in pesticide formulation and production plants are often exposed to relatively high levels of pesticides. This makes them prone to allergic reactions, respiration problems, eye and skin irritation and a variety of other health hazards including carcinogenic reactions. In order to minimize these risks, the workers should be encouraged to use protective clothings, gas masks, gloves and other devices. They should be trained to carefully handle the equipment during production as well as in packaging. No person should be allowed to work alone in the laboratory/plant. Proper warning labels should be put on the chemicals which are known carcinogens, mutagens and teratogens. Drinking and smoking should be strictly prohibited inside the plant. Proper ventilation to keep inhalation exposure within permissible limits must be maintained. Blood analysis of plant workers must be carried out annually. In addition to mamalian toxicity, protective measures should be taken against explosion.

a) **Fire.** The plant should have emergency exit plan, prohibition of smoking and proper marking for fire exit, and location of fire extinguishers. An efficient alarm system and functional emergency showers are indispensible.

b) **Explosion.** Workers must be familiar with explosive nature of the chemicals being handled and proper explosion shields and eye protectors should be located within easy reach.

c) **Storage.** All volatile chemicals must be stored in a covered area preferably in an underground enclosure and extra precautions should be taken during the use and storage of highly inflammable solvents.

d) **Disposal.** Solvents should be evaporated in fume hoods, and not discarded in drain.

Application of modern tools in pesticide formulation [17]

i) **Computers** are being employed very effectively for evaluation and optimization of pesticide formulations. Computer assisted correlation analysis of both physical and biological properties of any formulations and its economic viability based on types and concentrations of the ingredients can be carried out smoothly and rapidly. Statistical methods are utilized to measure the degree of correlation between responses and independent variables. The computer programmes can also be designed to operate equipment used to evaluate pesticide formulations and to carry out complex theoretical calculations to guide choice of ingredients. Computer aided technique is used for the evaluation of sedimentation rates/shelf life of flowable formulation and solubility parameter theory to optimize the choice of emulsifiers for EC formulation.

ii) **Laser.** The laser technique allows the application of basic colloidal theory to the practical development for pesticide formulation. It has been found that alachlor microcapsule shows enhanced bioactivity as compared to conventional formulation technique.

iii) **Fluidized bed granulation.** In this process granulation and drying occur concurrently in the same unit. This eliminates operator exposure, protects dust explosion and does not contaminate the environment. The process is ideally suited for the preparation of controlled release formulations. In the granulation process one can convert a liquid formulation containing a surfactant into a solid formulation (WG and SG). Besides these a few more equipments such as particle size analyser (PSA), electronic particle counter (EPC) and rheometer are being used in pesticide formulations.

Environmental aspects

Escaping gases, obnoxious smell along with solid, powdery and liquid exhausts from the formulation plants contribute vastly to air, ground water and soil pollution. Plants manufacturing dust formulations produce noise from jet or hammer mills. Emulsifiable concentrate formulation plants consume large quantity of solvents like xylene which are highly inflammable and toxic to human beings. These pollution problems can be reasonably controlled by employing appropriate engineering techniques and plant management.

CONCLUSION

All the applied pesticides ultimately reach the soil, water and air and influence the eco-system. The ever increasing sacrilege of air we breathe, water we drink and food we ingest, has become a matter of great concern. While the use of these chemicals cannot be completely stopped, however the judicial use of more effective, target specific and ecofriendly pesticides can provide a respite from the looming danger of extermination of life. All these aspects must be borne in mind at the time of formulation and screening of the product. The formulation chemists must be involved in the decision making process for product development to ensure that the final product is ecofriendly. Emphasis should be laid on data generation, toxicity to non-target organisms and effect on persistence residues. Perhaps the slow release formulations with effectively targetted kill and easy biodegradability hold out the promise for ecologically safe pest control measures.

REFERENCES

1. Cross, B. Trends in agrochemical formulations. *Regulation of Agrochemicals' A Driving Force in their Evolution.* (G.J. Marco, R.M. Hollingworth and J.R. Plimmer, eds.). American Chemical Society, Washington D.C., 89-100 (1991).
2. Cross, B. and Scher, H.B. (eds.). *Pesticide Formulations, Innovations and Developments.* American Chemical Society Symp. Sr. 371, Washington D.C. (1987).
3. Corty, C. *Developments in formulations and application technology for safe use of agrochemicals. Fifth Internatn. Congr. Pestic. Chem.*, (J. Miyamoto and P.C. Keanney, eds.), Oxford, Pergamon, **4** (1983).
4. Valkenburg, W.V. *Pesticide Formulations*, Marcel Deckker, New York (1973).
5. Mathews, G.A. *Pesticide Application Methods*, Longman Scientific and Technical, John Wiley & Sons, New York, 52-70 (1992).

6a. Buchel, K.H. *Chemistry of Pesticides*, Wiley, U.K. (1983).

6b. Trought, K. *Trends in Pesticide Formulations*. Agrow. Reports; Surrey, U.K. (1989).

7. Valkenburg, W.V., Sugavanam, B. and Khetan, S.K. *Pesticide Formulation,* New Age International (P) Ltd., New Delhi (1998).

8. Tungikar, V.B. and Das, K.G. Controlled release pesticide technology. *Chemical Age of India*. **30** (9), 827-30 (1979).

9. Wilkins, R.M. Controlled release systems : Prospects for the future. *Proc. Eighth Internatn. Congr. Pestic. Chem.*, Options 2000, (N.N. Ragsdale, P.C. Kearney and J.R. Pimmer, eds.) American Chemical Society, Washington, D.C., 96-103 (1995).

10. Shieh, T.R. Biopesticide formulations and their applications. *Proc. Eighth Internatn. Congr. Pestic. Chem.,* Options 2000 (N.N. Ragsdale, P.C. Kearney and J.R. Plimmer, eds.) American Chemical Society, Washington, D.C., 104-114 (1995).

11. Larson, R.O. *Focus on phytochemical pesticides, 1. The Neem Tree* (M. Jacobson, ed.) CRC Press, FL, USA (1989).

12. Shieh, T.R. Biopesticide formulation and their applications. In : *Formulation of pesticides in developing countries* UNIDO, U.N. Publications, Sales No. E. 83, 11.B.3; 1D.297 (1983).

13. *CIPAC Handbook*, Analysis of Technical and Formulated Pesticides. Collaborative International Pesticides Analytical Council, Cambridge, England, Black Bear Press, **1**, (1970); **1A**, (1980); **1B** (1983) and **1C**, (1985).

14. *CIPAC Handbook, Analysis of Technical and Formulated Pesticides.* (A Martin and G. Dorbat, eds.) Collaborative International Pesticides Analytical Council, Cambridge, England, Black Bear Press, **D**, (1988); **E** (1993) and **G** (1995).

15. *CIPAC Handbook, Physico-chemical Methods for Technical and Formulated Pesticides.* (W. Dorbat and A. Martin, eds.) collaborative International Pesticides Analytical Council, Cambridge, England, Black Bear Press, **F**, (1995).

16. *Official Methods of Analysis*, 16th Ed. (Patricia Cunniff, ed.), Virginia, USA, Association of Official Analytical Chemistry (1995).

17. *Advances in pesticide formulation technology* (H.B. Scher ed.). American Chemical Society, Symp. Sr. 254. Washington D.C. (1984).

CHAPTER 5

Biotechnology in Pest Management

- Introduction
- Predators
- Parasitoids
- Fungi
- Bacteria
- Viruses
- Nematodes
- Bioherbicides

- Botanical pesticides
- Resistance
- Antagonist
- Biotechnology in IPM
- Transgenic plants
- Advantages and disadvantages
- Protection of industrial rights
- *References*

Domestication of an organism for the benefit of mankind can be best described as biotechnology. The carefully developed and suitably tailored biotechnological systems have the merits of being efficient, economical and ecofriendly. With the application of biotechnology in pest control, disease management in crop plants is expected to become more efficient and cost effective with no adverse effect on the ecosystem. Use of biopesticides, botanical pesticides, parasites and predators of pests and antagonistic pathogens and development of genetically engineered crop varieties carrying the disease resistance traits hold promise for exploitation.

Indiscriminate use of broad spectrum pesticides has not only affected the non-target microbial denizens of the soil which play a vital role in maintaining the nutrient balance, but has polluted the ecosystem, particularly the aquatic environment. The situation is increasingly becoming more alarming because of the emergence of new diseases and acquired resistance in the causative organisms necessitating the formulation of stronger and broad spectrum chemicals.

Efforts are afoot to develop ecofriendly, pest-specific pesticides such as biopesticides, synthetic pyrethroids, sulfonylureas, juvenile hormones, pheromones and attractants. Various biotechnological contraptions are now being used to induce disease resistance in plants. The varying degree of spontaneous disease and pest resistance in plants has been widely exploited in the breeding programmes to develop resistant crop varieties. Biotechnology has opened up the new field of obtaining pesticidal chemicals from genetically modified cell cultures and enhanced production of growth limiting exudates from higher plants.

Predators

Many organisms have developed a taste for variety of pests. Predators such as dragon flies, ground beetles, lady bird beetles, spiders, toads, frogs and birds generally require several preys to complete their life cycle [1]. They are pivotal in maintaining the natural balance and limit their population. A variety of organisms belonging to orders Hymenoptera, Diptera, Coleoptera and Homoptera are being successfully used to control insect pests of crops like coffee, rice, sugarcane, pulses and tobacco. The control of cottony cushion scale, carried out by the predator, *Rodolia* beetle all over the world, is a classical example.

Parasitoids

Organisms parasitising the insect pests of crop plants are called parasitoids. Unlike the predators which consume the host, these organisms live within the host to complete a part of life cycle and in the process kill the host. They are generally specific and require only one host. The important parasitoids [1] are Trichogrammatid, Braconid, Eulophid and Chalcid wasps which are used to control insect pests of crops. The cabbage butterfly is well parasitized by *Apanteles glomeratus* at larval stage and *Pteromalus puparum* at pupal stage. *Trichogramma* spp. reduces spotted, pink and American bollworm incidence in cotton.

Pathogens

Pathogens are the disease causing organisms such as bacteria, viruses, fungi and protozoa which cause infection and kill the host. Some of the commercialized pathogenic pesticides are :

Fungi

The fungi that generally parasitise insects are Phycomycetes and Basidiomycetes. The mycelium of the fungus pierces through the integument and enters body cavity of the insect and finally kills the insect. A few commercially available pathogenic fungi which control insects pests are *Verticillum lecanii* and *Entomophthora* spp. (aphids, mites). *Nomurea releyi* (lepidopteran larvae, alfalfa weevil) and *Beauveria bassiana* (Colorado potato beetle, codling moth). Many fungi parasitise fungal hosts, *Coniothyrium minitans* invades the sclerotia of *Sclerotinia sclerotium* and *S. cepivorum*. *Fusarium roseum* and *Sporidesmium sclerotivorum* control the growth of *Cleviceps purpurea* and *Spirodesmium minor*. *Trichoderma,* is a much exploited fungus has been successfully used against *Sclerotium rolfsii, Rhizoctonia solani, Pythium aphanidermatum* and *Fusarium oxysporum* infection in lupine, tomato, sugarbeet, chickpea, brinjal and tobacco. *Trichoderma herzianum*, a fungus, protects Eucalyptus and teak plantations against *Rhizoctonia solani*.

Bacteria

It is known that bacteria produce bioinsecticides, bioherbicides, biofungicides and bactericides (antibiotics). Most of the bacteria used in biocontrol of insects belong to genera *Agrobacterium, Bacillus, Pseudomonas, Streptomyces* and *Serratia*. *Agrobacterium radiobacter* specifically controls the crown gall disease caused by *Agrobacterium tumefaciens*. *Bacillus subtilis* is used for seed treatment against *Sclerotium cepivorum* parasitising onion and *Fusarium roseum* which damages corn. *Pseudomonas fluorescens* and *P. cepacia* are active against *Rhizoctonia, Pythium* and *Fusarium,* causing damp off disease in crop plants, beans and clovers. The most popular and commercially exploited is the soil bacterium *Bacillus thuringiensis* (Bt) available in various formulations which are used to control a number of insect pests of crops. The protein delta endotoxin, formed during sporulation of the bacterium, also exhibits insecticidal activity. The bacterium enter the insect body through contaminated feed and multiply causing death of the insect either due to intoxication or septicaemia. Some known strains of Bt [2] controlling a number of insects, are *Bacillus thuringiensis* (Bt) (Japanese beetle, lepidopteran larvae), Bt. var. *israelensis* (dipteran); Bt var. *kurstaki* (Boll worms, fruit borer etc.) and *B. popillae* (Coleopteran insects). A large number of formulations of Bt like Dipel Bt (Caterpillar of fruit crops), Dipel 10G (Corn borer), M-one (Potato beetle), Vectobec (Mosquito and black fly larvae) are now available for use against lepidopteran and other insects. Another strategy is to transfer the entomotoxic gene 'Cry', responsible for producing the toxin, from *B. thuringiensis* to crop plants. Such transgenic plants have been developed in crops like soybean, maize, cotton, potato, tomato, squash and canola. These plants have already developed inbuilt resistance to variety of insect pests. Transgenic chickpea plants carrying the modified gene Cry IA from Bt, resist the attack by the pod borer, *Helicoverpa armigera*. Transfer of bar gene from *Streptomyces* spp. to tobacco, made the plant persistant to the herbicide 'bialaphos'. Genetically modified soybean grown in USA, carries a bacterial gene resistant to the herbicide 'glyphosate'.

Viruses

The viruses also control many lepidopteran insects including leaf eating caterpillars. The nuclear polyhedrosis virus (NPV) and granulosis virus (GV) are commonly used to control a number of insect pests [3]. NPV of *Helicoverpa armigera* are used for the control of cotton, pulses and vegetable crops. The granulosis virus of *Achaea janata* and *Chilo infuscatellus* are found useful to control semilooper and top shoot borer in sugarcane.

Nematodes

Beneficial nematodes provide below 'ground pest' control against root feeding insects like white grub, *Brahmina coriacee*, web worms and beetle grubs. Biovector is a nematode based biopesticide which is very active against cut worms, army worms, weevils, flea beetles, cucumber beetles and strawberry root weevils.

Bioherbicides

Many insects, fungi and bacteria are found useful to control a number of weeds. *Colletotrichum coccodes, C. malvarum* and *C. gloeosporoides* are found effective against weeds in cotton, rice and soybean while *Cercospora rodmanii* is useful against water hyacinth. *Phytophthora pulmivora* and *C. gloeosporoides* have now been registered in U.S.A. for commercial application [4, 5].

Botanical pesticides

Many products and preparations from plants competitively kill the insects and their easy biodegradability make them completely ecofriendly (cf. Chapter 2).

Resistance

The spontaneous resistance to insect pests can be developed in crop plants by modifying them at genetic level. This can be done through induced mutations or by picking up the desired chromosomal slice from the donor and introducing it into the gene sequence of the recepient.

Many excellent varieties of crops often have to be abandoned due to their high susceptibility to insect pests. Introduction of genes inducing resistance to insect pests in such varieties through genetic engineering is the only way to save the situation. Once suitable vectors are identified and the technique of transferring the desired trait is perfected, incidence of disease epidemics will become a rare phenomenon.

Antagonists

The micro-organisms used for suppression of plant diseases are known as antagonists. Several fungi like *Trichoderma, Gliocladiun* and bacteria like *Pseudomonas, Bacillus* and *Streptomyces,* have been found to be effective against number of plant diseases caused by soil borne fungi.

Farm yard manure and mulching applied as soil amendment may be beneficial because of their antagonistic effect on phytophagous nematodes. The fungus *Paecilomyces lilacinus* (Thom) and bacterium *Pasteuria penetrans* (Throne) also find useful application in biological control of phytophagous nematodes.

Biotechnology in Integrated Pest Management

Advanced techniques such as recombinant DNA, genetic engineering, modification of the enzyme processes, cell fusion, plant cell and tissue culture, clonal propagation, monoclonal and polyclonal antibodies, embryo and other germ cell manipulations, and process and system engineering have revolutionised agricultural production and diversified the agro based industry. These techniques, in addition to increase the production and improve the quality of the produce, also aim at making the pest resistance ability, a spontaneous attribute of the crop plant.

Another possibility is to transfer a gene segment coding for the synthesis of a pesticidal compound to a clone microbe and use this transformed system for large scale production of enzyme precursors, pesticides and repellent chemicals. The existence of human system like antigen/antibody response in plants may also offer a unique technique of protection through the development of immune system in the crop plants. Another promising area is the identification and action optimization of natural biostatic or biocidal products released or extracted from higher plants. The exudate from the roots of *Agropyran* spp. prevents the growth of weeds in its vicinity. Many pesticides show phytotoxicity in crop plants because of the presence of certain phytogens. To overcome this problem, the fusion of protoplasm technique using gene might be helpful.

Recombinant DNA technique, a tool for gene cloning and gene manipulation have made it possible to transfer genes from one organism to another by overcoming the species barrier [6]. This strategy has been successfully utilized to combat viral diseases. The transgenic tobacco plants show resistance to cucumber mosaic virus [7]. Another major breakthrough in nematology is the identification of a dominant *Meloidgyne incognita* gene to impart resistance to the root knot nematode [8]. The soil bacterium *Agrobacterium tumefaciens* causes aneoplastic disease in dicot plants called crown gall through the transfer of t-DNA to plant cells. This suggests that the bacterium can be used as a vector for the introduction of new genes into plants.

Transgenic plants

Presence of foreign genes in plants and their structure, expression, transfer and application have been comprehensively reviewed [9] with special reference to gene transfer in cereals. At present the only reliable transformation in cereals yielding transgenic plants is based on direct gene transfer using protoplasm based protocols. The production of transgenic crop plants is now combined with the construction of restriction fragment length polymorphism (RFLP) linkage maps for most major crop plants [10]. Integration of RFLP techniques into plant breeding procedures has facilitated the production of transgenic plants with increased disease resistance and stress tolerance through the transfer of the desired genes from related wild species. Genetically induced resistance to herbicides like glyphosate and sulfonylureas has already been developed [11, 12] using this technique.

Another approach is to develop herbicide resistance genes against detoxification pathways in susceptible plants. The difference in sensitivity is not always associated with a difference in susceptibility of the molecular target but an ability of the resistant plant to detoxify or break down the herbicide before it gets a chance to inactivate the susceptible target gene product. A microbial gene encoding an enzyme detoxifying herbicide has been modified into model species by transformation [13]. The detoxified enzymes show resistance to relatively high levels of phosphinothrin (Basta) and 3,5-dibrom-4 hydroxybenzonitrile (bromoxynil). The effectivity of the herbicide 2,4-D, which is degraded by a variety of microorganisms in the soil might be markedly enhanced by the detoxifying enzymes.

A few more potential applications of biotechnology are discussed in Table 5.1.

Advantages

The advantages of using biopesticides lie in their target specificity. They do not have adverse effect on non-target organisms including human beings and are readily biodegradable. The parasitising or predating organisms may provide a sustainably effective control against insect pests without disturbing the natural balance. With easy and cheap production protocols, these cost-effective and eco-friendly inputs will enable in minimising the production loss due to crop insect pests.

Disadvantages

While the efforts, both in terms of time and money, required to develop a genetically engineered system may be enormous, stability of the induced changes will greatly determine the long term efficacy of the biopesticides. Nevertheless, it will be a promising area for future. Systems like *Bacillus thuringiensis* carrying toxic genes are expected to provide enormous ground for developing effective biopesticides. However, the toxicity of the new chemicals or precursors produced by the engineered organisms has to be critically evaluated with reference to human health and ecological impact. Induction of intrinsic pest resistance through the implant of toxin producing genes in the crops should also be viewed with reference to productivity, nutritive and clinically safety value of the produce and stability of the change.

Protection of Industrial rights

Traditionally, processes and technologies developed are protected as industrial property by the patent law. It is doubtful however, whether the genetically engineered seeds microbes could be protected against 'piracy'. The hybrid seeds developed are protected under "International Conven-

Table 5.1. Some potential of application of biotechnology on agricultural crops

Crop	Principal problem	Area of improvement	Comments
Cereals			
Rice	(a) Sensitivity to drought stress (b) Poor tolerance to salinity and alkalinity (c) Severity of pest and disease damage (d) Poor utilization (loss)	1. Improvement of drought and salinity tolerance 2. Greater resistance to stem borers, brown hoppers, gall midge and leaf sheath blight 3. Improvement of nutritional and table quality of grains 4. Mechanical tissue improvement to prevent lodging of tall varieties	All improvements through gene transfer from other *Oryzae* species or the related plant varieties.
Wheat	(a) Peaking of fertilizer response in HY varieties (b) Excessive dependence on irrigation (c) Temperature related yield behaviour (d) Sensitivity to smut and rusts	1. High yield character with limited irrigation 2. Improved varieties for non-irrigated areas 3. Reduction in temperature sensitivity of yield. 4. Disease resistance	-do-
Pulse	(a) Severe pest damage (b) Low yields	1. Varietal development for marginal lands and un-irrigated areas 2. Resistance to caterpillars and pod borers	-do-
Oilseeds			
Mustard, Rapeseed	(a) Severe pest attack (b) Low yield (c) Low oil content	1. Varietal improvement for yield and suitability for growing in rainfed soils. 2. Improvement of oil content of seeds. 3. Induction/enhancement of pest resistance (specially against aphid)	-do-
Groundnut	(a) Sensitivity to 'Tikka' disease (b) White grub damage (c) Low oil content	1. Genetic improvement of yield potential and oil content 2. Induction of resistance to pests and diseases	-do-
Cash crops			
Cotton	(a) Severe incidence of bollworms (b) Fibre quality/oil content (c) Multiple flushes	1. Induction of resistance to bollworms/cotton pest complex 2. Change in flushing pattern/reduction of flushes 3. Improvement of staple and oil content of seeds	-do-
Sugarcane	(a) Borer attack (b) Low yields of the varieties grown in northern India	1. Resistance to pest attack 2. Varietal improvement for yield 3. Induction of early maturity	-do-
Tobacco	(a) Heavy photo respiration	1. Reduction of photo respiration	-do-

(Contd.)

Crop	Principal problem	Area of improvement	Comments
Jute	(a) Sensitive to drought (b) Preflowering and branching resulting in breakage of fibre (c) Retting quality and process	1. Induction of drought tolerance 2. Elimination of branching habitolus 3. Improvement of retting bacterial for speed and quality	Genetic engineering and tissue culture
Planatation crops			
Rubber	(a) High incidence of secondary leaf fall (disease)	1. Resistance to *Phytopthora palmivora*	-do-
Tea	(a) Sensitive to drought (b) Distinct "banjhi" period (rest period) (c) Low yields of hill tea varieties (d) Poor aromatic quality of plains tea	1. Induction of drought tolerance 2. Reduction/elimination of "banjhi" period (resting period) 3. Flavour and yield improvements 4. Modification of leaf disposition to improve photosynthesis	-do-
Fruits	(a) Biennial bearing habit e.g. mango (b) Poor keeping quality e.g. litchi, mangoes, banana etc. (c) Pest and disease damage e.g. hoppers in mango, scab in apple etc.	1. Induction of annual bearing habit 2. Disease resistance (apple) 3. Pest resistance (mangoes, litchie and others) 4. Improvement of keeping quality (banana, mangoes and litchi)	-do-
Cardamom	(a) Severe disease incidence (b) Low yields	1. Introduction of disease resistance 2. Genetic improvement for yield characteristics	-do-

Source : M.G. Srivastava, Pesticides, April 1987.

tion for the Protection of New Varieties of Plants" (1961) for a period of 18 years. Unless a system of protection is evolved such discovery will remain restricted. One can prevent the reuse of the hybrid seeds by introducing a terminator gene which robs the produced seeds of their hybrid vigour. The unwanted fall out of this strategy will be the transfer of terminator gene to the unintended crop plants during cross pollination due to which the farmer will be unable to produce and use even the traditional seeds of the crops. There are many other constraints of technical, economic and legal nature arising from biotechnological applications. International and national co-operation and commitment is essential for biotechnological applications in agricultural and horticultural crops. Although some problems have emerged in areas like crop protection however it is possible to inject a high degree of resistance in most crop varieties. It will be incorrect to ignore the biodynamics of pests and pathogens to successfully adapt to the changed environment through natural selection or mutation. Therefore, the pesticide industry should gear itself to utilise the new opportunities that are being opened up by biotechnology. Already a few Indian laboratories have made a beginning in this frontier area of biotechnological research but their efforts are very meagre in comparison to developed countries. It is necessary that these efforts are coordinated and channelised towards well defined objectives to avoid duplication and repetition and make them more relevant to our national needs.

REFERENCES

1. Singh, S.P. Bio-organisms : Use and potential. **In** : *Botanical and Biopesticides.* (B.S. Parmar and C. DevaKumar eds.), Westvill Publishing House, New Delhi, 131-164 (1993).

2. Kulshrestha, G. Biotechnology application in pest management. *Ann. Pl. Protec. Sci.* **3** (1), 1-14 (1995).

3. Battu, G.S. and Ramakrishnan, N. Efficacy of nuclear polyhedrosis virus against *Spilosoma obliqua* (Walker) on cowpea. *J. Biol. Control*, **3**, 99-102 (1989).

4. Hatzios, K.K. Biotechnology application in weed management—now and in the future. *Ad. Agron.* **41**, 325 (1987).

5. Senez, J.C. The new biotechnologies. *The Courier*, 4 (1987).

6. Obukowicz, M.G., Perlak., F.J., Kusano-Kretzmer, Mayer, E.J. and Watrud, L.S. Integration of the delta endoxin gene of *Bacillus thuringiensis* into the chromosome of root colonising strains of *Pseudomonas* using Tn 5 Gene, **45**, 327-331 (1986).

7. Harrison, B.D., Mayo, M.A. and Bulcombe, D.C. Virus resistance in transgenic plants that express cucumber mosaic virus statellite RNA. *Nature*, **328**, 799-802 (1987).

8. Rick, C.M. and Fobes, J.J. *Tomato Genetics Co-operatives*, **24**, 25 (1974).

9. Weising, K., Schell, J. and Kahl, G. Foreign genes in plants; transfer, structure, expression and application. *Ann. Rev. Genetics* **22**, 421-477 (1988).

10. Tanksley, S.D., Young, N.D., Paterson, A.H., Bonierable, M.W. RFLP mapping in plant breeding : new tools for old science. *Biol. Technol.*, **7**, 259-266 (1989).

11. Comai, S.L., Faccioth, D., Hiatt, W., Thompson, G., Rose, R., Stalker, D. Expression in plants of a mutant aro A gene from *Salmonella typhimurium* confers tolerance to glyphosate. *Nature* (London), **317**, 741-744 (1995).

12. Haughen, G., Smith, J., Mazur, B. and Sommerville, C. An Arabidopsis acetolactate synthase gene in tobacco confers resistance to sulfonylurea herbicides. Molecular and General Genetics (1988).

13. Stalker, D., McBride, K. and Malyj, L. Expression in plants of a bromoxynil-specific bacterial nitrilase that confers herbicide resistance. **In** : *Genetic Improvement of Agriculturally Important Crops : Progress and Issues* (R.T. Fraley, N.M. Frey and J. Schell eds.) Cold Spring Harbour Laboratory, Cold Spring Harbour, 37-40 (1988).

CHAPTER 6

Pesticide Residues and Their Environmental Implications

- Introduction
- Environmental implications of pesticide residues
- Residues in soil, water and water bodies
- Pesticide contamination of food
- Decontamination and detoxification
- Toxicological properties of pesticides and their safe use
- Minimization of pesticide residues in the environment
- Analysis of pesticide residues
- Multi-residue methods
- Design of experiment
- Sample preparation
- Collection of field sample and storage
- Extraction and clean up
- Supercritical fluid extraction
- Identification and quantification
- Instrumental methods of analysis
- a) Chromatography (TLC, HPTLC, GLC, HPLC)
- b) Spectroscopy (IR, MS, NMR)
- c) Biotechnology based analytical method (Immunoassays/ELISA)
- Validation and confirmation of residue identity
- Sanitation or phytosanitation
- *References*

The pesticide residue is a substance or a mixture of substances in food, feed, soil, water and air originating from the use of pesticides and includes the specified degradation and conversion products, metabolites, reaction products and impurities. These are considered to be of toxicological significance and have varying duration of persistence. Pesticides with longer persistence exert the desired effect for a longer duration but influence the non-targetted organisms also by entering their bodies through air, water and food. This problem becomes more acute with systemic pesticides which tend to accumulate in the system to attain toxic levels.

While the use of pesticides has become indispensible for protecting the agricultural produce from pest damage, their indiscriminate use has polluted almost all conceivable habitats. Some of the common contaminants with longer persistence comprise of the organochlorine group of compounds like DDT, BHC, aldrin and heptachlor. These can find their way into cereals, vegetables, fruits, pulses and oil seeds as well as in soil, water and animal tissues.

Incidences of pesticide poisoning in human beings and toxicity to other mammals, fish and wild life have been reported from time to time. Data on pesticide residues from supervised trials are being generated by various agricultural universities and Indian Council of Agricultural Research (ICAR) institutions in the country [1, 2, 3]. The All India Co-ordinated Research Project on Pesticide Residues sponsored by the ICAR, has further accelerated the pace of research in this area. However, still the quality and quantity of the data generated so far are not enough for formulating guidelines on the judicious and ecosafe use of pesticides.

Pesticides in different formulations such as liquids, dusts or granules are generally applied to the soil, plants and stored products. Animal and human bodies get exposed to these chemicals during their application. They become air borne through volatilization and wind action and ultimately reach the soil through precipitation, and ground water through percolation, leaching and run-off. Many chlorinated pesticides like DDT, BHC, aldrin and endrin are non-biodegradable and highly persistent. DDT and dieldrin appear as common contaminants in rivers and streams of temperate regions. Bioaccumulation of these persistent pesticides results in biomagnification along the ascending trophic chain and reach fatal levels [4, 5, 6].

Environmental implications of pesticide residues

The persistence of pesticides in the environment depends upon chemical nature, formulation, dose, method of application, prevailing temperature, sunlight, humidity, air movement and microbial complement of the soil. Except the chlorinated pesticides, other pesticide formulations like carbamates, organophosphorus and pyrethroids are easily biodegradable and highly susceptible to chemical hydrolysis and enzymatic breakdown.

Analysis of market commodities in various parts of India showed widespread contamination of vegetables with pesticide residues. In Bombay, out of 313 vegetable samples assessed, nearly half of the samples about 153 contained residues of HCH, lindane, DDT, aldrin and endrin and in many cases their levels were above the permissible limits. It was also found that leafy vegetables and potatoes had higher residue as compared to others. In Andhra Pradesh, 145 out of 350 samples were found to be contaminated with organochlorine insecticide residues. In Delhi 80 per cent of vegetable samples were contaminated with HCH and DDT residues. Similar contamination with HCH and DDT residues in vegetable samples were reported from Haryana, U.P. and Marathwada. The residue scenario was not different with edible oils, oil seeds, pulses, cereals, fruits, milk and milk products, meat, fish, crab and egg samples collected from different parts in India.

Residues in soil

Pesticides ultimately find their way to the soil either directly or indirectly through washing down by rain water or blown down by wind. Thus, soil constitutes a major environmental sink for most of the pesticides from where they are taken up by plant, move into bodies of invertebrates or contaminate the ground water [6]. The animals that feed upon those invertebrates may in turn concentrate these systemic and persistent toxins to levels that may kill them. These residues also interefere with the activities of microbial component of the soil which is vital for maintaining the soil fertility.

Residues in water and water bodies

Pesticides reach water bodies like rivers, lakes, and ponds through various routes like surface run off, aerial spray, spray drift, industrial wastes discharge, sewage effluents and accidental spillage

[7]. These contaminants lead to poisoning of aquatic invertebrates and fish mortality and may also affect fecundity in fish.

Pesticide contamination of food

Pesticide contamination of food commodities [2] is of paramount importance for the export market which contribute substantially to the national income. If the food commodities do not meet the importers specifications, the consignment is usually refused entry or fetches much lower price as it is put to unintended use. This means that any country which does not monitor its export commodities for pesticide residues, runs a grave risk of having them rejected and suffering heavy economic losses. Similarly, a country that does not have a system for proper monitoring and control of contaminated imports will be identified by other countries as dumping ground for sub-standard food commodities.

These compelled many countries to impose restrictions on the use of persistent pesticides since 1964. DDT was the first to be banned because it readily accumulates in the animal tissues. This was followed by others like BHC, aldrin, dieldrin, heptachlor and chlordane. Now these are completely banned in U.S.A., European countries, Japan, Australia and India for agricultural use. This restriction was prompted by the voluminous information collected on persistence of pesticides like DDT, BHC and cyclodiene insecticides such as aldrin and heptachlor in soil, water and plants, animals and food commodities in countries like USA, European countries and India [1, 2, 8, 9].

Decontamination and detoxication

Repeated use of pesticides leads to the accumulation of toxic residues on fruits, vegetables and cereal crops making them unfit for human consumption. Such food stuff must be decontaminated before consumption. A number of ways are now recommended for decontamination and detoxification, namely a) physical, b) chemical and c) biochemical methods.

The physical methods include peeling, washing, steaming, cooking and boiling which can reduce residues significantly from crop material. The chemical methods comprise of oxidation, reduction and hydrolysis and are applied according to the nature of crop and pesticide. The biochemical methods employ micro-organisms and enzymes for detoxification of pesticides.

Toxicological properties of pesticides and their safe use

One of the important pre-requisite of developing and commercializing a pesticide is toxicological evaluation of the product. Besides acute toxicity against rats and mice, the possible interference with the ecological equilibrium are examined very carefully.

The toxicity is usually recorded as the dose required to kill 50% of the population of test animals (LD_{50}) and is expressed as mg/kg of the body weight of the animal. The administration of toxicant can be oral, intravenous injection and skin absorption (dermal). The degree of toxicity is of the order intravenous oral dermal. In general, oral toxicity values against rats are quoted to indicate the mammalian toxicity of pesticides. The lower the LD_{50} value, the higher is the toxicity of the chemical and its gradation with respect to toxicity (cf. Chapter 11).

The following practices can be adopted to minimise the toxicity.

a) Maximum residue limit (MRL of tolerance)

Maximum residue limit (MRL) is the optimum safe concentration of the left-over pesticide or its degradation products according to "Good Agricultural Practices" (GAP). Good agricultural practices (GAP), in the use of pesticides, are defined as the officially recommended or authorised use of pesticides under practical conditions at any stage of production, storage, transport, distribution and processing of food and agricultural commodities and animal feed, keeping in mind variations in the requirements within and between regions and the maximum quantities necessary to achieve adequate control, so as to leave harmless quantity of residues [10a, 10b]. Care must be taken during application of agro inputs such as good seeds as well as dosages of fertilizers and pesticides and irrigation intervals with minimum application variation in requirements within and between different regions, so as to have good quality produce. The other important factors such as harvesting, processing, transport and storage are followed after proper standardization to minimize the variation. It has been observed that under GAP conditions, the pesticide residues become minimal and also toxicologically acceptable without any health hazards. The concentration is expressed as mg of pesticide/kg of commodity. The MRL (tolerance) of pesticides for different agricultural commodities have been fixed by the joint deliberations at a meeting of FAO/WHO. In India, the MRL is fixed by Ministry of Health, under Prevention of Food Adulteration Act, 1954. The most important feature of fixing these limits is to ensure that a commodity treated with pesticides is fit for human consumption. The tolerance provides an assurance to the general public that the food which is offered for consumption is safe. Some common terms used in calculation of MRL are discussed below.

b) Good Laboratory Practices (GLP)

Good laboratory practices originally applied to toxicology studies have now been extended to field studies with reference to residues and fate of pesticides in the ecosystem. The main objective of the GLP is the quality assurance system dealing with the organisational process and conditions under which non-clinical health and environmental safety studies are planned, conducted, monitored, recorded, achieved and reported.

There are many steps which need to be taken to ensure that the results reported reflect the true position. The Codex Committee on Pesticide Residues (CCPR) [11a] has recommended guidelines for good practices in pesticide residue analysis. All the three basic aspects such as analyst, basic resources and analysis have been duly emphasized. According to Carl [11b], the methods described in detail or even standardized are not sufficient to assure reproducibility and accuracy of the results. Analyst's experience and good laboratory handling play the most important part. Variations in results arise due to independent analysis protocols adopted. It is essential to carry out a blank determination to ensure that solvents, reagents or the equipment do not contribute to the inconsistency of the results. Analytical reference standards are without doubt the key to good or bad analytical results. Samples spiked with the parent compound or any known metabolite, ought to be analysed to determine the loss of these compounds during analysis. The periodic check on GLC determination step must be carried out. Introduction of internal laboratory control can prove useful in imparting reproducibility which can be monitored by introducing check samples at regular intervals.

To carry out quality scientific field trials, the most essential requirements are to have a well designed experiment, trained personnel and good analytical instruments, which can generate internationally acceptable, accurate, reliable and reproducible data following standard operating procedure. These data are to be stored in best possible conditions for regulatory authority.

Recently the GLP has gained tremendous importance after liberalisation of industrial policy and GATT agreement. In India, there is neither any GLP compliance authority nor GLP auditors for monitoring the system existing in the laboratories. These need to be looked into carefully in view of the WTO agreement.

c) Acceptable daily intake (ADI)

Acceptable intake is the daily exposure to a level of pesticide residue which does not manifest into an appreciable risk during the entire life time. This appreciable risk implies a practical certainty that injury will not result even after a life time exposure and is expressed in terms of mg of residue ingested per kg of body weight per day.

d) Waiting period

It is expressed as tol value 'T' and is defined as the period required for the residual level of the pesticide to decline to tolerance limit (MRL value). In other words waiting period is the time lag between pesticide application and harvest of the crop for consumption. Observance of waiting period becomes an important factor to determine the suitability of consumption of the treated samples. The crop should not be harvested until the pesticide residue level of the edible portion of the plant samples reaches below the MRL value. One can thus minimise the hazards of pesticide ingestion by observing the recommended waiting period.

Minimization of pesticide residues from the environment

Some suggestions for minimizing the pesticide residues

  i) Use pesticides only when absolutely necessary. Preferably use biopesticides and non-chemical methods of pest control.

 ii) Apply only appropriate pesticides at recommended dosages.

iii) Avoid using persistent pesticides like DDT, BHC, aldrin, chlordane, endrin on vegetables, fruits and fodder crops and directly on animals for veterinary purposes.

 iv) Never mix pesticides directly with stored grain or use them to disinfest storage premises.

  v) Pluck ripe fruits and vegetables before pesticide application. After pesticide spray harvest the crop only after the recommended waiting period.

 vi) Thorough washing followed by rubbing can help in reducing pesticide residues on the produce. Peeling of vegetables and fruits also results in lowering the levels of residues.

Pesticide residue analysis [12a, 12b, 12c]

Multi-residue methods

Pesticide residues in crops can be determined either by single or multi-residue methods when the application history is not known or crops are treated with more than one pesticide. The multi-residue method determines residue of each pesticide and its degradation products within a short period with minimum number of runs (multicomponent) and is applicable to a variety of environmental and food samples likely to contain a number of pesticide residues. The method of extraction of pesticide is so chosen that a large number of pesticides can be extracted, irrespective of low recoveries, and are then analysed by GLC method fitted with capillary column.

The extent of contamination of food commodities with pesticides residues is regulated by variety of factors including growth, dilution and adsorption on the surface. The residue resulting from a given method depends on timing and dosage of pesticide application which may vary with site and climate. The limits of such variations become important for the assessment of safety and particularly to the establishment of maximum residue limits (MRL). In order to determine the maximum residue level, the produce should be analysed after treatment with a known pesticide, following good agricultural practices (GLP).

Five basic steps are to be followed in order to obtain reliable and reproducible data in pesticide residue analysis. These are : i) design of the experiment, ii) sample preparation, iii) extraction and clean up, iv) detection and quantification and v) confirmation of residue identity.

i) Design of experiment

a) In crop residue analysis, the design of the experiment is very critical and important. The sample subjected to residue analysis should truly represent the crop treated with a given pesticide.

b) The trials should be carried out with the commercial formulations of the intended pesticides applied uniformly using conventional gadgets.

c) Since the objective of residue studies is to provide the basis for the estimation of maximum residue levels (MRL), the design of the experiment should be aimed at determination and evaluation of the conditions and factors that lead to the highest residue levels.

d) The trials to determine residue levels under various conditions should be conducted with different varieties at different growth stages under varying agroclimatic conditions and cropping patterns.

e) When applied on a standing crop, the prime objective should be to determine the amount of pesticide residue still persisting on the crop at the time of harvest. If significant residue is found in the harvested crop, it is necessary to assess its fate during storage and processing and likely effects on the consumers.

f) The size of the experimental plot should be large enough for taking representative samples. A significant buffer zone should be left between the experimental plots to prevent cross contamination.

g) At least two dosages should be included in each residue experiment, one at the maximum recommended dose and the other at double the rate.

h) An untreated plot close to the treated plot should be maintained to serve as source of control samples.

ii) Sample preparation

Sampling should be done so as to collect the true representative sample of whole of the population. Since it is normally not possible to take observations on the entire population, the data are recorded only on a fraction of randomly selected population representating the whole lot.

Collection of a field sample

Samples should be lifted from randomly selected spots in a plot. The number of spots is determined by the plot size. All the samples from each plot should be homogenised and pooled together. From this mixture, the required amount of sample should be saved for relevant analyses.

In an experimental field, it is necessary to do cleaning, pruning, trimming and selection of the crop components followed by mixing and subdividing, to bring the sample size to approximately 250 g without interfering with the residue level of the sample. General sampling procedure of crops and food commodities have been described by several workers but air samples require special care as variables like wind, rain, temperatures and snow are essentially unpredictable.

Storage of samples

The containers for storing samples should preferably be made of glass while the plastic and cardboard containers should be avoided. The samples must be stored in a cold room or deep freeze to prevent decomposition. The degradation is faster in chopped and homogenized samples due to higher enzymatic activity.

iii) Extraction and clean up

Extraction

Various extraction procedures such as solvent stripping, blending and Soxhlet extraction for pesticide residue from different substrates have been developed. While these methods are good for plant samples, extraction of pesticide residues from soil, water, air and animal tissues and contamination with fatty materials pose problems. The extracting solvent must be suitable for the extraction of compounds with wide range of polarity from various matrices containing different amount of water, fat, sugar and other substances. In most cases, absolute extraction is not possible and more often the observed residue values are found misleading. Fortification is considered as an excellent technique to evaluate the clean up procedure but does not prove to be an efficient extraction process. A peculiar situation arises when degraded pesticide molecules like phenols form conjugates during extraction. In order to deal with these problems, preparatory steps must be taken before analysis. The conjugates must be hydrolysed either by acid or enzyme. The extraction also depends upon the purity of the solvents and reagents which substantially affect the accuracy of the results.

Clean up

Extracts of pesticide residues are always contaminated with co-extractives like plant pigments, fats and other organic compounds which interfere with their estimation. The process of removal of these contaminants before analysis is known as clean up.

Residues and extractives from different environmental matrices exhibit similar physical and chemical properties and may be classified as i) lipophilic, ii) hydrophilic, and iii) amphilic. Clean up procedures extensively used by most analysts fall into two categories a) partitioning where the pesticides are partitioned utilizing the property of greater affinity of the residue to different solvents than the interfering substances, and b) liquid/solid column chromatography using various sorbents like florosil. However, column clean-up procedure using florosil does not yield reproducible results. If at all florosil is used, a standard solution of pesticide must be added in the control extract to determine the adsorption/elution pattern of the particular batch. Other clean-up procedures are c) thin layer chromatography, d) sweep co-distillation, e) cold precipitation, f) sublimation, and g) chemical methods. Based on the principle of sweep co-distillation, the Universal Residue Extractor (URE) and Accelerated Solvent Extraction (ASE) systems were introduced for the recovery of pesticides and organic residues from a wide range of samples like meat, fats, cheese and vegetable oils. Both the extraction procedures are fast and efficient with low operation and maintenance cost.

Supercritical fluid extraction (SFE)

Supercritical fluid extraction using CO_2 is an easily manageable and automated procedure which scores over the liquid based extraction method in eliminating the use of hazardous solvents and at the same time economising on time, cost, labour, glassware and laboratory space. SFE has now achieved a high degree of selectivity during the extraction process unlike liquid extraction, which requires clean up before analysis [13, 14]. Several studies on estimating multi-residues of pesticides in food stuff using different SFE conditions with minimal cost have been carried out successfully. In exceptional cases however, the recovery of residue may not be absolute but with modifications, SFE satisfactorily covers compounds with wide range of polarity. It can extract variety of pesticides from different matrices using water absorber Hydromatrix (HMX) [15], a diatomaceous earth material, for direct analysis by GLC and HPLC techniques. Even aqueous extraction of highly pH dependent pesticides gives satisfactory results [16].

(iv) Identification and quantification

Positive identification

Positive identification of pesticides is important in the samples from nature specially in such samples where no information on the source of residues is available. In addition, metabolism, photodegradation and breakdown products may further complicate the analysis as the clean up procedures do not remove all the interfering substances from the extractives.

For positive structural information at the residue level the best available method is GC-MS spectrometer. NMR also provides most of the structural information as MS but it is not very sensitive for residue level analysis. FT-IR and multicolumn GLC and HPLC are extensively used for identification of pesticides. Chemical derivatization now-a-days is commonly used for positive identification of pesticides. Commonly used procedures are addition, oxidation, rearrangement, dechlorination, reduction and dehydrochlorination. This method of chemical derivatization continues to play an important role in both qualitative and quantitative analysis of pesticides and metabolites because of i) increase in sensitivity of detection and selectivity for identification by GLC and HPLC detectors, and ii) increase in thermal stability of compounds with improved extractability.

Instrumental methods of analysis

The analysis of pesticide residue is a highly complex procedure with number of limitations all along the operation. Therefore, it is necessary to build up a set of controls designed to prevent erroneous conclusions. A continuous programme of analysis at regular intervals, of very carefully prepared homogenous samples of known a.i. is developed to monitor the performance called as 'Analytical Quality Assurance' [11b]. If analytical data is generated by a number of laboratories, the desired level of reliability can be attained and maintained only through such a programme.

Chromatography

Chromatography is an analytical procedure in which the flow of a liquid or a gas promotes the separation of substances through differential migration from a narrow zone in a porous sorptive medium. The chromatographic techniques like TLC, GLC, and HPLC using various types of sensitive detectors, capable of detecting up to nanogram, picogram and femtogram levels, are more commonly used in residue analysis.

Thin Layer Chromatographic (TLC)-enzyme inhibition technique

Thin layer chromatography (TLC) is one of the most widely used analytical technique for the separation and identification of organic molecules because of its simplicity, reliability and cost effectivity.

This technique is found useful for the analysis of organophosphorus (OP) and carbamate group of pesticides. OP pesticides, containing sulphur atom are first converted to oxygen analogues using NBS as oxidising agent, before treating it with esterase obtained from blood plasma or serum. The organophosphorus pesticide zones on the developed TLC plate is sprayed with solution containing cholinesterase enzyme and is left at ambient temperature to react with the phosphate zones for inhibition of the enzymatic activity. After spraying with chromogenic reagents like indoxyl acetate the phosphate zones can be located against fluorescent background. Zones containing as little as microgram quantities of phosphates can easily be detected by this procedure.

High performance thin layer chromatography (HPTLC)

High performance thin layer chromatography (HPTLC) is a very promising technique which can be made more efficient by selecting the plate coating material, thus improving sample application procedure and development of appropriate mobile phase. This technique can be used for separation and quantification of organic molecules including pesticide residues.

Silica gel of 60 micron size is mostly used as TLC sorbent to prepare HPTLC precoated plates, where separation of pesticides is much faster with high sensitivity. Quantification can be carried out using a spectrodensitometer or reflectance fluorescence spectrophotometer. In HPTLC, the recent improvements in spotting technique, plate technology and development of detection system have inducted automation and a high degree of sensitivity. Two-dimentional HPTLC is also used as an improved technique for separating components of a complex mixture.

Gas liquid chromatography (GLC)

The prime role of GLC is to resolve a sample mixture into its individual components and provide identification and precise quantification of each component which makes it an ideal choise for multi-residue analysis.

The gas liquid chromatography (GLC) is based on a principle of separation of volatile components which are distributed between a fixed stationary phase and a mobile inert gas phase. Here the stationary phase is a non-volatile liquid distributed on a solid support. The sample mixture upon vaporisation in the injection port at a definite temperature is carried away by the inert carrier gas into the packed column where partitioning of the components between two phases takes place. This is followed by diffusion and mass transfer operation leading to the sequential emergence of components from the column which are identified by different detectors.

For pesticide analysis glass column is preferred over metal column to avoid decomposition of certain type of pesticides. The types of columns generally used are :
  i) Packed 3-10 ft internal diameter (I.D.) 2-4 mm
 ii) Capillary 5-100 ft (I.D.) 0.25 mm
iii) Megabore
iv) Preparative (I.D.) 9 to 12 mm

The specificiation of the column affects the proper resolution of the mixture and the resolution increases with the square root of the column length. The detectors generally used are ECD, FID, NPD, HECD, FPD, MCD, PID and mass spectrometer.

Capillary GLC technique

This technique is of great advantage for analysis of pesticide residues of technical grade and their degradation products. It has a superior resolution capability and is very useful in multi-residue analysis. Here the individual isomer of any compound like synthetic pyrethroids, organochlorines, organophosphates and pheromones can be resolved from each other as well as closely related impurities and estimated quantitatively without prior derivatization. There is a low column bleed since superior coating techniques of small levels of liquid phase are used eliminating constant column rejuvenation, instrument adjustment and detector cleaning. The technique gives fast and accurate analysis using hydrogen as a carrier gas.

In the case of certain non-volatile pesticides, derivatization is needed to get as stable volatile component with increased sensitivity towards a particular detector. The detection limit of organic molecules by GLC is given in Table 6.1.

Table 6.1. Maximum detectable limits of the major GLC detectors

Abbreviation	Detector	Min. detectable limits (g)
TC	Thermal conductivity	10^{-7}
ECD	Electron capture	10^{-12} to 10^{-13}
FPD	Flame photometric phosphorus and sulphur	10^{-10} to 10^{-11}
FID	Flame ionisation	10^{-8}
NPD	Thermoionic for nitrogen and phosphorus	10^{-12} to 10^{-13}
MCD	Microcoulometric	10^{-7}
HECD	Electrolytic conductivity	10^{-11}

GC/MS also provide selective detection at subnanogram level mainly used for confirmation identity. Other selective detectors are FTIR spectrometer, Microwave Plasma Emission Detector (MPED), Atomic Absorption Spectrometer (AAS) and Chemiluminescent Detector (Thermal Energy Analyser).

High performance liquid chromatography (HPLC)

It is based on the principles propounded by Tswett who demonstrated the ability of an adsorbent column to separate a coloured component obtained from a plant extract. Today high efficacy packings in narrow bore columns, sensitive detectors and precise solvent delivery systems have inducted very high level of sensitivity in this basic technique. In LC the separation may be described as solute-solvent interaction where the solute molecules establish an equilibrium between stationary and mobile phases. Chemical derivatisation involves altering the chemical structure of the analyte making it suitable for determination by chromatographic technique. In both GC and LC, derivatization is applied to enhance the detector response.

For routine analysis LC technique appears to be the best if carried out under isocratic conditions as it allows the analysis for longer duration. Gradient elution often becomes necessary when a whole series of compounds are required to be separated over a wide range of polarities. Thus LC becomes a useful technique for separation and identification of wide range of polar metabolites, urea herbicides and systemic fungicides which are otherwise difficult to separate and characterise by any other analytical technique. Based on the choice of column and mobile phase three major ways of separation are employed in HPLC : a) normal phase, b) reverse phase and c) ion exchange chromatography. The normal phase is polar (hydrophilic) and silicagel is commonly used as sorbent. The reverse phase is non-polar (hydrophobic) and C_8 (octyl silane) and C_{18} (octadecyl silane) sorbent are commonly employed. In ion exchange either cation or anion exchanger is used. The column is made of stainless steel and generally has a diameter of 6 mm and length 25 to 100 cm. The proper separation and identification depends on the type of columns and solvents as well as detector systems. The common detectors are :

  i) R.I. (Low sensitivity)
 ii) U.V. fixed or variable wave length
iii) Photodiode Array (PDA—highly sensitive)
 iv) F.I.D.
  v) Electrochemical (selective for reduced compounds)
 vi) Fluorescence (limited to fluorescent compounds)
vii) Polarographic/potentiometric
viii) Mass spectrometer
 ix) Infrared spectrophotometer

The sensitivity of HPLC is not as high as GLC method. However, polar pesticides, their degradation products, non-volatile pesticides such as urea herbicides and thermolabile pesticides can be analysed only by HPLC without derivatization.

Spectroscopy

Infrared spectrophotometer is generally used for qualitative identification of organic molecules having a strong band of absorption but is now applied for quantitative analysis especially for pesticide residues. FT-IR has recently been found to be useful for trace quantity analysis of organic molecules with high sensitivity. It offers several advantages over normal IR in quantification at nanogram level.

Mass spectrometer (MS) is a tool for structural identification of complex organic molecules with quantitative detection limit of 10^{-6} to 10^{-9} g. It is now extensively used for identification of pesticides especially herbicides and their metabolites. Negative mass spectrometer dealing with negative ions (negative chemical ionization mode) is found very sensitive for pesticide analysis. Damico *et al.* [17a, 17b, 17c] reported the practical application of mass spectroscopy in residue analysis of a number of carbamate, organophosphate and chlorinated pesticides. Benson and Finocchiaro [18] also used mass spectrometer to characterise 14 pure carbamate pesticides. The coupling of GC-MS facilitates the study of volatile pesticide compounds in a complex mixture and is currently replacing GC because newer regulations require more specific and sensitive analytical data. For the non volatile, thermolabile and highly polar compounds such as urea and triazine herbicides, however, the GC-MS technique is found unsuitable. The liquid chromatography fitted with MS (LC-MS) can overcome this limitation. Currently LC-MS interfaces which show the greatest potential to accomplish these goals are particle beam thermospray [19] and electrospray [20]. A current sensitivity of 10-20 ng in the full scan mode suggests that as low as parts per billion (ppb) detection limits with particle beam can be achieved with proper sample preparation. Thermospray with a sensitivity of 0.1-10 ng range full scan, has been found to be most suitable for residue analysis of most herbicides and many other pesticides along with the metabolites which are polar and thermolabile in nature. This might be possible due to the presence of basic sites (amines) on some pesticides and herbicides resulting in high proton affinity. Electrospray technique also has similar high sensitivity and specificity in the analysis of pesticides in the femtogram detection limits.

GC-MS and LC-MS are now commonly used to locate and identify low level of unknown metabolites of pesticide. They provide molecular weight information instantly on minor components. These combined techniques conveniently provide definite and conclusive confirmation of residue identity. Another improved technique LC-MS-MS is found more useful to enhance sensitivity and selectivity in rapid analysis of hundreds of crude samples per day, under conditions that are extreme for other systems. It yields reproducible and precise results and does not involve daily maintenance or cleaning of the instrument. It also requires less clean up of soil, water and crop samples than analysed by LC-MS.

Nuclear magnetic resonance spectrometer

It is a non-destructive useful tool in the structural identification of organic molecules. With the introduction of FT-NMR, 1-2 µg samples can be analysed. ^{13}C NMR is also useful in determining the relative concentration of stereoisomers of an optically active compound. ^{31}P NMR has great potential in analysing organophosphorus pesticides and their degraded products. Photolytic degradation products of various pesticides have also been studied by NMR. Fukuto et al. [21] reported the use of ^{31}P NMR for detection and identification of fenthion in plants and its photodegraded products on the plant surfaces. LC-NMR is now-a-days used for the unambiguous structural elucidation of pesticide metabolites isolated from complex biological matrices which could not be identified by LC-MS.

Biotechnology based analytical methods (Immunoassays ELISA's) [22, 23]

ELISA called as enzyme linked immunosorbent assay is now becoming a rapid and reliable analytical method for the estimation of pesticide residues in numerous environmental samples. This is possible because the sample requires less clean-up than for GLC and HPLC methods and is also economical than the classical methods usually employed. Another advantage of this technique is that it can be conveniently used for many modern pesticides which show instability to chromatographic technique and degrade during analysis by classical methods. Immunoassays can also be used now as a complementary technique to GLC and has been seen to work satisfactorily for many polar pesticide metabolites. The technique can be coupled with supercritical fluid extraction (SFE), capillary electrophoresis (CE) or TLC procedures.

The method is based on the principle of binding of an analyte to an antibody generated for specific analyte such as pesticide. Due to the specificity of the antibodies, only micro sample preparation is necessary. Antibodies for specific antigens such as pesticides can be prepared by immunizing animals like rabbits, mice and goats. As most of the pesticides have much smaller mass and are unable to induce antibody formation, these need to be conjugated to a protein body such as bovine serum albumin (BSA) to make it large enough to be recognised as a foreign body by the immune system. Conjugation of any pesticide to a protein molecule is facilitated by adding functional groups of the pesticides such -OH, -COOH, -SH and -NH$_2$. called as spacer arms. This combination of an analyte with spacer arm is known as hapten.

Antibodies can be polyclonal or monoclonal type, depending upon the analytical needs. Monoclonal antibody produced by a single clone is more specific, unlike the polyclonal antibody produced by number of clones. After obtaining the antibody, a variety of formats [Fig. 6.1] may be followed comprising of three components i) target analyte (pesticide), ii) specific antibody and

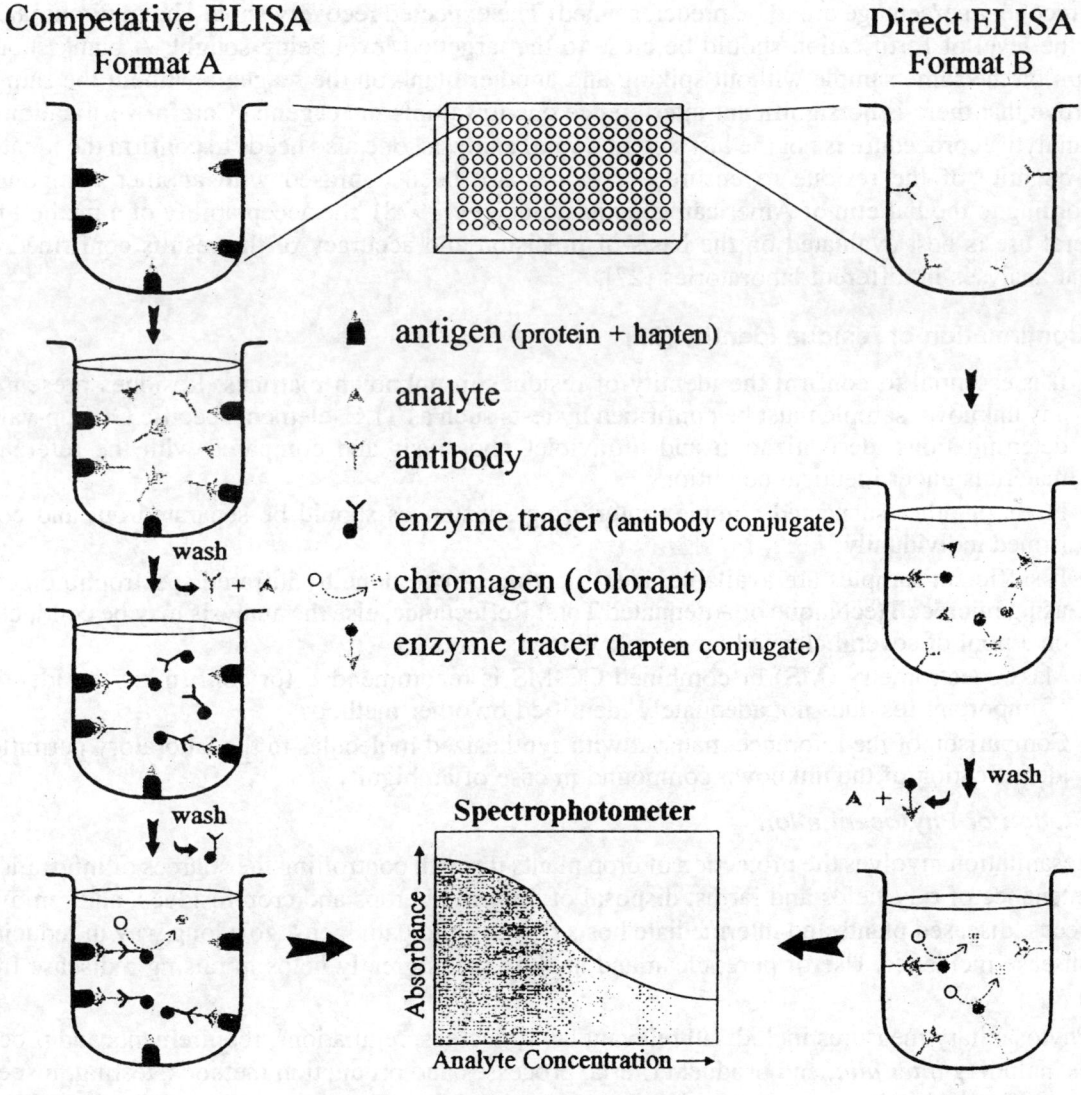

Competative ELISA
Format A

Direct ELISA
Format B

🔲 **antigen** (protein + hapten)
⚓ **analyte**
Y **antibody**
Y **enzyme tracer** (antibody conjugate)
○⤳ **chromagen** (colorant)
⚓ **enzyme tracer** (hapten conjugate)

wash

Y + ⚓

wash

A + ⚓

wash

Spectrophotometer

Absorbance →

Analyte Concentration →

Fig. 6.1. ELISA format.

iii) an enzyme tracer. The enzyme tracer is a hapten conjugated to an enzyme which reacts with chromogen such as *p*-nitrophenyl phosphate to produce a coloured complex whose intensity is measured by spectrophotometer against a standard curve. The ELISA technique has been extensively used for residue analysis of herbicides [24], fungicides [25] and insecticides [26].

Validation [27]

Proper validation of the analytical procedure with positive identification of the pesticide residue is imperative for collecting reliable data. The process involves several steps for the validation of an analytical procedure. The control samples must be free from any pesticide before fortification with the known quantity of pesticide. The steps for processing of samples like extraction, purification and recovery percentage are to be predetermined. The expected recovery should be minimum 80% and the level of fortification should be close to the targetted level being sought. A blank should be run on the same sample without spiking and another blank on the reagents without the sample to prove that there is no significant interference due to sample or reagents. Careful verification of the analytical procedure is not the last step in a valid analysis, one also needs to confirm the identity and quantity of the residue to ensure that it has not been confused with another compound. According to the bulletin of American Chemical Society (1978), the acceptability of a method for general use is best evaluated on the basis of precision and accuracy of the results confirmed by repeat analyses in different laboratories [27].

(v) Confirmation of residue identity [28]

- It is essential to confirm the identity of residues in unknown matrices. Residues present in any unknown sample must be confirmed by tests such as TLC, element specific GLC, p-value determinations, derivatization and ultraviolet photolysis and compared with the reference materials under identical conditions.
- PCBs or other suspected complex interfering substances should be separated out and confirmed individually.
- If sufficient samples are available, identification can be done by infrared spectrophotometry using microcell technique or Attenuated Total Reflectance, else the analysis may be conducted on a pool of several cleaned up sample extracts.
- Mass spectrometry (MS) or combined GC-MS is recommended for confirming the identity of important residues not adequately identified by other methods.
- Comparison of the reference material with synthesized molecules in the laboratory permitted identification of the unknown compound in case of ambiguity.

Sanitation or Phytosanitation

Phytosanitation involves the protection of crop plants through controlling the sources of infestation. Maintenance of tidy fields and farms, disposal of unwanted crops and crop residues, and removal of weeds, diseased plants and intermediate hosts of pests and pathogens, go a long way in reducing the disease incidence. Use of pure, clean and healthy seeds greatly helps in raising a disease free crop.

Phytosanitary measures include all relevant laws, decrees, regulations, requirements and procedures including *inter alia*, end product criteria, processes and production methods, testing, inspection, certification and approval procedures. Also quarantine treatments including relevant requirements associated with the transport of animals or plants or with the materials necessary for their survival during transport, provisions on relevant statistical methods of sampling procedures and methods of risk assessment and packaging and labelling requirements have proved to be directly relevant to food safety.

REFERENCES

1. Bindra, O.S. and Kalra, R.L. A review of the work done in India on pesticide residues. In : *Progress and Problems in Pesticide Residue Analysis* (D.S. Bindra and R.L. Kalra, eds.) PAU-ICAR 1-44 (1978).

2. *Pesticide Residues in the Environment in India* (C.A. Edwards, G.K. Veeresh and H.R. Kruger, eds.) *Proc. Symp.*, 1978 at U.A.S. Hebbal, Bangalore, India (1978).

3. Agnihotrudu, V. and Mithyantha, M.S. *Pesticides Residues—A review of Indian Work.* Rallis India Ltd., Mumbai 173 (1978).

4. *Pesticide Residues in Food.* Published by Consumer Education and Research Centre, Elisbridge, Ahmedabad, India, (1989).

5. *Studies on pesticide Residues and Monitoring of Pesticidal Pollution* (R.L. Kalra and R.P. Chawla, eds.), Department of Entomology, PAU, Ludhiana, 230 (1983).

6. Edwards, C.A. *Persistent Pesticides in the Environment.* C.R.C., Ohio, USA (1976).

7. Edwards, C.A. *Pesticides in Aquatic Environment* (M.A. Khan, ed.). Academic Press, New York, 11-138 (1977).

8. FAO/WHO Pesticide Residues in Food. *Wld. Hlth. Org. Rep. Sr.* **525**, 1973.

9. FAO evaluation of some pesticide residues in food. AGI 1974/M/11 WHO, Pesticide Residues Sr. **4** (1975).

10a. Kalra, R.L. Pesticide management in India—A critique on the current situation. In : *Pesticides, Crop Protection and Environment* (S. Walia and B.S. Parmar, eds.). Oxford & IBH Publishing Co. Pvt. Ltd., New Delhi, 213-232 (1995).

10b. Anonymous. Codex maximum limits for pesticide residues. FAO/WHO Food Standard Programme CAC **13**, Ed. 2, FAO Rome (1986).

11a. CCPR. Guidelines for good analytical practices in residue analysis and recommendations for the methods of analysis for pesticide residues. *Tech. Monograph 8*, GIFAP, Bruxelles, Belgram (1983).

11b. Carl, M. Internal Laboratory quality control in the routine determination of chlorinated pesticide residues. In. *Pesticide Residues* (H. Frehse and H. Geissbuhler eds.) Pergamon Press, New York, 19-28 and 46-49 (1979).

12a. Gunther, F.A. and Blinn, R.C. *Analysis of Insecticides and Acaricides*, Wiley Interscience, New York (1955).

12b. Roy, N.K. and Handa, S.K. Refinements in pesticide residue analysis. In : *Pesticides : Development, Toxicity and Safety* (R.B. Raizada and T.S.S. Dikshith, eds.) I.T.R.C. Lucknow, India, 75-87 (1992).

12c. Handa, S.K., Agnihotri, N.P. and Kulshrestha, G. *Pesticide residues : Significance, management and analysis.* Research Periodicals & Book Publishing House, India, 1-226 (1999).

13. Taylor, L.T. *Supercritical Fluid Extraction.* Wiley, New York (1996).

14. Wenelawia, B. (d.). *Analysis with supercritical fluid extraction and chromatography.* Springer-Verlag, New York (1992).

15. Pearce, K.L., Trenerry, V.C., and Were, S. Supercritical Fluid Extraction of Pesticide Residue from Strawberries. *J. Agric. Food Chem.* **45**, 153-157 (1997).

16. Lehotay, S.J. and Eiller, K.I. Development of a method of analysis for 46 pesticides in fruits and vegetables by supercritical fluid extraction and gas chromatography ion trap mass spectrometry. *J. A.O.A.C. Internatn.*, **78** (3), 821-830 (1995).

17a. Damico, J.N. and Benson, W.R. The mass spectra of some carbamate pesticides. *J.A.O.A.C.* **48**, 344 (1965).

17b. Damico, J.N. The mass spectra of some organophosphorus pesticide compounds. *J.A.O.A.C.* **49**, 1027 (1966).

17c. Damico, J.N., Barron, R.P. and Ruth, J.M. The mass spectral data of some chlorinated pesticidal compounds. *Org. Mass Spectrometry.* **1**, 331 (1968).

18. Benson, W.R. and Finocchiaro, J.M. Rapid procedure for carbaryl residues : modification of the official colorimetric method., *J.A.O.A.C.* **48** (3), 676-679 (1965).

19. Blakely, C.R. and Vestal, M.L. Thermospray interface for liquid chromatography/mass spectrometry. *Anal. Chem.* **55**, 750-754 (1983).

20. Dole, M., Mack, L.C., Hines, R.L., Mobley, R.C., Ferguson, C.D. and Alice, M.B. Molecular beams of macroions. *J. Chem. Phys.*, **49**, 2240-2249 (1968).

21. Fukuto, T.R., Hornig, E.O. and Metcalf, R.L. Nuclear magnetic resonance in the examination of the thermal decomposition of O,O-dimethyl 0-[4-(methylthio)-3-tolyl phosphorothioate. *J. Agric. Food Chem.*, **12**, 169-171 (1964).

22. Linde, C.D. and Gole, K.S. Immunoassays (ELISA's) for pesticide residues in environmental samples. *Pestic. Outlook,* **6** (4), 18-23 (1997).

23. Hammock, B.D. and Gec, S.J. Residues : Biotechnology based methods. **In** : *Proc. Eighth Internatn. Cong. Pestic. Chem.*, Options 2000. (N.N. Ragsdale, P.C. Kearney and J.F. Plimmer eds.), American Chemical Society, Washington DC, 204-214 (1995).

24. Yeung, J.M., Mortimer, R.D. and Collins, P.G. Development and application of altapid immunoassay for dibenzoquat in wheat and barley products. *J. Agric. Food Chem.* **44**, 376-380 (1996).

25. Bushway, R.J., Brandon, D.L., Bates, A.H., Leili, Larkin, K.A. and Young, B.S. Quantitative determination of thiabendazole in fruit juices and bulk juice concentrate using a thiobendazole monoclonal antibody. *J. Agric. Food Chem.* **43**, 1407-1412 (1995).

26. Jourdan, S.W., Scutellaro, A.M., Fleeker, J.R., Herzog, D.P. and Rubio, F.M. Determination of carbofuran in water and soil by rapid magnetic particle based ELISA. *J. Agric. Food Chem.* **43**, 2784-2788 (1995).

27. A chemical perspective (2nd Ed.). American Chemical Society, Washington D.C., 41 (1978).

28. Moye, H.A. *Analysis of Pesticide Residues*, John Wiley and Sons, New York, (1980).

CHAPTER 7

Alternative Methods of Insect Pest Control

<table>
<tr><td>

• Introduction
• Pheromones
• Juvenile hormones (JH)
• Anti-juvenile hormones (Anti-JH)
• IGRs and Chitin synthesis inhibitors

</td><td>

• Repellents
• Antifeedants
• Chemosterilants
• Microbial agents
• *References*

</td></tr>
</table>

Indiscriminate use of synthetic pesticides has created several environmental hazards like excessive toxic residues in food commodities, resistance development in insects to pesticides, outbreak of secondary or minor pests, and bioaccumulation of toxins in the environment. Thus the scenario created by the use of synthetic pesticides triggers the need to look for better alternatives which should be effective and eco-friendly. Recently some of the new alternative techniques/methods involving products like pheromones, hormones, IGRs, anti-juvenile hormones, attractants, repellents and chemosterilants (sterile male technique) are being used and discussed in brief.

Pheromones

Birch [1] and Shorey and Makelvey [2] defined pheromones as a chemical(s) released from one insect to induce a response in another individual of the same species. There are various types of pheromones such as sex, alarm, aggregation, trail, for maintaining chemical communication with the species. Sex pheromones (sex attractants) control the mating behaviour of insects. It is produced in small amounts by the females and is a mixture of chemical compounds, which are generally derivatives of unsaturated even carbon (C_{10}-C_{18}) straight chain alcohols, acetate derivatives, free alcohol, aldehyde etc. A large number of sex pheromones have been isolated but the quantity is too little for practical use. Pheromones can be used effectively against agricultural insects like pink bollworm moth, codling moth, tobacco, caterpillar, grape vine moth, cabbage looper. Some of such products are **gyptol**, 10 acetoxy-Z-7 hexadecanol (1) isolated from the female gypsy moth and

Z-7,8-epoxy-2-methyl octadecane, **disparlure** (2) which have been identified as sex pheromones. Another compound 10E-12-Z-hexadecanol, **bombykol** (3) of the natural silkworm moth, *Bombyx mori* was characterised as a sex attractant.

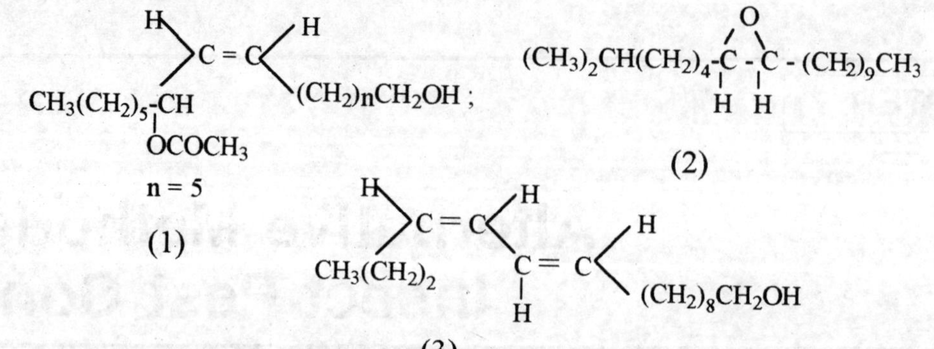

$$CH_3(CH_2)_5-CH \overset{\overset{H}{|}}{\underset{OCOCH_3}{C}}=C \overset{H}{\underset{(CH_2)_nCH_2OH}{}} ;$$

n = 5

(1)

$$(CH_3)_2CH(CH_2)_4-\overset{O}{\underset{\underset{H}{|}}{C}}-\overset{}{\underset{\underset{H}{|}}{C}}-(CH_2)_9CH_3$$

(2)

$$CH_3(CH_2)_2 \overset{H}{\underset{}{}}C=C \overset{H}{\underset{\overset{|}{H}}{}} \overset{|}{\underset{}{C}}=C \overset{H}{\underset{(CH_2)_8CH_2OH}{}}$$

(3)

A large number of sex pheromones with diverse chemical structures (4a, 4b,5,6) have also been isolated from American cockroach, the honeybee queen and the western pine beetle respectively.

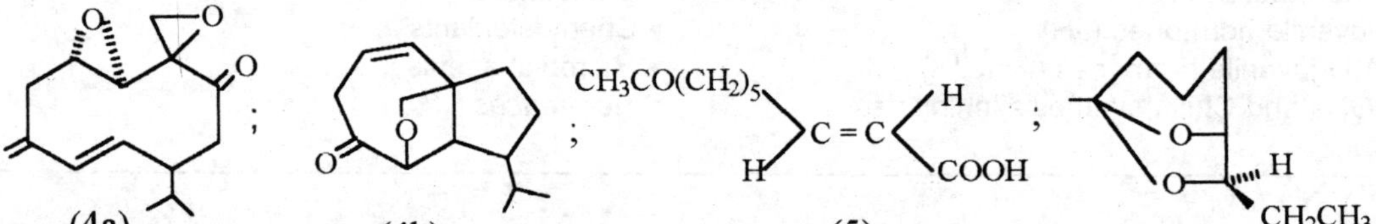

(4a) ; (4b) ;

$$CH_3CO(CH_2)_5 \overset{H}{\underset{H}{}}C=C \overset{H}{\underset{COOH}{}} ;$$

(5)

(6) CH₂CH₃

Besides sex pheromones, alarm pheromones (7a) and trail marking pheromones (7b) are produced to warn insects from ensuing danger.

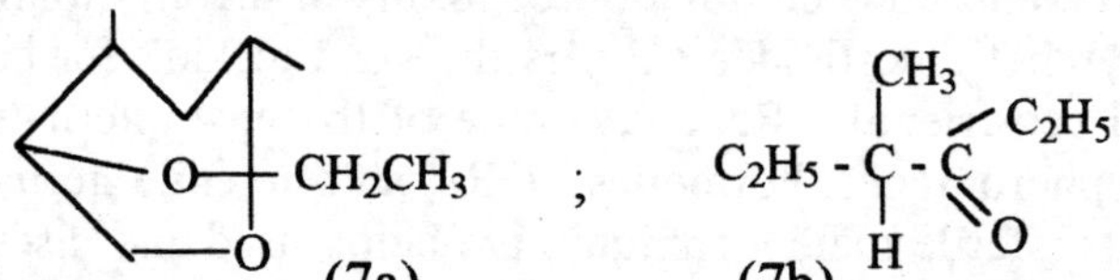

(7a) CH₂CH₃ ;

$$C_2H_5-\overset{\overset{CH_3}{|}}{\underset{\underset{H}{|}}{C}}-\overset{C_2H_5}{\underset{O}{C}}$$

(7b)

The pheromone (*E*)--farnsene (8) is found very effective in controlling aphids when used in combination with insecticides.

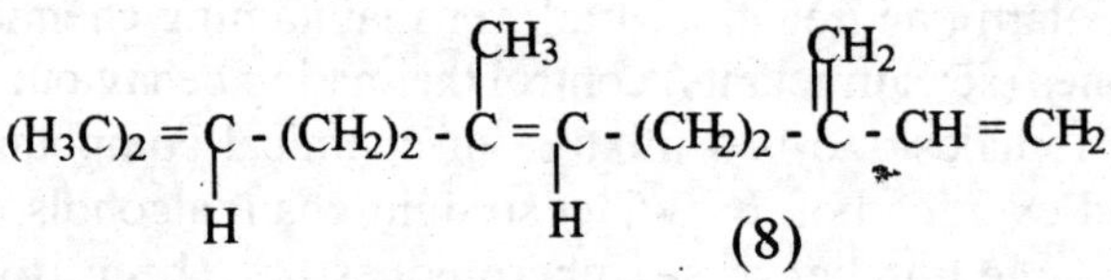

$$(H_3C)_2=\overset{}{\underset{\overset{|}{H}}{C}}-(CH_2)_2-\overset{\overset{CH_3}{|}}{\underset{\overset{|}{H}}{C}}=\overset{}{\underset{}{C}}-(CH_2)_2-\overset{\overset{CH_2}{||}}{\underset{}{C}}-CH=CH_2$$

(8)

It may be practised to minimise insect population to a low level using conventional insecticides and then maintained low by the use of pheromones.

The glue and kill strategy is found successful in case of mosquito oviposition pheromone (-) (5-R,6S)-6-acetoxy-5-hexadecanolide, if applied alongwith juvenile hormone (JH) type insecticide. The search for synthetic chemical lures has yielded several interesting compounds like methyl eugenol, (9) attracting the fruit fly and feeding stimulant, while siglure (10) is a synthetic attractants for the fruit fly. Attempts have been made to synthesise pheromones but they are not found so effective probably due to lack of stereochemical configuration. Although a few synthetic attractants, are found effective for insects other than fruit flies like butyl sorbate (11), an effective attractants for European cockchafer and methyl linolenate (12) for bark beetles. These are related to natural products and act as food or oviposition lures. It seems that the chemical released by male bark beetles act as both sex and food attractants.

$$CH_3CH = CH\text{-}CH = CHCO_2C_4H_9 \; ;$$
$$(11)$$

$$C_2H_5CH=CH\text{-}CH_2\text{-}CH=CH\text{-}CH_2\text{-}CH=CH(CH_2)_7COOCH_3$$
$$(12)$$

Most attractants, unfortunately affect the adults species only and immature insects are not attracted by the food or oviposition lures. The use of attractants in public health control programmes does not seem to be promising in integrated pest management (IPM) programme. However, the synthetic lures are found most useful in preliminary survey studies for monitoring purposes and the protein hydrolysate-insecticide preparations are still used for the control of some species of fruit fly. The unpopularity in its use may be due to some constraints like proper designing of traps, lure types, effective formulations and availability of quality control product. If these problems are sorted out the use of pheromones in future may become popular in insect pest control programmes.

Juvenile hormones (JH)

The first idea of a hormone based insecticide was founded after the discovery of Williams [3], that treatment of the pupal stage of an insect with a hormone extract would disturb morphogenesis during the pupa-adult moult and produce an intermediate form of insect incapable of survival.

The first chemical having JH activity, **farnesol** (13a) was isolated from the feces of the yellow meal worm, *Tenebrio molitor*. Among the several active analogues, commercially available **methoprene** (13b) is the most active and used as a larvicide and IGR. Later on, the methyl ether (13c) and the diethylamine (13d) derivatives of (13a) were developed which possessed greater biological activity. A few more derivatives of 13a were prepared and found to be active whereas a monocyclic sesquiterpenoid **juvabione** (14) showed a relatively high specific JH activity against pyrrhocoridae insects. Aliphatic and aromatic analogues of juvabione also exhibited similar activity. Besides, several synergists like PBO (15) and Sesamex (16), few synthetic related compounds (17, 18) also exhibit significant JH activity.

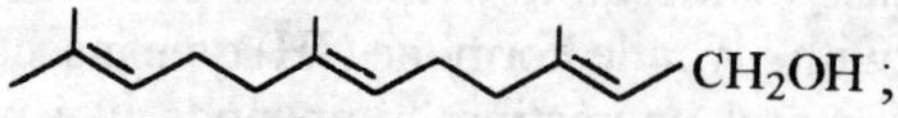

(13a)

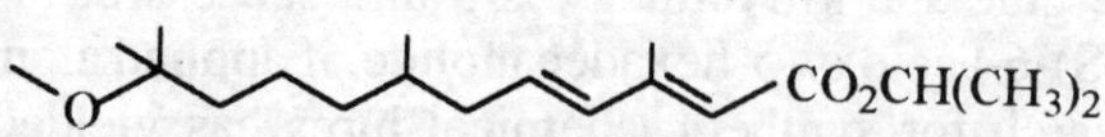

(13b)

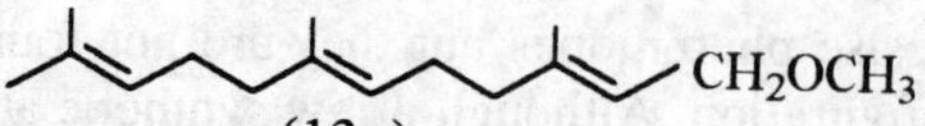

(13c)

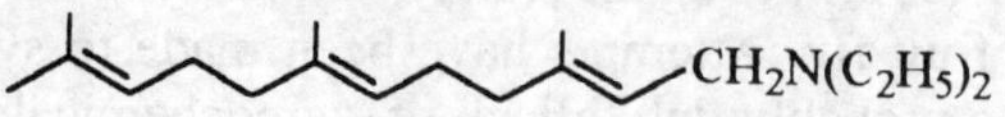

(13d)

Several non-methylenedioxy aromatic derivatives were prepared and found to be very active. The para substituted aromatic geranyl ethers (19) possess high JH activity and three naturally occurring moulting hormones (MH) like **ecdysones** (20) have also been identified. These polyhydroxy sterioids [4] interfere with moulting, ovarian development, embryogenesis and diapause. Some simpler and potent synthetic analogues like 22, 25-bisdeoxyecdysone (21) inhibits development, reproduction and diapause in a wide range of insect pests.

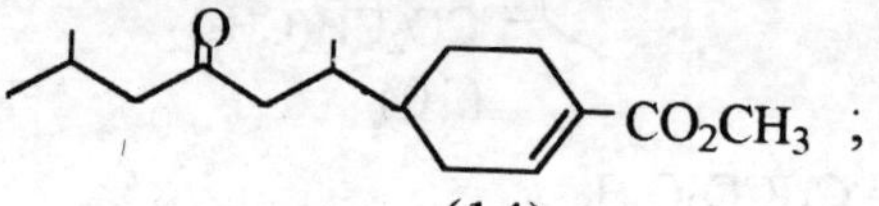

(14)

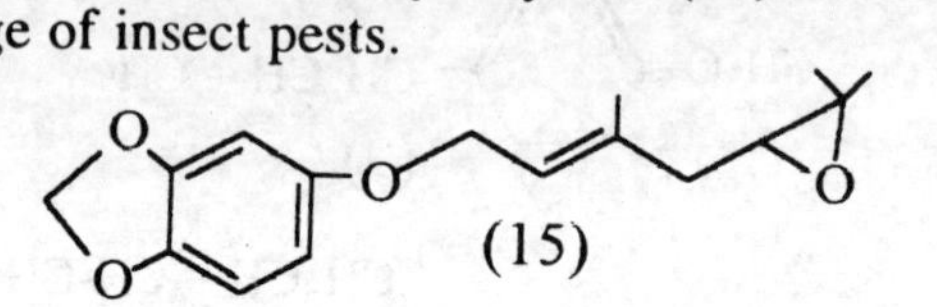

(15)

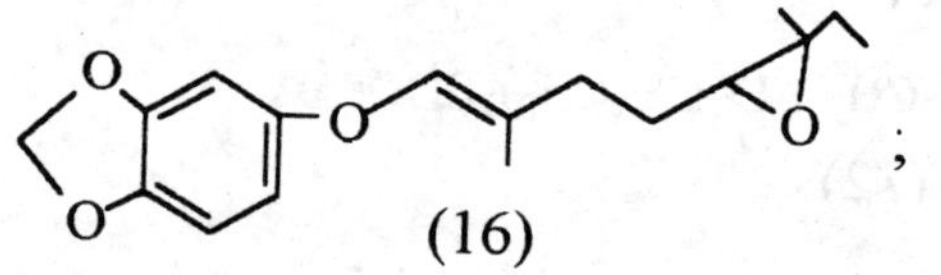

(16)

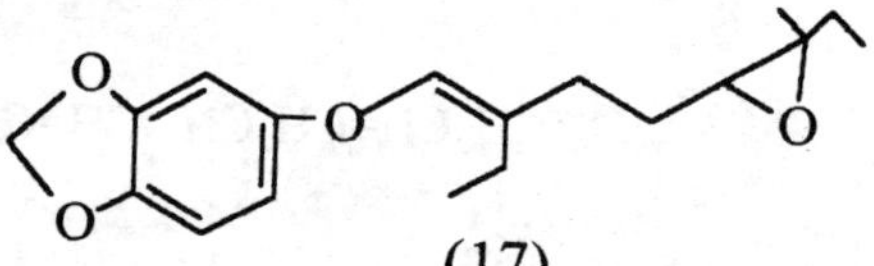

(17)

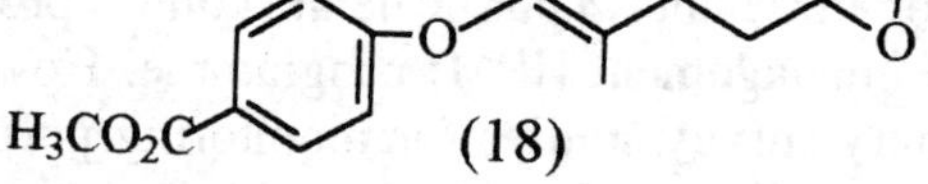

(18)

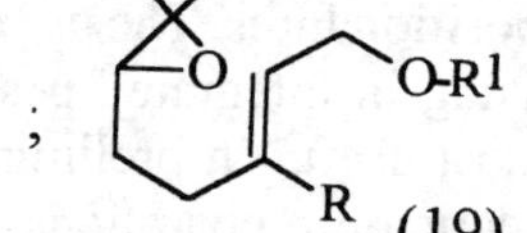

; R = CH₃, C₂H₅ &

(19)

R¹ =

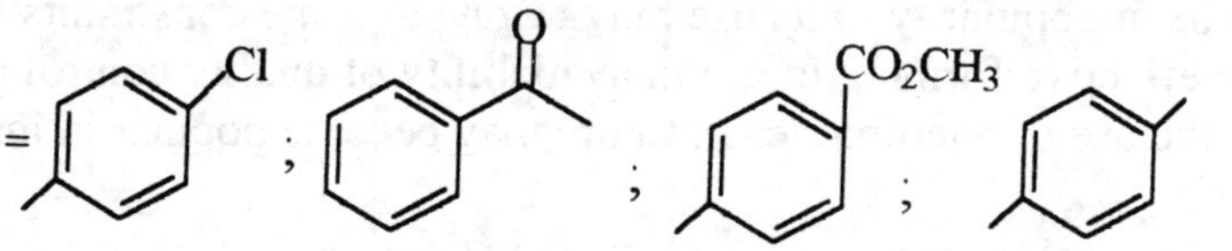

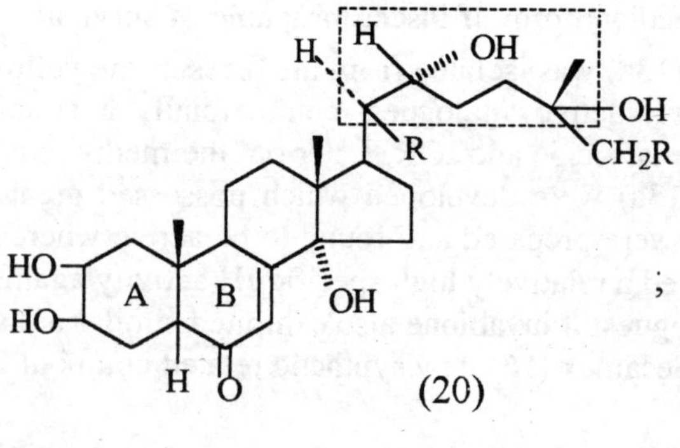

(20)

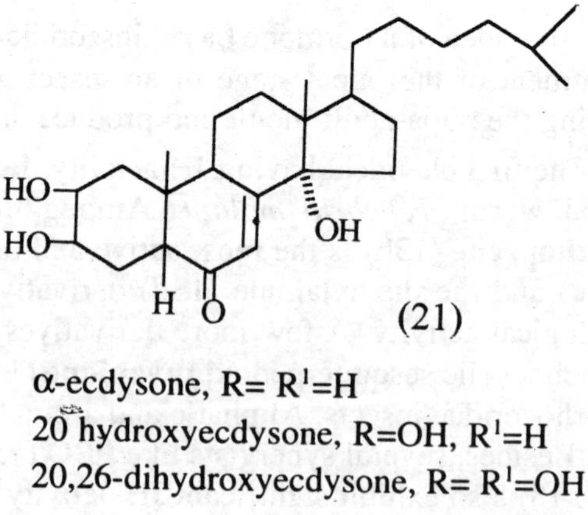

(21)

α-ecdysone, R= R¹=H
20 hydroxyecdysone, R=OH, R¹=H
20,26-dihydroxyecdysone, R= R¹=OH

Extensive studies on the biosynthesis of α-ecdysone [5] revealed that the insect is unable to build-up the steroid skeleton and depends on dietary sterol for normal growth. Comparative studies on the biological action of ecdysones of insect origin and of several ecdysoids of plant origin Slama et al. [6] suggested that high MH activity is due to the presence of Z-fusion of rings A and B; -hydroxylic function at position 3; keto-group at position 6 in conjugation with a Δ^7 double bond and steroid side chain with an appropriately R-oriented hydroxyl function at position 22.

Several active, related derivatives of ecdysterone structurally differ with respect to the side chain can be isolated from insects as well as from number of plants, and marine crab, Callinectes sapidus. A few steroidal compounds satisfying the structural requirements for MH activity were found to inhibit post-ecdysial hardening and sclerotisation of the insect cuticle.

Matolcsy et al. [7] screened for anti-ecdysone action of a number of compounds known to act as inhibitors of steroid biosynthesis in other organisms like hypcholesterolaemic agents used in human therapy and fungicides acting by the inhibition of ergosterol biosynthesis. Among these, the fungicidal molecules such as **triarimol**, 2,4-dichloro α-pyrimidin-5-yl-benzhydrol (22) were the most active. Based on the same hypothesis Hammock et al. [8a] bioassayed analogous compounds (23) and found that the acute symptoms on the test insects were similar to induced by precocene II.

Some new JH active compounds have also been reported recently [8b] as an alternative method of insect control.

Anti-juvenile hormones (Anti JH)

Slama et al. [6] screened a large number of compounds which are structurally related to juvenoids but lacking JH activity and considered to be antimetabolites of JH. Attempts were made by G. Matolcsy et al. [9] and others [10] to find anti-JH activity of a number of metabolite analogous of mevalonic acid and homomevalonic acid but found unsuccessful. However, fluorinated mevalonate analogue, 4-fluoromethyl-4-hydroxy-tetrahydro-2H-pyran-2-one, **Fmev** (24) acts as a potent inhibitor of JH biosynthesis. The most promising invention in this area was the characterisation of two active principles. 2,2-dimethyl-7 methoxy chromone, **precocene-I** (25) and 6,7-dimethoxy analogue, **precocene-II** (26). Matolcsy et al. [11] also proved that molecule (27) lacking chromone ring moiety possesses anti-JH action.

Several 3-alkylthio-1,1,1-trifluoro-2-propanones with JH-like side chains developed [12] with the idea of possible transition state analogue inhibitor of JH esterase. The most active compound was (28) which exhibited delaying pupation and inhibited JH esterase of the cabbage looper, Trichoplusia ni. Among various acetylenic esters as JH biosynthesis inhibitor a compound containing 1,3-benzodioxole moiety (29) was synthesized which showed strongest inhibitory action on both enzymes.

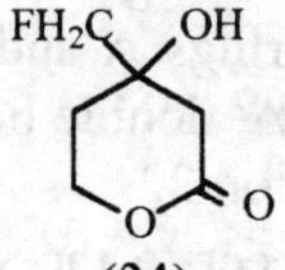

SAR studies on precocenes indicated that the presence of methoxy group in 7-position is responsible for activity. Replacement of methyl group by ethyl in position 2 reduced the activity.

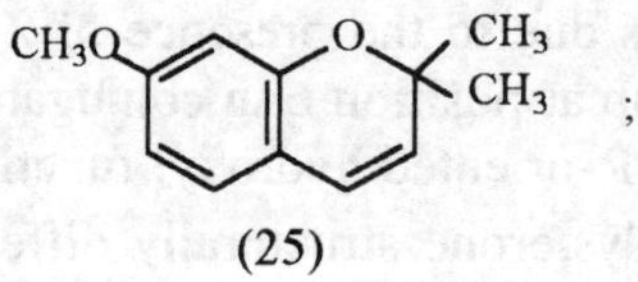

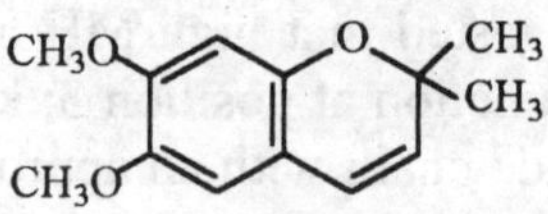

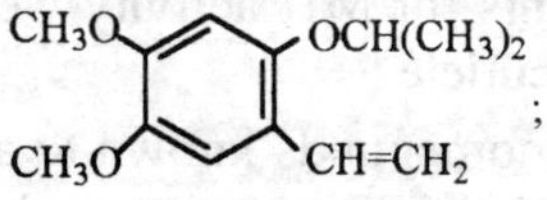

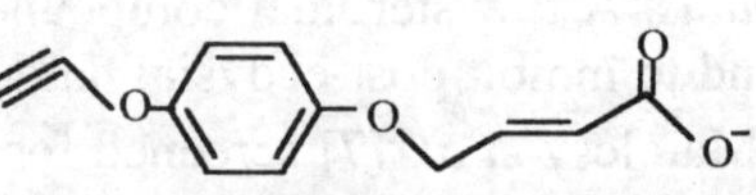

The action of precocenes on the molecular level is not understood, but it inhibits the final step in JH biosynthesis after epoxidation of farnesenic acid, following treatment of precocene II to the test insects. Bristol research scientists put forward a hypothesis that precocene and its derivatives are epoxidised to 3,4-epoxides showing alkylating properties. These epoxides then compete with the epoxidation step in JH biosynthesis destrupting the ability of *Corpora allata* to synthesise JH. The future use of these type of chemicals as insect control agent is difficult to predict.

IGRs and Chitin synthesis inhibitors

A large number of benzoyl urea compounds like **diflubenzuron**, dimilin (30), **triflumuron**, alsystim (31), **chlorfluazuron** (32), **buprofezin**, appland (33) exhibit new type of insecticidal activity reported first time by scientists [13a, 13b] at Philips-Duphar B.V. Netherlands and Mitsui [14]. These compounds are most effective to larvae, which failed to shed their exuviae and died if diet was mixed with these compounds. The new addition is **hexaflumuron** (34) having wide spectrum of activity with low toxicity and biodegradability.

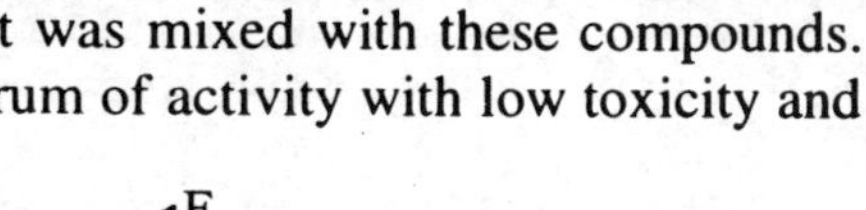

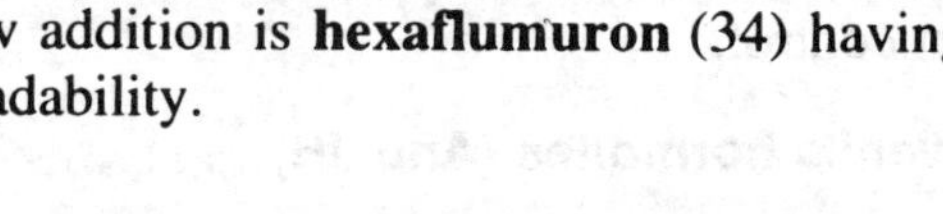

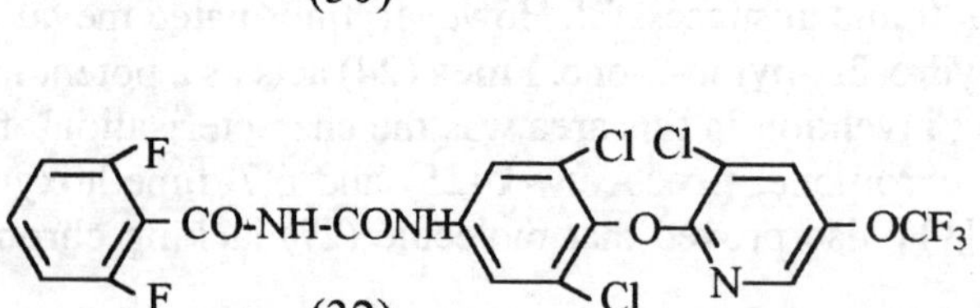

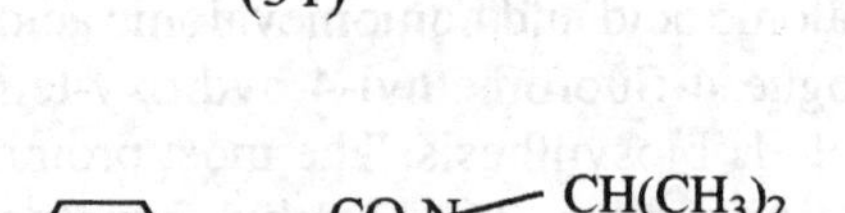

The compound (33) is quite different in chemical structure from benzoyl ureas and is also effective against brown plant hoppers and green house white fly. It was observed that sucking insects like aphids could not be controlled by these insecticides suggesting that this compound does not penetrate the insect cuticle or the plant epidermis. Biochemical and histological studies revealed that these insecticides prevented moulting of insects by interfering with formation of a new cuticle. These chitin synthesis inhibitors have two advantages (a) their mode of action is entirely different from conventional insecticides and are effective on strains resistant to these insecticides, (b) they have selectivity and low toxicity to non target beneficial organisms and are effective at low application rates. Their main application is on cotton and soybean crops in developed countries. The various species of flies and mosquitoes are effectively controlled by benzoylureas. Diflubenzuron degraded under aqueous condition mainly through hydrolysis at the benzamide moiety to give 4-chlorophenyl urea (CPU) and 2,6-difluorobenzoic acid (DFBA). Their stability in water depends on pH and temperature. In soil, the rate of degradation is strongly dependent on the particle size of diflubenzuron. In animal system, these are extensively metabolized and totally excreted. But in insects these are not degraded in most of the cases. These insecticides can be included in the IPM programme due to their safety towards environment. The preparation of a benzoylurea is given below (30, Fig. 7.1).

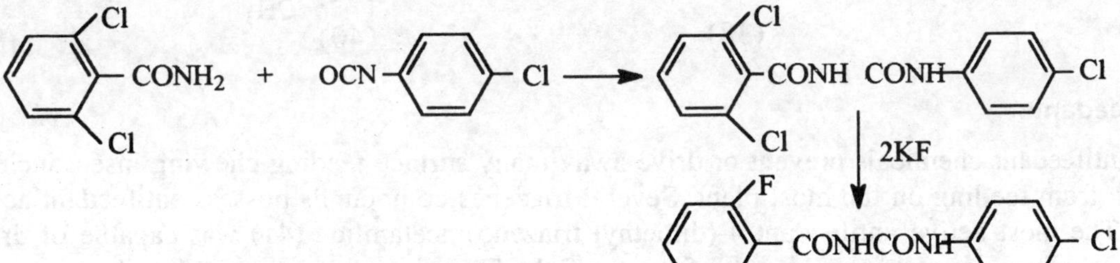

Fig. 7.1. Preparation of benzylurea.

Repellents

Insect repellent compounds can be either vapour or contact type and repel insect pests to move away from them. These compounds are harmless to human beings. The use of naphthelene and *p*-dichlorobenzene in moth bath is a common example of repellents [15]. In early days, a number of extractives from herbs like citronella oil containing various terpenes like geraniol, citronellol and borneol were used as insect repellent. But these are highly volatile and short lived. Later on dimethyl phthalate (35) [16] became prominent but was not effective against all species. Other synthetic repellents like 2,2-dimethyl-6-butylcarboxy-2,3-dihydropyran-4-one or indalone (36) and 2-ethylhexane-1,3-diol (37) or Rutzers (612) were very effective if mixed with dimethylphthalate.

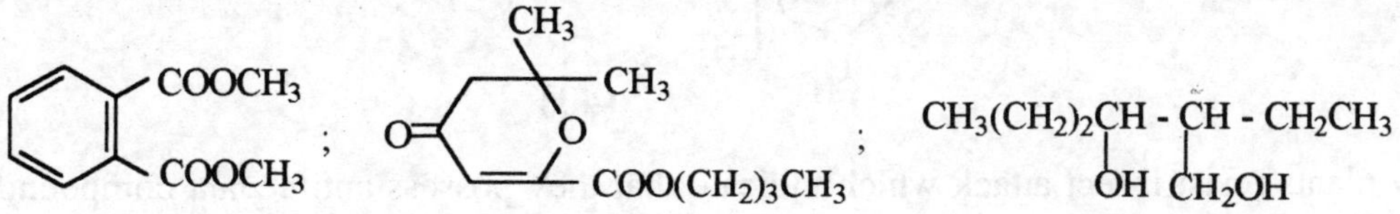

(35)   (36)   (37)

A more effective compound is N,N-diethyl-*m*-toluamide, **deet** (38) introduced in 1955 with broad spectrum of activity against all biting insects like mosquitoes, flies, chiggers. Another isomer diethyl phenylacetamide, **Depa** (39) is equally effective and can be synthesised [17] by the interaction of diethylamine with phenyl acetyl chloride (Fig. 7.2). Recently a *p*-menthane derivative (40) was found effective for insects and mites.

$$C_6H_5CH_2COCl + 2HN\ (C_2H_5)_2 \longrightarrow C_6H_5CH_2CON(C_2H_5)_2 + (C_2H_5)_2NH.HCl$$

(39)

Fig. 7.2. Reaction of diethylamine with phenyl acetyl chloride.

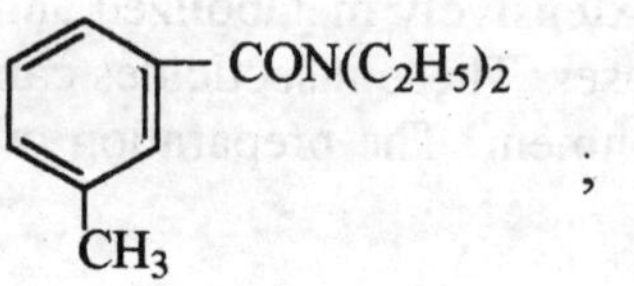

(38) 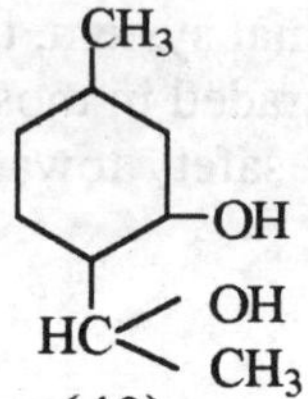   (40)

Antifeedants

The antifeedant chemicals prevent or drive away many surface feeding chewing insect including aphids from feeding on the most plant. Several triazenes compounds possess antifeedant activity [18]. The most active antifeedant 4-(dimethyl triazeno) acetanilide (41) was capable of driving away chewing caterpillars and beetles from the field. The triphenyltin fungicides like triphenyltin acetate (42) and quaternary derivatives of some heterocyclic amines (43) also possess antifeedant activity. Phenoxyethanols and acids like 2,4,6-trichlorophenoxy ethanol and 2,4,6-trichlorophenoxy acetic acid were investigated which exhibited the highest antifeedant activity in both *in vivo* and *in vitro* studies [19].

(41)   (42)

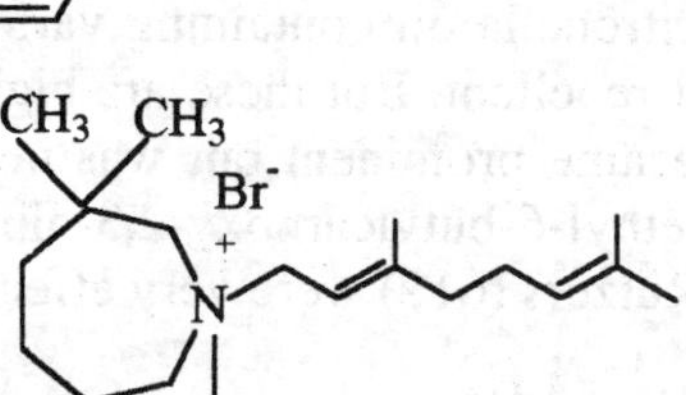

(43)

Many plants resist insect attack which indicate that they possess antifeedant compounds. The neem tree contains a complex tetranortriterpenoid molecule azadirachtin (cf. Chapter II) which is a most active antifeedant against lepidopteran spp. at very low concentration. Later on, simpler molecules like polygodial (44) and other related compounds warburganal (45) and ugandensidial

(46) having antifeedant activity were isolated from the bark of the plants, *Warburgia stihmannii* and *W. urgandensis* [20]. Schkuhrin I & II isolated from the African plant *Schkuhrina pinnata* possess high antifeedant activity [21].

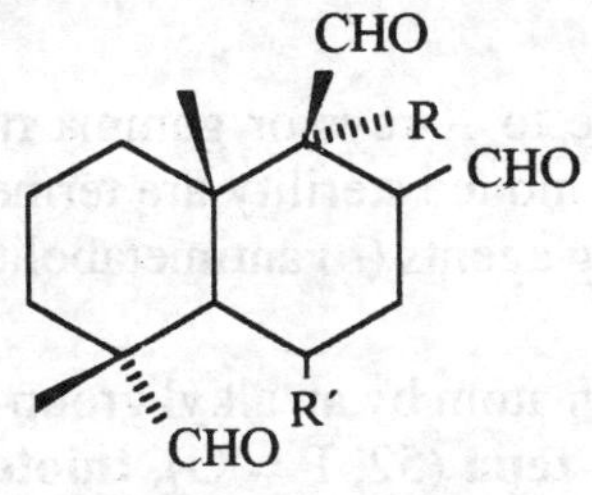

R = R′ = H, (44a)
R = OH, R′ = H (44b)
R = OH, R′ = OCOCH₃, (44c)

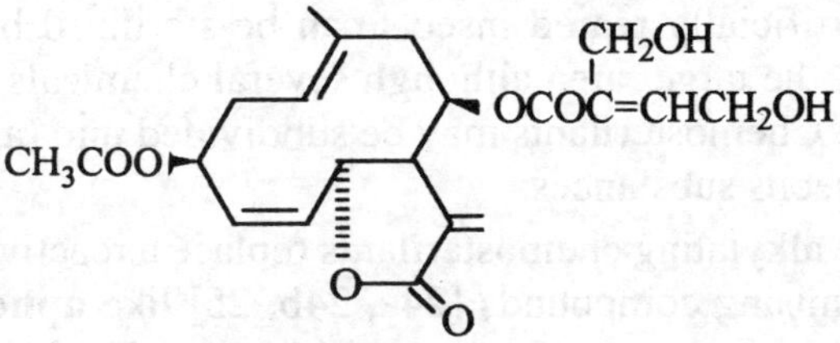

(45)

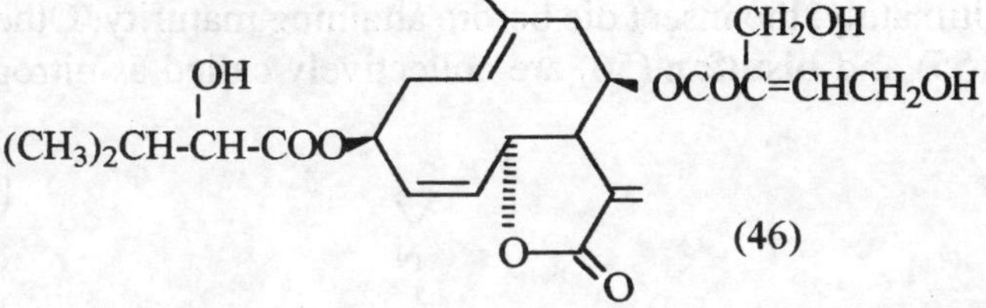

(46)

The activity of these molecules is due to the presence of methylene lactone moiety. A few sesquiterpenes isolated from plants [22] like isoalantolactone (47), **bakkenolide** (48) **Yatein** (49) and **bisabolangelone** (50) with or without methylene lactone moiety exhibit good antifeedant activity against several insects.

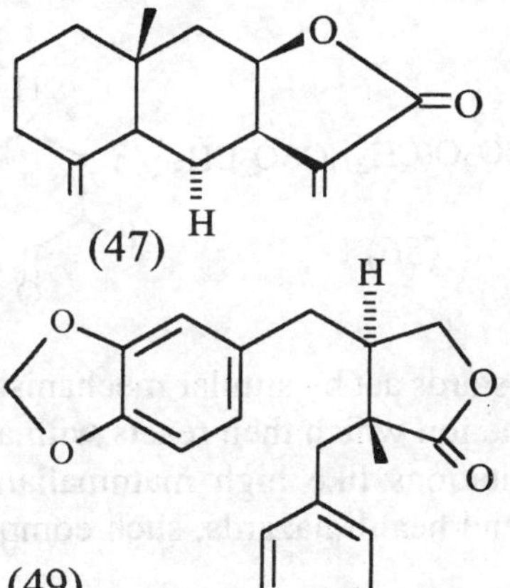

(47)

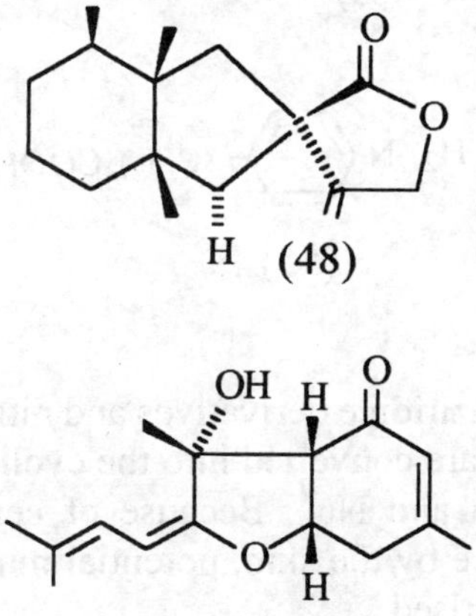

(48)

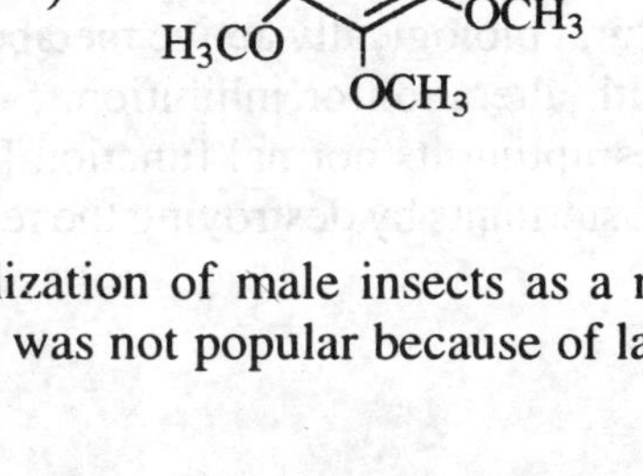

(49)

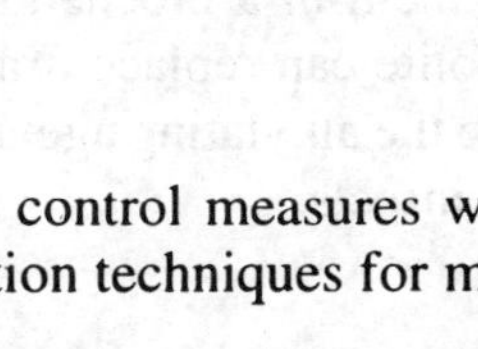

(50)

Chemosterilants

The idea of sterilization of male insects as a means of pest control measures was proposed by Knipling [23] but was not popular because of lack of application techniques for mass sterilization of insect pests. The method is completely safe, specific and without any environmental hazard because the sterile male only mates with females of his own species and will be very promising in future as exhibited in the elimination programme of screw worm carried out in USA. This technique can also be utilised to control Mediterranean fruit flies, melon flies, sheep blowflies and housefly.

The artificially reared insects can be sterilized by exposure to X-rays or gamma rays before release in the target area although several chemicals known to induce sterility are termed chemosterilants. Chemosterilants may be subdivided into (a) alkylating agents (b) antimetabolites and (c) miscellaneous substances.

a) The alkylating chemosterilants replace a reactive hydrogen atom by an alkyl group. The most promising compounds [24a, 24b, 25] like **apholate** (51), **tepa** (52, P = O), **thiotepa** (53, P = S) and **metepa** (54) are aziridinyl molecule containing nitrogen atom attached to electron withdrawing groups like P= O, C= O, CN, SO or SO_2. These chemicals are used at doses lower than required for general toxicity and act by causing the insect eggs not to develop or hatch into larvae. Ultimately, the insect die before attaining maturity. Other alkylating agents like **chlorambucil** (55) and **bisulfan** (56) are collectively called as nitrogen mustards [26].

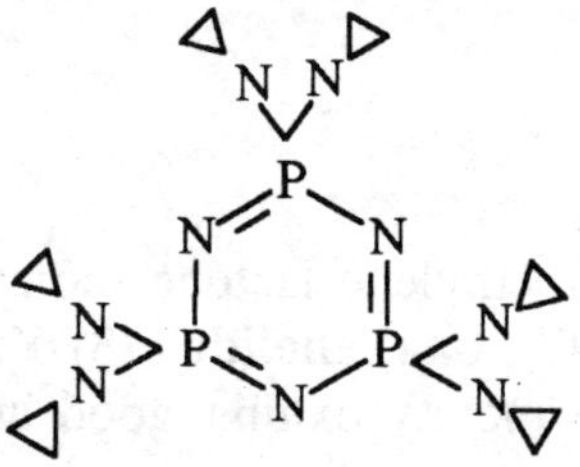

(51) 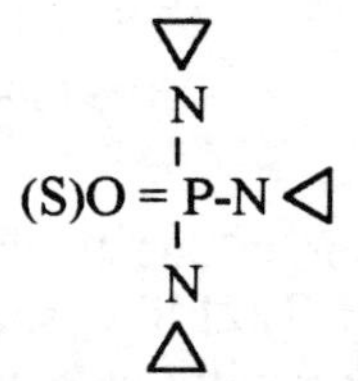   (52, P=O ; 53, P=S) 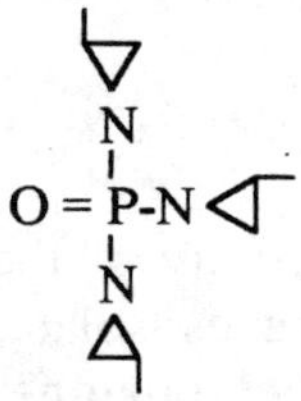   (54)

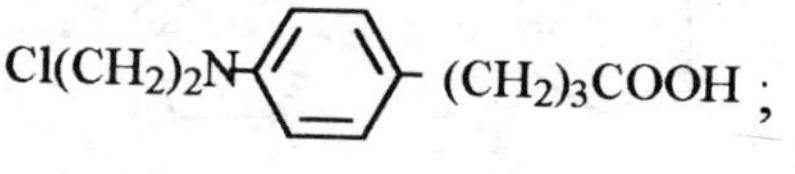

$Cl(CH_2)_2N$— (CH_2)_3COOH ; $CH_3SO_2O(CH_2)_4OSO_2CH_3$;

(55)   (56)   (57)

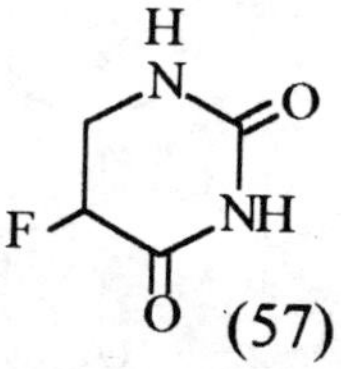

Both the aziridine derivatives and nitrogen mustards act by similar mechanisms. The nitrogen mustards are converted into the cyclic aziridine ion which then reacts with available nucleophiles Nu and Nu'. Because of certain limitations like high mammalian toxicity, quick absorbance by the skin, potential mutagenic and health hazards, such compounds could not be popularised.

b) Antimetabolites are chemicals that mimic natural biologically active metabolites, so they may substitute them in a biochemical process with alteration or inhibition. 5-Fluorouracil (57), antimetabolite can replace uracil in RNA desrupting its normal function [27]. Antimetabolites, unlike the alkylating agents act as female sterilants by destroying the reproductive system of female insects.

c) Chemosterilants of miscellaneous group include a wide range of compounds like mitomycin and cycloheximide, the alkaloid colchicine, hexamethylphosphoric triamide, triphenyltin, urea, thiourea and triazine derivatives.

Microbial agents

These are already discussed in Chapter 2.

REFERENCES

1. Birch, M.C. (ed.). *Pheromones*. North-Holland, American Elsevier, New York, 495 (1974).

2. Shorey, H.H. and McKelvey, J.J. (eds.) *Chemical control of insect behaviour, theory and application.* Wiley Inter-Sciences Publications, New York, USA, 414. (1977).

3. Williams, C.M. The juvenile hormones of insects. *Nature* (London) **178**, 212 (1956).

4. Bowers, W.S. Insect hormones and their derivatives as insecticides. *Bull. Wld. Hlth. Org.* **44**, 381-389 (1971).

5. Thompson, M.J., Svoboda, J.A., Kaplanis, J.N. and Robbins, W.E. Metabolic pathways of steroids in insects. *Proc. Roy. Soc. Sr. B.* **180**, 203 (1972).

6. Slama, K., Romanux, M. and Sorm, F. Insect hormones and bioanalogues, springer, Wein-New York, 261 (1974).

7. Matolcsy, G., Varjas, L. and Bordas B. *Acta Phytopath. Acad. Sci. Hung.* **9**, 161 (1974a).

8a. Hammock, B.D. and Mumby, S. Inhibition of epoxidation of methyl farnesoate to juvenile hormone III by cockroach corpus allatum homogenates. *Pestic. Biochem. Physiol.* **9**, 39-47 (1978).

8b. Chouwdhury, H. and Walia, S. Insect hormone mimics and antagonists, a novel group of insect control chemicals. **In :** *Pesticides crop protection and environment* (S. Walia and B.S. Parmar eds.). Oxford & IBH Publishing Co. Pvt. Ltd., New Delhi, India, 140-156 (1995).

9. Matolcsy, G., Varjas, L. and Bordas, B. Inhibitors of steroid biosynthesis as potential insect antihormones. *Acta Phytopath. Acad. Sci. Hung.* **10**, 455 (1975).

10. Bowers, W.S., Ohta, T., Cleere, J.S. and Marsella, P.A. Discovery of insect anti-juvenile hormones in plants. *Science* **193**, 542 (1976).

11. Matolcsy, G., Farag, A.I., Varjas, L., Belai, I. and Darwish, Y.M. In. *Juvenile hormone biochemistry* (G.E. Pratt and G.T. Broors, eds.), Elsevier, North-Holland, Biomedical Press, Amsterdam, 393 (1981).

12. Prestwich, G.D., Eng., W., Roe, A.M. and Hammeck, B.D. Synthesis and bioassay of isoprenoid 3-al-kylthio-1,1,1-trifluoro-2-propanones-potent, selective inhibitors of juvenile hormone esterase. *Arch. Biochem. Biophys.* **228**, 639-645 (1984).

13a. Wellinga, K., Mulder, R. and Van Daalen, J.J. Synthesis and laboratory evaluation of 1-(2,6-disulbstituted benzoyl)-3-phenyl ureas, a new class of insecticide I. 1-(2-6-dichlorobenzoyl)-3-phenyl ureas. *J. Agric. Food. Chem.* **21**, 348-354 (1973).

13b. *Ibid.* II. Influence of the acyl moiety on insecticidal activity. *J. Agric. Food Chem.* 21, 993-998 (1973).

14. Mitsui, T. Chitin synthesis inhibitors : benzoylurea insecticides. *Japan Pestic. Inform.* No. **47**, 1-17 (1985).

15. Kilgore, W.W. and Doutt, R.L. *Pest Control : Biological, physical and selected chemical methods.* Academic Press, New York, (1967).

16. Martin, H. and Woodcock, D. *The scientific principles of crop protection,* Arnold, London (1983).

17. Swamy, R.V. Development a new insect repellent—a review. **In :** *Pesticides crop protection and environment* (S. Walia and B.S. Parmar, eds.) Oxford & IBH publishing Co. Pvt. Ltd., New Delhi, 305-310 (1995).

18. Ley, S.V. The synthesis of some insect antifeedants. **In :** *Recent advances in the chemistry of insect control* (N.F. Jones ed.). Royal Society of Chemistry, London (1985).

19. Matolcsy, G., Saringer, G., Gaborjanyi, R. and Jarmy, T. Antifeeding effect of some substituted phenoxy

compounds on chewing and sucking phytophagous insects. *Acta, Phytopatho. Acad. Sci. Hung.* **3**, 275 (1968).

20. Kubo, I. and Nakanishi, K. In : *Host plant resistance to pest* (R. Hedin, ed.). *American Chemical Society Symp. Sr.* **62**, Washington DC (1977).

21. Pettei, M.J., Miura, J., Kubo, I. and Nakanishi, K. Insect antifeedant sesquiterpene lactones from *Schkuhrina pinnata* : the direct obetention of pure compounds using reverse phase preparative liquid chromatography. *Heterocycles*, **11**, 471 (1978).

22. Harmata, J. and Nawrot, J. *Biochem. Systematics and Ecol.*, **12**, 95 (1984).

23. Knipling, E.F. Internal treatment of animals with phenothiazine to prevent development of hornfly larve in the manure. *J. Econ. Entomol.* **31**, 315 (1938).

24a. Gaines, T.B. and Kimbrough, R.D. Toxicity of Metepa to rats—two other chemosterilants. *Bull. Wld. Hlth. Org.*, **31**, 737 (1964).

24b. *Ibid.* Sterilizing carcinogenic and teratogenic effects of Metapa. *Bull. Wld. Hlth. Org.* **34**, 317 (1966).

25. Ross, R.B. Recent developments in chemotherapy of cancer. *J. Wash. Acad. Sci.*, **52**, 209-216 (1962).

26. Fahmy, O.G. and Fahmy, M.J. *J. Genet.* **52**, 603 (1961).

27. Crystal, M.M. Chemical structure and sterilizing activity of N,N^{l}-alkylenebis-(1-aziridinecarboxamides) in screw worm flies. *J. Econ. Entomol.*, **60**, 1005 (1967).

Quantitative Structure Activity Relationship (QSAR) & Computer Assisted Molecular Modelling (CAMM) in Pesticide Design

- Introduction
- Application of QSAR
- Application of CAMD/CAMM
- Planning in Pesticide design
- Conclusion
- *References*

Pesticidal action may depend on one or more processes taking place in the target organism including permeation and transport of the pesticidal molecule, detoxification and interaction with the bioreceptor. These processes are conditioned by physico-chemical parameters exemplified hydrophobic, electronic and steric effects. Attempts to quantitatively correlate the pesticidal activity with physico-chemical parameters in the rational design of effective pesticides are often referred to as quantitative structure-activity relationship (QSAR) studies. Applying QSAR techniques, the complexity of synthetic programmes can be optimised thus facilitating the invention of safer and more effective pesticides.

The Hansch approach [1,2] has been the most widely and efficiently employed in correlating biological activity with physico-chemical parameters. In Hansch approach, linear free energy constants are used for regressing a quantitative fit in the most effective combinations of a set of mathematical functions of the substituents so as to modulate bioactivity in a congeneric series. The biological activity of the members of the series for instance can be expressed as

$$\text{Log } 1/C_s = a\pi^2 + b\pi + c\sigma + dE_s + e$$

where C_s is the concentration of the member S, which gives a standard biological response in a standard time interval (EC_{50}, LC_{50} etc.); a, b, c, d, e are coefficients and π, σ, E_s are hydrophobic, electronic and steric substituent variables respectively. The mathematical functions could be so choosen that certain physico-chemical properties influence the bioactivity and the structural modifications which enhance such properties would lead to generate potent compounds. A number of attempts have been made to apply the Hansch approach towards designing of compounds having an optimised structure among congeners [Wooldridge, 3 and Martin, 4].

Application of QSAR

The QSAR methods have now been widely used in the synthetic programme of insecticides, fungicides, herbicides in order to develop the best models for futuristic designs and as well to understand the mode of action. Since the subject is vast, it is not possible to cover all types of pesticidal molecules and only some salient points are discussed in this chapter to highlight its importance in pesticide design.

Some of the physico-chemical parameters for common substituents were initially developed in physical organic chemistry by the pioneering work of Hammett and Taft during the study of reaction mechanisms of organic compounds. Later on, other physico-chemical parameters were developed by Hansch, Fujita and Verloop. An exhaustive compilation of physico-chemical parameters for common substituents has been published by Hansch and Leo [2].

Hansch and Fujita [1] first applied QSAR studies on the activity of phenyl diethyl phosphate insecticides on houseflies and observed that electronic and hydrophobic parameters of the substituents on the phenyl ring were mainly responsible for the activity of the compounds. Later on Hansch and Deutsche [5] observed that cholinesterase inhibition of this series was dependent on the electronic effect of substituents and hydrophobicity. In the case of alkyl phosphonic esters, the steric factors played a major role, as found in the case of 2,4-trichloro N-alkyl phosphoramidates. Gozzo *et al.* [6] observed that insecticidal activity of O,O-di-ethyl O-(1-R-3-R-1,2,4-triazol-5-yl) phosphorothioates was solely dependent on steric parameters of R, expressed by the STERIMOL parameters L, B1 and B4.

Metcalf and Metcalf [7] correlated the toxic action of a series of O-alkyl-O-aryl alkyl and aryl phosphonates (1) and phosphonothionates on houseflies with Taft steric constants (E_s) and Hammett electronic constants (σ^*).

Sasaki *et al.* [8] studied the QSAR of a series of O,O-dimethyl O-(4-substituted-2,6-dichlorophenyl) phosphorothioates and O-alkyl O-methyl O-(2,4,6-trichlorophenyl) phosphorothioates for their antifungal activity against *R. solani*. The activity was found to be parabolically related to the length (Sterimol L) and hydrophobicity (π) of the substituents.

In the field of herbicides Yang *et al.* [9] first studied the QSAR of a series of O-aryl-O-ethyl N-isopropyl phosphoramidothioates for their herbicidal activity on the barnyard grass, *Echinochloa crussgalii*, using physico-chemical parameters of the aryl substituents. The activity was found to be dependent on the hydrophobicity of the whole molecule as well as on the position-specific steric effects of the substituents.

The QSAR methods have been applied by Roy *et al.* [10] on series of organophosphorus pesticides to obtain the best model for the design of most potent pesticides. Gupta *et al.* [11] reported QSAR studies on O,O-diaryl S-ethyl phosphorothioates for their fungicidal activity against five fungi and observed that both hydrophobic (R_M) and electronic (σ) parameters were responsible for the activity. Roy *et al.* [12] studied the QSAR of bisaryl *iso*propyl phosphonates for their antifungal activity against six fungi and found that the activity was dependent mainly on lipophilicity, proximity polar effect, and molar refractivity. QSAR technique was also applied to a series of methyl diaryl phosphonates and 1,1-dichloroethyl diarylphosphonates for their antifungal activity against four fungi [12]. The results showed that hydrophobicity and electronic character of substituents were responsible for the fungicidal activity.

The QSAR study of O-alkyl S,S-diaryl phosphorotrithioates for their fungicidal activity shows that activity was dependent on alkyl groups attached to oxygen atom [10]. However, the presence of electron withdrawing substituents in the benzene ring were favourable for fungicidal activity.

The QSAR studies of S-alkyl S,S-diaryl phosphorotrithioates for their nematicidal activity against *Meloidogyne incognita* showed that the STERIMOL parameter B4 of the alkyl groups attached to sulphur atom along with the electronic nature (σ) and the lipophilicity (π) of the phenyl ring substituents were found to be influencing the activity [13].

Gupta and Roy [10] studied the QSAR of O-methyl O,O-diaryl phosphorothionates for their fungicidal activity against *Sclerotium rolfsii* and observed that increase in hydrophobicity and minimum width of ortho substituents and decrease in minimum width of meta substituents enhanced the activity of compounds towards *S. rolfsii.*

Chakravarti [14] applied the QSAR studies on a series of O-aryl-N-*n*-propyl 1,1-dichloroethylphosphonamidates and reported that the herbicidal activity was dependent on the hydrophobic and electronic character of the substituents. There was a marked preference for electron withdrawing substituents with low hydrophobicity primarily at the para.position of the phenyl ring. In addition, smaller size of the substituents was generally preferred for higher activity and selectivity was inversely correlated with hydrophobicity.

A comprehensive QSAR study of sixty-two O,O-bisaryl alkyl phosphonate fungicides was carried out by Hansch approach [12]. The variation in fungicidal activity among members of the series against four fungi e.g. *Helminthosporium oryzae, Pyricularia oryzae, Alternaria alternata* and *Rhizoctonia bataticola* was dependent mainly on changes in hydrophobicity of the phenyl substituents and molar refractivity of the alkyl substituent. Molar refractivity of the substituents at the para position of the phenyl ring and electronic factors also played significant roles in two cases individually. Principal component analysis of the fungicidal activity data revealed that the first principal component had 79.2% of the information content. Physico-chemical interpretation of the scores corresponding to the first principal component gave results consistent with those obtained by Hansch analysis. It appears that the fungicidal molecule interacted with the bioreceptor through two para positions of the phenyl rings while the alkyl substituent and the steric factor of the substituents play a dominant role during the interaction (Fig. 8.1).

The application of CAMD/CAMM

The application of Computer-Assisted Molecular Designing (CAMD)/Modelling (CAMM) tool can now be applied in the discovery of pesticides. This also helps in providing analytical methods for rationalizing or predicting molecular properties related to molecular conformation and config-

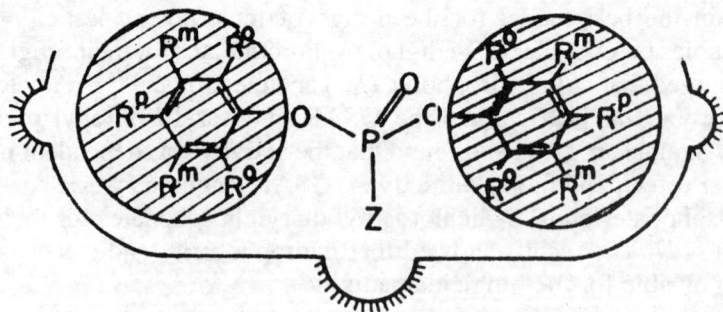

Fig. 8.1. Receptor diagram of O,O-bisaryl alkyl phosphonates. Substituents at the *ortho, meta,* and *para* positions of the phenyl rings are designated as R^o, R^m and R^p, respectively, and the alkyl substituent is designated as Z. The shaded area shows hydrophobic sites that facilitate binding and transport to the bioreceptor. Lines with fringes represent three-dimensional steric interaction sites of the receptor to which the molecules must fit. [Roy *et al.* J. Agric. Food Chem. 44, 3971 (1996)]

uration, using quantum chemistry to predict the structures of proteins and biorational discovery process, and also for structuring the experimental design of the synthesis programme.

A three-dimensional model to develop ergosterol biosynthesis inhibitors using 8-7-isomerase, an enzyme present in fungi has recently been constructed (Fig. 8.2). The model is based on three compounds : (S)fenpropimorph, 1-[3-(4-tert.butylphenyl)-2,3-dimethyl-2-propenyl-pyridine and the natural substrate for the $\Delta 8$-$\Delta 7$-isomerase [15].

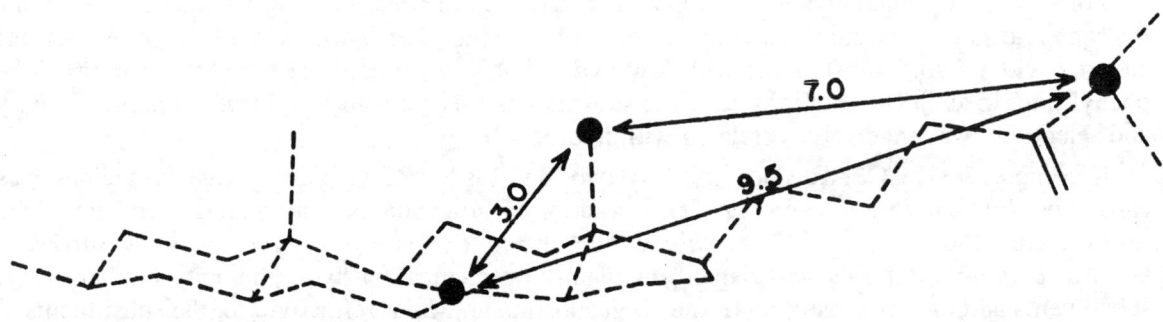

Fig. 8.2. 3 D-model for inhibitor of ergosterol biosynthesis. [Jansen J.S. *et al. IUPAC Seventh Internatn. Congr. Pestic. Chem. Hamburg, abs.ID-05 (1990)]*

The QSAR study on sulfonylurea herbicides provided the following finding based on the experimental data obtained in the laboratory, whereby a Caliper model was postulated to contain 3 independent planar conformations, 3 negative centres and a concave centre within the active sulfonylurea structures. Due to the difficulties in obtaining the X-ray crystal structure of ALS target enzyme, an approach using CoMFA and Leapfrog software, a preliminary 3-D contour map of ALS was established. With the help of Disco software and data search, novel structures of potential ALS inhibitors have been designed [16] (Fig. 8.3).

The QSAR conformational analyses and computer graphics study of triflumizole analogues of fungicidal N-(1-imidazol-1-yl-alkylidene) anilines (3) showed that the minimum energy structure

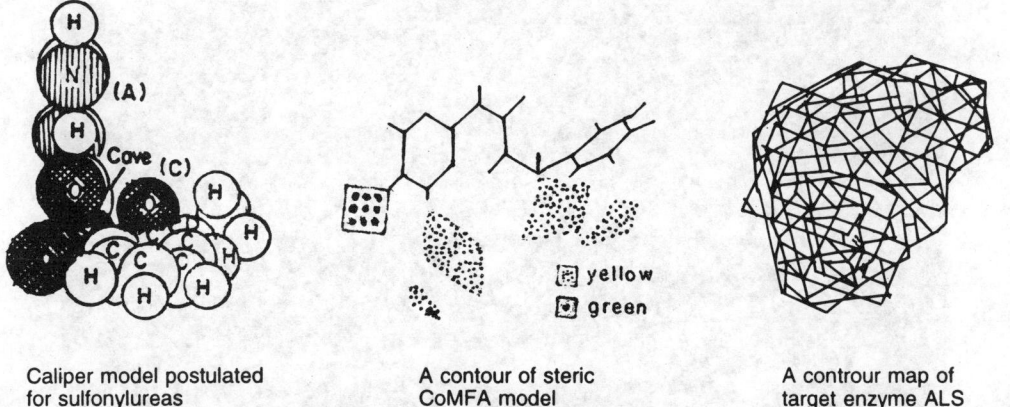

Caliper model postulated   A contour of steric   A controur map of
for sulfonylureas   CoMFA model   target enzyme ALS

Fig. 8.3. 3 D-QSAR study on conformation of sulfonyl urea herbicide. [Li, Z.M. *et al.*, IUPAC *Ninth Internatn. Congr. Pestic. Chem.*, London, abs. 1, E-018 (1998)]

of triflumizole fits the receptor site of cytochrome P-450 blocking the oxidation of lanosterol to ergosterol. Fig. 8.4 represents a schematic model of the interaction of triflumizole with the cavity of P-450 indicating the sites of importance of steric and hydrophobic properties of the imidazole derivatives [17].

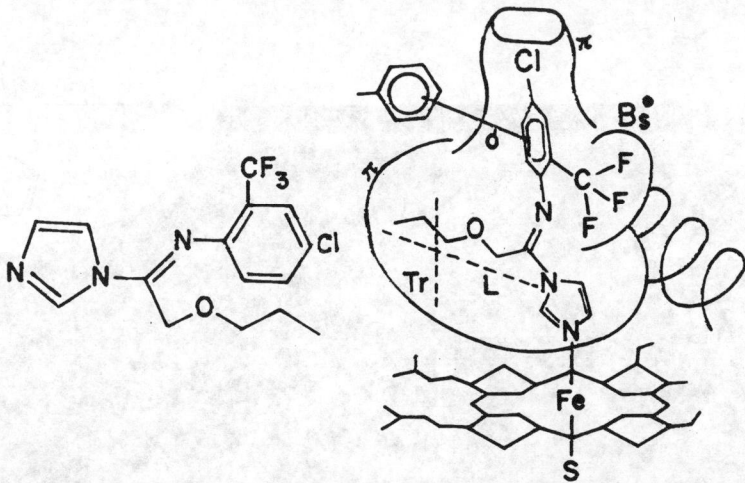

Fig. 8.4. Model of triflumizol fungicide. [Nakayama, A. *et al.*, *J. Pestici. Sci.* 14 (1), 23-37 (1989)]

The QSAR study on fungicidal substituted *cis*-2(1H-1,2,4-trizol-1-yl) cycloalkanols (4) was carried out [18], whereby conformational analyses and molecular graphics study showed, that the active compounds in their minimum energy conformation favourably interact with the cytochrome P-450 receptor site (Figs. 8.5a, 8.5b). The study also revealed that meta-substituent on the phenyl residue hindered free rotation which was detrimental to the activity. On the other hand, the 5-chlorothiophen-2-yl residue can rotate freely resulting in a good fit to the receptor site while derivatives with benzyl substituent cannot fit the receptor site well.

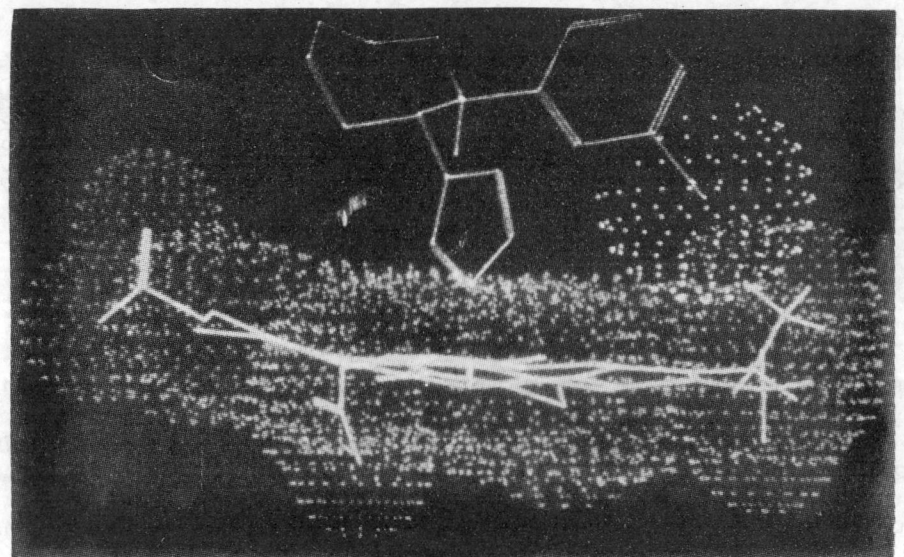

Fig. 8.5a. Computer graphics of the Vander Waals surface-stick model of the binding of conformer I of compound (upper) to cytochrome P-450 porphyrin (lower, partial structure).

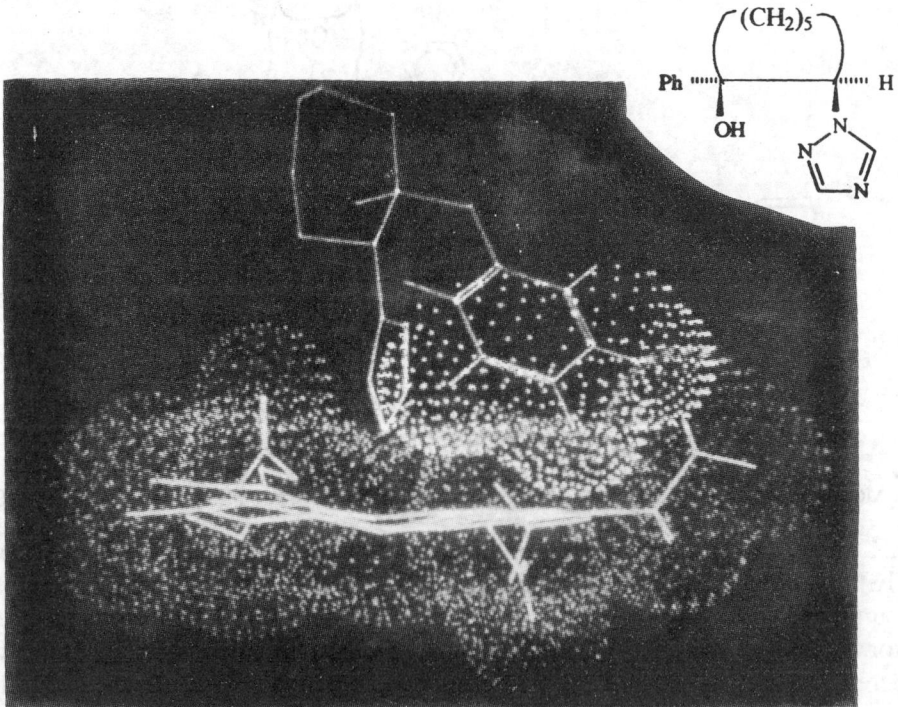

Fig. 8.5b. Computer graphics of the Vander Waals surface-stick model of the binding of the most stable conformer I of compound (upper) to cytochrome P-450 porphyrin (lower, partial structure). [Kataoka, T. *et al., Pestic. Biochem. Physiol.* 34(3), 228-239 (1989)]

Planning in pesticide design

Certain important features are discussed and must be borne in mind before planning a pesticide discovery using this approach.

i) Lead compound

The lead compound may be a natural product isolated from a plant or microbial broth, a compound active after screening of randomly synthesized molecules. If the mode of action of the compound is known it will be easier to establish rules based on QSAR model.

The QSAR sub-goal may itself be broken down into several sub-groups comprising identification of toxophore, spacers, secondary binding sites and the ballast positions in a molecule.

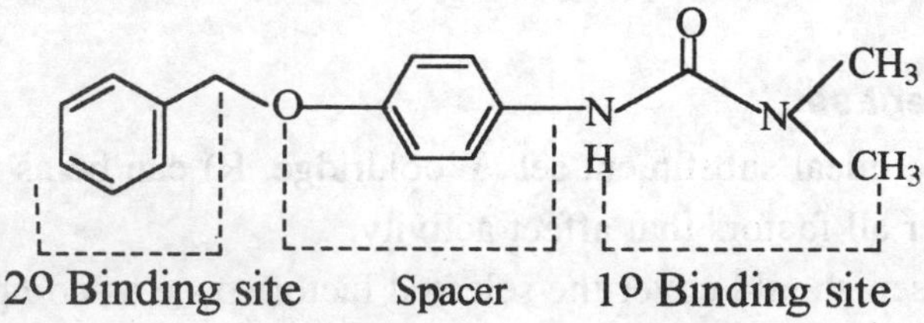

$$2^0 \text{ Binding site} \qquad \text{Spacer} \qquad 1^0 \text{ Binding site}$$

For example in diuron, the herbicide urea moiety is identified as toxophore while the phenyl ring serves as a spacer between two binding sites. The benzyloxy group serves as the secondary binding site held to the target site by hydrophobic binding.

The ballast site is one which can accomodate very large groups with diverse physico-chemical properties and its identification is very important in agrochemical design. An excellent example of this is the benzoyl urea insect chitin inhibitor whose para substitution on the aniline moiety may be very large and is used to adjust the properties of the molecule.

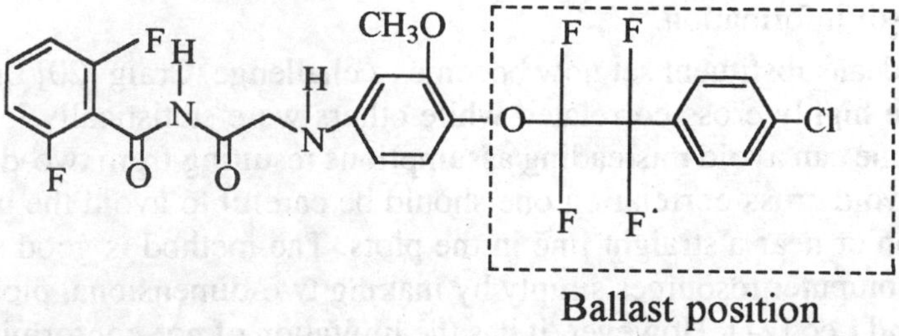

Ballast position

Once the activity is optimised by derivatization of the parent molecule, most programmes set out a new sub-goal for the investigation of biosteric analogues. Reviews of this approach are given by Lipinski [19] and others. Strategies based on the QSAR model are best used. The activity optimization and analogue synthesis may be carried out by intrinsic or green house testing termed as agrokinetic optimization properties such as vapour pressure, water solubility, soil and plant stability. The selectivity must be adjusted to actual requirements found in the field. Again the QSAR strategies will be the major contributors to the programmes if experiments are designed for that purpose. For every lead there are literally millions of analogues and derivatives that could be pursued. As an example, if there are three non-equivalent positions for substitution in a lead and if one considers only the small set of substituents for which complete information on lipophilicity (π), electronic (σ or F and R) and size/shape (MR or the STERIMOL parameters L, B1 and B4) are available i.e. around 100 substituents, then there would be 100^3 or one million possible derivatives of the lead. There is no guarantee that a commercial compound might exist in the lead area, however one must pursue on probabilistic knowledge.

The biological test may be used as a guide in a project which has a profound effect on the selection of a strategy for optimization. Experience has demonstrated that activity optimization and establishment of sound QSAR is relatively straight forward in pure isolated enzyme systems. It is an established fact that a poor series design can be the cause of a failure to find best QSAR model for activity. Conversely, adoption of strategies for the proper selection of substitutent sets may be required to increase the efficiency of the design process and ensure the development of QSAR models.

ii) Selection of substituent set

The major criteria to select ideal substituent set (Wooldridge, R) can be as follows :

  a) The set should cover all factors that affect activity.
  b) The substituent chosen should cover the selected factor space as completely as possible
  c) The set should represent those properties with substituents whose parameters vary independently that is they should be orthogonal.
  d) The number of substituents chosen will depend on the design strategy but should be the minimum necessary to avoid chance correlations.
  e) Target compounds should be chosen to preserve synthetic resources but not that they are easily synthesised.
  f) The synthesised compounds must be stable under condition of bioevaluation.

The Wooldridge's set seems to cover lipophilicity (π), electronics (F&R) and steric size/shape (E_s and MR) well. However, E_s and MR would not normally be included in the same analysis since they provide same type of information.

The selection of an ideal substituent set now becomes a challenge. Craig [20] first demonstrated that certain factors were highly cross correlated while others were statistically least independent. He then suggested that one can avoid misleading assumptions resulting from two-dimensional plots of the parameters. To avoid cross-correlation one should be careful to avoid the use of only those substituents which lie on or near a straight line in the plots. The method is good since it could be accomplished without computer resources simply by making two-dimensional plots on paper from data given by Hansch and Leo [21]. However, it has the limitation of not-conforming to the criteria of representing all of the potentially pertinent factors in an initial set. An alternate approach to substituent selection was given by Hansch and his co-workers [21] using Cluster sets, in initial set of compounds for synthesis. The substituents are well spread in parameter space and appear to be orthogonal except parameter π and MR, based on the correlation matrix.

Although Cluster analysis is found to be very successful as a tool to design a set of substituents well spread in parameter space but it fails to guarantee complete orthogonality. Another approach of Wootten and his co-workers [22] suggested the ease of synthesis by limiting their database choice to 35 substituents which are easy to introduce into an aromatic ring. The method for choosing substituents from this set has become known as multi-dimensional non-linear mapping

Table 8.1. Wooldridge's ideal substituent set

Substituent	π	E_s^a	MR	F	R
H	0.00	0.00	1.03	0.00	0.00
n-C_6H_{13}	3.05	−1.54	28.87	−0.06	−0.09
SCH_3	0.61	−1.07	13.82	0.20	−0.18
O-n-C_6H_{13}	2.55	−1.04	30.90	0.25	−0.55
CN	−0.57	−0.51	6.33	0.51	0.19
Br	0.86	−1.16	8.88	0.44	−0.17
OH	−0.67	−0.55	2.85	0.29	−0.64
t-C_4H_9	1.98^a	−2.78	19.62	−0.07	−0.13
$NHC=OCH_3$	−0.97	−2.98	14.93	0.28	−0.26
$N(CH_3)_2$	0.18	−2.84	15.55	0.10	−0.92
SO_2CH_3	−1.63	−2.63	13.49	0.54	0.22
CF_3	0.88	−2.40	5.02	0.38	0.19
$N(C_3H_3)_2$	2.18	−2.84	34.03	−0.02	−0.92
NH_2	−1.23	−0.61	5.42	0.02^a	−0.68
SO_2NH_2	−1.82	−2.62	12.28	0.41	019^a

Values changed to conform with Hansch and Leo.

Table 8.2. Substituents chosen for synthesis

Substituent	Cluster group	Substituent	Cluster group
$C(CH_3)_3$	18	$NHC(=O)CH_3$	7
C_6H_5	18	$NHC(=O)C_6H_5$	14
$CH_2C_6H_5$	18	OC_6H_5	18
$CH(CH_3)_2$	17	$OC(=O)C_6H_5$	15
$CH=CHC(=O)CH_3$	12	$P(CH_3)_2$	1
CH=CHCOOH	3	SO_2CF_3	11
H	2	SO_2CH_3	4
$N(CH_3)_2$	9	$SO_2C_6H_5$	13

pattern and is based on maintaining a pre-set distance between substituents in multi-dimensional space. A compound with hydrogen atom usually is chosen. The Euclidean distance of all points in the data set from the seed is calculated, a distance criterion is established and the substituent that meets this criterion is thus chosen as the first substituent and the new seed. The same guidelines are followed to choose other substituents.

iii) Combinations of the chosen parameters through linear regression analysis

The parameters that are used in the design should be the ones used in the initial evaluation of the biological data. Any good regression programme namely, SAS REGRESS [23) can be used for analysis. As an absolute minimum, there should be three observations per variable in the final model. However, four to six observations per variable is the usual rule of thumb. The output of

Table 8.3. Statistics for each variable

Variable	Mean	Standard deviation	Smallest value	Largest value
π	0.64	1.11	−1.63	2.08
F	0.18	0.26	−0.15	0.73
R	−0.02	0.42	−0.92	1.04
MR	20.99	9.11	1.03	34.64

	Correlation matrix			
	π	F	R	MR
π	1.000			
F	−0.358	1.000		
R	−0.175	−0.064	1.000	
MR	0.449	0.034	−0.083	1.000

this programme contains specific information that can be used to decide the best model. These are r (or R^2), s, F and the Students 't' test for the coefficients of each model. The rule for choosing the best model is to seek the simplest model that is not improved by addition of further terms. The criteria used to define best model are .

R^2 : can be interpreted on the percentage of biological variation which can be explained by the model and R^2 should be greater than 0.50. Although R^2 of unity would be a worthy goal in an error free system, such a value would be suspect in a complex system, such as green house testing or to a lesser extent *in vitro* testing. An estimate of the pure experimental error for the test system used provides how large R^2 theoretically could be. The following data derived from Doweyko *et al.* [24] would be illustrative.

Standard error of estimate for biodata (log 1/c)	Expected R^2 range
0.20	0.839 - 0.947
0.25	0.772 - 0.914
0.30	0.704 - 0.853

Here the lack of accurate reproducibility of biological test data results in an unavoidable standard deviation of about 0.20 to 0.25 log1/C (pc) unit *in vitro* tests, and under greenhouse with much controlled condition, a standard error of 0.30 pc units is observed. Therefore, a useful rule is to accept a model which could explain 50-85% of the variance in *in vivo* testing and thus expect values of 70-90% *in vitro* tests.

F-test : The F-test for any acceptable model must be significant at the 95% level or stated another way, the probability that the data does not fit a linear model must be 0.05 or less (p = 0.05 or less).

Students's 't' test : When adding a parameter to a model, the Students' t test for significance, should be greater than 95% (p > 0.05).

s : When comparing models a significant decrease in the standard error of estimate, s, should be used as a guide. A decrease of 0.05-0.1 in s is an estimate of a significant change.

r : When comparing models, a significant increase in the correlation coefficient r should be realised to justify the addition of a new term or the rejection of a model. An increase of 0.05-0.1 in r is an estimate of a significant increase.

The best model of QSAR becomes a hypothesis that must be tested by choosing a new set. The concept of an ideal substituent set must be maintained to fill in gaps in observed activity and extend the tested area towards higher activity. One should reconsider the factors to be included in the original design set to find out whether it is important to include them in the new set.

CONCLUSION

The successful application of QSAR and molecular modelling techniques requires effective experimental design and selection of substituents to represent the factors important in determining the property being optimised.

The key element in the successful application of quantitative design of the structure requires the hands of scientists knowledgeable in pesticide design, and laboratory chemists and biologists responsible for the design project. Software is available for statistical analysis and data base management, but software customized to fit local practices and data representation (e.g. biological and physico-chemical property database) are likely to be more useful.

It is expected that QSAR and molecular modelling will become an important tool in future towards designing efficacious and safe pesticides which are currently in high demand in plant protection science.

REFERENCES

1. Hansch, C. and Fujita, T. ρ-σ-π analysis a method for the correlation of biological activity and chemical structure. *J. Am. Chem. Soc.* **86**, 1616-1628 (1964).

2. Hansch, C. and Leo, A. *Substituent constants for correlation analysis in chemistry and biology*, Pamona College, California, USA (1979).

3. Wooldridge, K.R.H. A rational substituent set for structure activity studies. *Eur. J. Med. Chem.* **15**, 63 (1980).

4. Martin, Y.C. A pactioner's perspective of the role of quantiative structure activity analysis in medicinal chemistry. **24**, 230 (1981).

5. Hansch, C. and Deutche, E.W. The use of substituent constant in the study of structure activity relationships in cholinesterase inhibitors. *Biochem. Biophys. Acta* **126**, 117-128 (1966).

6. Gozzo, F., Masoero, M., Quadrelli, L. and Zagni, A. A novel series of O,O-diethyl O-(1,3-disubstituted-1,2,4-triazol-5-yl) phosphorothioates, part 11, Quantitative sturcture activity relationships, *Pestic. Sci.* **11**, 344-323 (1989).

7. Metcalf, R.L. and Metcalf, A.M. Toxic action of O-alkyl-O-phenyl phosphonothioate insecticides. *Pestic. Biochem. Physiol.*, **22**, 169-177 (1984).

8. Sasaki, M., Kato, T., Ooshi, T., Takayama, C. and Mukai, K. Synthesis and antifungal activity of O,O-diakyl O-aryl phosphorothioates and related compounds. *J. Pesti. Sci.* **9**, 737-744 (1984).

9. Yang, H.Z., Zhang, Y.J., Wang, I.X., Tan, H.F., Cheng, MR X.D., Xing, Chen R.Y. and Fujita, T. Quantitative structure activity study of herbicidal O-aryl O-ethyl N-isopropyl phosphoramidothiate. *Pestic. Biochem. Physiol.*, **26**, 275-283 (1986).

10. Roy, N.K. Chloroalkyl phosphonates and phosphorothioates a new group of fungicides. *Proc. Indian natn. Sci. Acad.* **B56 (3)**, 305-310 (1990).

11. Gupta, R.L. and Roy, N.K. Recent developments on organophosphorus pesticides and their environmental impact. In : *Agrochemicals and sustainable agriculture* (N.K. Roy, ed.), APC Publication Pvt. Ltd., New Delhi, pp. 123-140 (1996).

12. Roy, N.K., Nidiry, E.S.J., Vasu, K., Bedi, S., Lalljee, B. and Singh, B. Quantitative sturcture-activity relationship studies of O,O-bisaryl alkyl phosphonate fungicides by Hansch approach and principal component analysis. *J. Agric. Food Chem.*, **44**, 3971-3976 (1996).

13. Gupta, R.L., Roy, N.K. and Prasad, D. Evaluation and structure activity relationship of S-alkyl S, S-daryl phosphorotrithioates for the nematicidal activity against root knot nematode. *Fundam. appl. Nematol.*, **16 (2)**, 97-102 (1993).

14. Chakravarti, S. Ph.D. Thesis, P.G. School, IARI, New Delhi (1992).

15. Jensen, J.S., Petterson, I., Jorgensen, F.S. and Klemmensen, P.D. Molecular modelling and QSAR in pesticide design. *Seventh Internatn. Congr. Pestic. Chem.* Hamburg, abs.ID-05 (1990).

16. Li, Z.M., Linn, J., Lai, C.M. and Ma, Y. Research on target enzyme ALS in structure activity study in sulfonyl urea herbicides. *Ninth Internatn. Congr. Pestic. Chem.*, London, abs. **1**, E-018 (1998).

17. Nakayama, A., Ikura, K., Katsuura, K., Hashimoto, S. and Nakata, A. Quantitative structure-activity relationships, conformational analyses and computer graphics study of triflumizole analogs; fungicidal N-(1-imidazol-1-yl-alkylidene) anilines. *J. Pestici. Sci.* **14** (1), 23-37 (1989).

18. Kataoka, T., Hayase, Y., Hatta, T., Mayashi, Y., Murabayashi, A., Mukisumi, Y. and Fujita, T. Quantitative structure activity study of fungicidal 1-substituted *cis*-2-(1H-1,2,4-triazol-1-yl) cycloalkanols. *Pestic. Biochem. Physiol.* **34** (3), 228-239 (1989).

19. Lipinski, C.A. Bioisosterism in drug design. *Annu. Rep. Med. Chem.* **21**, 283 (1986).

20. Craig, P.N. Comparison of the Hansch and the Free-Wilson Approaches to Structure-Activity Correlations. In : *Advances in chemistry, Biological correlation—The Hansch Approach.* (R.F. Gould, ed.), Sr 114, American Chemical Society, Washington, U.S.A., 115-129 (1992).

21. Hansch, C., Leo, A., Unger, S.H., Kim, K.H., Nikaitani, D. and Lien, E.J. Aromatic constants for structure activity correlations. *J. Med. Chem.* **16**, 1207 (1973).

22. Wootten, R., Cranfield, R., Sheppey, G.C. and Goodford, P.J. Physicochemical-Activity relationships in practice.2. Rational selection of benzenoid substituents. *J. Med. Chem.* **18**, 607 (1975).

23. SAS Programs are available through the SAS Institute, Inc. Box 8000, Cary, N.C. USA 7411-8000.

24. Doweyko, A.M., Bell, A.R., Minatelli, J.A. and Relyea, D.I. Quantitative structure-activity relationships for 2[(phenylmethyl)sulfonyl} pyridine oxide herbicide. *J. Med. Chem.* **26**, 475 (1983).

CHAPTER 9

Suggestions for Future Development, Intellectual Property Rights and Indian Patent System

- Present scenario in the area of pesticides
- Use of new effective insecticides, fungicides, herbicides and PGR
- Newer approaches to control insect pests
- Need to discover and develop new agrochemicals using QSAR and CAMM
- Chemical hybridizing agents
- Importance of biotechnology on crop protection
- New improved formulations and application technology
- Intellectual property rights
- Indian patent system
- Conclusion
- *References*

Plant protection chemicals are an essential input in modern agriculture. Their use is likely to continue in the near foreseeable future. As Prof. N.E. Borlaug, Nobel Laureate, once quoted *"The people of the world must decide that we either use agricultural chemicals (pesticides) and use them wisely and in the right amounts and the right kinds to produce the food we need or we starve"* Thus the choice is to use pesticides judiciously in an integrated manner.

In older times, a few inorganic chemicals like sodium chlorate, copper sulphate Bordeaux mixture, hydrogen cyanide, lead arsenate, sulphur and mercuric compounds were used as plant protection chemicals. In the year 1930, synthetic pesticides were for the first time introduced. After the second World War, optimism for high agricultural production combining high yielding varieties, chemical fertilizers and pesticides was great, and efforts all over the world concentrated on the development of broad spectrum pesticides which could be used against a complex range of insect pests, diseases and weeds. Synthetic organic insecticides (DDT, BHC aldrin, dieldrin, endrin, parathion and carbaryl) were introduced {cf. chapter 3(I, IV, V)}. However, it was realised soon that the use of these pesticides caused several problems. It was observed that many phytophagous pests had developed resistance to these insecticides, requiring heavier doses and more frequent application and newer chemicals. In addition, the use of these pesticides caused the break down of the natural ecosystem and destroyed quite a few beneficial organisms with concomitant resurgence of many secondary and minor pests of economic importance.

The levels of DDT, BHC and their metabolites in human fat tissues and human milk in Nicaragua were found the highest in the world. The chronic effects of pesticides and its misuse led to the ban of DDT and related compounds in developed countries. Recently, the Government of India has banned the use of DDT, BHC and related compounds like aldrin, endrin, chlordane in agriculture sector. However, these are still being used in some developing countries.

Sustained R&D efforts have given way to safer insecticides like malathion, and synthetic pyrethroids {cf. chapter 3(II)} which are environmentally more acceptable. Some of these, like decamethrin obtained as a single enantiomer is remarkably potent and effective at lower application rate (2-10 g/ha). The pyrethroids possess low mammalian toxicity and are safer than organophosphates and carbamates. The scientists are now working to develop ecofriendly chemicals which are safer and more selective in action against the target pest, and do not persist longer than necessary to achieve its objective but without harming the ecosystem. In the field of herbicides, development of resistance in common broad leaved weeds in cereal crops towards phenoxyacetic acid is posing a serious threat to world cereal production.

It is known that the discovery and development of a new agrochemical is a long multidisciplinary process involving high cost. The synthesis of a new molecule invariably involves the following routes :

a) Emperical approach—random synthesis

b) Analogue synthesis with optimisation of lead molecule

c) Molecule leads from natural pesticides

d) Bio-rational approach

e) Mathematical modelling and computer programmes aided by mode of action/physico-chemical approach.

f) Stereospecific synthesis and manufacture

The golden thumb rule is that the new product if commercialized should be eco-friendly, economical, easy to handle and should not pose any health hazard.

The modern synthetic insecticides are highly effective against flies, forestry and timber pests and are capable of checking outbreaks of typhus, bubonic plague and malaria, but repeated applications and misuse are known to induce resistance which is becoming a serious problem. So far most of the conventional insecticides act on the insect nervous system, pyrethroids on the Na^+ ion channel whereas organophosphates and carbamates inhibit the enzyme acetylcholinesterase. Although, new types of insecticide acting on other target sites were developed, their use is restricted because of inherent constraints. Alternate methods of pest control using semiochemicals and other infochemicals such as pheromones, insect growth hormone mimics, repellents, antifeedants, chitin synthesis inhibitors and chemosterilants are explored (cf. chapter 7) during the last twenty years. Some of such novel insect behaviour modifying chemicals possess high selective action against the target insect pests and would not hopefully induce resistance as frequently as conventional insecticides. However, the conventional insecticides like OPs, carbamates, pyrethroids can be made selective by moderating the rates of penetration and metabolism in various species. Biopesticides (both botanical and bio-organisms) are increasingly becoming popular in insect pest management programmes. The availability of natural products in small quantity however is viewed as a serious disadvantage. Attempts have, therefore, been made to develop synthetic analogues of lead molecules though with little success. The total synthesis of azadirachtin, the active component of neem, is still in progress. However, a few simpler analogues of azadirachtin prepared as simpler templates did not show the desired level of insect antifeedant activity (cf. chapter 2). Efforts must be made to explore many unexploited natural products which might provide new selective and safer pesticides with different biochemical modes of action [1, 2] to enable us to overcome problem of insect resistance development. The *iso*butylamides synthesised by modelling on natural insecticides like **piperine** (1), appear to be promising active analogue (2), effective against insects resistance to pyrethroids. Significant research work has been carried out on chemicals affecting insect neurotransmitters and neurohormones.

The pentapeptide, cockroach muscle neurotransmitter, **proctolin** (3) is one such example which along with the toxin generated by insect predators like spider venom could provide alternate future insect control chemicals.

Arg - Tyr - Leu - Pro - Thr

(3)

In case of herbicides, with different modes of action following selectivity and persistence {cf. chapter 3(IX)}, one can achieve selective control of weeds with various crops. The recently developed and the most effective sulfonylurea herbicides can selectively control broad leaved weeds in cereals at a very low dose (20 g/ha). Now-a-days more emphasis is being given on plant growth regulators (PGRs) to improve crop yields and quality of produces as well as tools for easy harvesting. Ethephon, a PGR is being widely used now-a-days on rubber plantations in microgram level to improve quality and enhancing the yield of rubber material. Similar research efforts in this area could also provide improvements in both quality and yield of other important crops. The use of cycocel for dwarfing wheat crop in the temperate climate, dessicants as an aid in harvesting, NAA in pineapple growing and use of defoliants for cotton harvesting are some known examples. Their use will certainly increase and research efforts are needed to develop new chemicals having these properties. Some other uses of PGR are the methoxyamine acid derivatives which interfere with release of ethylene and delay ripening of fruits in cold storage, improving economy of fruit growers. Certain chemicals can impart strength to seedlings during transplanting. Two of the these transplant shock inhibitors Hymexazole and BASF 10604 are now under commercial trials in rice transplantations.

Induction of male sterility is an indispensable tool in the hybrid programmes of small grain cereals like wheat and rice. Chemical induction of male sterility using chemical hybridizing agents is gaining importance all over the world. Based on the proline level in both sterile and fertile gametes, chemists of Shell Research Chemical Company (Kent), have deployed proline analogues such as methanoproline (4a) and others like (4b) as effective sterilants which allow the production of commercial F_1 hybrids of wheat with enhanced traits such as higher yields and improved disease resistance.

(4a) (4b)

Another idea of using chemical hybridizing agents (CHA) is to restore fertility in genetic male sterile variety of cereals which has been of interest for a long time [3]. But this approach of fertility restoration over male sterility has only recently come in use. This was first time developed in the context of maize hybrids but later on extended to barley and wheat crops. There are a number of advantages of chemical restoration of genetic male sterility. For example, the chemical may not be required to be 100% effective since its purpose is to increase supplies of female parent seed. Once an adequate male sterile female parent seed supply is available, hybrid seed can be produced using the same production techniques through heterosis. However, the chemical is not to be used in the hybrid seed production, a lower volume of chemical would be adequate as compared to amount of CHA needed. It would have a favourable impact on crop and environmental residues.

Amongst the various effective fungicidal chemicals, the azole group of compounds are very effective {cf. chapter 3(VIII)}. A few more effective and eco-friendly azole compounds commercialized recently like a) **tetraconazole** (5) (for the control of powdery mildew and brown rust, at low doses), b) **bromoconazole** (6) (rust, septoria, scab), c) **fenbuconazole** (7) (rust of different types) whose dose is 75 g/ha exhibit systemic, preventive and curative action.

(5) (6) (7)

In the area of insecticides, although a large number of synthetic pyrethroids have been developed as safe and effective ones but still a few of them are toxic to fish. Recently developed non-ester synthetic pyrethroids like **ethofenprox** (MTI-800), and some related compounds {cf. chapter 3(II)} are not only safe to fish but also exhibit broad spectrum of activity. Pyrethroid, **lamdacyhalothrin** is another potential substitute for DDT in malaria control programme and has already been found useful in countries, like Nepal, Brazil and Indonesia.

So far, not much attention has been paid to investigate the effects of the chemicals on the host plant and the environment. A few chemicals act indirectly either by increasing the resistance of the crop plant towards insect pests or by modifying the environment to make it less conducive to the support of the insect pests. These might be beneficial as compared with those exerting a direct toxic action on the pest organisms.

The conventional systemic fungicides when sprayed on plants generally move towards upward direction (apoplastic movement). There is hardly any chemical having downward (symplastic) movement. If such fungicides are developed they could be effective against not only root pathogenic fungi, but also against many nematodes causing serious damage to potatoes, sugar beet and tropical crops which do not have adequate chemical protection. A few synthetic plant growth regulators like **daminozide** (8), owes its activity against potato scab due to downward translocation. **Pyroxychlor** (9), when sprayed on leaves is also translocated down to the roots to control the infecting phycomycetes such as foot-rot in peas. Similarly, the herbicide **glyphosate**, is known to translocate from the leaves down to the rhizomes.

$$HOOC\text{-}(CH_2)_2CONH\text{-}N(CH_3)_2 \quad ;$$

(8)   (9)   (10)

This downward movement from leaves to root zone is attributed to the presence of a weakly acidic moiety ensuring biphasic solubility.

Certain diseases can be minimised by alteration of the nutrition of the host plant such as stabilizing nitrogen nutrition by the addition of nitrification inhibitors, like **nitrapyrin** (10). Research efforts are also needed to develop spore germination inhibitors like 3,4-dimethoxy cinnamate isolated from plants. Viral and bacterial diseases can cause considerable crop damage. Since no chemical is available to combat such diseases, there is an urgent need and scope to discover such molecules. Plant breeding programmes have been successful in developing new races of crop plants resistant to root pathogenic fungi. However, much of the success is short lived as the resistant variety in a short time becomes again susceptible to a new pathogenic strain. Research efforts to solve this complex problem must be strengthened.

Inhibitors of other enzyme systems like those involved in oxidative phosphorylation and inhibition of nerve impulses by blocking GABA-ergic transmission with compounds like pyrotoxin, bicyclic ketones and benzodiazepenes is another potential area to be looked in. A break-through in the development of these products can definitely lead to a new generation of future insecticides.

Biotechnology and genetic engineering provide ample opportunities for crop improvement through tissue culture and manipulation of plants and organisms [4, 5] (cf. chapter 5). Biopesticides (microbial pesticides) like Bt, NPV are commercialized, though in limited forms. The technique for the introduction of foreign genes into crop plants has been used to enhance herbicide, fungal and viral resistance in crops. Transgenic cotton plant developed by Monsanto, USA has been tested successfully in field. The recent Economic Research Service (ERS) of USDA report cites that use of certain crops like cotton, corn and soybean improved through biotechnology is associated with significantly higher yields, significantly reduced herbicide treatments and fewer insecticide treatment for target pests. The introduction of Bt cotton has reportedly reduced the use of over 900,000 kg of chemical insecticides in the U.S. Similarly the Bt and virus resistant potatoes have significantly reduced chemical use. Plants tolerant to glyphosate and sulfonyl urea herbicides have also been discovered. The introduction of a maize plant mutant, resistant to rootworm and borer would decrease the over reliance for synthetic pesticides on this crop. The application of molecular techniques such as DNA cloning *in vitro* has extended our knowledge on the mechanism of resistance to insect pests and herbicides. Plant biotechnology has also provided spectacular information towards design and construction of plants with induced resistance to insect pests and herbicides. It might be possible in future to isolate the genes involved in stress responses in plants and to manipulate the synthesis and accumulation of specific plant secondary metabolites having pest control potential. In the twenty first century, biotechnology and biopesticides would probably play a major role in integrated pest management (IPM) programmes as they do not damage natural regulatory forces operating in the ecosystem and are helpful in minimising the use of synthetic pesticides. The development of safe formulations is another potential area of research to be looked into carefully so that there is no danger to beneficial insect pests (cf. Chapter 4). Research efforts must also be directed to develop effective pesticide delivery systems for those products which are active at g/ha level so that they can be target specific like a gun. The use of synergists to enhance bioactivity of pyrethroids might in future be extended to other classes of insecticides. To prevent injury to crops from herbicide use, the combination of herbicide safeners is also becoming popular, however, sufficient number of safeners are not available for all group of herbicides {cf. Chapter 3(IX)}. Research effort in this direction thus needs to be augmented.

Quantitative structure activity relationship (QSAR) and other elements of statistics with computer graphics referred to as computer-assisted molecular modelling (CAMM) or designing (CAMD) is becoming an important tool in synthesing efficacious and safe pesticides (cf. chapter 8). It is a powerful method in the molecular architecture of biologically active materials. The real breakthrough may come when an organic chemist uses the methods of CAMM as a guide in the synthetic programme rather than just to rationalise the results that have already been achieved. It is a tool to encourage a divergent approach rather than the convergent thinking normally associated with modelling.

The pesticide application technology should also be precise, accurate and target oriented. The application efficacy and accuracy of recently developed electrostatic sprayers where the charged droplet adhere better to the leaves of the crop is much more than a hydraulic sprayer. This delivery system can also be effectively employed for the application of pheromones. The application of multiple pest control measures will be preferable in future than a single method in order to reduce the emergence of resistant pests. Thus optimum pest management can be achieved if the complex interrelationships between crop, pest and environment is better understood. The Agricultural Advisory Service (AAS) alongwith pesticide manufacturers must train the farmers and advise them on the advantages in the use of newer group of pesticides, IPM technologies and safety measures using high infotech. The farmers should be supplied not only with quality pesticides but a complete protection package. Such pest control measures would essentially play a vital role for sustainable, nutritional and adequate food production for mankind.

Intellectual Property Rights (IPR)

The IPR provides for the legal ownership of a discovery or an invention by a person or a party, prohibiting others from unauthorised copying or imitation, thereby giving the inventor the exclusive right for commercial exploitation over a prescribed period. This decision was taken in the Uruguay

Round of GATT in 1994, establishing the Trade Related Intellectual Property Rights (TRIPS) regime. The IPR was given the name of TRIPS due to its trade ramifications. The TRIPS Agreement no longer permits the free exchange of genetic resources as it brings into its fold the issues pertaining to the breeders and farmer's rights, benefits sharing among the nations and patent protection. The World Trade Organisation (WTO) was created under the TRIPS guidelines of the Marrakesh Agreement. The Agreement came into effect on January 1, 1995. It would be implemented in phases by developed, developing, least developed countries and by economies in transition. Article 65.4 provides a relaxation period up to January 1, 2005 in case of the developing countries which could not cover product patent protection in certain areas.

The Annexure-C of the WTO makes the product and process patents obligatory. It states that patents shall be available for any inventions, whether products or processes, in all fields of technology provided they are new and capable of industrial application. Article 27.3b is important in this regard as it spells out the patent and *sui generis* obligations for plants, animals, microorganism and other biological and non-biological products and processes. The deed did not contain provisions for patentability of the community knowledge. India and several other WTO signatories have a large pool of intellectual wealth that does not necessarily belong to any one individual and the TRIPS Agreement makes no provision to protect this kind of community intellectual wealth. This is one of the major causes of concern for India. The WTO provides that the plant varieties must be protected either by patents or by *sui generis* system such as the plant breeders rights (PBR) provided in the Union for the protection of new varieties of plants (UPOV). UPOV' 91 has a strong bias in favour of protection for plant breeders but has none on the rights of farmers. Eighty seven per cent of the seeds sown in the country are provided by the farmers and their rights should be well protected.

Indian patent system

In India, patent rights accrue under the purview of the Patent Act, 1970, which seeks to give a fair share to the interest of the party that invented the process or products and the public that benefits from it. All products except food, chemicals and drugs or substances produced by chemical processes can be patented for a period of seven years from the date of filing of the application or five years from the date of acceptance whichever is earlier.

The major hallmark of the Indian Patent Act is that it provides for the process patents in place of the product patents in the field of chemicals, food and drugs. The implementation of the revised patent Act has become inevitable as the WTO has set October 22, 1999 as the dead line by which India has to provide information about introduction of either product patents or exclusive marketing rights (EMR) for drugs and agrochemicals.

In India, medicines and many agrochemicals are available at a far lesser price than most other countries. The lower price is due to application of reverse engineering method applied by the industries. Indian companies mostly copy the product with some additions or improvements to it, thus doing away with preclinical studies. On the other hand, the developed countries make huge investments in terms of both money and time in developing a new molecule. Patent protection should also be extended to the innovating company so as to give it a window for exclusive marketing of about seven to ten years in case of a twenty year patent grant.

In the absence of patent protection, the competitor gets a chance of reverse engineering the molecule to develop it at a far lesser price within a relatively shorter time. The apprehensions on the Indian side however, is that the product patents would give monopolistic leverage to transnationals to flood the Indian market with a plethora of crop protection molecules and drugs leaving them at a major disadvantage.

Benefits that may accrue to India through IPR

a) Availability of new and world standard products without delay.

b) Easier inflow of the latest technology

c) Availability of co-licensing opportunities for the Indian sector.

d) Better utilization and employment of scientific manpower.

e) Fresh investment and expansion of domestic industries

f) An automatic check on spurious products.

g) Greater protection to Indian products.

Post-amendment scenario

The new IPR stipulations may make India pay royalty for the use of inventions more directly to the countries having a monopoly in the field of drugs and agrochemicals [6]. But at the same time it will enable the country to take advantage of the existing knowledge in technology to produce world standard products, which may give a boost to the whole scientific community and put India on an equal footing with the rest of the world. There is a concern in certain sections that the new patent regime might curb the spirit of Indian scientists but this is rather unfounded. On the contrary, a better patent protection in the long run would encourage inventions and inspire scientists to become more competitive, charting India to the path of newer challenges in science and technology. This is, however, possible only when we adopt western work-culture and ethos. The inertia and lethargic bureaucratic attitude must give way to more entrepreneural mode. The patent office must become more receptive to new challenges and adopt world standards and specifications in their functioning.

CONCLUSION

The awareness of Indian scientific community to issues of IPR is very vague at present. The science and technology system should i) stimulate and encourage the creativity and inventiveness of researchers, promote awareness of IPR and need to protect inventors, ii) develop skills to understand and use the techno-legal business information contained in patents and thus manage and exploit IPR nationally and internationally. It should also facilitate global access to IPR information data base, capture and assess the intellectural property generated at an early stage, recognise and reward inventiveness. The educational system presently needs to (iii) devise appropriate curricula on IPR and introduce these as formal courses at the graduate, post-graduate levels in law, science, engineering and management, iv) train people to draft and interpret patents and other IPRs and also provide refresher training course to IPR professionals and encourage research undertakings to update IPR knowledge and application.

REFERENCES

1. Cremlyn, R.J. Future development in : *Agrochemicals*. John Wiley & Sons, 361-372 (1990).
2. Mukerjee, S.K. Agrochemical Research in India : Priorities must change. In : *Pesticides, Crop Protection and Environment* (S. Walia and B.S. Parmar eds.), Oxford & IBH Publishing Co. Pvt. Ltd., New Delhi, 11-18 (1995).
3. McRae, D.H. Advances in chemical hybridization. Plant Breeding Reviews. **3**, 169-191 (1985).
4. Lindsey, K. and Jones, M.G.K. *Plant Biotechnology in Agriculture*, Open Univ. Press, Milton, (1989).
5. Keeley, L.L. and Hayes, T.K. *Insecticide Biochem.*, **17**, 639 (1987).
6. Devakumar, C. and Walia, S. GATT trade deal : Grave implications on Indian agriculture and pesticide industry. *Pestic. Res. J.* **5** (2), 14 (1993).

CHAPTER 10

Regulatory Aspects of Pesticides : Insecticides Act, 1968

- Insecticides Act
- Composition of Central Insecticide Board (CIB) .
- Insecticide analysts, inspectors and their duty
- Central Insecticide Laboratory (CIL)
- Licensing officers and their functions
- Function and composition of Registration Committee

- Salient features of registration of pesticides and guidelines
- Data requirements on various types of materials to be registered
- *References*

Insecticides Act

The Insecticide Act was enacted in 1968 and its implementation started in 1971 with the framing of the Insecticides Rules. The main preamble of the Insecticides Act is safety oriented legislation. This is an Act to regulate the import, manufacture, sale, transport, distribution and use of insecticides with a view to prevent risk to human beings or animals and for matters connected therewith. Different provisions have been made under this Act and rules framed thereunder to ensure safety of insecticides to human beings and animals. The registration committee constituted under the Insecticide Act registers a pesticide only after satisfying itself on various safety parameters. On the basis of toxicity, the committee has refused permission to the use of about 21 pesticides in the country and restricted the use of another 10 pesticides. It endeavours to register safer pesticides which are highly toxic to the target pests with minimal or no harmful effects on non-target organisms. Pesticides being toxic in nature can cause hazards to human health and environment if used indiscriminately without proper precautions.

The various regulatory provisions made in the Act include compulsory registration, compulsory licensing, inspection, drawal, analysis of samples, detention, seizure, confiscation of stocks, suspension and cancellation of licenses. The enforcement of this Act is the joint responsibility of the Central and State Governments.

A. Central Insecticides Board (CIB)

The Central Government constituted a Board to be called, the Central Insecticides Board to advise the Central Government and State Governments on technical matters arising out of the administration of this Act and to carry out the other functions assigned to the Board by or under this Act. The Board may advise on matters relating to :

a) the risk to human beings or animals involved in the use of insecticides (pesticides) and the safety measures necessary to prevent such risk.

b) the manufacture, sale, storage, transport and distribution of pesticides with a view to ensure safety to human beings or animals.

c) The Central Government may after consultation with the Board and subject to the condition of previous publication, by notification in the Official Gazette, make rules for the provisions of this Act.

Composition of the Board

The Board members shall consist of the following :

i) The Director General of Health Services, *Ex-officio* Chairman

ii) The Drugs Controller, India, *Ex-officio*

iii) The Plant Protection Advisor to Government of India, *Ex-officio*

iv) The Director of Storage and Inspection, Ministry of Food, Agriculture, Community Development and Co-operation (Deptt. of Food), *Ex-officio*

v) The Chief Advisor of Factories, *Ex-officio*

vi) The Director, National Institute of Communicable Diseases, *Ex-officio*

vii) The Director-General, Indian Council of Agricultural Research, *Ex-officio*

viii) The Director-General, Indian Council of Medical Research, *Ex-officio*

ix) The Director, Zoological Survey of India, *Ex-officio*

x) The Director-General, Bureau of Indian Standards, *Ex-officio*

xi) The Director-General of Shipping or in his absence the Deputy Director General of Shipping, Ministry of Transport & Shipping, *Ex-officio*.

xii) The Joint Director; Traffic (General), Ministry of Railways (Rail Board), *Ex-officio*.

xiii) The Secretary, Central Committees for Food Standards, *Ex-officio*

The following additional members are to be nominated by the Central Government :

a) One person to represent the Ministry of Petroleum and Chemicals.

b) One pharmacologist

c) One medical toxicologist

d) One person, who shall be incharge of the department dealing with public health in a State, to be nominated by the Central Government.

e) Two persons who shall be Directors of Agriculture in States.

f) Four persons, one of whom shall be an expert in industrial health and occupational hazards.

g) One person to represent the Council of Scientific and Industrial Research.

The functions of the Board include :

   i) to advise to the Central Government on the manufacture of insecticides under the Industries Development Regulation Act, 1951.

   ii) to specify the uses or classification of insecticides on the basis of their toxicity as well as their being suitable for aerial application.

   iii) to advise on the tolerance limits for insecticides residues and establishment of minimum intervals between the application of insecticides and harvest in respect of various commodities.

   iv) to specify the shelf life of insecticide formulations.

   v) to suggest colourisation, including colouring matter which may be mixed with concentrates of insecticides particularly those highly toxic in nature, and

   vi) to carry out other functions such as resupplemental, incidental and consequential.

The Board is also competent to constitute committees. To execute power, the Board constituted six expert panels to finalise the approved usages of different group of insecticides. These are as follows :

   i) Panel on organophosphorus insecticides

   ii) Panel on carbamate insecticides.

   iii) Panel on chlorinated hydrocarbon insecticides

   iv) Panel on fungicides

   v) Panel on herbicides and plant growth regulators.

   vi) Panel on fumigants and rodenticides.

These panels finalise their recommendations which are accepted by the Board and adopted by the Registration Committee.

B. Insecticide (pesticide) analysts

The Central Government or a State Government may, by notification in the official gazette, appoint persons in such number as it thinks fit and possessing such technical and other qualifications as may be prescribed for insecticide analysts in such areas and in respect of such insecticides or class of insecticides (pesticides) as may be specified in the notification.

Insecticide (pesticide) Inspectors and their duty

The Central Government or a State Government may, by notification in the Official Gazette, appoint persons in such number as it thinks fit and possessing such technical and other qualifications as may be prescribed to be Insecticide Inspectors for such areas as may be specified in the notification.

An insecticide inspector shall have the following power :

a) To enter and search at all reasonable times and with such assistance if any, as he considers necessary any premises in which he has reason to believe that an offence under this Act or the rules made thereunder has been or is being or is about to be committed or for the purpose of satisfying himself that the provisions of this Act or the rules made thereunder or the conditions of any certificate of registration or licence issued thereunder are being complied with.

b) To enquire the production of and to inspect, examine and make copies of or take extracts from registers, records or other documents kept by the manufacturer, distributor, carrier, dealer or any other person in pursuit of the provisions of this act or the rules made thereunder and seize the same, if he has reason to believe that all or any of them, may furnish evidence of the commitment of an offence punishable under this Act or the rules made thereunder. In case of seizure, a magistrate has to be informed about it and his orders be taken for custody.

c) To make such examination and enquiry as he thinks fit in order to ascertain whether the provisions of this Act or the rules made thereunder are being complied with and for that purpose stop any vehicle.

d) To stop the distribution, sale or use of an insecticide (pesticide) which he has reason to believe is being distributed, sold or used in contravention of the provisions of this Act or the rules made thereunder, for a specified period not exceeding twenty days or unless the alleged contravention is such that the defect may be removed by the possessor of the insecticide (pesticide), seize the stock of such insecticide.

e) To take samples of any insecticide (pesticide) and send such samples for analysis to the Insecticide Analyst for test in the prescribed manner. However, samples are to be drawn in the presence of the person from where stocks samples are being taken unless that person is wilfully absent.

f) To exercise such other powers as may be necessary for carrying out the purposes of this Act or the rules made thereunder.

D. Analysis of samples and reports

The Central Government and the State Governments are competent to appoint Insecticides Analysts for any defined area and for particular type of insecticides or class of insecticides. A copy of the analyses report will be delivered to the person from whom the sample was taken. If the person intends to challenge the report he can inform the Inspector or the Court. The court can then direct re-analysis of samples at CIL where report would be finalised.

E. Central Insecticides Laboratory (CIL)

The Central Government by notification in the Official Gazette, has established a Central Insecticides Laboratory at Faridabad, Haryana with its branches at Bombay and Hyderabad under the control of a Director to be appointed by the Central Government to carry out the functions entrusted to it by or under this Act. The Central Government is also in the process of setting up of five Regional Pesticides Testing Laboratories at Bombay, Calcutta, Chandigarh, Hyderabad and Kanpur. Besides these various State Governments have set up a total of 35 pesticides testing laboratories with capacity to analyse about 37,015 samples per annum.

F. Licensing Officers and its functions

The State Government may by notification in the Official Gazette, appoint such persons as it thinks fit to be licensing officers for this specific purpose of this Act and define the areas in respect of which they shall exercise jurisdiction. Any person desiring to manufacture or to sell, stock or exhibit for sale or distribute any insecticide may make an application to the licensing officer for the grant of a license as per norms prescribed.

G. Registration Committee

The Central Government has constituted a Registration Committee whose composition is as follows :

The registration committee comprises a Chairman nominated by the Central Government. There will be a Secretary to be appointed by the Central Government Besides there are five members (technical) including the Drug Controller, India and the Plant Protection Adviser. The committee may also co-appoint number of experts for a fixed period it may deem fit without voting right for their specific advice. The Registration Committee shall regulate its own procedure and the conduct of business to be transcated by it.

H. The function of Registration Committee

The function of the Registration Committee is :

i) to register insecticides (pesticides) after scrutinising their formulae and verifying claims made by the importer or the manufacturer regarding their efficacy and safety to human beings and animals.

ii) to specify the precautions to be taken against poisoning through the use of handling insecticides.

iii) to perform such other functions as are assigned to it by or under this act.

Three types of registration have been contemplated under this Act such as **provisional registration, regular registration** and **repeat registration**. Provisional registration is granted when the product is introduced for the first time in the country. It is a limited registration granted for two years for data generation. Commercialization is generally not allowed under this type of registration. Regular registration is granted when complete data has been generated with the insecticide to the satisfaction of the committee. Once the product has been granted regular registration, any subsequent applicant for the same product is granted as repeat registration on establishment of chemical identity. Repeat registration is also full registration but data requirement for this is very less. Both regular and repeat registration certificates have indefinite validity period and there is no need to renewal.

The guidelines on different data required for the grant of registration under the Insecticides Act, 1968 are to be submitted along with the application, after detailed studies on efficacy and safety of the product to human beings are shown to be supported by necessary scientific data.

Some salient features of registration of pesticides and guidelines are discussed in brief and are mentioned [1,2] here.

Chemical composition and allied data to be required for registration

i) Source of supply : The name and address of the supplier

ii) Chemical composition : The minimum purity of technical grade along with impurities profile.

iii) Specifications : Complete specifications for the product quality, the pesticides and their formulations where BIS specifications are not available.

iv) Analytical test report : Detailed analysis of a particular batch as per specification.

v) Shelf-life data : Expected shelf-life claim for the product in support of this claim.

vi) Methods of analysis : The methods of analysis of pesticides and formulations are to be submitted.

Packaging and labelling requirements

i) Type and manner of packaging. As per approved by the Registration Committee, if specifications are not covered earlier, new specifications are to be submitted.

ii) Leaflet requirements : As per Insecticides Act and Rules.

iii) Instructions for storage and use including first aid precautionary matters : As per Insecticides Act and Rules.

iv) Information on disposal of used packages, surplus materials and washing of pesticides : As per Insecticide Act and Rules.

Prohibition of import and manufacture of certain insecticides (Pesticides)

i) No person shall himself or by any person on his behalf import or manufacture any misbranded insecticide. An insecticide to be called misbranded if its label contains any statement, design or graphic representation relating *thereto,* which is false or misleading in any material particular or its package is otherwise deceptive in respect of content if it is intimated or sold under the name of another insecticide, wrong packing, registration number or higher toxicity level than the prescribed limit due to mixing or adulteration.

ii) Any insecticide, the sale, distribution or use of which is for the time being prohibited under **Section 27**.

iii) Any insecticide except in accordance with the condition on which it was registered.

iv) Any insecticide in contravention of any other provision of this Act or any rule made thereunder.

Data requirements on bioefficacy and residues

The detailed protocols regarding residue data needed for registration of pesticides as per Gaitonde Committee report are :

i) Experimental data on bioefficacy of formulations to be registered under different agroclimatic conditions in India; bioefficacy data generated under different agroclimatic conditions in India.

ii) Phytotoxicity data on some key crops to be generated.

iii) Translocation study within plant or animal by data generated under Indian or foreign conditions.

iv) Persistence study in soil, water and plants and metabolism of pesticides in soil, water and plants together with the nature of metabolites and their degree of toxicity.

v) Directions concerning the dosage or concentration to be applied

vi) Compatibility with other chemicals; data to show that if mixed, do not exhibit phytotoxicity.

vii) Time of application and type of application equipment and the manner in which it is to be used.

Data requirements on various types of materials to be registered are :

1. Data requirement for the registration of technical grade for indigenous manufacture of approved pesticides.

2. Data requirement for the registration of formulation of the approved products.

3. Data requirement for registration of new formulation of the approved pesticides.

4. Data requirement for the registration of formulation for use in public health programme for indoor applications—bioefficacy data on two household insect pests based on repeated trials.

5. Data requirement for the registration of combination products
   - i)   Bioefficacy data on combination products.
   - ii) Cost-benefit ratio of combination products.

6. Experimental data on residues for submission for registration purpose (to be carried out under Indian conditions)
   - i)   Methods of samplings and residue analysis in food and feed, water, soil and wild life according to protocol prescribed.
   - ii)   Expected residue level in edible crops and soil under good agricultural practices (GAP).
   - iii) Worker hazards after application.

7. Data requirements on safety/toxicology on technical as well as on different types of formulations to be used in the field. Acute toxicity studies in mammals (oral, dermal, inhalation, irritation), birds, fish, bees to be generated. Supplementary information from toxicological studies like neurotoxicity, teratogenicity, carcinogenicity, mutagenicity and effects on reproduction are to be provided.

8. Data requirements for imported technical grade pesticides having a.i. content below and above 90%.

9. Data requirements for granting provisional registration on pesticide (technical) and its formulation.
   - i)   Chemistry : Source, purity, identity, product quality.
   - ii)   Packing and labelling : Proposed/existing levels with manner of packaging.
   - iii) Efficacy : Bio-effectiveness
   - iv)   Residues : Results of trials on translocation within the plant or animal, metabolism of the pesticide in the plants, animal, soil, residue data on crops under foreign conditions.
   - v)   Toxicity : Acute toxicity and sub-acute toxicity studies in mammals and information on birds, fish and human toxicity data from foreign countries.

10. Data requirements for registration of neem based products.

11. Data requirements for registration of household pesticides (bioefficacy and residues) as carried out.

12. Data requirements for registration of formulation of pesticides on bioefficacy and residues for regular registration under section 9(3).

13. Data requirements on bioefficacy and residues for formulations of pesticides for provisional registration under section 9(3b) of the Act.

14. Data requirement for registration of *Bacillus thuringiensis* (Bt) and *B. sphaericus* (Bs) for pest control and vector control respectively (chemistry : technical, formulation, bioefficacy, toxicity, processing, packaging and labelling (cf. H).

15. Data requirements for registration of pesticides for export, purpose : As per prescribed guidelines of the Act.

REFERENCES

1. *The Insecticides Act, 1968* (46 of 1968), Government of India, Ministry of Law and Company Affairs, Government of India Press Coimbatore, India, (1970).
2. Registration of Pesticides and Guidelines. **In :** *Pest Management and Pesticides : Indian Scenario* (B.V. David, ed.), Namutha Publication, Madras, India, 276-325 (1992).

CHAPTER 11

Safety Measures in Handling and Uses of Pesticides

- Introduction
- Toxicity versus hazards of pesticides
- Safe handling and safety measures
- Disposal of empty pesticide containers
- First-aid treatment against pesticide intoxication.

- Symptoms, antidotes and guidelines for emergency medical treatment in case of pesticide intoxication
- *References*

Use of pesticides in crop protection and post harvest care of the produce in the country is necessary to boost crop production and minimise storage loss. Pesticides are toxic substances with varied chemical configuration marketed in different formulations for use as placement, spray and fumigation. Poisoning from pesticides often happens as a result of negligence or misuse during handling. There is always a possibility of accidental exposure through ingestion, inhalation and contact also. Ingestion of pesticides through contaminated food and drinking water may also produce adverse effects on human body while prolonged and excessive exposure may lead to intoxication. No chemical is entirely without risk but there are safe ways of using them. Therefore, one should 'READ ALL LABELS' carefully and handle them with precautions, store and apply only as recommended on the label. It is easier to prevent pesticide poisoning than to treat it.

Pesticide marketing involves educating the farmers and distributors not only about the merits and attributes of the formulation, methods and time of application but also about the handling risks and safety measures to be taken in the event of accidental spillage, contact inhalation and ingestion. The toxicity potential is indicated by the colour of the triangle on the pesticide pack as given in Table 11.1.

Toxicity versus hazards of pesticides

Toxicity is the inherent capacity of a substance to cause damage. Hazard refers to the risk of poisoning involved in actual practice. Thus the most toxic compounds if handled carefully will

Table 11.1. Categories of warning symbols

Category	LD$_{50}$ (oral) mg/Kg	Warning Symbol	Message for the medical practitioner
Extremely toxic	1.50	Poison Red	Most dangerous, requires best medical help
Highly toxic	51-500	Poison Yellow	Less dangerous but requires good medical help
Moderately toxic	501-5000	Danger Blue	Not an emergency situation
Slightly toxic	5000	Caution Green	Not at all an emergency situation

pose little or no hazard. This also means that a chemical with low mammalian toxicity may prove hazardous if used without precautions. The relationship between hazard and toxicity can be represented in the form of following equation (Table 11.2).

$$Hazard = Toxicity \times Contamination \times Time$$

where, hazard is the risk of poisoning, toxicity is the ability to cause damage, contamination is the extent of exposure to the pesticides, and time involves the duration of contact with the pesticide. Thus hazard or the possibility of poisoning can be reduced by minimizing the variables on the right side of the equation.

Table 11.2. Toxicity version hazards of pesticide

Toxicity	Contamination	Time
1. Choose pesticides with low dermal or oral mammalian toxicity. 2. Use the least toxic formulation. 3. Use the lowest effective concentration.	1. Use appropriate protective clothing. 2. Avoid direct contact with pesticides. 3. Master the technique of application.	1. Do not exceed the recommended working time. 2. Wash skin immediately if body gets contaminated during pesticide application. 3. After pesticide application, immediately take bath with soap and plenty of water and change into clean clothing. 4. Wash protective clothing frequently.

Safe handling and safety measures

Pesticides are usually safe if properly transported, stored and handled with necessary precautions.

A. **Before application**, the following precautions are to be taken.

   i)   Read the labels on the containers before opening them in order to avoid accidental injury.

   ii) Do necessary calculations before dilution of the formulation with water and the dilution should be carried out in the open or in well ventilated places.

   iii) Make sure that the pesticide application equipment is in working condition before using it. Do not blow or suck with mouth the nozzle of the spraying equipment during use.

   iv) During handling of dangerous pesticides use appropriate protective clothings such as handkerchief, cap, goggles, gloves, shoes to cover different parts of the body.

   v)   Do not tear open the pesticide bags but cut them with a knife.

   vi) Alert the neighbourhood about the pesticide application programme as a precautionary measure. Never work alone while handling/spraying and do not allow children to the site of application.

   vii) Keep plenty of water, soap and towels at the site of operation in case the operator gets contaminated.

   viii) Pluck ripe fruits, vegetables and edible plant parts before application of pesticides.

   ix) If body gets contaminated, wash immediately with soap and water.

   x)   Never rub your eyes or face and drink water, food, tobacco during application.

   xi) If sprayer has been used earlier for applying herbicides, wash it with washing soda (2%) solution before using it for other pesticides.

   xii) Adequate supervision is necessary if handling for first time to avoid getting harmed.

   xiii) Operators should not be allowed to work more than eight hours a day.

   xiv) For mixing the pesticides it is desirable to use a long handled stirrer and a deep vessel to protect the operator from splashings. Use funnel while pouring into the spray pump/ spraying equipment.

B. **After application**, the following precautions need to be taken.

   i)   Return the unused pesticide formulation to the labelled container only and after closing it tightly, store in a room where children can not enter and play.

   ii) Empty containers should be disposed off as per instructions disposal of empty containers.

   iii) Never leave diluted pesticide formulation in the applicator; if anything is left over, pour it on a barren land where animals cannot reach but not in canals, ponds, wells or streams.

   iv) After application of pesticide, clean the applicator first with detergent and then rinse it thrice with plenty of water. Decontaminate the drums, sticks, and measuring cups used for making dilutions in a similar way. All washings obtained should be disposed off over the barren land.

   v)   Take bath with soap and plenty of clean water and wash clothes with soap separately. Don't mix up contaminated clothes with other clothes during washing.

   vi) Wash caps, handkerchief, shoes, belts, goggles or spectacles worn during pesticide application for future use.

Disposal of empty pesticide containers

   i)   Empty containers should never be left unattended. Children and animals are reported to have been poisoned because of careless disposal of pesticide containers.

ii) Never dump empty pesticide containers in water bodies like streams, canals and ponds.

iii) Do not put empty pesticide containers to any other use.

iv) Decontaminate the metallic or plastic containers in the following manner :

    a) Drain off the container by keeping it in an inverted position.

    b) Fill about half of the container with water, cap it tightly and shake gently.

    c) Discard the water and drain the container by keeping it inverted.

    d) Repeat the above process few times.

    e) Dispose of the water used for rinsing (if necessary, contaminated water can be treated with acid/alkali before being discarded).

v) Bury the containers, away from human dwellings and animal sheds.

vi) Avoid burning of containers. If burnt, do not stand near the smoke.

First aid treatment against pesticide intoxication

Basic information about the first aid treatment against pesticide intoxication will help the sales and marketing personnel to ensure the safety of the user.

a) Poisoning through the skin by contact

    i) Remove off the dress

    ii) Wash the exposed portion with plenty of water using soap

    iii) To alleviate skin irritation, apply vitamin E cream or inert preparations such as vegetable oil and vaseline.

b) Eye contact

    i) Open the eye lids and wash eyes with clean running water

    ii) Do not apply any lotions

c) Poisoning by inhalation

    i) Bring the patient immediately to fresh air.

    ii) Loosen the dress

    iii) Wrap the patient in a blanket

    iv) Avoid jarring or noise

    v) If necessary give artificial respiration

d) Poisoning by oral ingestion

    i) Move the patient to fresh air

    ii) If the patient is conscious make him to vomit by gently stroking or touching the throat with finger.

    iii) Administer emetics like mustard powder, one teaspoon of common salt in a glass of warm water.

    iv) Clean water and milk dilute the poison and help counteract acid or alkali poisons.

    v) Administer universal antidote.

Universal antidote preparation

A mixture containing 80 g of activated charcoal, 4g of tannic acid and 4g of magnesium oxide in warm water will absorb and neutralise the poison. Charred *papad* can act as a substitute for activated charcoal and strong tea decoction for tannic acid.

Guidelines to be followed in case of pesticide intoxication are mentioned in Table 11.3.

Table 11.3. Guidelines for emergency measures with medical treatment in case of pesticide intoxication

Group	Symptom	First aid/ Antidote	Special medical treatment	Precautions
Organochlorines (DDT, BHC, Aldrin, Endo-sulfan, Heptachlor, Chlordane, Kelthane, Lindane)	Nausea, severe vomiting, weakness in arms and legs, headache, diarrhoea, tremors or convulsions.	Removal of poison with gastric lavage or induce vomiting with universal antidote.	Control convulsions with diazepam paraldehyde or soluble barbiturates. Atropinize with repeat doses of 2-4 mg at 5 to 10 min intervals (max. 25 to 50 mg/kg). 10% calcium gluconate to be given by I.V. (in severe cases use steroids and dialysis). Feed the patient with rich protein carbohydrate and calcium diet.	2-PAM, Protomam Chloride morphine, theophyline are not recommended. Diet should be fat free; Oil laxatives should be avoided.
Organophosphates (Dimethoate, Methyl Parathion, Acephate, Ethion, Monocrotophos, Quinalphos, Phosphamidon, DDVP, Phorate, Chlorpyriphos, Fenitrothion)	Headache, nausea, contraction of pupil, sweating, slow pulse rate, diarrhoea, muscular twitching, pulmonary edema, quivering of tongue, convulsions, coma, heart block and death.	Should drink 30 g of magnesium sulphate or sodium sulphate in one cup of water. Administer oxygen if difficulty is noticed in breathing.	Artificial respiration is necessary if the respiration is irregular or shallow. Inject atropine sulphate by I.V. in a dose of 2-4 mg for adults and 0.04 to 0.08 mg/kg body weight for children, every 5-10 min until adequately atropinised as shown by dilated pupils and dry mouth, and maintain atropinisation for atleast 24-48 h and carefully observe the patient as the drug is withdrawn. Recommence treatment if signs of poisoning return. Also administer 2-PAM, 1000-2000 mg IM or IV. IM of Diazepam at 5-10 mg for treating convulsion and anxiety. If available Toxugain (Merck) 25 g can be administered to reactivate cholinesterase enzyme.	Do not use morphine, theophyllin, aminophyllinor, barbiturates.
Carbamates (Carbofuran, Carbaryl, Aldicarb, Propoxur)	Symptoms tend to be of quicker onset and shorter duration. Headache, nausea, salivation, constriction of pupil of the eye blurred vision, sweating, slow pulse rate, diarrhoea, muscular twitching, convulsion, coma, heart block and death.	Administer universal antidote.	Artificial respiration is necessary if the respiration is irregular or shallow. Inject atropine sulphate by I.V. in a dose of 2-4 mg for adults and 0.04 to 0.08 mg/kg body weight for children, every 5-10 min, until adequately atropinised as shown by dilated pupils and dry mouth and maintain atropinisation for at least 24-48 h, and carefully observe the patient as the drug is withdrawn. Recommence treatment if signs of poisoning return. Also administer 2-PAM, 1000-2000 mg IM or IV. IM of Diazepam at 5-10 mg for treating convulsion and anxiety. If available Toxugain (Merck) 25 g may be ingested to reactivate cholinesterase enzyme.	Oximes such as 2-PAM should not be given. Morphine, barbiturates, phenothiazines are also contradicted.

(Contd.)

Group	Symptom	First aid/Antidote	Special medical treatment	Precautions
Synthetic Pyrethroids (Fenvalerate, Cypermethrin, Decamethrin, Permethrin)	Facial sensation such as tingling, burning or numbness. Irritation of oral and nasal mucosa, salivation, convulsive seizures.	Not associated with systemic poisoning. The effects are reversible and no specific treatment is necessary.	Control seizures with injectable diazepam or barbiturates.	CNS stimulants should not be given.
Dithiocarbamates (Maneb, Mancozeb)	Exposure to dithio-carbamates followed by alcohol ingestion may produce headache, palpitation, nausea, vomiting and flushed face.	Give gastric lavage. Administer activated charcoal or universal antidote.	No specific treatment is available only symptomatic therapy is possible.	
Bipyridyliums (Paraquat, Diquat)	Within h irritation of mouth and throat with nausea, vomiting abdominal pain and diarrhoea with blood. 1-3 days later signs of kidney and liver damage, intensive lung damage become visible. Diquat causes profuse watery diarrhoea while paraquat causes respiratory failure 5-14 days after poisoning.	Give gastric lavage and leave the gastric tube *in situ*. Give 1 litre of 15% aqueous suspension of Fuller's earth together with a suitable purgative until Fuller's earth is seen in stool. (4-6 h post treatment) or give 50 g of charcoal in 150 ml or water of 8 beaten eggs.	Induce hemodialysis. Avoid the use of oxygen therapy for the first 48 h. Give potassium ferrocyanide 0.5 g in water.	
Copper Compounds (COC, Oil based Copper formulations)	Violent retchin muscular spasm and collapse.	Give gastric lavage with large quantities of water containing one ounce of milk of magnesia, potassium ferrocyanide solution.	Administer other demulcants like egg white. Dimercaprol or pencillamine.	
Carbophenoxy compounds	Loss of appetite and weight, vomiting, depression, muscular weakness may be produced by repeated exposures. Twitching of muscles, urinary incontinence and coma may occur in large doses.	Quinidine sulfate or sodium sulfate useful as a cathartic. In severe poisoning forced alkaline diuresis is required.	Symptomatic treatment.	

(Contd.)

(Contd.)

Group	Symptom	First aid/ Antidote	Special medical treatment	Precautions
Zinc phosphide and Aluminium phosphide	Vacant look, gastroenteritis, severe abdominal pain, frequent vomiting followed by respiratory distress and death.	Give gastric lavage with one tea spoonful of mustard powder in a glass of warm water and make the patient to drink it. After vomiting has stopped, give 5g of potassium permanganate dissolved in a glass of warm water. Ten min later give a solution of half of a teaspoonful of copper sulphate in a glass of water. Fifteen min later give solution made by dissolving one table spoonful of magnesium sulphate (Epsom salt) in a glass of water.	Administer morphine for the relief of abdominal pain, administration of 100% oxygen for pulmonary edema, anticonvulsant theraphy, high doses of corticosteroids and blood transfusion for the treatment of shock and hemorrhage.	
Anticoagulants (Bromadiolone, Bradifacoum, Warfarin, Coumafuryl, Chlorophacinone)	Back pain, abdominal pain, nausea, vomiting and diarrhoea, bleeding from nose, gums, gastrointestinal and urinary tracts, internal bleeding leading to shock and coma.	Give gastric lavage.	Give agnamephyton 5-10 mg to adults or 1-5 mg to children IM or IV. Dose of vitamin K1 intramuscularly @ 1 mg/kg/day or inject large dose of vitamin K1 intravenously (65 mg first day, smaller doses following until prothrombin level normal). Orally administer vitamin K1 and water soluble forms of menadione. If the patient is seriously ill blood transfusion is necessary along with the administration of large doses of vitamin K1.	
Mercurials	Blisters on skin, severe nausea, vomiting, abdominal pain, diarrhoea, salivation, thirst, headache, loss of peripheral vision, loss of coordination in speech and gait prostration.	Give universal antidote. Give milk or white of egg beaten with water, then a table spoonful of salt in a glass of warm water and repeat till the vomit fluid is clear. Repeat milk or white of egg beaten with water or give 6 to 8 tea spoonfuls of a mixture of activated charcoal 2 parts, 1 part magnesium oxide and 1 part tannic acid in a glass of water.	High colonic irrigation with freshly prepared injection of 100 to 200 ml of sodium formaldehyde sulphoxylate solution I.V. For later treatment give sodium citrate alone 1 to 4 g every 4 h by mouth. Give 100 ml of 10% calcium gluconate solution by I.V. for muscular spasm.	

Group	Symptom	First aid/Antidote	Special medical treatment	Precautions
Organo-phosphorus herbicides (Glyphosate)	Irritation, sensitization, photo-irritation and photo-sensitization.		Symptomatic treatment.	
Triazines (Atrazine, Imazine)	Skin irritation.	Symptomatic treatment.		
Organic arsenicals (Sodium arsenate, Paris green)	Loss of appetite, weight loss, weakness, nausea and constipation. Colic, peripheral neuropathy and headache. Some loss of hair and giddiness.	Stomach should be emptied by vomiting or lavage with warm water and activated charcoal followed by saline cathartic.	Inject BAL (dimercaprol) intramuscularly. Dehydration should be combated by saline infusion. Diet should be liquid, supplemented with vitamins.	
Cyanides (Hydrogen cyanide, Calcium cyanide)	Headache, weakness, confusion, nausea and vomiting, slowed respiration blood becomes dark, the patient suffers cyanotic collapse and cessation of respiration.	Displace cyanide from cytochrome oxidase by converting hemoglobin to methaemoglobin, ensure adequate supply of sulfur to permit conversion of cyanideion to relatively non-toxic thiocyanates.	Artificial respiration, amyl nitrate through inhalation. Inject sodium nitrate (or) sodium thiosulfate intravenously.	
Carboxides	Second degree burns resembling that as produced by mustard gases. Vesication formation of large blisters, marked desquamation and formation of residual pigment.	Symptomatic treatment.	Dimercaprol (BAL) may be given before symptoms appear. Administer barbiturates for convulsions.	
Methyl Bromide	Unconsciousness leading to anesthetic death, malaise, headache, visual disturbance and vomiting, ataxia, tremor.	Clonezepam at 3-4 mg/kg.	Symptomatic treatment.	
EDB (Dibromoethane)	Comatose, vomiting burns of eyes and throat, chemical odour, barchycardia.		Symptomatic treatment.	

REFERENCES

1. Association of basic manufactures of pesticides, C/O BASF, May and Baker House, 4th Floor, S.K. Ahire Marg, Bombay, India. "Pesticides Poisoning, First aid and treatment".

2. Crop Protection Chemicals Reference (CPCR). Chemical and Pharmaceutical Press a joint venture of John Wiley and Sons Inc., New York and Chemicals and Pharmaceutical Publishing Corporation, 8th Ed. (1992: 1894; Suppliments 77).

3. ICMA reprints on Safety and Environment. Reprinted in India with kind permission of Chemical Industries Association, London, U.K. (1979).

4. International group of National Association of manufacturers of Agrochemical Products. Guidelines for emergency measure in cases of Pesticides poisoning. Avenue Albert Lancaster 79a, 1180 Brussels, Belgrade.

5. Marshall Sittig. Hand book of Toxic and Hazardous Chemicals and Carcinogens, 2nd Ed. Noyes Publications, Park Ridge, New Jersey, USA, 950 (1985).

6. Robert E. Gosselin, Roser P. Smith, and Heroid C. Modge, Clinical Toxicology of Commercial Products, 5th Ed., William and Wilkins Baltimore, USA (1984).

7. U.S.E.P.A. Noyes Data Corporation. Park Ridge, New Jersey, USA, **1**, 827, **2**, 666 (1990).

8. U.S.E.P.A. Pesticides Safety for Farm Workers, Occupational Safety Branch (H7 506 C). Office of Pesticides Programme, U.S. Environmental Prot. Agency Washington DC.

9. Wayland, Ed., Hays Jr., J. and Laws Jr., Edward R. Hand book of Pesticide Toxicology, Academic Press Inc., New York, USA, **1-3**, 1-1576, (1991).

10. Instruction for the safe use of pesticides, published by Directorate of Plant Protection Quarantine and Storage, Ministry of Food and Agriculture, Govt. of India, Faridabad, Haryana (1962).

11. Safe use of pesticides, published by Publications and Information Division, ICAR, Krishi Bhawan, New Delhi, India (1987).

Appendix

Important pesticides registered for use in the country with their common name including warning symbol, chemical name, trade name and manufacturers

	Common name	Chemical name	Trade name	Manufacturers
1.	Acephate (I) [Blue]	O,S-Dimethylacetylphosphoramidothioate	Orthene, Asataf, Starthene	M/s Chevron Chemical Co., USA M/s Rallis India Ltd. M/s Shaw Wallace Co.
2.	Alachlor (H) [Blue]	2-Chloro-N-2′,6′ diethyl-N-methoxymethyl-acetanilide	Lasso, Attack 10G, Catch 50EC	M/s Monsanto Chemical Co. Ltd. M/s Searle (I) Ltd.
3.	Aldicarb (I) [Red]	2-Methyl-2-(methylthio) propionaldehyde O-methylcarbamoyl oxime	Temik	M/s Union Carbide Corp. USA M/s Rhone-Poulenc Agrochemicals (I) Ltd.
4.	Aldrin (I) [Red]	1,2,3,4,10,10-Hexachloro-1,4,4a,5,8,8a-hexahydro-exo-1,4 endo-5,8-dimethanonaphthalene	—	Banned in India
5.	Allethrin (I) [Blue] (pallethrine)	(RS)3-Allyl-2-methyl-4-oxocyclopent-2-enyl-(1RS, 3RS,1RS,3SR)-2,2-dimethyl-3-(2=methylprop-1-enyl) cyclopropane-carboxylate	Pynamin, Jet, Goodknight	M/s Sumitomo Chem. Co. Japan
6.	Alphacypermethrin (alphamethrin), (I) [yellow]	A racemate comprising (S)α-cyano-3-phenoxybenzyl (1R, 3R)-3-(2,2-dichlorovinyl)-2,2-dimethylcyclo-propane carboxylate and (R)α-cyano-3-phenoxy-benzyl(1S, 3S)-3-(2,2-dichlorovinyl)-2,2-dimethyl-cyclopropane carboxylate	Fastac, Fendona Renegade, Bestox Alphaguard	M/s Cyanamid India Ltd. M/s Gharda Chemicals Co.
7.	Alpha-Naphthyl acetic acid (PGR) [Green]	(2-Naphthyloxy) acetic acid	Fruitone-N, Celmone	M/s Rhone-Poulenc Agrochemicals (I) Ltd. M/s Excel Industries Ltd.
8.	Aluminium phosphide (Fumigant) [Red]	—	Celphos, Quickphos	M/s Excel Industries Ltd. M/s United Phosphorus Ltd.
9.	Atrazine (H) [Blue]	6-Chloro-N²-ethyl N⁴-isopropyl-1,3,5-triazine-2,4-diamine	A Atrex, Gesaprim Atrataf	M/s Hindustan Ciba-Geigy Ltd. M/s Rallis India Ltd.
10.	Anilophos (H) [Blue]	S-4-Chloro-N-isopropyl carbaniloylmethyl O,O-dimethyl phosphorodithioate	Arozin, Aniloguard	M/s Hoechst Schering AgrEvo Ltd. M/s Gharda Chemicals Ltd.

(Contd.)

	Common name	Chemical name	Trade name	Manufacturers
11.	BHC (HCH), Lindane (I) [Yellow]	Hexachlorocyclohexane	Gamma Col., Lintox	M/s Zeneca-ICI Agrochemicals Ltd. Banned in India (BHC)
12.	Benomyl (F) [Green]	Methyl 1-(butylcarbamoyl)benzimidazol-2-ylcarbamate	Benlate, Benomyl	M/s EI Du Pont India Pvt. Ltd. M/s EID Parry (I) Ltd.
13.	Butachlor (H) [Blue]	N-Butoxymethyl-2-chloro-2',6'-diethyl acetanilide	Machete, Bilchlor	M/s Monsanto Chemicals of India Ltd. M/s Bayer (India) Ltd.
14.	Benthiocarb [Blue] (Thiobencarb)	S-4-Chlorobenzyl diethyl thiocarbamate	Saturan	M/s Kumai (Japan) M/s Chevron Chemical Co. USA
15.	Bromadiolone (R) [Red]	3-[3-(4'-Bromobiphenyl-4-yl)-3-hydroxy 1-1-phenylpropyl)-4=hydroxycoumarin	Deadline, Maki	M/s Lipha S.A.
16.	Bitertanol (F) [Green]	1-(Biphenyl-4-yloxy)-3,3-dimethyl-1-(1H-1,1,2,4-triazol-1-yl) butan-2-ol	Baycor, Sibutol	M/s Bayer (India) Ltd.
17.	Captafol (F) [Blue]	N-(1,1,2,2-Tetrachloroethylthio) cyclohex-4-ene-1,2-dicarboximide	Difolatan, Folcid	M/s Rallis (India) Ltd.
18.	Captan (F) [Green]	N-(Trichloromethylthio) cyclohex-4-ene-1,2-dicarboximide	Captaf, Hexacap	M/s Rallis (India) Ltd. M/s Zeneca-ICI Agrochemicals Ltd.
19.	Cartap hydrochloride (I) [Yellow]	S,S'(-2-Dimethylaminotrimethylene)bis (thiocarbamate) hydrochloride	Padan, Caldan 4G & 50 SP	M/s Takeda M/s Dhanuka Pesticides Ltd.
20.	Carbaryl (I) [Yellow]	1-Naphthylmethyl carbamate	Sevin 50 WP Sevidol 4 : 4G	M/s Rhone-Poulenc Agrochemicals (I) Ltd. M/s Union Carbide (I) Ltd.
21.	Carbendazim (F) [Green]	Methyl benzimidazol-2-yl carbamate	Bavistin, Akozim	M/s BASF (India) Ltd. M/s EI Du Pont India Pvt. Ltd. M/s Rallis (India) Ltd.
22.	Carbofuran (I) [Red]	2,3-Dihydro-2,2-dimethylbenzofuran-7-yl N-methyl carbamate	Hexafuron, Furadan, Curaterr	M/s EID Parry (I) Ltd. M/s Rallis (India) Ltd.
23.	Carboxin (F) [Blue]	5,6-Dihydro-2-methyl-1,4-oxathine 3-carboxanilide	Vitavax, Hiltavax	M/s Uniroyal Chem. Co. M/s Hindustan Insecticides Ltd.
24.	Chlorfenvinphos (I) [Red]	2-Chloro-1-(2,4-dichlorophenyl) vinyl diethyl phosphate	Birlane, Apachlor	M/s Cyanamid India Ltd. M/s Rhone-Poulenc Agrochemicals (I) Ltd.
25.	Chlormequat chloride (CCC), Chlorocholine chloride (PGR) [Blue]	2-Chloroethyl trimethyl ammonium	Cycocel, CCC 700, Lihocin	M/s BASF (India) Ltd. M/s Cynamid India Ltd.

	Common name	Chemical name	Trade name	Manufacturers
26.	Chlorothalonil (H) [Green]	2,4,5,6-Tetrachloro-isophthalonitrile	Bravo, Kawach 75 WP	M/s Diamond Alkali Co. M/s Sandoz (I) Ltd.
27.	Chlorobenzilate (I) [Blue]	Ethyl 4,4'-dichlorobenzilate	Akar, Folbex	M/s Hindustan Ciba Geigy Ltd.
28.	Chlorpyriphos (I) [Blue]	O,O-Diethyl O-(3,5,6 trichloro-2-pyridyl) phosphorothioate	Dursban, Tricel, Rader	M/s Excel Industries Ltd. M/s De-Nocil Crop Prot. Ltd. M/s Searle India Ltd.
29.	Coumatetralyl (R) [Red]	4-Hydroxy-3-(1,2,3,4-tetrahydro-1-naphthyl) coumarin	Racumin	M/s Bayer (I) Ltd.
30.	Coumachlor (R) [Red]	3-[1-(4=Chlorophenyl-3 oxobutyl]-4-hydroxycoumarin	Ratitan	M/s Hindustan Ciba-Geigy Ltd.
31.	Cyfluthim (I) [Yellow]	(RS)α-Cyano-4-fluoro-3-phenoxybenzyl-(1RS,3RS,1RS,3R)-3-(2,2-dichlorovinyl)-2,2-dimethylcyclopropane carboxylate	Baythroid, Baygon spray	M/s Bayer (I) Ltd.
32.	Cypermethrin (I) [Yellow]	(RS)α-Cyano-3-phenoxybenzyl (1RS,3RS,1RS,3SR)-3-(2,2=dichlorovinyl)=2,2 dimethylcyclopropane carboxylate	Basathrin, Ripcord, Ankush	M/s Cyanamid (I) Ltd. M/s Gharda Chemicals Co. Ltd. M/s BASF India Ltd.
33.	Dalapon (H) [Green]	2,2-Dichloropropionic acid	Dalacide, Dalapon 85% WP	M/s BASF India Ltd. M/s Herbicides (I) Ltd.
34.	Deltamethrin (I) [Yellow] (decamethrin)	(S)α-Cyano-3-phenoxybenzyl-(1R,3R)-3-(2,2-dibromovinyl)-2,2=dimethylcyclopropane carboxylate	Decis, Butox, K-othrin, Decakill	M/s Roussel Uclaf, USA M/s Hoechst Schering AgrEvo Ltd.
35.	Dichlorvos (I) (DDVP) [Yellow]	2,2-Dichlorovinyl dimethyl phosphate	Vapona de de vap. didivane uniphos	M/s Hindustan Ciba-Geigy Ltd. M/s Bayer (I) Ltd.
36.	Dicofol (A) [Blue]	2,2,2-Trichloro-1,1-bis(4-chlorophenyl) ethanol	Kelthane Colonel-S	M/s Rhom & Haas Co., USA M/s Hindustan Insecticides Ltd. M/s Indofil Chemicals Co. Ltd.
37.	Diflubenzuron (I) [Blue]	1-(4-Chlorophenyl)-3,2-(2,6-difluorobenzoyl) urea	Dimilin	M/s Philips-Duphar, B.V.
38.	DDT (I) [Yellow]	1,1,1-Trichloro-2,2-bis-(4-chlorophenyl) ethane	DDT, WP, DP	M/s Hindustan Insecticides Ltd.

(Contd.)

	Common name	Chemical name	Trade name	Manufacturers
39.	Dinocap (F, A) [Blue]	2,6-Dinitro-4-octylphenylcrotonates and 2,4-dinitro-6-octylphenyl crotonate	Karathane	M/s Rhom & Haas Co., USA M/s Indofil Chemicals Co. Ltd.
40.	Dithianon (F) [Blue]	5,10-Dihydro-5,10-dioxanophtho [2,3-b]1-4-dithi-in-2,3-dicarbonitrile	Delan	M/s Cyanamid India Ltd.
41.	Diuron (H) [Blue]	3-(3,4-Dichlorophenyl)1-1-dimethyl urea	Cyprex, Dodene	M/s Cyanamid India Ltd. M/s Rhone-Poulene Agrochemicals (I) Ltd.
42.	Dodine (F) [Blue]	1-Didecylguanidinium acetate	Karmex WP, Direx	M/s Bayer (I) Ltd. M/s El Du Pont India Pvt. Ltd.
43.	Diazinon (I) [Yellow]	O,O-Diethyl O-2-isopropyl 6-methylpyrimidin-4-yl phosphorothioate	Basudin, Dianon, Ditaf	M/s Hindustan Ciba-Geigy Ltd. M/s Rallis India Ltd.
44.	2,4-D (H) [Blue]	2,4-Chlorophenoxy acetic acid and ester-salts	Erbitox, Combi, Agrodone-48	M/s Bayer (I) Ltd. M/s Zeneca-ICI Agrochemicals Ltd. M/s Agromore (P) Ltd.
45.	Edifenphos (F) [Yellow]	O-Ethyl S,S-diphenyl phosphorodithioate	Hinosan	M/s Bayer (I) Ltd.
46.	Endosulfan (I) [Yellow]	6,7,8,9,10,10-Hexachloro-1,5,5a,6,9,9a hexahydro-6,9-methano 2,4,3-benzodioxathiepin-3-oxide	Thiodan, Endocell	M/s Excel Industries Ltd. M/s Hoechst Schering AgrEvo Ltd.
47.	Ethofenprox (I) [Green] (Etofenprox)	2-(4-Ethoxyphenyl)-2-methylpropyl 3-phenoxybenzylether	Trebon, Nukil 10% EC	M/s Mistu Toatsu (Japan) M/s Dhanuka Pesticides Ltd.
48.	Ethephon (PGR) [Blue]	2-Chloroethylphosphonic acid	Ethrel, Terpal	M/s Rhone-Poulenc Agrochemicals (I) Ltd. M/s Rallis India Ltd. M/s BASF India Ltd.
49	Ethion (I) [Red]	O,O,O',O'-Tetraethyl S,S'-methylene bis (phosphorodithioate)	Cethion, Ethiol, Tafethion	M/s FMC, USA M/s Rallis India Ltd.
50.	Fenitrothion (I) [Blue]	O,O-Dimethyl O-4-nitro-m-tolyl phosphorothioate	Folthion 50EC, Sumithion	M/s Sumitomo Chemical Co., Japan M/s Bayer (I) Ltd.
51.	Fenarimol (F) [Blue]	(+)-2,4'-Dichloro-α-(pyrimidin-5-yl) benzhydryl alcohol	Rubigan 12% EC	M/s Dow Elanco Ltd. M/s De-Nocil Crop Prot. Ltd.
52.	Fenobucarb (I) [Blue] (BPMC)	2-sec-Butylphenylmethyl carbamate	Bassa, Bipvin 50 EC	M/s Sumitomo Chemical Co., Japan M/s Rhone-Poulene Agrochemicals (I) Ltd.
53.	Fenthion (I) [Yellow]	O,O-Dimethryl O-4-methylthio-m-tolyl phosphorothioate	Lebaycid 1000	M/s Bayer (I) Ltd.

(Contd.)

	Common name	Chemical name	Trade name	Manufacturers
54.	Fenvalerate (I) [Yellow]	(*RS*)-α-Cyano-3-phenoxybenzyl(*RS*)-2-(4-chloro-phenyl)-3-methylbutyrate	Sumicidin, Fenkil-20	M/s Sumitomo Chemical Co., Japan M/s Rallis India Ltd. M/s United Phosphorus Ltd.
55.	Ferbam (F) [Blue]	Ferric tris-dimethyl dithiocarbamate	Trifungol	M/s Elf Atochem.
56.	Fluchloralin (H) [Blue]	*N*-(2-chloroethyl)-2,6-dinitro-*N*-propyl-4-(triflyoro-methyl) aniline	Basalin	M/s BASF (India) Ltd.
57.	Fluvalinate (I) [Blue]	(*RS*)-α-Cyano-3-phenoxybenzyl *N*-(2-chloro-α,α,α-trifluoro-*p*-tolyl)-D-valinate	Mavrik	M/s Sandoz (I) Ltd.
58.	Formothion (I, A) [Yellow]	S-[Formyl (methyl) carbamoylmethyl] O,O-dimethyl phosphorodithioate	Anthio	M/s Sandoz (I) Ltd.
59.	Fosetyl-Al (F) [Green]	Al-salt of ethylhydrogen phosphonic acid	Aliette 80 WP	M/s Rhone-Poulenc Agrochemicals (I) Ltd.
60.	Glyphosate (H) [Blue]	*N*-(phosphonomethyl) glycine	Round up, Glycel	M/s Monsanto Chemicals of India Ltd. M/s Excel Industries Ltd. M/s Zeneca-ICI Agrochemicals Ltd.
61.	Hexaconazole (F) [Blue]	(*RS*)-2-(2,4-Dichlorophenyl)-1-(1*H*-1,2,4-triazol-1-yl) hexan-2-ol	Anvil, Planete, Contaf	M/s Zeneca-ICI Agrochemicals Ltd. M/s Rallis (India) Ltd.
62.	Iprodione (F) [Blue]	3-(3,5-Dichlorophenyl)-*N-isopropyl*-2,4-dioxo-imidazolidine-1-carboxomide	Rovral	M/s Rhone-Poulenc Agrochemicals (I) Ltd.
63.	Isoproturon (H) [Blue]	3-(4-*Isopropylphenyl*)-1,1-dimethyl urea	Arelon, Dhar, Isoguard 75 WP	M/s Hoechst Schering AgrEvo Ltd. M/s Rhone-Poulenc Agrochemicals (I) Ltd. M/s Gharda Chemicals Ltd.
64.	Kitazin (F) [Blue]	S-Benzyl O,O-diethyl phosphorothioate	Kitazin	M/s Kumiai Chemical Industry (Japan)
65.	Lambda cyhalothrin (I) [Yellow]	An equal mixture of (*S*)α-Cyano-3=phenoxybenzyl-(*Z*)-(1*R*,3*R*)-*cis*-3-(2-chloro-3,3,3-trifluoroprophenyl)-2,2-dimethyl-cyclopropane carboxylate and (*R*)-α-cyano-3-phenoxybenzyl(*Z*)-(1*S*,3*S*)-3-(2-=chloro-3,3,3-tri-fluoroprophenyl)2,2-dimethyl cyclopropane carboxylate	Karate	M/s Zeneca-ICI Agrochemicals Ltd.
66.	Malathion (I) [Blue]	S-1,2-Bis (ethoxycarbonyl)ethyl O,O-dimethyl phosphorodithioate	Celthion, Malatox, Malaphos	M/s Cyanamid India Ltd. M/s Excel Industries Ltd. M/s United Phosphorus Ltd:

(*Contd.*)

	Common name	Chemical name	Trade name	Manufacturers
67.	Maleic hydrazide (PGR) [Green]	6-Hydoxy-2-H-pyridazin-3-one	MH, Regulox, Mazide	M/s Rhone-Poulenc Agrochemicals (I) Ltd. M/s Uniroyal Chemical Co.
68.	Mancozeb (F) [Green]	Manganese ethylene bis (dithiocarbamate) polymeric complex with Zn-salt	Dithane M-45, Indofil M-45	M/s Bayer (India) Ltd. M/s United Phosphorus Ltd. M/s Indofil Chemicals Co. M/s Hindustan Insecticides Ltd.
69.	Metaldehyde (M) [Yellow]	r-2, c-4, c-6, c-8-tetramethyl, 1,3,5,7-tetroxocane	Halizan, Metason, Snailkil	M/s Lonza M/s Pesticides India
70.	Methabenzthiazuron (H) [Green]	1-(1,3-Benzothiazol-2-yl)-1,3-dimethylurea	Tribunil	M/s Bayer (India) Ltd.
71.	Methyl chlorophenoxy acetic acid (MCPA) (H) [Blue]	4-Chloro-2-methyl phenoxy acetic acid	Agroxone, Blagal	M/s Rhone-Poulenc Agrochemicals (I) Ltd. M/s Zeneca-ICI Agrochemicals Ltd.
72.	Metalaxyl (F) [Blue]	Methyl N-(2-methoxyacetyl)-N-[2,6-xylol)-DL-alaninate	Ridomil, Apran	M/s Hindustan Ciba-Geigy Ltd. M/s De-Nocil Crop Prot. Ltd.
73.	Metoxuron (H) [Blue]	3-(3-Chloro-4-methoxyphenyl)-1,1-dimethyl urea	Dosanex, Deftor	M/s Sandoz (I) Ltd. M/s Hoechst Ltd.
74.	Methomyl (I) [Red]	S-Methyl N-(methylcarbamoyloxy) thioacetimidate	Lannate	M/s EI DuPont India Ltd. M/s Rhone-Poulenc Agrochemicals (I) Ltd.
75.	Metolachlor (H) [Blue]	2-Chloro-6-ethyl-N-(2-methoxy-1-methyl ethyl) acet-o-toluidide	Dual	M/s Hindustan Ciba-Geigy Ltd.
76.	Metribuzin (H) [Blue]	4-Amino-6-tert.-butyl-4,5-dihydro-3-methylthio-1,2,4-triazin-5-one	Sencor, Lexone	M/s Bayer India Ltd. M/s EI DuPont India Ltd.
77.	Monocrotphos (I) [Red]	Dimethyl (E)-1-methyl-2-(methylcarbamoyl) vinyl phosphate	Nuva, Cron, Azadrin, Monocil	M/s Cyanamid India Ltd. M/s Hindustan Ciba-Geigy Ltd. M/s De-Nocil Crop Prot. Ltd.
78.	Myclobutanil (F) [Blue]	2-p-Chlorophenyl-2-(1H-1,2,4-triazol-1-ylmethyl) hexanenitrile	Systhane	M/s Indofil Chemicals Co. M/s Rohm & Haas, USA
79.	Methyl parathion (I) [Red]	O,O-Dimethyl O-4-nitrophenyl phoshorothioate	Folidol, Metacid	M/s Bayer India Ltd.

(Contd.)

	Common name	Chemical name	Trade name	Manufacturers
80.	Neem products (I) (azadirachtin) [Green]		Godrej Achook, Neemzal, Margocide OK, Neem Gold, Neembecidin	M/s EID Pary Chemicals M/s S.T. Stanes and Co. Ltd. M/s SPIC
81.	Oxadiazon (H) [Green]	5-Tert. butyl-3-(2,4-dichloro-5-*isopropoxyphenyl*-1,3,4-oxadiazol- =2(-3*H*)-one	Ronstar	M/s Rhone-Poulenc Agrochemicals (I) Ltd.
82.	Oxycarboxin (F) [Blue]	2,3-Dihydro-6-methyl-5-phenyl-carbomyl-1,4-oxathin 4,4-dioxide	Plantvax	M/s Uniroyal Chemical Co.
83.	Oxydemeton-methyl (I) [Yellow]	S-2, Ethyl sulfenylethyl O,O-dimethyl phosphorothioate	Metasystox, Metamol-250	M/s Bayer India Ltd. M/s Konkan Pesticides
84.	Oxyfluorfen (H) [Green]	2-Chloro-α,α,α-trifluoro-*p*-totyl-3-ethoxy 4-nitrophenyl ether	Oxygold	M/s Indofil Chemicals Co.
85.	Paraquat (H) [Yellow]	1,1′-Dimethyl-4, 4′-bipyridinium	Scythe, Speedway	M/s Cyanamid India Ltd. M/s Zeneca-ICI Agrochemicals Ltd.
86.	Penconazole (F) [Blue]	1-(2,4-Dichloro-β-propylphenethyl)1-*H*-1,2,4-triazole	Topas	M/s Hindustan Ciba-Geigy Ltd.
87.	Pendimethalin (H) [Blue]	*N*-(1-ethylpropyl)-2,6-dinitro-3,4-xyledine	Stomp	M/s Herbadox M/s Cyanamid India Ltd.
88.	Permethrin (I) [Yellow]	3-Phenoxybenzyl (1*RS*, 3*RS*; 1*RS*, 3*SR*) 3-(2,2-di-chlorovinyl-2,2 =dimethyl cyclopropane carboxylate	Ambush	M/s Zeneca-ICI Agrochemicals Ltd.
89.	Phenthoate (I) [Yellow]	*N*-α-Ethoxycarbonyl benzyl O,O-dimethyl phosphorodithioate	Cidial, Papthion	M/s Sumitomo Chemicals Co., Japan M/s Nissan Chemical Co., Japan
90.	Phorate (I) [Red]	O,O-Diethyl S-ethylthio-methyl phosphorodithioate	Thimet, Phoril	M/s Cyanamid India Ltd. M/s Rallis India Ltd.
91.	Phosalone (I) [Yellow]	S-6-Chloro-2, 3-dihydro-2-oxo-benzoxazol-3-ylmethyl O,O-diethyl phosphorodithioate	Zolone 35 EC	M/s Rhone-Poulenc Agrochemicals (I) Ltd.
92.	Phosphamidon (I) [Red]	2-Chloro-2-diethyl carbamoyl-1-methyl vinyl dimethyl phosphate	Sumidon, Dimecron	M/s Sudarshan Chemical Industries M/s Hindustan Ciba-Geigy Ltd.
93.	Pirimiphos-methyl (I,A) [Blue]	O-2-Diethylamino-6-methyl pyrimidin-4-yl O,O-dimethyl phosphorothioate	Acetellic	M/s Zeneca-ICI Agrochemicals Ltd.

(Contd.)

(Contd.)

	Common name	Chemical name	Trade name	Manufacturers
94.	Prallethrin (I) [Blue]	Roth:*S* (2-methyl-4-oxo-3-prop-2-ynyl cyclopent-2-enyl(1*R*)-cis-trans-2,2- = dimethyl 3-(2 methyl-prop-1-enyl) cyclopropane carboxylate	Jet (coil), Etoc	M/s Sumitomo Chem. Co., Japan
95.	Pretilachlor (H) [Blue]	2-Chloro,2′, 6′-diethyl-N-(2-propoxyethyl) acetanilide	Rifit	M/s Hindustan Ciba-Geigy Ltd.
96.	Profenphos (I) [Yellow]	O-4-Bromo-2-chlorophenyl O-ethyl S-propyl phosphorothioate	Curacron	M/s Hindustan Ciba-Geigy Ltd.
97.	Propanil (H) [Blue]	3′, 4′-Dichloropropionanilide	Stomp F-34, Surcopur	M/s Cyanamid India Ltd. M/s Bayer India Ltd.
98.	Propiconazole (F) [Blue]	(+)-1-[2-(2,4-Dichlorophenyl)-4-propyl-1,3-dioxolan 2-ylmethyl]-1*H*-1,2,4- = triazole	Radar, Tilt	M/s Zeneca-ICI Agrochemicals Ltd. M/s Hindustan Ciba-Geigy Ltd.
99.	Propuxur (I) [Yellow]	2-*Isopropoxyphenyl methyl carbamate	Baygon 20%	M/s Bayer India Ltd.
100.	Quinalphos (1) [Yellow]	O,O-Diethyl O-quinoxalin 2-yl phosphorothioate	Ekalaux, Quinaltaf	M/s Sandoz India Ltd. M/s Rallis India Ltd.
101.	Simazine (H) [Green]	6-Chloro-*N*²,*N*⁴-diethyl-1,3,5-triazine-4-diamine	Gesatop, Calibar, Weedex, Aquazine	M/s Hindustan Ciba Geigy Ltd.
102.	Temephos (I) [Blue]	O,O,O′,O′-Tetramethyl O,O′-thiodi-*p*-phenylene bis (phosphorothioate)	Abate	M/s Cyanamid India Ltd.
103.	Thiometon (I) [Yellow]	*S*-2-Ethylthioethyl (O,O-dimethyl phosphorodithioate)	Ekatin 50 ZP	M/s Sandoz (I) Ltd.
104.	Thiophanate methyl (F) [Green]	Dimethyl 4, 4′-(o-phenylene) bis (3-thioallophanate)	Topsin, Alert, Baynate	M/s Nippon Soda Co. Ltd. M/s Bay Organics Ltd. M/s Nagarjuna Agrochemicals Ltd.
105.	Thiram (F) [Blue]	Tetramethyl thirum disulfide	Furam	M/s SMP Pvt. Ltd.
106.	Triadimefon (F) [Yellow]	1-(4-Chlorophenoxy)-3,3-dimethyl-1-(1*H*-,2,4-triazol-1-yl) butan-2-one	Bayleton	M/s Bayer India Ltd.
107.	Tri-allate (H) [Blue]	*S*-2,3,3-Trichloroallyl di-*isopropyl (thiocarbamate)	Avadex BW, Far-Go	M/s BASF India Ltd.
108.	Triazophos (I) [Yellow]	O,O-Diethyl O-1-phenyl-1-*H*-1,2,4-triazol-3-yl-phosphorothioate	Hostathion	M/s Hoechst Schering AgrEvo Ltd.
109.	Trichlorfon (I) [Yellow]	Dimethyl 2,2,2-trichloro-1-hydroxy ethyl phosphonate	Dipterex	M/s Bayer India Ltd.

	Common name	Chemical name	Trade name	Manufacturers
110.	Tricyclazole (F) [Yellow]	5-Methyl-1,2,4,-triazolo[3,4-b] (1,3)benzothiazole	Trooper, Bean	M/s De-Nocil Crop Prot. Ltd.
111.	Tridemorph (F) [Yellow]	2,6-Dimethy-4-tridecylmorpholine	Calixin	M/s BASF India
112.	Trifluralin (H) [Green]	α,α,α-Trifluoro-2,6-dinitro *N,N*-dipropyl-*p*-toluidine	Treflon 48%, Tri-4	M/s De-Nocil Crop Prot. Ltd. M/s Cyanamid India Ltd.
113.	Warfarin (R) [Red]	(*RS*)4-Hydroxy-3-(3-oxo-1-phenylbutyl) coumarin	RB, GB, TP, CB	All India Medical, Hopkins
114.	Zinc Phosphide (R) [Red]	Tri-zinc phosphide	Commande, Ratol	M/s Excel Industries Ltd. M/s United Phosphorus Ltd.
115.	Zineb (F) [Green]	Zinc ethylenebis (dithiocarbamate) (Polymeric)	Dithane Z-78	M/s Rohm & Hass, USA M/s Bayer India Ltd. M/s Rhone-Poulenc Agrochemicals (I) Ltd.
116.	Ziram (F) [Green]	Zinc bis (dimethyldithiocarbamate)	Cuman, Dhanuka Z-27	M/s Hindustan Ciba Geigy Ltd. M/s Hoechst Schering AgrEvo Ltd.

Biopesticide

	Common name	Chemical name	Trade name	Manufacturers
117.	*Bacillus thuringiensis* (Bt) (I) [Green]		Biobit, Bioasp, Bollgard, Certan Dipel-8, Halt	M/s Rallis India Ltd. M/s Biotech. International M/s Sandoz (I) Ltd. M/s Monsanto Chemicals of India Ltd.

A - Acaricide, F - Fungicide, H - Herbicide, I - Insecticide, M - Molluscicide, R - Rodenticide, [ ] Warning symbol

Subject Index

A

AC-64, 134
Acaricides, 1, 45, 60, 99
Acephate, 109
Acetal fragment, 20
Acetogenins, 24
Acifluorfen, 204
Acrinathrin, 60
Actidione, 31
Acyl phosphonate, 160
ADI, 269
Adjuvants, 252
Advantage, 261
Affinin, 21
Agvitor, 88
Ajugarin, 23
Alachlor, 202
Aldicarb, 119, 126
Aldrimorph, 177
Aldrin, 57
Alkyl thiocyanate, 2
Allelochemicals, 38
Allethrin, 58
Allidochlor, 225
Allodan, 52
Allyloxydim sodium, 223
Ametryne, 212
Amidosulfuron, 219
Aminophon, 217
Aminprofos-methyl, 217
Amobam, 151
Anabasine, 10
trans-Anethole, 24

Annonin, 24
Antagonist, 260
Antibacterial activity, 31
Antibiotics, 2
Anti-caking agents, 252
Antidotes, 225, 325
Anti-dusting agents, 253
Antifeedants, 9, 13, 22, 288
Anti-foaming agents, 253
Antifungal activity, 31
ANTU, 139
Apholate, 290
Aqueous concentrates, 243
Aroylphosphonate, 160
Arylthioureas, 146
Asimisin, 24
Asparagusic acid, 27
Atrazine, 211, 229, 230, 238
Attractants, 144
Auxins, 37
Avermectins, 31
Azadirachtin, 10, 12, 13, 15, 16
Azadirone, 12
Azinphosmethyl, 88, 106, 107
Aziprotryne, 212
Azoxystrobin, 186

B

Bacteria, 258
Bakkenolide, 289
Barban, 206
Barium salts, 2

Barthrin, 60
Basagran, 237
Basta, 37
Baysan, 182
Benodanil, 162
Benomyl, 2, 166, 189, 193
Bensulide, 216
Bensultap, 26
Bentazone, 223, 237
Benthiocarb, 207, 229
Benzamizole, 216
Benzilic acid, 45
Benzimidazole, 2, 165
Benzophenone, 45
Benzophenone hydrazones, 72
Benzylanilines, 50
Benzylphenylethers, 50
BHC (HCH), 2, 51, 55, 144
Bicyclo orthocarboxylates, 72
Bifenfoxmethyl, 204
Bifenthrin, 60
Bioherbicides, 225, 259
Bioorganisms, 1, 31
Biopesticide formulations, 246
Bipyridylium, 2
Bisabolangelone, 289
Bisulfan, 290
Blasticidin S, 34
Bombykol, 282
Botanical pesticides, 259
Brassica, 203
Briquitte BR, 246
Brodifacoum, 142

Bromadiolone, 143
Bromoconazole, 308
Bromoxynil, 205
Broussonin A&B, 30
Bulan, 45
Bupirimate, 175
Buprofezin, 286
Butachlor, 202
Butamifos, 217
Buthidazole, 215
Buthiobate, 171
Buturon, 210
iso-Butyl phosphonate, 160
sec-Butyl phosphonate, 160

C

Caffeine, 39
Calcium arsenate, 2
CAMD/CAMM, 295
Capillin, 29
Captafol, 156, 188
Captan, 2, 155, 188
Carbamates synthesis, 118, 119, 123
Carbaryl, 119, 125
Carbendazim, 166
Carbofuran, 127, 133
Carbophinothion, 107
Carbosulfan, 133
Carboxamide, 162
Carboximide, 162
Carboxin, 162
Caronaldehyde, 66
Cartap, 26
CDAA, 201
CDEA, 201
Cellocidin, 31
Ceredazine, 213
Cerezin, 84, 157, 188
Cevadine, 21
CGA-123407, 203
CGA-173506, 25
CGA-80000, 173
Chemosterilants, 145, 289
Chitin, 3, 286
α-Chloralose, 140
Chloramben, 198
Chlorambucil, 290
Chloraniformethan, 168
Chloranil, 153, 154
Chlorazine, 211
Chlorbromuron, 210
Chlorbufam, 206
Chlordane, 52
Chlordene, 52

Chlorfenvinphos, 92
Chlorfluazuron, 286
Chlorimuron-ethyl, 220, 239
Chlorinated cyclodiene, 2
Chlorobenzilate, 45
Chlorohydrins, 146
Chlorophacinone, 142, 143
Chloropropham, 229
Chlorosilatran, 140
Chloroximes, 185
Chloroxuron, 209
Chlorpyrifos, 100
Chlorsulfuron, 218
Chlorthalonil, 153, 194
Chlorthion, 98
Cholecalciferol, 139
Chromatrography, 272, 273
Chromenes (benzopyrans), 23
(trans)-Chrysanthemic acid, 9, 65
Chrysanthemum cinerariaefolium, 9
CIB, 315
CIL, 317
Cinerin-I & II, 9
Clean up, 271
Clerodane, 23
Clitocine, 24
Clofop-isobutyl, 200
Clomiphene, 145
Clotrimazole, 179
Cocaine, 39
Compounds, 177
Computers, 254
Concentrated emulsions, 246
Connen, 157
Consumption, 3
Contamination of food, 267
Controlled release formulations, 245
Copper compounds, 149
Coroxon, 96
Cotton, 5
Coumachlor, 142
Coumafuryl, 142
Coumaphos, 101
Coumestrol, 29
Crimidine, 139
Crotonamides, 74
Cyanatryn, 212
Cyanazine, 212
Cyanofenphos, 89
Cyclafuramid, 163
Cyclic phosphates, 24, 102-103
Cyclodienes, 52-54, 56
Cycloprothrin, 60
Cyclosulfamuron, 220
Cycluron, 208

Cyfluthrin, 60
Cyhalofop-butyl, 200
Cyhalothrin, 60
Cymoxanil, 186
Cyometrinil, 203, 225
Cypendazole, 167
Cypermethrin, 59, 60, 62, 64
Cypothrin, 60
Cytokinins, 38

D

2,4-D, 199, 227, 232
Dalapon, 198
Dalbergia retussa, 24
Daminozide, 309
Dazomet, 132, 151
DDA, 48
DDD, 45
DDE, 45, 48
DDT, 2, 45-48, 50, 55, 145
DDVP, 77, 90
Deactivators, 252
Decalin synthesis, 19
Decamethrin, 60, 64, 66
Deciquam, 170
Decontamination, 267
Deet, 288
Dehydroacetic acid, 172
Demeton, 101
2,4-DEP, 86
Depa, 288
Derris, 2, 9
2,4-DES, 200
Destruxins, 31
Detoxification, 267
DFDT, 45
DFP, 76
Diallate, 207
Diamidafos, 132
Diazinon, 99, 100
Dicamba, 199
Dichlobenil, 205
Dichlobutrazole, 182
Dichlofluanid, 156
Dichlormate, 205
Dichlormid, 225
Dichlorobenzil, 45
Dichlorofenthion, 132
Dichloromethyldiphenyl, 158
Dichlorophen, 154
Dichlorovinyl chrysanthemic acid, 66
Dichlozoline, 165
Diclofop-methyl, 200

Dicofol, 45
Dicoumarin, 141
Dicrotophos, 92
Dieldrin, 57
Difolpet, 153, 156
Dihalovinyl acids, 65
Dihydroazadirachtin, 12, 13
Dimefox, 76
Dimethirimol, 174
Dimethoate, 107
Dimethrin, 60
Dimetilan, 121
Dimilin, 286
Dinoben, 198
Dinocap, 153, 154
Dinoseb, 203
Dinoterb, 203
Dioxapyrolomycin, 31
Diphacinone, 142, 143
Diquat, 214, 236
Diram, 152
Disadvantages, 261
Disparlure, 288
Dispersing agents, 252
Disposal, 254, 324, 325
Disulfoton, 106
Ditalimfos, 158
Dithiocarbamate, 150
Dithireanitrile, 25
Diuron, 209, 229
DMPA, 216
DNOC, 203
Dodemorph, 177
Dodine, 170
Domoic acid, 26
2,4-DP, 199
Drazoxolon, 178
Drimane sesquiterpenoids, 22
Dry formulation, 243
Dry lubricants, 252
Dust, 243
Dust bases, 243
Dust driftless, 245

E

Ecdysones, 12, 284
Ecdysterone, 12
Ecofriendly insecticide, 2, 8
Edifenphos, 84, 157, 188, 190
ELISA, 3, 276, 277
Empenthrin, 62
Emulsifiable concentrates, 243
Emulsifiers, 252

ENB, 50
Encecalin, 23
Endosulfan, 52, 57
Endrin, 52
ENP, 50
EPN, 86, 89
Epoxide model, 13
EPTC, 207
Ergocalciferol, 139
Estragole, 24
Etaconazole, 183
Ethephon, 37
Ethion, 108, 216
Ethirimol, 174
Ethofenprox, 308
Ethoxyquin, 176
Ethyl hydrogen propylphosphonate, 37
Ethylene dibromide, 131
Etofenprox, 63
Etridiazole, 179
Eugenol, 24
Explosion, 254
Extraction, 271
Eye contact, 325

F

Farnesol, 283
FDN, 170
Fenaminosulf, 171
Fenamiphos, 132
Fenarimol, 176
Fenbuconazole, 308
Fenfuram, 163
Fenitrothion, 97
Fenopanil, 181
Fenoxaprop-ethyl, 201
Fenpiclonil, 35
Fenpropathrin, 60, 70
Fenpropidin, 178
Fenpropiomorph, 177
Fenpyrithrin, 62
Fenthiaprop-ethyl, 201
Fenthion, 88, 98
Fenuron, 209
Fenvalerate, 60, 68
Fire, 254
First aid treatment, 325
Floating granules, 244
Flowable, 246
Fluazifop-butyl, 200
Fluchloralin, 204
Flucythrinate, 60

Fluidized bed granulation, 255
Flumethrin, 60
Flumeturon, 209
Flumipropyn, 224
Fluoro, 63
Fluoroacetate, 146
Fluorochloridone, 223
Fluorodifen, 204
Fluotrimazole, 181, 193
Flupropadine, 141
Flurazol, 226
Fluridone, 237
Flusilazol, 184
Flutriafol, 184
Fluvalinate, 60, 69
FMC 67825, 133
Folpet, 155, 188
Fonofos, 89
Food grain, 5
Food production, 3
Fuberidazole, 167
Fumigants, 130, 137
Fungi, 258
Fungicides, 1, 162
Furadantin, 146
Furalaxyl, 172
Furathiocarb, 122
Furcarbanil, 163

G

GC-MS, 3, 272, 275, 276
Genistein, 29
Gibberellins, 38
GLC, 48, 52, 273, 274
GLP, 268
Glyphosate, 37, 216, 217, 309
Glyphosine, 37, 216
Gophacide, 140
Granules, 244
Griseofulvin, 34, 35
Guazatine, 170
Gyptol, 281

H

Hadacidin, 35
Haedoxan-A, 25
Haloxon, 82
Harmane, 22
Hemizonia fitchii, 23
HEOD, 53
Heptachlor, 52

Herbicidin A&B, 36
Herbimycins A&B, 36
Hexaconazole, 185
Hexaflumuron, 286
HHDN, 53
Hollofibres, 245
HOO34, 157
HPLC, 3, 274
HPLC-MS, 3, 274, 275
HPLC-MS-MS, 3, 275
HPTLC, 273
Hydantoin phosphonate, 24
Hymexazol, 178

I

IAA, 37
IBA, 37
Imazalil, 180
Imazamethabenz, 222
Imazapyr, 222
Imazaquin, 222
Imazethapyr, 222
Imidacloprid, 74
Imidan, 108
Indanediones, 142, 146
Indian patent system, 311
Indole derivatives, 75
Industrial rights, 261
Insect growth regulators (IGR), 22
Insecticides Act, 314
Insecto, 27
Intellectual property rights, 310-311
Invert emulsifiable concentrates, 243
Ioxynil, 205
IPC, 206
IPM, 5, 260
IPR, 310, 312
Iprodione, 164, 165
Isodrin, 52
Isolan, 121
Isomalathion, 104
Isophos-3, 217
Isoprothiolane, 172
Isoproturon, 209, 233, 234
Isovepol, 12
Isoxazolidines, 224
Isoxazolines, 224

J

Jasmolin I&II, 9
Juglone, 29, 38

Juvabione, 283
Juvenile hormones, 3, 283

K

α-Kainic acid, 26
Kasugamycin, 33, 34
Kitazin, 157, 158
Kitazin-p, 157, 188, 189

L

Laminated structures, 245
Larvicides, 1
Laser, 255
Lead compound, 299
Limbocinin, 12
Limonoids, 14
Linuron, 209
Liquid formulations, 243
Lupulone, 29
Lutenone, 29

M

Malathion, 104
Maleimidoanilides, 164
Mancozeb, 152
Maneb, 151
Marine insecticides, 25, 26
Matrine, 28
MCPA, 2, 199
Mepanipyrim, 187
Mepronil, 194
Mercury compounds, 149
Merigolds, 27
Merphos, 87
Mesulfan, 156
Metabolism, 48, 67, 110-114, 124-
     128, 188-190, 227-230
Metalaxyl, 172
Metazachlor, 203
Metepa, 290
Metflurazon, 213
Methabenzthiazuron, 229
Metham sodium, 131
Methamidophos, 108
Methazole, 215, 236
Methfuroxam, 163
Methiuron, 211
Methomyl, 119, 128
Methoprene, 283

β-Methoxyacrylates, 35, 186
Methoxychlor, 45
6-Methoxyeuparm, 22
Methyl bromide, 131
Methyl isothiocyanate, 131
Metolachlor, 202
Metoxuron, 209
Metsulfuron-methyl, 221
Meturin, 210
Michaelis-Arbuzov reaction, 78
Microbial origin, 26
Microcapsules, 245
Microemulsions, 246
Milbemycins, 31
Mocap, 132
Mode of action, 48, 52, 55, 67, 77,
     78, 124, 134, 187, 188, 231
Molluscicides, 1
Mon-4606, 202
Monocrotophos, 91
Monolithic matrices, 245
Monuron, 209, 234
Morpholine, 2, 174
MRL, 267, 268
MTBO, 103
Multi-residue methods, 269
Muzigadial, 23
Myclobutanil, 185

N

Naphthalic anhydride, 225
α-Naphthyl acetic acid, 37
Natural resistance, 149
Neem, 10-12
Neem formulations, 247
Nemagon, 131
Nematodes, 259
Nereistoxin, 25
Neviram, 151
Niburon, 210
Nicotinoids, 9
Nitrapyrin, 309
Nitrofen, 204
Nitromethylene heterocycles, 71
NMR, 276
Norbormide, 140
Nuarimol, 176

O

Odoracin, 28
Odoratrin, 28

Oil concentrates, 243
Oil solutions, 243
OMPA, 76, 110, 111
Organomercuric compounds, 153
Organosulphur, 149
Organotin compounds, 149
Oudemansin A, 34, 35
Oxadiazon, 215
Oxadixyl, 190
Oxathin, 2
Oxetane derivative, 184
Oxime O-ethers, 62, 225
Oxine copper, 176
Oxycarboxin, 163
Oxyfluorpen, 204

P

Paclobutrazol, 37
Paraquat, 2, 214, 236
Parasitoids, 258
Parathion, 96, 97
Pathogens, 258
Patulin, 38
Pellitorine, 21
Pendimethalin, 205, 229, 235
Penicillic acid, 38
Pentachloroaldrin, 57
Perkow reaction, 80-82
Permethrin, 60, 61
Perthane, 45
Pesticide residues, 266
Pesticides banned, 7
Pesticides registered, 6-7
Pesticides review, 7
Phenothrin, 26
Phenoxy acetic acid, 2, 227
3-Phenoxy benzyl alcohol, 63, 64
Phenoxy sulfonylurea, 220
Phenthoate, 105
Phenyl alkyl phosphonate, 159
Pheromones, 1, 281
Phorate, 106
Phosalacine, 36
Phosalone, 106, 115
Phosdiphen, 157
Phosdrin, 91
Phosphamidon, 92
Phosphinothricylalanylalanine, 36
Phosphite chemistry, 82-85
Phosphorothioates, 95, 96, 157, 160
Phosphorotrithioates, 160
Phosphorylated monoterpenoids, 161
Photoaldrin, 57

Photochemistry, 55, 57, 67, 70, 114, 116-117, 128-129, 190-194, 232-238
Photodieldrin, 57
Photoheptachlor, 57
Photoinduced toxins, 22
Phthalic anhydride, 225
Phthalide, 172
Physostigmine, 39
Phytoalexins, 30
Piericidin-A, 31
Pinosylvin, 29
Piperazine, 194
Pipercide dihydro, 21
Piperine, 307
Piperophos, 217
Pirimicarb, 121
Pirimiphosmethyl, 99, 100
Pisatin, 29
Pisiferic acid, 35, 36
Plant families with insecticidal activity, 11
Plant growth regulators (PGR), 1, 37
Pluridane, 24
Poison dusts, 144
Poisoning, 325
Polygodial, 23
Polyoxins B&D, 34
Population growth, 3
Prallelthrin, 62
Precocene-I&II, 23, 285
Predators, 258
Pretilachlor, 203
Primisulfuron methyl, 220
Prochloraz, 180
Proctolin, 307
Procymidone, 164, 165, 194
Prolan, 45
Promotryne, 212
Propachlor, 201
Propamocarb, 171
Propanil, 233
Propazine, 211
Prophos, 132
Propineb, 152
Propioconazole, 183, 190, 192
Propoxur, 119
Prosafeners, 226
Protective collids, 252
Prothiocarb, 171
Protolimonoids, 12
Proximpham, 206
Prynachlor, 201
Pterocarpan, 29
Purine, 38

Pyracarbolid, 163
Pyramat, 121
Pyramin, 213
Pyranocoumarin, 27
Pyrazole derivatives, 73
Pyrazon, 213
(trans)-Pyrethric acid, 9
Pyrethrin-I&II, 9
Pyrethrum extracts, 2
Pyrichlor, 213
Pyridine, 2, 174
Pyridinitrile, 174
Pyridino-furanone, 174
Pyridinyl pyrimidines, 174
Pyrido indoline triones, 73
Pyridylthiophenene carbamates, 187
Pyrifennox, 174
Pyrimidine, 2, 73, 174, 187, 224
Pyroxychlor, 309
Pyrrolnitrin, 35

Q

QSAR, 3, 294-297
Quality control, 250
Quinalphos, 98
Quinine, 39
Quinofop-methyl, 200
Quinomethionate, 176

R

Rat sterility, 145
Red squill, 146
Registration committee, 318
Repellents, 144, 287
Reserpine, 138
Residue identity, 278
Resistance, 260
Resmethrin, 60
Rice, 5
Rodenticide, ideal, 136
Ronnel, 98
Rotenoids, 10
Rugby, 133
Ryania, 16
Ryanodine, 16
Ryanodol, 16

S

Sabadilla, 21
Safeners, 225
Safety aspects, 254

Salicylanilide, 2, 162
Salithion, 103
Sanitation, 278
SAR, 3, 39, 93, 94, 143, 184, 226
Sarin, 77
Sativan, 29
Schradan, 76
Scilliroside, 137
Sebuthylazine, 211
6α-Senecioyloxychaparrinone, 25
SFE, 272
Shelf-life, 253
Siduron, 210
Silaflunofen, 63
Simazine, 211
Smoke, 246
Sodium fluoroacetate, 139
Soman, 77
Sophocarpine, 28
Spectroscopy, 275
Stereochemistry, 52
Stickers, 253
Storage, 254
Streptomycin, 31, 33
Strigol, 40
Strobilurin-A, 34-35
Strychnine, 138, 146
Sugarcane, 5
Sulfadiazole, 210, 211
Sulfallate, 208
Sulfometuron-methyl, 221
Sulfonylpyrroles, 75
Sulfonylureas, 218, 238
Sulfotep, 110
Sulphur compounds, 2, 149
Surfactants, 251
Suspoemulsions, 246
Synthetic pyrethroid, 3, 59
Systemic fungicides, 162

T

2,4,5-T, 199
Tabun, 77
TCMTB, 179
Tebuthiuron, 210, 211
Tefluthrin, 62
Tepa, 290
TEPP, 76, 110
Terbuthylazine, 211

Terbutol, 205
α-Terthienyl, 22, 27
Tetrachloroaldrin, 57
Tetrachloroethanes, 45
Tetrachlorvinphos, 93
Tetraconazole, 308
Tetrahydroazadirachtin, 12, 13
Tetramethrin, 60
Tetranactin, 31
Tetranortriterpenoids, 12
Teutoxin, 36
Thallium sulfate, 146
Therapeutants, 149
Thiabendazole, 167, 189
Thiadiazines, 72
Thiafensulfuron, 219
Thiazafluron, 210, 211
Thiocyclam, 26
Thiodicarb, 119
Thioester, 62
Thiolutin, 34, 35
Thiono-Thiolo rearrangement, 95
Thiophanate-methyl, 168, 189, 193, 194
Thiotepa, 290
Thiourea derivatives, 72
Thiram, 153
TLC, 273
Tobacco, 2
Toyocamycin, 36
2,4,5-TP, 199
Transgenic plants, 261
Triadimefon, 182, 192
Triadimenol, 182
Trialaphos, 36
Triallate, 207
Triamphos, 158
Triarimol, 176, 285
Triazbutil, 183
Triazole, 3, 181
Triazolyl O,N-acetals, 181
Triazophos, 99
Tribenuron-methyl, 219
Tricamba, 40
Trichlorfon, 88, 89
Trichloronate, 90
2,4,5-Trichlorophenyl derivative, 158
Tricyclazol, 179
Tridemorph, 177
Triethylene melamine, 146

Triflumuron, 286
Trifluralin, 204, 228, 238, 239
Triflusulfuron methyl, 219, 220
Triforine, 168, 169, 194
Triketone compounds, 224
Trimorphamide, 169, 177
Trithiophosphate, 103

U

Ultra low volume, 246
Unsaturated amides, 21

V

Validamycin-A, 34
Validation, 278
Vegetables, 5
Vepaol, 12
Vepinin, 12
Veracevine, 21
Veratridine, 21
Vignafuran, 29
Vinclozoline, 164, 193
Viruses, 259

W

Waiting period, 269
Warfarin, 141, 147
Water dispersible granules, 244
Water emulsifiable gels, 244
Wettable powder, 244
Wetting agents, 252

Y

Yatein, 289

Z

Zeatin, 38
Zinc phosphide, 137, 146
Zineb, 151
Zinophos, 132
Ziram, 153